VD

Conceptual Basis of the Classification of Knowledge

Proceedings of the
Ottawa Conference
on the Conceptual Basis
of the Classification
of Knowledge
October 1st to 5th, 1971

Les fondements de la classification des savoirs

Actes du
Colloque d'Ottawa
sur les fondements
de la classification
des savoirs
du 1er au 5 octobre 1971

edited by/edité par
Jerzy A. Wojciechowski
University of Ottawa

Verlag Dokumentation, Pullach/München 1974

Printed by Anton Hain KG, Meisenheim/Glan, West Germany
Bound by Thomas-Buchbinderei GmbH, Augsburg
ISBN 3-7940-3649-2

Acknowledgments

The Ottawa Conference on the Conceptual Basis of the Classification of Knowledge was financed by grants from the Canada Council and the University of Ottawa. The publication of the Proceedings was made possible through an additional grant from the University of Ottawa.

The Organizing Committee wishes to express its gratitude for the financial assistance which it has received from these two institutions.

The Organizing Committee wishes also to thank the Department of Documentation, Libraries and Archives of UNESCO and in particular Drs. O.A. Mikhailov and P. Lazar for advice and encouragement.

Foreword

The present volume contains the Proceedings of the Ottawa Conference on the Conceptual Basis of the Classification of Knowledge. The Conference took place at the University of Ottawa, October 1st to 5th, 1971. The Organizing Committee was composed of Professors Guy Lafrance, William Shea and Jerzy A. Wojciechowski, Chairman, all from the Faculty of Philosophy of the University of Ottawa.

A few words of explanation are in order concerning the reasons for the organization of the Conference. In Newsletter No 1 issued by the Organizing Committee the reasons are explained as follows:

> Generally speaking, the aim of the Conference is to discuss the epistemology of the Classification of Knowledge in the light of the present state of the problem.
>
> As far as philosophers are concerned, the Conference seems to be justified precisely because of the present state of the Classification of Knowledge. For two thousand years, the Classification of Knowledge has been a major philosophical problem. Lately, it has evolved from a philosophical preoccupation into a specialized technique, a species of the general techniques of classification, far removed from philosophical preoccupations and equally ignored by philosophers. At the present time there exist powerful technical models of classification and highly developed classifications of various branches of knowledge. What is lacking is a satisfactory general classification of knowledge and an adequate clarification of the philosophical problems underlying such classification. The need of clarification is important from the theoretical as well as practical point of view because the general systems of classification of knowledge depend on theories of knowledge. It is therefore highly desirable that philosophers and specialists in classification get together and discuss problems of common interest.
>
> A subsidiary aim of the Conference is to discuss the project of the establishment at the University of Ottawa of a Research Center for the Classification of Knowledge.

It seems to the present writer that the results of the Conference have borne out the Organizing Committee's assessment of the need for holding this Conference. It is to be hoped that the readers of the Proceedings will arrive at a similar conclusion.

The conference was unique in this that for the first time philosophers and classificationists from the world over were brought together to discuss the problem of the overall classification of knowledge from a fundamental point of view. In the course of the Conference it became evident how large is the gap between these two groups. The aims, angles of approach, means and methods in the case of each group were different. This was to be expected. The more disturbing fact was the estrangement of each group from the other, lack of knowledge of what the other was doing and the difficulty of communication. In this sense one may say that the Conference was a very imperfect one. But its very imperfection was perhaps its most revealing characteristic. That the Conference brought it to light, has been, in the eyes of the present writer at least, its greatest value.

The Conference amply proved that from the philosophical point of view the general classification of knowledge is a wide open problem. Not only does there not seem to exist a satisfactory classification, but, moreover, we do not seem to have adequate methodological tools to deal with this problem. Over and above this difficulty there is the question of the very feasibility of a philosophically satisfactory classification. This is a serious problem indeed, for if no satisfactory classification can be devised, then it looks as if we are faced with the impossibility of forming a coherent and total view of the universe of knowledge and consequently of the universe of being itself. If this is the case, then two very important consequences can be drawn. In the first place, the belief in the possibility of forming a coherent world-view was the basic assumption of philosophical systems as well as of the pre-philosophical thinking and of science. This belief seems to be open to question. If this is the case, then a fundamental revision is needed in our understanding of the nature and purpose of knowledge. The second consequence concerns the value of the progress of knowledge. This, in its present form, involving an ever-increasing proliferation of branches of knowledge, and a marked absence of synthesis may turn out to be a modern, more sophisticated version of the building of the tower of Babel.

Something of the tower of Babel spirit is present in the Proceedings: All papers read at the Conference are included. The aim of the editor was to bring forth a volume which would, as faithfully as possible, represent the actual proceedings of the Conference and capture its spirit. In view of the special character of the Conference, no attempt was made to select or to alter the papers so as to produce a more coherent and uniform book. As it is, the volume expresses the diversity of interests of the participants. The editor believes that this in itself is instructive and indicative of the problems described above. The present volume will achieve its aim and therefore justify its publication, if it arouses the interest of philosophers and classificationists in the problem of conceptual basis of an overall classification of knowledge. May it become an invitation to philosophers and classificationists to become better acquainted with each other and try to bridge the gap which exists at present between them. Cooperation and mutual understanding will benefit not only these scholars but knowledge in general.

Jerzy A. Wojciechowski
Chairman, Organizing Committee

Contents

Acknowledgments ... 5
Foreword ... 7
Contributors ... 11

Jerzy A. Wojciechowski
The philosophical relevance of the problem of the classification of knowledge ... 13

Eric de Grolier
Le système des sciences et l'évolution du savoir ... 20

Ganesh Bhattacharyya / S.R. Ranganathan
From knowledge classification to library classification ... 119

Alwin Diemer
L'ordre (classification) universel de savoirs comme problème de philosophie et d'organisation ... 144

R. Mölgaard-Hansen
On the problem of universality in knowledge classification ... 161

William R. Shea
The classification of scientific terms as "theoretical" and "observational" in contemporary philosophy of science ... 172

Zygmunt Dobrowolski
Secteurs scientifiques autonomes et leur rôle dans la classification des sciences ... 186

Frederick Suppe
Some philosophical problems in biological speciation and taxonomy ... 190

Réjane Bernier
Comments upon Prof. Suppe's paper on "Some philosophical problems in biological taxonomy and speciation" ... 244

Guy Lafrance
Le problème épistémologique de la classification dans les sciences humaines sociales ... 252

Patrick A. Heelan
The logic of changing classificatory frameworks ... 260

E. J. Ashworth
Classification schemes and the history of logic ... 275

C. A. Hooker
The impact of quantum theory on the conceptual bases for the classification of knowledge ... 284

Suman Datta / Jason E.L. Farradane
A psychological basis for general classification ... 319

Phyllis A. Richmond
Review of: Suman Datta and Jason E.L. Farradane.
A psychological basis for general classification 332

J.L. Jolley
The Holotheme: A paper submitted to the First Ottawa Conference
on the Conceptual Basis of the Classification of Knowledge 335

Addendum to "The Holotheme": Presentation.
By Robert Fairthorne 367

Derek Austin
A conceptual approach to the organisation of machine-
held files for subject retrieval 371

J.M. Perreault
Commentary 399

Robert A. Fairthorne
Temporal structure in bibliographical classification 404

Phyllis A. Richmond
Review of: R. A. Fairthorne, Temporal structure in
bibliographical classification 413

Ejnar Wåhlin
The AR Complex – Adapted systems used in combination with
a common reference system 416

Ingetraut Dahlberg
Principles for the construction of a universal classification system.
A Proposal 450

Б.М. Кедров /Москва/
КЛАССИФИКАЦИЯ НАУК. ЕЕ ПРИНЦИПЫ. ЕЕ ЦИКЛИЧЕСКАЯ
ФОРМА. 472

Normand Lacharité
Questions about "Classifiables" and "Classifyers" submitted
as a commentary to various papers 494

Contributors
to the Ottawa Conference on the
Conceptual Basis of the Classification of Knowledge

Prof. Dr. E. Jennifer Ashworth
University of Waterloo, Dept. of Philosophy
Waterloo, Ont., Canada

Derek Austin, Principal Investigator,
OSTI/PRECIS Project,
British National Bibliography,
7 Rathbone St, London WIP 2AL, U.K.

Prof. Dr. Réjane Bernier,
Université de Montreal, Dept. de Philosophie
Montreal, Que., Canada

Ganesh Bhattacharyya, Senior Lecturer,
Documentation Research and Training Centre,
Bangalore, India

Dr. Ingetraut Dahlberg,
D-6000 Frankfurt/M. 50, Woogstr. 36a

Dr. Suman Datta (Mrs.)
The City University, Centre for Information Science,
23 Goswell Rd. London ECIM 7BB, U.K.

Prof. Dr. Alwin Diemer,
Universität Düsseldorf, Philosophisches Institut,
D-4000 Düsseldorf-Nord, Cecilienallee 40

Prof. Dr. Zygmunt Dobrowolski, Emeritus Professor,
University of Warsaw,
ul. Zjednoczenia 33, Warszawa, Poland

Robert A. Fairthorne, Professor Emeritus,
30 Clockhouse Rd, North Farnborough, Hamps., U.K.

Prof. Jason E.L. Farradane, Director,
The City University, Centre for Information Science,
23 Goswell Rd, London ECIM 7BB, U.K.

Prof. Eric de Grolier,
68 rue de la Tour, F-75016 Paris

Prof. Dr. Patrick A. Heelan, Professor of Philosophy,
State University of New York,
Stony Brook, NY, USA

Prof. Dr. C.A. Hooker,
The University of Western Ontario, Talbot College,
Dept. of Philosophy,
London, Canada

J. L. Jolley
"Jays", New Road,
Great Missenden, Bucks., U.K.
Academician Bonifatij Michailowič Kedrow, Director,
Institute of History of Natural Sciences and Technology
of Academy of Science of the USSR,
Moskwa K-12, Staropanskij per. 1/5, USSR
Prof. Dr. Normand Lacharité,
Université de Québec, Dépt. de Philosophie,
Montreal, Canada
Prof. Dr. Guy Lafrance,
Université d'Ottawa, Faculté de Philosophie,
Ottawa, Ont. K1N 6N5, Canada
R. Mölgaard-Hansen,
Danish Centre for Documentation,
Committee on Classification Research,
Oster Voldgade 10, DK-2000, København
Jean M. Perreault,
Director of the Library,
University of Alabama,
Huntsville, Ala. USA
S.R. Ranganathan,
National Research Professor in Library Science
and Honorary Professor,
Documentation Research and Training Centre,
Bangalore, India
Prof. Dr. Phyllis A. Richmond,
6628 Airtree Park Drive, apt. 202,
Mayfield Village, Ohio 44143, USA
Prof. Dr. William R. Shea,
McGill University, Dept. of Philosophy,
Montreal, Que., Canada
Prof. Dr. Frederick Suppe,
University of Maryland, Dept. of Philosophy
College Park, MD 20742, USA
Ejnar Wåhlin, Konsultbyrå
Birger Jarlsgatan 102, S-114 20 Stockholm, Sweden
Prof. Dr. Jerzy A. Wojciechowski,
Université d'Ottawa, Faculté de Philosophie,
Ottawa, Ont. K1N 6N5, Canada

Jerzy A. Wojciechowski (Ottawa)

The philosophical relevance of the problem of the classification of knowledge

The problem of the classification of knowledge and of classification in general is as old as philosophy and as open and unsettled as a philosophical system. Two opposite factors play a dominant role in the sphere of classification, and counterbalance each other. There is, on the one hand, the imperative desire of the intellect to understand the overwhelming richness of the phenomena with which it has to cope, and on the other, the varied and varying, seemingly endless multiplicity of data. Confronted with this situation, the thinker has two choices. He may attempt to reach for the apparently supreme intellectual ideal of the total conceptual horizon, and try to conceive a global and exhaustive classification, or he may renounce this lofty aim and opt for a less grandiose but one which is more attainable and less controversial, which is that of partial classification.

When we look objectively at these two endeavors we realize that in fact they are complementary. And indeed they have been pursued simultaneously, but with varying degrees of intensity and success. That much is clear. What is much less obvious is the relation between partial classifications and a total one, irrespective of whether these classifications pertain to the external world (things, phenomena) or to the order of knowledge. The incertitude is due to the fact that a satisfactory total classification does not result from the addition of particular classifications. If it did, we would not lack an adequate total classification, even in our times of rapidly extending knowledge. All we would have to do, would be to add new partial classifications to the existing ones. This however is not the case. The conclusion which must be drawn concerns the nature of total classifications. The differences between them are not simply bigger, more of the same type of things, they are in some important way qualitatively different. This is why they raise different kinds of questions. These questions we usually refer to as philosophical.

It is only natural that philosophy has, since its beginnings, been concerned with the problem of total classifications. One cannot inquire into the nature of reality or of knowledge without considering the modes, or forms of existence. Diversity is more evident than unity, and no rational process can overlook it. If thinking is directed to the knowledge of what there is, it must not only notice the extreme variety of data but do somethings about it, something other that is, than arguing it away. To take the problem of classification seriously is, and has always been, a sign of intellectual maturity and a prerequisite for effective investigation of reality. One is tempted to say that the richer the knowledge becomes, the more elaborate become the classifications, and the more sophisticated the techniques of classification. I said "one is tempted to say" because nowadays knowledge seems to run ahead of classifications and we are in danger of being defeated by the wealth of data. Their very multiplicity of variety is the cause of a new kind of unintelligibility and of ignorance, namely that produced by the wealth of available information. I am not certain whether calling this state of affairs "learned ignorance" will do much to alleviate our present predicament.

Traditionally, philosophy was concerned with the problem of what to classify, and how to do it. In our age of highly – not to say hyper – developed critical awareness, it is necessary to step back from these straightforward questions and ponder what questions could be or should be asked concerning classification. Without attempting to predetermine the solution of this problem one may point out that the answer will depend, among others, on whether one envisages the problem from a philosophical or stricly methodological point of view. Let us clearly distinguish between the philosophical endeavors of the classification of reality and/or of knowledge, and those of the classificationist, librarian or other, whose main concern is a neat classificatory pattern regulated by the idea of pragmatic usefulness. For the philosopher, the overriding concern is the adequacy of the classification, i.e. the conformity of the classification with the object classified. Consequently, the classification should be subordinated to the nature of that which is classified. The classificationist instead must serve two masters: the object classified and the users of the classification. The usefulness of classification is to him of prime importance. This, in turn, forces him to consider the problem of information retreaval and to adapt the classification to the conditions which would facilitate efficient retreaval. These remarks lead to another important question, namely what actually happens when we classify, i.e. what is the nature of the classificatory process. This is a second order question. It has to be distinguished from the first order questions and dealt with according to its epistemological status. Our ability to distinguish between first and second order questions is a measure of progress accomplished by critical thinking. In order to fully grasp this progress and to understand the nature of problems which we have to face, it is worthwhile to consider a philosophical system eminently concerned with total classification and dating from the pre critical era. Aristotle's philosophy seems to be the natural choice, both because of its internal rigor and its historical importance.

"To think is to order" claimed Aristotle and built his system on two fundamental classifications, namely those of ten categories and five predicables. It is important to stress that the ten categories were meant to be at the same time the supreme classifications of the order of being and of the fundamental elements of thought. According to Aristotle, the categories of being, i.e. of objective reality, and those of concepts were rigorously the same. To arrive at this conclusion Aristotle had to adopt a realistic theory of knowledge, the hylomorphic theory of nature, and the metaphysical theory of act and potency as explanation of the fundamental nature of being. At the basis of his system lies the belief that nature is a form-giving force, orderly, and ordering, the universe – a well organized whole, essentially intelligible, and knowledge, an image of reality reflecting nature's harmonious structure.

Aristotle's ideas have exercised lasting influence on Western thought and he may rightly be called the father of all classificationists. Twenty centuries later Kant speaking of Aristotle's categories remarks that "This rhapsody must be considered a mere hint for future inquirers, not as an ideal developed accordingly to rule . . ." Having said this he proceeded to draw his own list of categories which although produced in a much more critical spirit than that of Aristotle, exercised much lesser and shorter-lived influence. Kant's dictum places on me the onus of explaining why present-day thinkers interested in the philosophical problem of classification may find the study of Aristotle's classifications worthy of his attention. There are, in my opinion, two main reasons for the interest in aristotelian classifications. First of all, they offer a perfect example of fundamental classifications both of the realm of being and of the order of knowledge. Secondly when seen in the context of Aristotle's philosophy, they allow a good insight into the role which classification plays in the elaboration of a philosophical system and, inversely, into the role which philosophical presuppositions play in the development of a fundamental classification pretending to be universal. In other words the aristotelian philosophical construct should be for us not a mere hint, as Kant wanted it to be, but an instructive lesson about the kind of problems associated with attempts of total classification. Whatever our opinion about aristotelian philosophy, one conclusion resulting from the study of this philosophical system is inescapable: to classify in a grand manner is to philosophize.

For Aristotle the problem was not fundamentally if it was possible to classify, but rather, how to classify. To-day, with all the highly developed and specialized techniques of classification the fundamental question which baffles us is whether the very intention of total classification of the outside world and of the order of knowledge is not only feasible but intellectually justifiable in the first place. The great thinkers of the past accepted as a foregone conclusion, that an individual mind can encompass the whole province of knowledge. Out of this conviction grew the belief that a thinker could and should provide a total classification as an intrinsic element of his system. Presently, our factual knowledge, has by far outdistanced the capacities of learning and of synthesizing of any individual intellect. Two kinds of consequences of this situation are of importance for our discussion. First of all, thinkers are less interested in, and more critical towards projects of total classification. Secondly, the necessary work of classification passed into the hands of specialists and became more and more a team effort. Whether a return to the one-man classifications is possible remains to be seen.

Let us return once more to Aristotle. According to him the universe of being was not only eternal but in addition static in its fundamental determinations provided by a hierarchy of generic and specific forms. The universe was, as it were, a classification in act. Since knowledge was meant to be a faithful image of the objective world, the attempts of classification were objectively justified, and the ideal of universal classification within the reach of the intellect. For us the situation appears, alas much more complicated. Let us try to explain why this is so. Possibly, several reasons may be given of this situation but, as far as I can tell, it results essentially from two factors, one more obvious, the other less evident. The obvious one, which I have already mentioned earlier is the expansion of knowledge. The other reason is the change in our understanding of the nature of knowledge. The Aristotelian mirror model of knowledge has been replaced by more complex models taking into account the subjective factors and the interplay of subjective and objective elements in the process of cognition. Consequently, the aristotelian idea of the object of knowledge, i.e. of the objective reality as something "being" there, in front of us, all ready made and waiting to be looked at and understood, had to make place for a more sophisticated but, in a sense, much less satisfying notion of a reality essentially bound to the act of knowledge. Reality appears therefore as being neither purely subjective nor objective. Moreover it seems to be dynamic in two different ways: because of the process of evolution taking place in nature, and because of the expansion of knowledge. Thus, reality reveals itself as essentially evolving and expanding.

If the above description is correct, it invites important questions. When we embark on the project of total classification, what is it really that we classify: that what is, as it is, or that what we know, as we know it? Are we dealing with the objective world or with our image of it? We know that the answer cannot be simple and that it depends, among others, on the size of things. The situation is different in the case of macro-objects from that of micro-objects. Micro-physics has made us aware of the fact that observation is far from being a simple act, and that data supplied by it are not the result of a direct subject-object relation. We have learned that the relation involves at least three factors: the subject, the means of observation and the object. Consequently the resulting datum is not a mirror image of the object but the result of an interplay of these three factors. The lesson which micro-physics has taught us is of great importance for the theory of knowledge and has a direct bearing on our problem. It can be stated simply thus: all observation is deformation.

Although the statement is universal, it must be qualified to avoid oversimplification. The deformation in question is, for all practical purposes, negligible in the case of macroscopic objects, directly perceptible, existing in the spatial and temporal vicinity of the observer. Beyond this immediate and preferential, as far as perception is concerned, group of objects, the deformation of the datum becomes greater and greater. The image of a sub-atomic particle or of a distant nebula is far from being the direct image of the object itself. Since a total classification has to include all the kinds of things from the most easily observable, to the only marginally perceptible ones, the classes which it establishes, cannot have the same relation to the external world. They simply cannot be all equally objective. The situation may be described as being inverse from that which obtains in cartography. Reality, which for the sake of simplicity, may be assumed to be flat is projected on a sphere, i.e. on the geometric place of points equidistant from the observer. It is logical to infer that the sphere and the flat surface come closest together in the region of most adequate observability. Unfortunately it would be rather difficult to obtain unanimous agreement on what is exactly most adequately observable. Therefore it is not clear whether one could map the area of least deformation, i.e. of the most adequate classification, to the general satisfaction.

As I see it, the above conclusion raises questions pertaining to three aspects of the problem of total classification, namely the objectivity of classification, the intellectual frameworks of classification and the communicability of information about classification. With regard to the first problem one may well wonder whether the objectivity of classification is such an important issue after all. To inquire about the objectivity of classification and to worry about the lack of it, may be a sign and a consequence of a philosophical option, namely a predilection for a realistic theory of knowledge. For a Kant or a Hegel, this problem would be solved a priori and in confirmity with their subjectivistic theories of knowledge, while for a Hume it would be a meaningless question. Of course, these philosophical divergencies of opinion may appear to a professional classificationists seriously and fruitfully engaged in furthering the techniques of classification, rather confusing, not to say meaningless. Chances are, that explicity or implicity, he is a realist too, convinced about the reality of things which he tries to classify and the meaningfulness of his classification, unwilling to be bothered by philosophical disputes. Unfortunately the situation is not as simple as that, and the philosophical problems cannot be brushed aside. They come to the fore whenever one reflects about the meaning of a classificatory system, no matter how adequate this system may be from the technical point of view.

In recent years philosophers of science became increasingly aware of the role of intellectual frameworks, i.e. paradigms, in the shaping and changing of scientific theories. The understanding of data is a function not only of their nature but of the general state of knowledge and of the prevailing opinions. It is the two latter factors which determine if and how a datum is received, whether it is judged meaningful and important or not, requiring a change in the established explanations or not. Similar considerations about the role of intellectual frameworks should be made with regard to total or even partial classifications. The role of the paradigm is clear in the case of aristotelian classifications and the change of the intellectual framework from Aristotle to, say, Kant is also obvious. Every classification bears the unmistakable imprint of the intellectual horizon of its time and of its creator. In this sense classifications are, one is tempted to say, organic elements of those intellectual horizons, and share in their value and in their fate. In other words they are always relative with regard to their respective paradigm.

If this conclusion is correct one may ask whether classifications are equally, to paraphrase an expression à la mode, intellectually degredable as the paradigms, or whether there are some elements of it which retain their value after the change of the paradigm. An example from the field of physics may illustrate the problem. The replacement of classical mechanics by relativity physics did away with the Newtonian paradigm, but the mathematical structure of classical mechanics remained as a first approximation to the more complex construct of Einsteinian equations. Moreover Newtonian equations retained all their practical value in the field of medium sized forces and objects, because in this case calculations made by means of Einsteinian equations do not give appreciably different results. The question therefore is whether in classifications there exists some parts of more permanent value than others.

One could argue perhaps that this question should be reworded and expressed as follows: is there any formal, mathematical or logical element which can be separated from the body of classifications and thus saved from oblivion as retaining its value in the face of more adequate classifications? The answer depends, no doubt, on the nature of classifications. In view of the fact that classifications differ widely as to the degree of their formalization, no one general answer is possible and each classification should be studied separately. It would be tempting to draw the following conclusion: The "survival value" of a classification with regard to the change of the paradigm is proportional to the degree of formalization or mathematization of the classification. This formula sounds good, and conforms to the prevailing general trend of mathematization and formalization of knowledge. If it is correct, then it is plausible to infer that the future of the science of classification lies in the successful application of logic and mathematics to this field of research, just as three or four centuries ago the future and the very development of modern physics depended on the successful mathematization of that science. It would seem therefore that the science of classification is in a situation comparable to that of physics in the early seventeenth century.

I expressed the preceeding statements in the conditional tense because I am not at all convinced whether drawing analogies with physics is plausible, and if it is, to what extent these analogies are valid. It seems to me that a distinction has to be made again between partial and total classifications. Formalization and mathematization present great advantages but they presuppose one important condition, namely a certain homogeneity of elements to which they are applied. Physics could be mathematized because the data were expressed in the form of number-measures and the degree of mathematization has been determined by the degree to which the data could be quantified. The success of this method presupposed a judicious choice of the subjectmatter of physics, namely the most general and most adequately measurable aspects of the material world. Analogical choices of homogenous elements could probably be made in the case of some partial classifications. But certainly, this could not be the appropriate way of dealing with total classifications.

The above reasoning presupposes an essential diversity in the realm of being, and the necessity for an adequate classification to recognize and to reflect this diversity, i.e. to be essentially a classification of objective reality. This point of view can be challenged at least in two different ways. To the philosophical pluralism one may oppose a metaphysical monism be it of the materialistic or spiritualistic variety. On the other hand one could argue that metaphysical conundrums not withstanding, it is possible to produce a working classification by limiting oneself to the realm of language and by classifying words.

As to the philosophical options, they will remain a matter of personal preference and choice. A purely linguistic classification will certainly be an interesting tool for the study of the nature of language and of conceptual structures, provided that it is done in an empirical manner without an attempt to force on the language preconceived and rigid patterns. The question which in the case of such classification will remain open is the problem of the relation of this classification to the objective reality, and its value for the study of that reality.

As I mentioned earlier, total classifications were the work of individual thinkers, and they expressed their "Weltanschauung". The question is whether in the present situation of knowledge, teamwork can achieve results analogical to those which were produced by individuals. The main problem here is that of communication. As we know communication of information is a complex problem and perfect communication is a myth. We can communicate signs but we cannot communicate meanings in the same way. Again the question of formalization of knowledge comes to the fore. The more formalized a body of knowledge, the more precisely the symbols by means of which it is expressed are communicable, the smaller the deformation of meanings. We know that, at least theoretically, a formalized classificatory system may be conceived. What we do not know, is the extent to which such a system is a procrustean bed. Consequently, it remains an open question whether we can formalize the science of classification as such, to what extent this is possible, and how meaningful will this formalization be? These are the final questions which I would like to raise.

Philosophers and classificationists have not talked to each other a great deal in the past and have not investigated problems which should be of common concern to them.

May this conference be the beginning of a fruitful dialogue.

Eric de Grolier (Paris)

Le système des sciences et l'évolution du savoir

0 Introduction

01 Préliminaires terminologiques

Le "Bulletin No 1" du Colloque d'Ottawa assigne comme objectif à celui-ci "une réflexion sur le problème épistémologique de la classification des savoirs". Observons qu'en anglais le libellé est quelque peu différent, et qu'il y est question, non d'épistémologie mais de "base conceptuelle" – ce qui ne nous paraît pas tout à fait identique, et non de classification des savoirs, mais de "classification of knowledge" – ce que, traditionnellement, on traduit plutôt par classification des connaissances. Ailleurs, il est vrai, le texte français dit "classification *du* savoir". Ces variations soulèvent quelques questions: savoir et connaissances sont-ils identiques? y a-t-il *un* savoir ou *des* savoirs? L'ambiguité de l'anglais "knowledge" invite à la réflexion: tantôt un singulier (theory of knowledge = théorie de *la* connaissance), tantôt un pluriel (classification of knowledge = classification *des* connaissances). Mais nous retrouverons ces problèmes plus loin.

Notre rapport s'intitule, lui, *le système* des *sciences* – suivant en cela une tradition bien établie en France depuis Goblot (1922). Ces termes paraissent impliquer: (1) une restriction à la connaissance scientifique (excluant, par conséquent, le recours à quelque forme que ce soit de dualisme ou de pluralisme des modes de connaissance, que la philosophie occidentale a élaborée, depuis Jean Scot Erigène, soit plus d'un millénaire); (2) la pluralité des sciences (par opposition aux tentatives pour constituer une "science unifiée" – par ex. celle du "Cercle de Vienne"); (3) la possibilité d'ordonner les sciences en *un* système structuré, possibilité qui est fermement niée par beaucoup de théoriciens actuels des problèmes de classification, par ex. Jean-Claude Gardin, qui suivent en cela, probablement sans s'en douter, une tendance philosophique relativement récente – à ce qu'il nous semble – en Europe, ne remontant guère au delà des néo-kantiens de l'école de Bade (Windelband, Rickert, 1894–96) et systématisée par Jaspers en 1932.

Un autre problème terminologique est encore soulevé par le "Bulletin No 1" du Colloque. En son paragraphe 3a, celui-ci mentionne l'existence de "techniques générales de classification", dont la classification des connaissances (ou "des savoirs") ferait partie comme "technique spécialisée", et qui utiliserait des "puissants modèles techniques" (dans le texte français, le mot "techniques" est ici omis, mais il est présent dans le texte anglais: "powerful technical models"). La classification est ici présentée comme une pure *technique* – opposée par ailleurs à un problème *philosophique.* Notre propre position est qu'il s'agit actuellement, tout d'abord, d'une science, à laquelle on peut laisser, si l'on veut, sa dénomination ancienne de taxonomie, ou la remplacer, comme l'avait déjà fait Durand (de Gros) en 1899 par taxinomie, ou bien encore adopter pour elle un terme différent: classologie, taxilogie. Bien sûr, comme toute science, celle-ci a des aspects et des méthodes techniques, et il est fort vrai que ces méthodes se sont renouvelées depuis une dizaine d'années, notamment avec l'emploi de modèles mathématiques, mais le problème de la classification ne se réduit pour nous ni à des spéculations philosophiques, ni à de pures techniques.

02 La science de la classification (taxinomie) comme discipline ou comme champ de recherches interdisciplinaires

Nous pensons, toutefois, que cette science de la classification (taxinomie, taxilogie) n'est pas encore véritablement constituée comme discipline "autonome": elle n'est qu'en voie de constitution. A notre connaissance, il n'y a dans le monde aucune chaire de science de la classification, aucun traité général qui en embrasse tous les aspects. Il est vrai que, par contre, il existe dans quelques pays un petit nombre de groupes de recherche qui – au moins théoriquement – sont spécialisés dans ce domaine d'études: le Classification research group de Londres (depuis 1952) est le plus ancien et le plus actif; des groupes analogues ont été créés aux Etats-Unis (1959) et en Inde (à Bangalore, en 1962). On peut également noter l'existence d'une association spécialisée: la Classification society, fondée en Grande Bretagne en 1964, divisée en deux branches, européenne et nord-américaine, depuis la création de cette dernière en 1970, comptant à peu près 500 membres et publiant un Bulletin. Elle a pour objet de promouvoir la coopération et l'échange d'informations entre les personnes intéressées par les principes et la pratique de la classification dans n'importe quelle discipline. Jusqu'à présent, cependant, pour autant que nous sachions, elle s'est surtout intéressée aux nouvelles techniques de classification automatique, et n'a pas du tout abordé les aspects "philosophiques" de la taxinomie. Si la Conférence d'Ottawa réussit donc à provoquer la création d'un "Centre de recherche sur la classification des savoirs" à l'Université d'Ottawa, comme il est indiqué au point 3b du "Bulletin No 1", et si ce Centre est organisé sur des bases réellement scientifiques, avec un programme suffisamment exhaustif (*comprehensive*, en anglais), alors peut-être pourra-t-on dater d'octobre 1971 la constitution effective de la taxinomie générale en tant que discipline . . .

Pour le moment, il faut la considérer comme un champ de recherches interdisciplinaires (ou plutôt multidisciplinaires, si l'on accepte la terminologie proposée par Pierre de Bie, 1970, p. 121). Il est donc important de voir quelles sont les contributions que peuvent apporter les diverses disciplines, antérieurement constituées, à la fondation et au développement de cette petite soeur cadette. Nous le ferons en suivant – approximativement – l'ordre historique dans lequel ces différents points de vue sont apparus.

03 Les différents points de vue pour l'étude de la classification

031 Le point de vue "technique"

Jacques Monod (1970, p. 166) écrit "Que les animaux soient capables de classer des objets et des relations entre objets selon des catégories abstraites, notamment géométriques, ne fait aucun doute". *Homo faber,* tout comme ses précurseurs, cousins plus ou moins lointains – abeilles, souris ou chimpanzés – a réalisé des opérations de classification dans un but pratique depuis quelque cinq millions d'années. La complexification croissante de ses techniques s'est reflétée dans les taxinomies qui leur sont associées, jusqu'à ce que Brisch et ses collaborateurs aient, pour la première fois sans doute, créé une entreprise réalisant des profits en établissant des classifications d'outillages pour l'industrie. On sait toute l'importance que F. W. Taylor attachait à de bonnes classifications pour l'organisation scientifique dans les usines; à sa suite, les services et associations de normalisation comme les bureaux d'étude des temps (en France, par ex., le Bureau des temps élémentaires) ont consacré beaucoup d'efforts à l'établissement de classifications de machines, de matériaux, de produits, et – plus récemment – les principaux pays industriels ont créé des organismes chargés d'étudier la normalisation des nomenclatures (codifications) dans l'industrie (National Computing Center, 1968). On peut aussi évoquer ici l'activité taxinomique intense, au plan international, du Comité électrotechnique international, mais celle-ci est liée à des efforts de normalisation terminologique et se rattache donc de très près au point de vue linguistique, que nous examinerons plus loin (034).

032 Le point de vue psychologique

Jean Piaget et ses disciples de "l'école de Genève" (Centre international d'épistémologie génétique) ont apporté une contribution, dont on ne saurait sous-estimer l'importance, à la taxinomie générale en montrant la genèse des opérations de sériation, d'inclusion de classes et leur caractère primitif par rapport aux structures linguistiques (cf. Piaget, 1970, p. 612–14). Il ne nous paraît pas nécessaire d'insister ici sur ces recherches, bien connues.

033 Le point de vue logique

Yehoshua Bar-Hillel (1957, 1962, 1964) a attiré l'attention sur les abus qui ont été faits de la terminologie logique en matière de "langages documentaires" et des systèmes de recherche d'information (*information retrieval*), dans des contextes où l'emploi de cette terminologie ne pouvait que conduire à des discours vides de sens. Il a en particulier dénoncé l'emploi illégitime de la notion de "class-concept" (1964, p. 322) et le recours exclusif à la logique des classes, alors que s'imposerait plutôt celui à la logique des relations (idem, p. 354). Ses sévères condamnations ne peuvent cependant viser que l'application inconsidérée, pseudo-scientifique, des résultats des recherches des logiciens au domaine taxinomique, et non leur utilisation rationnelle.

034 Le point de vue linguistique

"Nommer, c'est classer" écrivions-nous naguère (Grolier, 1956, p. 2), et nous trouvons une pensée identique sous la plume de François Jacob (1970, p. 57): "Nommer une plante, c'est déjà la classer" – mais Linné dès 1751 (cité par Jacob, 1970, p. 58) et Lavoisier en 1789 (idem, p. 50–51) exprimaient déjà la même idée, et Friedrich Engels, un peu moins d'un siècle plus tard, leur faisait écho en intitulant un des fragments de *Dialectique de la nature* "Importance des noms" (trad. P. Naville, 1950, p. 304–5). La linguistique contemporaine conduit dans plusieurs directions des recherches qui ont d'importantes implications taxinomiques: qu'il s'agisse de la base linguistique des "catégories" (Benveniste, 1958), du "fondement naturel" du genre, de la correspondance entre classification grammaticale et classification sémantique (Lyons, 1968; trad. française, 1970, p. 128–130), de la détermination par la structure linguistique du découpage de la réalité, de la "catégorisation de l'univers" (Sapir-Whorf) ... Des notions comme celles d'hyponymie (Lyons, p. 346–348), de "structure hiérarchique du vocabulaire" (idem, p. 349–350), de champs sémantiques (Trier, 1931; Matoré, 1953; Ohman, 1953; etc.), d'analyse componentielle (Lyons, p. 359–366; Bendix, Conklin, Goodenough, Lounsbury, Mounin, etc.), sont d'une extrême importance en taxinomie. Et il n'est pas douteux que les tentatives de structuration du vocabulaire, depuis – au moins – Roget (1852), en passant par R. Hallig et W. von Wartburg (1952), U. Weinreich (1963), et bien d'autres, ont encore insuffisamment attiré l'attention des spécialistes de la classification. On peut mentionner encore ici la tentative décriée par les linguistes "traditionnalistes", mais cependant très intéressante, d'A. Juret (1960) pour structurer le vocabulaire indo-européen suivant une classification des idées abstraites. Enfin, il faut signaler le grand intérêt des recherches poursuivies, depuis 1955, par Jean-Claude Gardin et ses collaborateurs ou disciples pour la constitution de langages artificiels visant à la description plus rationnelle d'objets archéologiques (artifacts), voire même à l'analyse sémantique de corpus littéraires (M. Allard *et al.* 1963). Il n'est guère douteux que la linguistique, dans son ensemble, et sa branche sémantique en particulier, doivent figurer parmi les principales disciplines de base sur lesquelles devra s'appuyer une future taxinomie générale.

035 Le point de vue "scientologique"

On nous permettra ce néologisme, adjectif correspondant à la "science de la science": cette discipline nouvelle, dont on peut à peu près dater la naissance, il y a moins de quatre décennies, avec l'article de S. Ossowski et M. Ossowska (1936) – bien qu'elle ait eu des précurseurs en Russie et que l'on puisse, sans doute, faire remonter ses lointaines origines jusqu'à Fichte. Elle s'est développée en englobant la philosophie et l'histoire des sciences, bien plus anciennement constituées, la "scientométrie", antérieure d'une dizaine d'années (Lotka, 1926; W. Hulme, 1922), et qui devrait – au moins en principe – servir de base théorique à une application pratique, la politique de la science *(science policy)*. Comment se sont constituées les différentes sciences; comment elles ont organisé leur "base institutionelle" – réseau de centres de recherche, de chaires professorales, d'associations professionelles, de circuits d'information et de diffusion; quels rapports elles entretiennent avec les disciplines voisines, quelles anastomoses les y rattachent: telles sont quelques-unes des questions scientologiques qui ont de profondes implications pour la taxinomie, car celle-ci peut trouver, dans les réponses qui y sont données, une base *objective.* Aux confins de la scientologie et de la logique, se trouvent les recherches concernant la classification à l'intérieur de chaque science: l'étude de la manière dont la systématique (pour employer le terme utilisé par les biologistes) de la discipline a évolué, s'est transformée au fur et à mesure du progrès des connaissances et dont elle doit actuellement être conçue, voire formalisée. Ce sont surtout les classifications biologiques qui ont été l'objet des travaux de cette nature les plus nombreux (par ex. Daudin, 1926; Souèges, 1938–1951; Simpson, 1945, 1961; Gregg, 1956) mais on en trouve aussi pour d'autres domaines (chimie, notamment: Vickery, 1961).

036 Le point de vue pédagogique

Henry E. Bliss a été l'un des premiers à attirer l'attention sur le point de vue pédagogique dans ce qu'il a appelé l'organisation des connaissances (1929). Sont à considérer, de ce point de vue, les programmes d'études des établissements d'enseignement (principalement les universités) et la division de ceux-ci en facultés, départements, etc.; les plans des encyclopédies systématiques – dont l'histoire remonte à la fin du 2e millénaire avant notre ère (Amanemope, en Egypte).

037 Le point de vue bibliographique, bibliothéconomique et documentaire

Les classifications utilisées dans les bibliothèques, dans les bibliographies et – beaucoup plus tardivement – dans les centres de documentation spécialisés ont été l'objet de toute une littérature, depuis Naudé (1627) et Montfaucon (1739), jusqu'à Šamurin (1955–1959), en passant par Peignot (1802), Petzholdt (1860), Richardson (1901), Bliss (1933) . . . Leur histoire est presque aussi ancienne que celle des classifications pédagogiques, dont d'ailleurs elles ont souvent dépendu – plus que des classifications scientifiques et/ou philosophiques, qui les ont, sans doute, influencées, mais avec un retard souvent très considérable.

038 Le point de vue mathématique et informatique

Robert A. Fairthorne a inauguré, en 1947, l'étude mathématique des classifications, mais celle-ci ne s'est vraiment développée qu'à partir du moment où l'emploi des ordinateurs sur une plus vaste échelle a provoqué la recherche d'algorithmes de classification automatique, soit dans l'optique de la traduction mécanique, soit dans celle de la recherche documentaire, soit enfin en vue d'aboutir à ce que l'on a ensuite appelé la taxonomie numérique (Sokal, 1966; Cole, 1968), pour faciliter les opérations de classification des organismes vivants (surtout en bactériologie), puis dans d'autres domaines archéologie et linguistique, notamment: cf. Gardin, 1971, Carvell et Svartvik, 1969. Ce champ de recherches, actuellement en pleine expansion, demanderait à lui seul un rapport spécial, et nous ne l'aborderons pas ici (voir Sparck Jones, 1971). Bar-Hillel (1962; cf. 1964, p. 346–351) avait très sévèrement jugé les premiers essais de théorie mathématique de la classification appliquée à la recherche documentaire, allant jusqu'à affirmer qu'il s'agissait d'un objectif (presque) impossible à atteindre (1964, p. 351). Cette mise en garde ultra-pessimiste n'a guère découragé les chercheurs et, s'il est vrai que beaucoup de méthodes proposées pour mesurer les "distances" entre sujets se sont révélées fallacieuses, il s'en faut de beaucoup que toutes les études réalisées aient été vaines. Il semble que Bar-Hillel, néo-positiviste de l'école de Carnap, se soit laissé ici entraîner dans la même voie où Auguste Comte lui-même s'était si imprudemment aventuré quand il avait voulu "établir en principe que jamais l'analyse mathématique ne saurait étendre son domaine au-delà de la physique des corps bruts" (1822; cf. 1854, p. 123). – En passant, il convient de signaler que des méthodes graphiques ont été utilisées avec succès pour réaliser des "classifications automatiques" dans différents domaines (archéologie: V. Elisseeff, 1965; diverses sciences humaines, P. Bertin au Laboratoire de cartographie de l'Ecole pratique des hautes études).

039 Le point de vue sociologique et ethnologique

Dans un mémoire célèbre de *l'Année sociologique* (1902) Durkheim et Mauss avaient inauguré ce qu'on pourrait appeler "l'ethnologie de la classification", ou la taxinomie ethnologique. Celle-ci n'a pas cessé d'être cultivée depuis, par des chercheurs aussi éminents, que Claude Lévi-Strauss, G. Dumézil, et par beaucoup d'autres (par ex., G. Dieterlen, 1952) et l'on pourrait lui rattacher aussi bien la plupart des travaux que nous avons évoqués ci-dessus (034) à propos de l'analyse componentielle.

04 Philosophie et taxinomie

Sans doute aurions-nous pu considérer le "point de vue philosophique" au § 03 ci-dessus – mais, s'agissant d'un Colloque organisé par une faculté de Philosophie, le point de vue philosophique méritait bien un paragraphe spécial . . . Selon que l'on donne au mot "philosophie" une acception plus ou moins large, incluant ou excluant la pensée mythique et religieuse – et, au moins pour ce qui est de la classification philosophique des connaissances, il nous semble impossible de ne pas adopter la définition la plus extensive – il s'agit d'une tradition encore plus ancienne que les deux millénaires dont parle le Bulletin No 1 du Colloque. Les Védas comme la Genèse contiennent une systématisation des connaissances, et les glorieux 6e et 5e siècles grecs ont déjà posé les bases de presque toutes les formes modernes de classification du savoir. Par ailleurs, comme pour ce qui concerne les classifications documentaires, les guides historiques ne manquent pas: qu'il suffise de citer ceux de Shields (1888–1889), Wundt (1889), Mariétan (1901), Flint (1904), Bliss et Ostwald (1929 tous les deux), Kedrov (1961–1965). Cependant, mis à part ce dernier (que nous n'avons pu étudier), tous ces auteurs nous sont apparus comme ayant plus ou moins manqué aux règles de l'historiographie moderne. Plusieurs ont conçu l'histoire des taxinomies philosophiques comme Hegel l'avait fait de l'histoire de la philosophie: celle d'une évolution aboutissant à leur propre système. La plupart n'ont guère tenté de mettre en relation la systématisation des savoirs, telle que les philosophes l'avaient conçue, avec les transformations survenues dans les théories scientifiques et dans l'organisation de la science. A en juger d'après les sous-titres de ses tomes (et aussi d'après ce que son auteur avait écrit précédemment, de 1946 à 1954), l'ouvrage de Kedrov n'est peut-être pas exempt d'un certain parti-pris, que l'on pourrait définir en parodiant Boileau: "enfin Engels vint" . . . Par ailleurs, il semble que toutes ces histoires aient fait preuve de ce que l'on pourrait appeler un "européocentrisme" excessif, selon une tradition qui remonte d'ailleurs loin (voir la phrase de Hegel: "Ce qui est oriental doit [. . .] s'exclure de l'histoire de la philosophie", p. 20 des *Leçons sur l'histoire de la philosophie,* tome II, Gallimard, 1970).

05 Objet du présent rapport

Il ne saurait être question de reprendre ici en sous-oeuvre les travaux de nos prédécesseurs, afin de parvenir à une conception plus objective de l'évolution du système des sciences, en fonction du progrès de la connaissance et de la complexité croissante de l'organisation de la recherche scientifique (elle-même fonction de l'évolution technologique, économique et sociale). Ce devrait être là un des objectifs du futur Centre de recherches de l'université d'Ottawa! Nous voudrions seulement, dans une première partie, dégager quelques tendances de cette évolution qui nous paraissent présenter une importance particulière, puis, dans une deuxième partie, étudier quelques aspects des répercussions, des conséquences, qu'elle a pu avoir dans le passé – ou qu'elle pourra avoir dans un avenir plus ou moins proche, compte tenu du phénomène d' "hystérésis" que l'on observe à cet égard – sur les systèmes d'organisation des connaissances utilisés en documentation et en analyse de l'information.

1 L'évolution du système des sciences

Dans le chapitre IX de son *Introduction à la philosophie,* qui résume d'ailleurs son ouvrage antérieur *Vom Ursprung und Ziel der Geschichte* (1949), Jaspers présente un tableau philosophique de l'histoire universelle qui comporte deux grands "tournants": celui de ce qu'il appelle la "période axiale" (entre –800 et–200), et celui de "notre âge technico-scientifique" qui, dit-il, "se préparait en Europe depuis la fin du Moyen Age", s'est "constitué spirituellement au XVIIe siècle [. . .] largement déployé depuis la fin du XVIIIe siècle" (1965, p. 105, 109, 104) et qui "est comme un second commencement, comparable seulement à la première découverte des outils et du feu" (p. 109). Quelque contestable que puisse être ce schéma, selon certains points de vue (notamment celui de l'histoire économique et sociale), il convient assez bien à notre propos présent. En effet, nous vivons encore, dans une large mesure, en matière de système des sciences, sur des conceptions fondamentales qui se sont formées entre la fin du 7e et le 3e siècles avant notre ère, mais il y a de fortes raisons pour penser que nous sommes les témoins de ce que – utilisant une conception due à Kuhn (1962) – nous devons bien appeler une "révolution scientifique", c'est-à-dire un changement de paradigme, rendu nécessaire par les très profonds changements qui se sont produits dans les différentes sciences (celles de la nature comme celles de l'homme) depuis – à peu près – la décennie 1895–1905.

11 De Thalès à Engels

Sans pour autant donner raison à Kedrov, pour qui Engels est le fondateur du système des sciences moderne, on peut admettre qu'il marque assez précisément le point de jonction entre l'ancien et le nouveau "paradigme" de la taxinomie. Il annonce la conception actuelle par plusieurs remarques de *Dialectique de la nature,* et en même temps il se rattache expressément à l'ancienne tradition. C'est une coïncidence remarquable que l'année de sa mort (1895) soit aussi celle où la découverte des rayons X par Roentgen marque le véritable début du bouleversement des théories physiques qui va amener, par voie de conséquence, celui des classifications des sciences. Cette longue période d'environ 2 500 ans peut être divisée en trois sous-périodes d'importance inégale: du –7e au–ler siècle; du 1er siècle à 1638; de 1644 à 1895.

111 Des physiologues ioniens à Geminus

Avec Thalès, Anaximandre et Anaximène, la cosmologie se dégage pour la première fois du mythe et recherche une explication globale rationnelle de l'univers; avec les Pythagoriciens, les mathématiques, bien qu'encore mêlées à des spéculations mystiques, accumulent assez de connaissances objectives pour qu'une première systématisation de leur contenu devienne possible, sous une forme qui sera reprise par Platon, puis maintes fois après lui et qui deviendra la *quadrivium* des écoles médiévales. Xénophane et Parménide posent en pleine clarté les problèmes de l'être, Héraclite ceux du devenir. Empédocle propose une théorie "chimique" – celle des quatre éléments – qui va subsister pendant près de deux millénaires et, assez probablement influencé par une pensée égyptienne qu'il dédivinise, une explication de l'évolution par la lutte de deux principes contraires (Amour et Haine) qui n'est pas sans présenter quelques résonances très modernes (il est cependant fort exagéré de dire, avec Voilquin (1964, p. 119), qu'il annonce Hugo de Vries). Leucippe et Démocrite introduisent l'atomisme. Les sophistes – il est vrai plus d'un siècle après Pânini, et bien moins savamment que lui – posent les premiers éléments d'une théorie linguistique comme ceux, par ailleurs, de la rhétorique et de la morale. Ainsi les présocratiques ont-ils mis en place toutes les pièces que les grands constructeurs de systèmes – Platon, Aristote, Epicure, les stoïciens – allaient pouvoir utiliser. Sur ceux-ci, trop connus, il n'est guère nécessaire de s'étendre. Il suffira de noter leur influence persistante jusqu'à l'époque moderne, que nous retrouverons dans les paragraphes suivants et qui, pour la triade stoïcienne, s'étend jusqu'à l'*Anti-Dühring* de 1877–78. Il est plus intéressant pour l'histoire de la classification des sciences de rappeler comment les progrès de la recherche scientifique alexandrine se sont reflétés dans la classification de Geminus (Rey, 1933, II, 180). Il faudra dix-huit siècles pour dépasser ce sommet.

Presque contemporains de Geminus, Varron à Rome puis Philon à Alexandrie donnent deux énumérations peu différentes des disciplines établies qui incluent non seulement les mathématiques pures et appliquées, mais aussi la grammaire, la rhétorique, la dialectique, plus (pour le premier) la médecine et l'architecture et (pour le deuxième) la logique. Tout le moyen âge respectera cette systématisation du *trivium* et du *quadrivium.* La compilation de Pline fournira d'autre part aux encyclopédistes médiévaux des matériaux de qualité fort inégale et une classification des sciences naturelles inférieure à celle d'Aristote. S'inspirant du Stagirite, mais sans tenir compte des restrictions qu'il apportait à son usage, Porphyre généralise la méthode dichotomique, qui trouvera parmi les constructeurs de classifications, jusqu'à Ampère, et plus tard encore, tant d'adeptes. Martianus Capella, et à sa suite Cassiodore, reprennent l'ordre de Varron. Mille ans vont se passer à d'innombrables variations sur le même thème: accorder, en la lui subordonnant, la raison à la foi. Pour ce qui est du système des sciences, cela donne depuis Isidore de Séville et Bède le Vénérable toute une série de rééditions, plus ou moins modifiées et complétées, du plan des *Noces de Mercure et de la Philologie* où transparaît, sous-jacente, la triade stoïcienne – la théologie remplaçant l'éthique. Si intéressante qu'en soit l'étude, nous la négligerons ici, mais il faut cependant marquer le lent enrichissement qu'apporte au contenu des schémas traditionnels le progrès des techniques et des sciences positives: le moyen âge n'est pas pour elles la "longue nuit" que l'on décrivait naguère: déjà Engels, (voir *Dialectique de la nature,* notes de 1875, éd. 1952, p. 191–193), puis Duhem et Koyré ont fait justice de cette légende. Mais il a fallu longtemps pour que, grâce aux "collèges" de traducteurs syriaques (7e–8e siècles), puis arabes (Bagdad, 832), enfin chrétiens (12e–13e siècles), ce qui restait de l'acquis scientifique grec puisse être réintégré dans la pensée occidentale; la philosophie rationnelle arabe (au sens large: y compris la persane et la juive) éveillera d'ailleurs la spéculation de celle-ci, d'Al Farâbi et Israëli à Avicenne, Avicébron, Al Gazali, Averroès. Il a fallu aussi que la "Renaissance du 12e siècle" soit rendue possible par le développement des villes, l'établissement des communes, la libération (relative) de l'enseignement se dégageant, au moins partiellement, de la tutelle épiscopale: c'est alors, comme l'écrit Bréhier (1937; éd. 1971, p. 103–104) que "les sciences du quadrivium qui avaient eu autrefois une fonction religieuse [. . .] sont en quelque manière libérées de cet office: mathématiques et astronomie s'apprêtent à exister pour elles-mêmes". Mais, chez Hugues de Saint-Victor, si l'on voit la mécanique ajoutée à la division aristotélicienne, celle-ci est, pour le reste, conservée et, finalement, il n'y a pas grande novation (cf. Bréhier, 1971, p. 169–171; Chatillon, 1968, p 155–157). Ce n'est qu'au siècle suivant que la *Somme de philosophie* faussement attribuée à Robert Grosseteste posera en principe l'autonomie des sciences naturelles vis-à-vis de la théologie. Ce 13e siècle est une époque de contrastes, où se mèlent l'ancien et le nouveau. Bonaventure peut bien reprendre, avec plus de méthode, la tentative "syncrétiste" d'Hugues de Saint-Victor et faire dériver toutes les connaissances des quatre degrés de la lumière divine, en même temps qu'il obtient de l'évèque de

Paris la condamnation d'une philosophie soupçonnée d'hérétisme; Albert le Grand et Thomas d'Aquin peuvent élever des monuments, que l'on retrouvera au 19e siècle contre le matérialisme, pour étayer la foi catholique par l'aristotélisme: celui-ci servira à Siger de Brabant pour dissocier la philosophie de la théologie. Et en Roger Bacon coexistent les deux tendances opposées: les trois *Opus* doivent fournir les moyens d'une rénovation intellectuelle propre à assurer la domination mondiale de la papauté, mais aussi la fondation d'une science expérimentale et mathématisée – pourtant, cette science elle-même doit utiliser l'illumination divine pour découvrir les secrets de la nature. On ne peut s'étonner de voir sa classification des sciences mêler des éléments très "modernes" à un schéma ancien (cf. Bréhier, 1971, p. 315–324; Šamurin, 1964, p. 83–92). L'occamisme du 14e siècle achève le divorce entre la foi et la raison, et le scepticisme radical d'un Nicolas d'Autrecourt joue un rôle "progressiste" dans la démolition de l'édifice monolithique des grands Docteurs scolastiques: la Renaissance n'est pas loin; elle est d'ailleurs déjà commencée à Padoue. Toutefois, si les 15e et 16e siècles voient se changer les conditions du travail intellectuel, avec la fondation des premiers musées (Laurent de Médicis), des premières bibliothèques publiques depuis la destruction de celles de Rome, d'Athènes ou d'Alexandrie, d'observatoires (Cassel, Uraniborg), d'une demi-douzaine de jardins botaniques (depuis celui de Padoue, 1545), d'institutions d'enseignement d'un type nouveau (la *Casa giocosa* à Mantoue; le Collège de France), des académies platonicienne à Florence et aldine à Venise, puis de toute la série de leurs soeurs dans les principales villes italiennes (Padoue, Sienne, Naples, Rome, Parme, Bologne); si – parallèlement, et concurremment au "décollage" économique – se multiplient les résultats de la "R. & D." d'alors (Copernic, 1543; la *Pirotechnia* de Biringuccio, 1540; le *De re metallica* d'Agricola, 1556; les oeuvres mathématiques de Tartaglia, Cardan, Viète, Stevin, à partir de 1537) – les effets de ces profonds changements sont lents à se faire sentir sur les systèmes de classification des connaissances. Angelo Ambrogini, dit Poliziano, en 1491, s'en tient, dans l'ensemble, à la version aristotélicienne de la fin du Moyen Age; en 1553, Mario Nizolio remet au goût du jour la triade stoïcienne, mais quand il subordonne l'éthique à la politique, on sent l'influence du *Prince* de Machiavel; l'Ardennais Christophe de Savigny, ami de Ramus, donne en 1587 des *Tableaux accomplis de tous les arts libéraux* qui, sans doute, doivent beaucoup à Agrippa de Nettesheim (et, par celui-ci, à Roger Bacon) et à la *Biblioteca universalis* de l'humaniste suisse Gesner, mais témoignent d'un effort rationaliste. Le chancelier d'Angleterre Francis Bacon est pris bien à tort par beaucoup d'historiens comme le fondateur de l'empirisme moderne; sa classification des sciences d'après les "facultés" de l'esprit humain (mémoire, imagination, raison) se trouve, trente ans avant *The Advancement of Learning* chez l'Espagnol Juan Huarte, qui l'a lui-même trouvée chez Avicenne – et cette source conduit, finalement, de nouveau, à Aristote . . . Cette période se termine avec l'utopiste Campanella qui, en plusieurs oeuvres de la fin de sa vie, reprend en somme le schéma de Nizolio, en rejetant la théologie du cercle des sciences, mais en admettant la théorie des deux voies de la connaissance.

113 L'âge d'or du rationalisme

A sept ans d'intervalle (1644–1651), Descartes et Hôbbes inaugurent véritablement la période "classique" du système des sciences. Chez eux, pour la première fois, l'esprit de la science moderne – celle de Galilée, de Kepler, de Harvey – inspire la classification des connaissances, sans référence aux anciens schémas médiévaux maintenant bien dépassés. Pourtant, dans sa dichotomie fondamentale Histoire/ Science, Hobbes suit partiellement Francis Bacon, et l'on pourrait retrouver, dans le détail de ses subdivisions, quelque influence de Savigny. Chez lui, comme chez Descartes, la théologie a disparu, mais il conserve l'astrologie. Dans ses plans pour une nouvelle encyclopédie (qu'il ne réalisera pas), le jeune Leibniz, en 1679, modernise lui aussi la triade stoïcienne en la complétant; Locke (1690) en donne une autre version, où émerge pour la première fois la sémiologie. D'Alembert (1751) reprend et complète le schéma baconien; Kant (1781) élabore une construction complexe, où la connaissance rationelle est opposée à la connaissance empirique (ou historique): celle-ci se divise en physique et anthropologie; celle-là est divisée et subdivisée, par dichotomies successives – la première opposant la connaissance philosophique (par concepts) à la connaissance mathématique (par construction de concepts). Ce système, où l'on retrouve au premier niveau la première dichotomie de Hobbes, sera maintes fois remanié pendant cent-cinquante ans par les néo-kantiens. Il représente un tournant dans l'histoire des classifications des sciences, tournant qui n'a pas toujours été apprécié à sa juste valeur. D'un côté, avec l'opposition qu'il marque entre deux logiques, celle des mathématiques d'un côté, celle des sciences de la matière de l'autre (cf. Goldmann, 1945; éd. française 1967, p. 13–14 et 67–68), Kant est à l'origine de tous les systèmes qui mettent à part les sciences abstraites, sciences formelles, sciences de l'ordre, etc. (Spencer, Wundt, Pearson, Husserl, Ostwald, Carnap . . .). D'un autre côté, le Kant de l'*Histoire universelle de la nature* (1755) et de l'*Essai d'introduire en philosophie la notion de quantités négatives* (1763) est un précurseur de la dialectique moderne (voir Engels, *Dialectique de la nature,* éd. 1952, p. 34–35, 52–53; Goldmann, 1967, p. 91). Enfin, la distinction que fait Kant, dans le donné, de trois domaines différents: mécanique, biologique, spirituel (Goldmann, 1967, p. 178–191) se retrouvera bien au-delà des écoles néo-kantiennes au sens strict. Un autre courant, nettement matérialiste, part de Diderot et du cercle holbachien: par le Dr Burdin, il aboutit aux différentes esquisses encyclopédiques du comte de Saint-Simon (1802–1813); le *Système analytique des connaissances positives de l'homme* de Lamarck (1820) reflète la même idéologie, dont on trouve une version affadie et dogmatique chez Auguste Comte: c'est un des problèmes de l'histoire du système des sciences d'expliquer pourquoi c'est précisément le comtisme, et non l'holbachisme ou le saint-simonisme qui a été adopté par la majorité des épigones du 19e et même du 20e siècle. Très peu de temps après Saint-Simon, entre 1808 et 1817, Hegel reprend une fois de plus le vieux schéma triadique stoïcien, mais en cherchant à lui donner une base rationnelle: il doit désormais exprimer le développement dialectique de l'Idée absolue. On nous taxera sans doute d'irrévérence, mais le monument hégélien nous inspire le même sentiment que ceux d'Albert von Bollstädt ou de saint Thomas:

il s'agit d'oeuvres achevées aux deux sens du terme, dont la belle ordonnance ne peut qu'être ou bien admirée en silence, ou bien soumise à une critique destructive. Que Hegel soit resté bien en arrière de la science avancée de son temps, on n'en voudra pour preuve que deux détails de sa classification: elle place ensemble, sous le titre de "nature organique" la géologie comme la botanique et la zoologie, alors que Lamarck, deux ans avant l'*Encyclopédie des sciences philosophiques*, a déjà reconnu l'unité et la spécificité de la biologie; elle néglige l'économie politique, cependant constituée comme science depuis un demi-siècle (la Société des économistes a été fondée en France en1767) et reconnue comme la première des sciences sociales par Cambacérès dans son *Discours sur la science sociale* de l'an VI. Avec Ampère, nous rencontrons pour la sixième fois (après Aristote, Roger Bacon, Descartes, Leibniz, Lamarck) un véritable "scientifique" auteur d'une classification des sciences. Aux triades hégéliennes, Ampère préfère la dichotomie porphyrienne, qu'il applique d'un bout à l'autre de son plan; celui-ci, avec l'opposition primordiale qu'il introduit entre la nature et la pensée (opposition qui sera reprise, après lui, par Du Bois-Reymond, Dilthey, et bien d'autres), est en régression sur celui des holbachiens. En 1851, un autre scientifique, Cournot, remanie la classification ampérienne, mais la transforme en appliquant (pour la première fois, sauf erreur) le principe du parallélisme de trois séries de sciences: série théorique, série cosmologique et historique, série technique et pratique. On y voit apparaître, sous le nom de "chrématologie", l'économétrie que Cournot lui-même a fondée avec son *Essai sur la théorie des richesses* en 1838. C'est l'époque, par ailleurs, où s'élabore la théorie du matérialisme historique, dont on décèle les germes chez Montesquieu, Turgot, Barnave et Cambacérès, qui se précise chez Guizot en 1822 et, peu après, chez Mignet et Augustin Thierry, prend forme avec Proudhon (1843) et se trouve enfin clairement exprimée par Marx et Engels dans leur manuscrit de 1845–1846 cependant "livré à la critique rongeuse des souris". C'est aussi le temps où, vingt ans après le mémoire de Carnot, resté à peu près ignoré, le principe d'équivalence et le concept d'énergie permettent d'intégrer toute une série de phénomènes physiques, où la théorie cellulaire, celle du développement embryonnaire et celle de l'évolution jouent le même rôle en biologie. Changement sans doute plus fondamental, les méthodes de l'analyse statistique, que Maupertuis, Buffon, Condorcet, Malthus, Laplace avaient tenté dès le 18e siècle d'appliquer aux phénomènes biologiques et sociaux, commencent à se généraliser: la portée de cette "attitude radicalement nouvelle" a été bien marquée par François Jacob (1970, p. 190–192, 212–220). Les années 1858–59 voient, presque simultanément, paraître la *Cellularpathologie* de Virchow, l'*Origin of species* de Darwin, *Zur Kritik der politischen Oekonomie* de Marx, les *Illustrations of the dynamical theory of gases* de Maxvell. En 1859–60, avec les visites des géologues anglais chez Boucher de Perthes à Abbeville, la préhistoire – après six décennies d'incompréhension, depuis les découvertes de John Frere (1797) – était enfin reconnue. Et c'est en 1858 (lettre à Marx du 14 juillet) qu'Engels commence à s'intéresser au problème de la classification des sciences, qu'il reprendra plus sérieusement entre 1873 et 1882. Les auteurs soviétiques (Kedrov, 1946–47; Šamurin, 1967, p. 8–15) ont exagéré l'originalité d'Engels: lui-même reconnaissait sa dette vis-à-vis de Saint-Simon et de

Hegel (Grolier, 1956, p. 132–133). Avant lui, Gorjaninov (1834, 1843) et Dove (1850) avaient proposé des classifications évolutives, basées sur la notion d'émergence de propriétés nouvelles, pour ranger les phénomènes naturels (et humains); Isidore Geoffroy Saint-Hilaire, en 1854, avait cru pouvoir affirmer que "les sciences se développent [. . .] précisément dans l'ordre même où leurs objets se sont produits". Et, comme la *Dialectique de la nature* n'est parue qu'en 1925, c'est H. M. Stanley qui, en 1884, a publié le premier un plan analogue à ceux de Gorjaninov et de Dove, mais tenant compte des progrès intervenus entre temps dans les sciences physico-chimiques. Il faut, certes, mettre au crédit d'Engels ses idées sur la nécessité de baser le système des sciences, non sur des constructions philosophiques a priori, mais sur "les connexions des phénomènes eux-mêmes", sur l'indispensable abandon des "oppositions diamétrales représentées comme inconciliables et insolubles", des "lignes de démarcation nettes et infranchissables", des "différences de classes fixées de force" (voir les références dans Grolier, 1956, p. 135–136). Il se rendait compte qu'il fallait "montrer [. . .] non seulement l'enchaînement entre les phénomènes de la nature dans les différents domaines pris à part, mais encore la connexion des différents domaines entre eux [. . .] au moyen de faits fournis par la science empirique de la nature ellemême" (texte de 1888; cf. Grolier, 1956, p. 142). A cet égard, d'ailleurs, il faisait écho au discours de James Clerk Maxwell à Cambridge, onze ans plus tôt, qui évoquait le processus de "cross-fertilization" comme l'une des caractéristiques de la science moderne. Mais il avait aussi écrit, en 1874, ces lignes révélatrices de l'esprit de l'époque: "Maintenant que la conception nouvelle de la science de la nature est *achevée dans ses traits fondamentaux* . . ." (souligné par nous). Cet optimisme "scientiste", qu'apparemment il devait conserver durant le quart de siècle qui lui restait à vivre, n'allait pas tarder à être cruellement démenti.

12 Crise et restructuration: 1895–1971

Presque cent ans après cette phrase d'Engels, aucun savant n'oserait plus affirmer qu'il n'y a plus rien à changer dans les "traits fondamentaux" de notre conception de la nature. Il a suffi d'une génération pour faire succéder, aux orgueilleuses certitudes de la science et de la philosophie rationaliste "fin de siècle" un scepticisme qui rappelle par certains aspects l'époque des sophistes grecs du 5e siècle avant J.-C. ou celle des occamistes parisiens de la deuxième moitié du 14e siècle. La science, aujourd'hui, a pris conscience de la profonde vérité de la notion pascalienne des "deux infinis" comme de la pensée d'Hamlet: "There are more things in Heaven and Earth, Horatio, Than are dreamt of in our philosophy" (Acte I, Scène 5). En son "troisième âge", la physique "n'en finit pas de s'enfoncer au coeur de la matière" (P. de Latil, 1971, p. 709): des quarks aux partons, une nouvelle réalité sous-jacente aux particules "élémentaires" – qui ne le sont plus – se découvre (Kendall et Panofsky, 1971). Quant aux cieux, après les quasars et les pulsars, nul ne sait encore combien de catégories d'objets inconnus, et pour l'instant encore inexpliqués, vont y apparaître. Cependant, "le microcosme se révèle plus complexe que jamais, mais pas chaotique" (Bitsakis, 1971, p. 83) – et il en est de même du macrocosme. Roger Caillois (1969, p. 24), dans un article remarquable où il dégage la portée – disons "philosophique" – de la table périodique de Mendeleev, écrit qu'il n'est "pas indifférent qu'à l'idée pascalienne, romantique et impressioniste d'un univers inépuisable du ciron à la nébuleuse, soit opposée l'épure limitée et, comment dire? orthogonale qu'une pensée moins précipitée et une recherche plus méthodique ont su amener à la lumière [. . .] L'univers sans doute est immense et labyrinthique. Il reste que les brumes, les nuées qui constamment s'y effilochent et s'y recomposent, dissimulent un plan quadrillé que balisent des échos, des rappels, des repères périodiques." François Jacob a construit son "histoire de l'hérédité" sur les quatre étapes de la découverte progressive, depuis le 17e siècle, de structures de plus en plus profondes, cachées et simples: d'une "série d'organisations emboîtées les unes dans les autres comme des poupées russes". Il écrit – et cette phrase pourrait s'appliquer à bien d'autres domaines que celui de la biologie: aux mathématiques comme à la physique, à la logique comme à l'astronomie – "Au delà de chaque structure accessible à l'analyse, finit par se révéler une nouvelle structure, d'ordre supérieur, qui intègre la première et lui confère ses propriétés. On n'accède à celle-ci qu'en bouleversant celle-là, en décomposant l'espace de l'organisme pour le recomposer selon d'autres lois" (Jacob, 1970, p. 24). Et, au dernier paragraphe de son livre, constatant que les organisations, les systèmes, les hiérarchies de la nature s'expliquent "en fin de compte" par "les propriétés des atomes que décrivent les lois de l'électromagnétisme de Maxwell", il termine, non par des affirmations, comme l'aurait fait sans doute un de ses pareils à l'époque d'Engels, mais par des interrogations: "peut-être y a-t-il d'autres cohérences possibles dans les descriptions. Mais enfermée dans son système d'explications, la science ne peut s'en évader. Aujourd'hui le monde est messages, codes, information. Quelle dissection demain disloquera nos objets pour les recomposer en un espace neuf? Quelle nouvelle poupée russe en émergera? " (p. 345).

On a décrit bien des fois, et il n'est donc pas nécessaire de le faire de nouveau ici, la "crise" de la physique, ouverte par l'expérience de Michelson et Morley dès 1887, puis étendue en 1895–98 par les découvertes successives de Roentgen, de Becquerel et des Curie et qui allait aboutir à une révolution majeure, aussi profonde (ou plus) que celle de Galilée, avec la théorie de la relativité et celle des quanta. Aux modèles déterministes de Newton et de Laplace succédaient des modèles probabilistes ou stochastiques. Parallèlement, les deux plus vieilles sciences, mathématique et logique, connaissaient elles aussi une "crise", ouverte en fait dès 1829 pour la première par Lobachevsky et 1847 pour la deuxième par Boole et De Morgan, mais qui se manifestait avec éclat avec les "paradoxes" de la fin du siècle (Burali-Forti, 1896). Transformée, élargie et unifiée par la théorie des ensembles et la topologie, la mathématique n'a sans doute pas achevé sa révolution: il n'y a guère plus d'une décennie qu'une "nouvelle structure, d'ordre supérieur", celle des catégories, a fait son apparition. Quant à la logique, elle n'a pas non plus fini de digérer les problèmes de la pluralité des valeurs, posés par Lukasiewicz et Post en 1920–21, ni ceux des limites de la formalisation, nés des théorèmes de Goedel dix ans plus tard. Une discipline que, depuis Euclide, l'on croyait à jamais assurée de sa pérennité, la géométrie, perd le droit à l'existence autonome – mais la logique elle-même peut-elle encore revendiquer un domaine qui lui soit propre, et ne doit-elle pas être fusionnée avec des mathématiques qui ne sont plus perçues comme science de la quantité? Piaget (1970, p. 575) en fait "un cas particulier d'algèbre générale".

Si spectaculaires qu'aient pu être, vers ces deux décennies cruciales qu'englobe la "belle époque", les transformations de l'astronomie qui, de science mathématique (mécanique céleste: titre encore, de 1892 à 1899, du livre d'Henri Poincaré) devient une science physique, celles de la biologie qui, comme la physique (et même avant elle), découvre avec De Vries les "sauts quantiques", fait disparaître, avec Büchner et Loeb, les barrières qui la séparaient de la chimie (déjà bien abaissées par Pasteur dès 1848, mais les conséquences de la *Dissymétrie moléculaire* ou des *Etudes sur la bière* ne seront aperçues que plus tard), et élabore la théorie du gène, celles même d'une psychologie explorant l'inconscient avec Breuer et Freud (1895–1900), découvrant les réflexes conditionnés avec Pavlov (1903) et, avec Spearman et Thurstone, se muant en psychométrie – ces changements n'ont pas eu sur l'évolution du système des sciences le même retentissement que les restructurations douloureuses des sciences dites exactes.

Il n'est que d'ouvrir quelque texte philosophique récent de vulgarisation pour voir comment les constructeurs de systèmes métaphysiques se sont saisis avec empressement des armes que leur fournissait une science doutant de ses propres fondaments et ne pouvant plus croire au "rêve d'une réductibilité intégrale tel que l'univers entier en ses manifestations innombrables se réduirait à une équation de base d'où l'on pourrait tirer toutes les autres" (Piaget, 1970, p. 52). C'est, par exemple, Jaspers (1964; texte français, 1970, p. 11–14) écrivant qu'aucune des "conceptions abstraites, mathématiques, de l'univers dans son ensemble ne peut être prouvée scientifiquement de façon définitive" et que, par conséquent, "le chemin qui mène à la connaissance de l'univers se perd pour ainsi dire dans l'infini"; qu'"il n'y a plus de particules élémentaires que l'on puisse imaginer" et que, donc, "l'essence de la matière ne saurait être définie"; que les savants "ne pourront jamais créer la vie"; qu'enfin les sciences modernes "avec leurs résultats stupéfiants sur le plan particulier [. . .] ne font qu'approfondir le mystère sur le plan général". Il ne reste plus alors qu'à "comprendre le monde à partir de lui-même", à l'aide d'une "expérience possible d'un amour du monde" qui "voit le monde réel en chiffres" – c'est-à-dire en symboles (p. 18). C'est Bochenski décrivant (1967, p. 17–26) "la crise de la physique newtonienne", la "critique de la science", la "crise des mathématiques", et, parallèlement, l'apparition de "l'irrationalisme vitaliste", la "renaissance de la métaphysique réaliste", pour conclure que "Si le 19e siècle fut une époque surtout moniste et matérialiste, la crise de 1900 annonce la prédominance, solidement établie, d'un personnalisme spiritualiste".

A vrai dire, dans ce mouvement qui débute à peu près avec l'empiriocriticisme de Mach et d'Avenarius, dont les premiers écrits philosophiques sont de 1872 et 1876, mais dont les oeuvres principales se situent entre 1883 (*Mécanique* de Mach), 1888–90 (*Critique de l'expérience pure* d'Avenarius) et 1905 (*Connaissance et erreur,* de Mach), le bergsonisme (1889–96) et la phénoménologie de Husserl (1900–01), il y a peu d'originalité véritable. C'est l'époque où fleurissent les "néo-. . .": néo-kantisme, néo-thomisme, néo-positivisme, sans compter toutes les philosophies qui se veulent neuves et ne font qu'habiller de vêtements à la mode quelque très ancienne idéologie. Même l'existentialisme, seul courant vraiment novateur, ne fait guère que continuer Kierkegaard, dont l'oeuvre essentielle se situe entre 1841 et 1845. Nous n'avons pas à faire ici de l'histoire de la pensée, et il importe peu de savoir si, comme l'art moderne et presqu'en même temps que lui, la philosophie de notre temps n'a pas eu surtout pour rôle d'exprimer l'angoisse humaine devant des bouleversements techniques et sociaux qu'elle a pressentis avant qu'ils adviennent réellement et qui peuvent faire douter de la survivance même de l'humanité: *Sein und Zeit* et *L'Etre et le Néant* encadreraient *Guernica.*

121 Les classifications philosophiques des connaissances

Il nous semble – mais peut-être n'est-ce qu'une illusion: seules des études statistiques précises permettraient de confirmer ou d'infirmer ce que nous n'avançons guère que comme une impression subjective – que, depuis le tournant du siècle, les philosophes comme les savants eux-mêmes sont beaucoup moins ardents qu'auparavant (et tout spécialement qu'au siècle précédent) dans leur recherche du système des sciences "idéal". Alors que tout philosophe qui se respectait croyait devoir, au 19e siècle, publier son schéma de l'ordre des sciences, et que maint "scientifique" en faisait autant, leurs émules sont plus rares depuis 1900. Seuls peut-être font exception les auteurs soviétiques qui, il est vrai, se situent dans la stricte tradition d'Engels et, par conséquent, échappent à ce discrédit général qui frappe ailleurs la taxinomie théorique (une note bibliographique dans Šamurin, 1967, II, p. 751–752, énumère onze travaux dans ce domaine, dont deux thèses, parus entre 1960 et 1964 en U.R.S.S.). Il est assez symptomatique de voir P. Auger, dans son rapport pour l'UNESCO sur les *Tendances actuelles de la recherche scientifique* (1961) se contenter de reprendre le vieux schéma comtien, un peu "rafistolé", sans chercher à proposer une systématisation nouvelle. Nous pouvons, d'autre part, négliger toute l'école néo-thomiste, dont le système des connaissances se résume en dernière analyse à une mise à jour de l'hylémorphisme de son maître.

Tout au début de la période qui nous occupe présentement, nous trouvons les néo-kantiens de l'école de Bade: développant une idée déjà présentée par Dilthey en 1883 (et qu'il a peut-être lui-même reprise à Ampère), Windelband (1894) et Rickert (1896–1902) introduisent une coupure fondamentale entre les sciences de la nature, nomothétiques, posant des lois, généralisantes, et celles de l'esprit, idiographiques, individualisantes. La fortune de cette conception est étrange: elle sera reprise, avec des variantes, par Stumpf (1906) – qui découvre, en dehors de ces deux séries, une série de "sciences neutres", ni de la nature, ni de l'esprit, dans lesquelles, sous l'influence de Husserl, il range la phénoménologie et l'eidologie, par Becher (1921), Oppenheim (1926), et se retrouve jusque dans la grande enquête de l'UNESCO sur les "Tendances principales de la recherche dans les sciences sociales et humaines", où elle a été adoptée sur les suggestions de Jean Piaget (1970, p. 2–8) qui, il est vrai, la modifie plus profondément, puisqu'il admet dans le groupe des sciences de lois une demi-douzaine de sciences humaines.

Entre 1900 et 1913, Husserl élabore la phénoménologie. Du point de vue qui nous occupe ici, son système comporte une distinction fondamentale, de nouveau absolument tranchée, entre sciences des faits et sciences de l'essence ou eidétiques, ces dernières comprenant les mathématiques – et la philosophie phénoménologique, procédant d'abord par "réduction eidétique" puis arrivant à la conscience pure par réduction transcendantale, au moyen des expériences intentionnelles. On trouvera une critique, à notre avis fort topique, de ce système qui aboutit, comme le bergsonisme, à la recherche, par delà les frontières assignées à la connaissance scientifique, "d'un autre type de savoir revenant [. . .] à doubler la science elle-même", dans Piaget (1970, p. 41–43).

Presque en même temps que l'idéalisme transcendantal de Husserl, fort proche du néo-kantisme, Croce développe un autre type d'idéalisme, néo-hégélien, avec sa tétralogie de la *Filosofia dello spirito* (1902–17). Il accepte les idées de Mach, Poincaré, Duhem, James sur la nature des théories scientifiques et les pousse à leur conséquence logique extrême: les sciences de la nature et les mathématiques, reposant entièrement sur des conventions arbitraires, ne sont que des pseudo-sciences. Seule réalité: l'esprit, et ses deux activités, théorique/pratique. Une deuxième dichotomie – d'ailleurs double – concerne l'objet de cette activité, individuel ou universel, et la nature de la connaissance, intuitive ou rationnelle. D'où finalement une division quadripartite de la science de l'esprit, qui est aussi la philosophie (et l'histoire):

		objet:	
		individuel	universel
activité:	théorique	esthétique	logique
	pratique	économique	morale
		connaissance:	
		intuitive	rationnelle

On ne saurait dénier à cette construction, dans sa rigueur symétrique, une certaine harmonie, qui en fait une sorte d'objet esthétique. Elle exclut toute métaphysique, qui ne peut fournir de connaissance réelle, faute d'intuition intellectuelle, et la religion, ensemble de mythes intéressants à étudier et de règles morales moins superficielles que celles de la morale laïque. Ceci dit, le système repose entièrement sur une interprétation de l'activité théorique de la science qu'aucun véritable scientifique, fût-il par ailleurs le plus acharné des adversaires du matérialisme, ne saurait considérer que comme une caricature: il a au moins ce mérite, d'ailleurs, de montrer où conduisent les conceptions de la théorie physique comme construction arbitraire, uniquement pratique, mises à la mode vers 1900, sous le coup de la crise de la physique classique.

Entre 1908 et 1929, le chimiste W. Ostwald publie divers ouvrages de "philosophie de la nature" exposant un système des sciences néo-positiviste; reprenant des idées exposées sous diverses formes avant lui par Neil Arnott en 1828, Pačoskij en 1891, Stumpf en 1906, il cherche à représenter graphiquement les relations des sciences entre elles et leurs interférences, et choisit pour cela la forme pyramidale.

Alfred North Whitehead mit encore plus longtemps à passer de la science à la philosophie; nommé à Harvard à 63 ans, il combat le matérialisme avec une étrange mixture néo-platonicienne-leibnizienne-spinoziste-bergsonienne additionnée de son interprétation personnelle de la théorie quantique et grâce à une notion ultra-extensive de "l'événement" qui en fait une sorte de monade revue et corrigée avec des souvenirs d'Héraclite et des exemples pris de la physique atomique. On retrouve chez lui, comme plus tard chez Teilhard, une conception que le leibnizien Maupertuis avait exposée dans son *Système de la nature* et dont on peut déceler les origines lointaines jusque chez Marsile Ficin (*Théologie platonicienne;* cf. Gusdorf, 1967, p. 443): la matière est douée de mémoire, d'intelligence, de désir ou d'aversion – Whitehead dit, ce qui revient au même, que tout "événement" est bipolaire, à la fois matière et conscience. De ce vaste système composite ressort-il une classification des savoirs? Sans doute: à côté des connaissances scientifiques existent les intuitions concrètes de l'art et de la religion; les dépassant toutes se trouve la philosophie. Il y a une hiérarchie entre les êtres, selon que, dans leur "bipolarité", prédomine ou non la conscience, et tout aboutit à Dieu.

Max Scheler, lui aussi, reconnaît plusieurs sortes de savoir: inductif, de la structure essentielle, du salut. D'où une triade: sciences-philosophie-religion. Nous n'entrerons pas ici dans le détail de ses conceptions sur les valeurs, la personne (isolée ou plurale), l'amour de celle-ci. Au surplus faudrait-il distinguer entre le Scheler personnaliste et théiste d'avant 1926 et le Scheler pessimiste des deux dernières années de sa vie. Indiquons en passant sa classification des unités sociales: foule-communauté-société, église/nation, dans laquelle il a élaboré des idées de Le Bon, de Tönnies, etc.

La question de savoir si les diverses philosophies existentialistes comportent une classification des connaissances, à défaut d'un système des sciences, demanderait une étude spéciale, que nous n'entreprendrons pas ici. L'opposition tranchée qu'elles établissement entre le *Dasein* et les autres étants, le pour-soi et l'en-soi, l'*Existenz* et le *Dasein,* au-delà des différences (ou même des oppositions: entre Heidegger et Jaspers) de terminologie, revient finalement à séparer l'homme du monde, dans lequel l'holbachisme avait voulu le réintégrer. Mounier (1962, p. 179–188) a raison de mettre au premier plan de tous les existentialismes la notion de transcendance, de montrer que pour toute l'école "la raison n'est pas le seul mode d'appréhension de l'existence", que "Le rapport originel d'une transcendance à une liberté ne peut [. . .] être du modèle de la connaissance rationnelle", que donc ce modèle doit être la foi, dont le "mode propre est l'incompréhensible" (p. 185) et que "chrétienne ou athée", la "réaction existentialiste [. . .] marque un retour du religieux dans un monde qui a tenté de se constituer dans le pur manifeste" (p. 188). C'est peut-être chez Jaspers que l'on trouve la conception la plus cohérente – et la plus antiscientifique: "les ensembles systématiques que les sciences ont pour tâche d'établir sont essentiellement et radicalement différents et irréductibles entre eux [. . .] Une faille sépare le monde physique du monde de la vie, celui-ci du monde de l'âme, celui-ci du monde de l'esprit. Il se trouvent hiérarchiquement ordonnés, de telle sorte que la réalité du degré supérieur implique celle du degré précédent, tandis que la réalité du degré inférieur paraît possible sans celle du degré supérieur [. . .] La totalité unique du monde, à quoi appartiennent tous les ensembles que les sciences explorent, ne constitue pas elle-même un ensemble susceptible de faire l'objet d'une théorie universelle [. . .] Il n'existe pas de représentation de l'univers; il n'y a qu'un système des sciences" (1965, p. 78–79). Pour arriver à la révélation de l'être, il faut donc renoncer à une représentation de l'univers: "le sens caché du savoir scientifique" est "d'arriver par la recherche à la limite où s'ouvre au savoir le plus lucide l'espace du non-savoir" (idem, p. 80). On reconnaît ici le leit-motiv de la philosophie contemporaine, que saint Bonaventure doit contempler avec plaisir du haut du Paradis: la connaissance scientifique n'est qu'une connaissance inférieure; c'est par une autre voie – celle de la lumière supérieure, pour reprendre le terme de Bonaventure (cf. Jaspers, p. 83: "la lumière instantanée, mais inoubliable, qui jaillit lorsqu'on essaie de penser philosophiquement l'être du monde") – que l'on atteint l'être universel. Cette expérience mystique, commune aux Victorins, à Bonaventure, à Roger Bacon, relie fondamentalement, de nos jours, des philosophies aussi différentes par ailleurs que celles de Bergson, Jaspers, Marcel, Hartmann.

Chez ce dernier, on retrouve une conception des quatre degrés, ou couches de l'être réel: matière, vie, conscience, esprit. Mais l'être réel n'est qu'une des deux sphères primaires de l'être, l'autre étant l'être idéal; à chacun de ces sphères primaires sont reliées une des deux sphères secondaires: connaissance et logique. La grande construction ontologique de Hartmann est de 1933–50: soutiendrait-il encore que l'élément "métaphysique" apparaît dans les sciences de la nature avec le "prodige de la vie", inexplicable soit mécaniquement, soit téléologiquement? Moins nettement peut-être que les existentialistes, Hartmann sépare aussi l'homme du reste du monde, puisque dans son système, si la conscience est donnée aussi aux animaux, l'être spirituel est, lui, nettement distinct de l'être psychique. Pour élucider cet élément catégorial nouveau, irrationnel, il s'inspire de Hegel (tout en le contredisant sur tel ou tel point) avec ses trois formes d'esprit: personnel, objectif, objectivé.

Avec Carnap (1934, 1938), nous retombons sur terre, bien loin de ces sphères éthérées; toutefois, la distinction tranchée du leader du Cercle de Vienne entre science formelle (logique et mathématiques) et science empirique (tout le reste, divisé par dichotomies successives en physique et biologique, celui-ci en biologie proprement dite et science du comportement, cette dernière en science du comportement individuel et science du comportement social) ne laisse pas de rappeler la distinction hartmanienne entre être réel et être idéal. Par ailleurs, on retrouve chez Carnap, à l'aboutissement de ses dichotomies, les cinq catégories de sciences de la pyramide ostwaldienne et, au delà, la série comtienne quelque peu modifiée:

nous sommes chez les néo-positivistes . . . Malgré les critiques qu'adresse Piaget à certains aspects de la "philosophie empiriste" (1970, p. 38–40) ou au "positivisme logique" (idem, p. 60–61), lui reprochant principalement "la coupure radicale introduite entre les faits d'expérience et le langage logico-mathématique", le fondateur de l'épistémologie génétique a proposé à plusieurs reprises (1950, I, 38–45; II, 237–238; 1970, 43–53, 329, 560–563) un plan de classification des sciences qui, finalement, diffère peu de celui de Carnap, sauf à se présenter, non sous forme linéaire, mais comme un "système circulaire ou spiroïdal".

Piaget veut éviter le "réductionnisme" et lui oppose ce qu'il appelle un "constructivisme" (1970, p. 286–288, 299–301, 52–53); il pense que "dans tous les domaines psychosociologiques [. . .] on se trouve actuellement en présence de trois sortes de niveaux [. . .] : l'organique, le mental et le social" (idem, p. 300 – cf. cependant la note p. 338, où il signale qu'une autre tendance – chez les Soviétiques – s'en tient "au dualisme plus simple des facteurs organiques et des facteurs sociologiques" et que "l'avenir départagera entre ces deux tendances"). Avec son insistance sur la "spécificité des phénomène d'échelle supérieure" (1970, p. 53) et la "hiérarchie des échelles de phénomènes" (idem, p. 561), ses références louangeuses à Ch. E. Guye (idem, p. 52 et 561) et à Bertalanffy (idem, p. 49, 61, 575, 592), Piaget se trouve d'autre part rapproché du dernier courant que nous voudrions considérer dans ce bref aperçu, à vol d'oiseau, des essais contemporains de taxinomie générale: celui des partisans, à quelque degré que ce soit, de l'"*emergent evolution*". Les origines de ce mouvement sont difficile à préciser. Derek Austin (1969, p. 81) fait remonter la "theory of integrative levels" (autre dénomination, synonyme) "au moins au positivisme de Comte", et, en tout cas, aux *Premiers principes* de Spencer (1862). En fait, on peut en déceler une première forme chez Bonnet en 1764, une version déjà nettement "moderne" chez Lamarck en 1809 (cf. Jacob, 1970, p. 42–43, 99–101) et nous avons vu ci-dessus (p. 32) les généralisations qu'en avaient proposé Gorjaninov et Dove dans la première moitié du 19e siècle. En tant que mouvement philosophique (plus ou moins) autonome, la théorie de l'émergence (ou des niveaux d'intégration) "émerge" elle-même lentement – dès Pascal et Leibniz sans doute, puis tout au long du 18e siècle, apparaît plus clairement chez Kant et chez Hegel, et trouve ce qui est probablement une de ses formes les plus remarquables, à la fois par quelques intuitions que l'on peut qualifier de géniales (sur les particularités du cristal, notamment) et par des affirmations dogmatiques qui nous paraissent aujour d'hui absurdes (comme sur le fait qu'il soit "ridicule de demander quelle est la cause de la pesanteur ou de l'électricité") dans *Le Monde comme volonté et comme représentation* de Schopenhauer (voir livre II, § 26–27, p. 175–201 de la trad. Burdeau-Roos, 1966). Plusieurs passages de la *Dialectique de la nature* (trad. française 1952, p. 260–261, 176–177, 301) peuvent faire ranger Engels parmi ses partisans, mais elle ne devient le signe de ralliement d'une véritable" école " que bien plus tardivement, à partir du Bergson de *L'évolution créatrice* (1907) et surtout avec l'Australien F. Samuel Alexander (1920) et l'Anglais Convy Lloyd Morgan (1923). Bien qu'elle ait compté des représentants en Suisse (Guye) et en France (Georges

Matisse, Louis Glangeaud, Teilhard de Chardin), l'école semble surtout avoir prospéré en pays anglo-saxons (Joseph Needham, A. B. Novikoff, J. K. Feibleman, L. von Bertalanffy . . .); aux Etats-Unis, Shields (1882) et Richardson (1901) peuvent lui être rattachés à titre de précurseurs. Chez ceux-ci, comme chez Alexander et Teilhard, la série des formes émergentes aboutit au divin: Dieu, "anges" ou point Oméga; d'autres s'arrêtent avant ce dernier stade: c'est le cas de François Jacob, dont le chapitre de "conclusion" de *La logique du vivant* s'intitule, précisément, "L'intégron" (1970, p. 320–345) et propose une version rationaliste de la théorie: "l'évolution biologique continue sans faille une longue évolution chimique" (p. 326); "il n'existe pas d'entité métaphysique pour se cacher derrière le mot de vie. Le pouvoir de s'assembler, de produire des structures de complexité croissante, de se reproduire même, appartient aux éléments qui composent la matière. Des particules à l'homme se rencontrent toute une série d'intégrations, de niveau, de discontinuités. Mais aucune rupture [. . .] Aucun changement d'essence" (p. 327). Ensuite, refusant pour qualifier "l'orientation que la sélection naturelle a imprimée au hasard" les termes de progrès, progression, perfectionnement, comme ceux de complication, complexité, il y voit une "tendance à l'assouplissement dans l'exécution du programme génétique", à la "constitution de nouveaux systèmes de communication" (p. 329), "à accroître les interactions de l'organisme et de son milieu" (p. 333), à développer "les moyens de recueillir l'information du dehors, de la traiter, d'ajuster en conséquence les réactions de l'organisme" (p. 336), ce qui "aboutit à de nouveaux systèmes de communication, de régulation, de mémoire, qui fonctionnent à un niveau plus élevé que l'organisme": dès lors apparaissent "les intégrons culturels et sociaux" qui "fonctionnent selon des principes inconnus aux niveaux inférieurs" (p. 341), et "débordent les schémas explicatifs de la biologie [. . .] Mais, si l'on voit à nouveau des paliers, des discontinuités de phénomènes et de concepts, on ne trouve aucune rupture avec les niveaux de la biologie" (p. 342). Si l'on veut, tout ceci rappelle Engels, et quand Jacob souligne que "Les concepts de démocratie, de propriété, de salaire sont aussi dépourvus de signification pour une cellule ou un organisme que ceux de reproduction ou de sélection naturelle pour une molécule isolée" (p. 341), il fait écho à une remarque tout à fait analogue du compagnon de Marx (*Dialectique de la nature,* 1952, p. 319). Mais, le séparant de celui-ci, il y a tout l'acquis des sciences de la nature depuis un siècle. Et Engels, qui parle dans l'*Anti-Dühring* (éd. 1950, p. 41, 55) de "l'apparente contingence des événements" ou des "contingences apparentes", n'aurait sans doute pas souscrit aux idées indéterministes de Jacob (p. 344–345): "un intégron particulier n'a qu'une certaine probabilité d'apparaître. Toute prévision concernant son existence ne peut être que d'ordre statistique [. . .] C'est donc sur la contingence que repose aujourd'hui l'unité d'explication". Jacques Monod (*Le hasard et la nécessité,* 1970) les approuverait par contre certainement, et la classification des "objets" qui ressort de son livre est compatible avec celle proposée par Jacob. On peut la résumer dans un tableau de "traits distinctifs", à l'instar de ceux proposés pour les phonèmes par l'école de Jakobson:

	objets naturels sauf C et V	cristaux (C)	êtres vivants (V)	artefacts
reproduction invariante & morphogenèse autonome	–	+	+	–
téléonomie	–	–	+	+

étant entendu que par téléonomie, chez les êtres vivants, il faut entendre "plus exactement les mécanismes morphogénétiques qui construisent les structures téléonomiques" (p. 38); alors que pour tout artefact, produit de l'activité d'un être vivant, l'adaptation fonctionnelle est obtenue "par application aux matériaux qui le constituent de forces extérieures à l'objet lui-même" (p. 22–23). C'est un "critère purement quantitatif", la "quantité d'information inférieure de plusieurs ordres de grandeur" représentée par les structures cristallines qui "permet de distinguer les êtres vivants de tous les autres objets, y compris les cristaux" (p. 25). Ceux-ci sont des "objets capables de reproduction invariante, mais dépourvus de tout appareil téléonomique" (p. 29); "l'invariance précède nécessairement la téléonomie" (p. 37), dont la théorie sélective fait "une propriété secondaire, dérivée de l'invariance considérée comme seule primitive" (p. 38). Monod (p. 98, 121) comme Jacob (p. 274, 295, 301–302, 324–325) marque les analogies entre cristal et organisme; l'un et l'autre insistent sur les processus d'information et de régulation communs aux engins automatiques et aux êtres vivants (Monod, 78, 81–82; Jacob, 271–274, 208). Par contre, si l'on ne trouve chez Jacob aucune attaque contre telle ou telle position philosophique proprement dite (la citation de Diderot en exergue de son livre, et d'autres passages le montrent cependant adversaire de tout déisme comme de tout vitalisme), Monod est à la fois vigoureux adversaire d'Engels (et du matérialisme dialectique en général: p. 46–59), de Teilhard (p. 44–46) et des écoles "organicistes ou holistes" (p. 92–94) comme de Bertalanffy (il parle dédaigneusement de la "vague théorie générale des systèmes"). Il n'y a pas lieu de discuter ici la question de savoir si, par certains côtés, il ne se rattache pas à une autre philosophie de l'émergence: celle de Bergson, qu'il cite avec quelque éloge (p. 39–40, 130), ni d'examiner si la "thèse du *pur* reflet, du *parfait* miroir" qu'il attribue au marxisme sur le problème des rapports entre le monde extérieur et la pensée humaine (p. 48–50) correspond bien à la vérité: une lecture même superficielle de *Matérialisme et empiriocriticisme* (cf. p. 86, 115–117, 136, 155, 166, 212, 238–239, 242, 282–285 de la trad. française, éd. 1951) conduirait à le réfuter sur ce point, ni enfin d'élucider si vraiment, comme il le soutient, une théorie reposant sur un "postulat d'invariance" est bien "par définition une théorie idéaliste" (p. 52: *a contrario,* voir Goldmann, 1967, p. 184, note). Ce sont là des problèmes importants pour la construction d'un "système des sciences" réellement scientifique, mais ils dépassent le cadre du présent exposé. Pour la discussion d'autres classifications des sciences contemporaines, basées ou non sur des philosophies de l'émergence et des niveaux d'intégration, nous nous permettrons de renvoyer à deux de nos études antérieures (Grolier, 1961, p. 100–106; 1965, p. 176–187). Il serait, sans doute, nécessaire de les reprendre pour les cinq ou six dernières années. Retenons seulement, parmi les tentatives récentes présentant une certaine originalité, celle de Georges Gusdorf (1966, p. 188–189, 209–219) qui, distinguant quatre "attitudes": scientifique (ou spéculative), religieuse, technique et poétique (ou esthétique), chacune "génératrice d'une philosophie", voudrait voir reconnaître la "nécessité d'inverser le sens de lecture de la classification des sciences" et propose comme "science des sciences, en lieu et place des mathématiques", l'anthropologie qui, écrit-il, "doit se voir reconnaître la priorité en droit et en valeur".

122 Structures organisationnelles et système des sciences

De tout temps, le système de "classification des savoirs" a dépendu des infrastructures organisationnelles de l'enseignement et de la recherche, comme il les a d'ailleurs influencées. Une par une, les sciences particulières, les "disciplines" se sont séparées, détachées du complexe magico-religieux d'abord – avec l'astronomie, la mathématique, la médecine, et, à un moindre degré, la zoologie, la chimie, déjà à Sumer et en Egypte – de la "philosophie" ensuite, à partir des Alexandrins. Il est exagéré de soutenir, comme l'a fait Kedrov (1956, p. 83–84), que "dans l'antiquité, il n'existait qu'une science: la philosophie" et que "entre la philosophie et toutes les autres branches du savoir, il n'existait qu'un seul type de rapports: la subordination de toutes les branches du savoir à la philosophie". C'était peut-être la prétention des philosophes, mais elle était déjà démentie par les faits. Il est vrai que des savants étaient philosophes, plus peut-être qu'aujourd'hui, et que chez les Ioniens on ne peut guère distinguer leurs thèses scientifiques et leurs thèmes philosophiques, mais Hippocrate n'est que médecin, Euclide géomètre, Geminus mathématicien, Archimède physicien, Ptolémée astronome. De ce point de vue, le Moyen Age a été une époque de régression, où non seulement les sciences, mais encore la philosophie ont été étroitement subordonnées à la théologie, au moins en Occident. Pourtant, les disciplines du trivium et du quadrivium avaient gardé une certaine autonomie: comme l'écrit Bréhier (1971, p. 10) "un mince filet coule de connaissances positives qui, par leur nature et leur essence, échappent à la vie religieuse". C'est beaucoup plus évident à partir du 12e siècle en Europe occidentale; en pays d'Islam, comme en Inde (où la linguistique est pleinement constituée comme science dès Panini) et sans doute en Chine, l'émancipation des sciences a été obtenue, au moins partiellement et malgré de multiples tentatives de résorption, plus rapidement, quoique de manière moins durable. Ce mouvement se précise à la Renaissance (cf. ci-dessus, p. 29); il devient irréversible aux 17e et 18e siècles: celui-ci voit, comme l'écrit Jacob (p. 83), apparaître une catégorie de "professionnels d'un type nouveau, les naturalistes"; il s'accélère au 19e siècle: on peut suivre, depuis le dernier tiers du 18e siècle, ce processus qui pousse à la création d'écoles spéciales (les vieilles universités restant fort longtemps plus ou moins en dehors et tendant – en France jusqu'à 1968 – à maintenir des cadres antiques), d'associations professionnelles et "sociétés savantes" spécialisées, d'abord nationales, puis internationales (à partir du moment où le progrès des communications permet la tenue de congrès internationaux de plus en plus nombreux). Depuis la parution en 1665, presque simultanément en France et en Angleterre, des deux premiers périodiques scientifiques, le nombre de ceux-ci s'est accru de manière spectaculaire et les périodiques "primaires" ont été doublés depuis la fin du 18e siècle par des périodiques "secondaires" (de résumés analytiques) ou "tertiaires" (revues annuelles de mises au point): les uns et les autres se sont de plus en plus spécialisés. Tout ce mouvement a donné une base organisationnelle, non plus à des dizaines de disciplines, mais à des milliers de sous-sous-sous-disciplines. Cependant, il ne s'est nullement agi de simple fragmentation: il y a eu, bien plus fréquemment, application de ce processus de "fertilisation croisée" que Maxwell évoquait en 1877; il s'est formé, entre les spécialités, des anastomoses de plus en plus nombreuses – de sorte que le schéma des relations entre "disciplines" n'évoque plus du tout une sage carte administrative, où les limites de chaque district sont bien tracées et où personne n'empiète sur le ressort du voisin, mais bien plutôt ces tableaux de connexions des circuits d'un ordinateur, où le profane n'aperçoit qu'un inextricable imbroglio. En outre, il y a eu des regroupements. Prenons-en un seul exemple, dans un champ qui intéresse particulièrement ce Colloque: en 1900–1903 s'étaient tenues deux conférences internationales d'histoire des sciences, mais dans le cadre du Congrès d'histoire comparée; à partir de la 3e Conférence (1904), ce cordon ombilical avec l'histoire générale a été coupé; en 1928, un Comité d'abord rattaché au Comité international des sciences historiques prit le nom d'Académie internationale d'histoire des sciences; l'Union internationale d'histoire des sciences fut enfin créée en 1947. Par ailleurs, après des congrès tenus depuis 1929 (à l'initiative du *Wiener Kreis*), une Union internationale de philosophie des sciences s'etait constituée en 1949; en 1955, les deux Unions fusionnèrent sous le nom d'Union internationale d'histoire et de philosophie des sciences. Pourtant, il est à noter que la sociologie de la science, d'une part, la politique de la science *(sientific policy)* d'autre part, restent en dehors de l'UIHPS: peut-être un nouveau regroupement interviendra-t-il encore quelque jour.

L'étude de cet aspect institutionnel du développement scientifique n'est pas très avancée – bien que l'on trouve d'assez nombreux matériaux plus ou moins dispersés, surtout sur les universités (Denifle, 1855; Paulsen, 1919–21; Irsay, 1933–35; etc.), à un moindre degré pour l'enseignement technique (Artz, 1966) et les sociétés savantes (Ornstein, 1913, etc.). Les essais de synthèse sont rares (Ben-David, 1971; Taton, 1970). Et, depuis les deux livres d'Eijkman (1910–11), il n'y a pas eu de travail complet sur les organisations scientifiques internationales – les enquêtes de l'Unesco en 1950, de la Library of Congress en 1962 et de l'OCDE en 1965 restant partielles et ayant plutôt le caractère de répertoires ou d'annuaires et les sciences sociales, et/ou humaines, demeurant négligées (Grolier, 1969b). Toutes ces recherches sont pourtant importantes pour établir les connexions et interférences entre les "disciplines". Par exemple, il est intéressant de préciser comment les associations professionnelles de scientifiques se sont fédérées pour constituer des conseils ou instituts interdisciplinaires: aux Etats-Unis, on trouve entre autres l'American institute of physics, le Social science research council, l'American council of learned societies, l'Engineers joint council; au plan international, après la période "unitaire" de l'Association internationale des académies (1899–1914) et celle de l'entre-deux-guerres où existaient deux organisations, l'une pour les sciences de la nature – Conseil international des recherches, puis Conseil international des unions scientifiques – et l'autre pour les "humanités" – Union académique internationale – continuant en somme les sections B et C de l'ancienne A.I.A., il y a maintenant six fédérations: CIUS, CIOMS, CISS, CIPSH, UATI et FISI: les trois premières pour les sciences (sciences "tout court"), les sciences médicales, les sciences sociales, la quatrième pour la philosophie et les "sciences humaines", les deux dernières, plus ou moins concurrentes, dans le domaine technique, mais de nombreuses associations restent en dehors de ces regroupements. Intéressante aussi est l'étude de l'évolution des organismes centraux de la recherche scientifique, au plan national: Académies des sciences, dans les pays socialistes; Conseils nationaux de la recherche en Grande-Bretagne; National academy of science and engineering/National research council et National science foundation, aux Etats-Unis; Centre national de la recherche scientifique en France. Des matériaux importants ont été publiés à cet égard, au cours des dernières années, dans le cadre des enquêtes sur la politique de la science menées par l'OCDE et l'UNESCO, mais ils n'ont pas – à notre connaissance – été utilisés de manière exhaustive du point de vue de l'étude des organigrammes de la recherche: de la façon dont sont répartis les thèmes de recherche, groupés les chercheurs par disciplines ou domaines, organisées les compétences respectives des diverses commissions . . . (pour les sciences sociales, on trouvera quelques indications suggestives dans le rapport d'Eric Trist, 1970; cf. aussi Lefer, 1971). Il y a là encore de vastes champs à défricher pour voir comment le "système des sciences" est *objectivement* conçu et réalisé, dans la pratique de la recherche scientifique des différents pays aussi bien que par les groupements et regroupements des scientifiques eux-mêmes dans leurs associations, congrès, etc., au plan national, au plan international régional (par exemple, en Europe, dans les Amériques, pour la région du Pacifique . . .) et, enfin, au plan mondial. On notera que cet aspect du "cadre institutionnel" de la recherche commence à être reconnu comme important du point de vue de l'épistémologie: Pierre Thuillier le signale dans une étude récente, fruit de son cours à l'université de Nanterre (1971, p. 553), et Victor Karady, présentant un projet de recherche de l'Ecole pratique des hautes études (1971, p. 105), écrit que "le procès d'organisation d'un secteur du champ intellectuel [. . .] seul permet d'éclairer l'histoire 'interne' des idées".

123 Quelques problèmes de délimitation et de recoupement

Il n'est pas question de traiter ici tant soit peu complètement l'ensemble des problèmes que soulèverait aujourd'hui une tentative de constitution d'une classification "des savoirs" en accord avec la théorie et la pratique de la science contemporaine. Ce serait là matière à un travail d'équipe de longue haleine – à mettre au programme de recherches du (futur) Centre d'Ottawa. Nous voudrions seulement indiquer, à titre d'exemples, quelques-unes des difficultés que l'on rencontrerait dans cette tâche.

Jacob écrit (1970, p. 11): "Les deux points de rupture de l'évolution, l'émergence du vivant d'abord, celle de la pensée et du langage plus tard, correspondent chacun à l'apparition d'un mécanisme de mémoire, celui de l'hérédité, celui du cerveau". Cette phrase, à laquelle souscriraient sans doute de nombreux biologistes, impliquerait une division tripartite de l'ensemble des sciences: sciences de la nature inorganique; sciences biologiques; sciences psychologiques (ou psychosociales). Nous avons vu d'ailleurs que c'est, en somme, le schéma de Carnap pour les sciences empiriques (ci-dessus, p. 25). On retrouverait d'ailleurs sans peine la même tripartition chez beaucoup d'autres auteurs: c'est, par exemple, celle de Julian Huxley (1967, p. 75): "la nouvelle représentation de l'évolution montre celle-ci à l'oeuvre au cours de trois phases consécutives, les phases inorganique, biologique et psychosociale". Mais, si l'on examine de plus près chacun de ces "points de rupture", on aperçoit que les choses ne sont pas si simples qu'elles le paraissent au premier abord. En premier lieu, où commence "le vivant"? Jacob se montre hésitant, et presque en contradiction avec lui-même, en ce qui concerne les virus. Page 287 de son livre, il les exclut du domaine proprement biologique: "si les virus manifestent certaines propriétés des organismes, ils sont loin de les posséder toutes. Et le manque d'autonomie qui en résulte empêche da les considérer comme des êtres vivants". Page 300, il adopte le même point de vue: "un virus possède certaines propriétés d'un système vivant, mais pas toutes [. . .] Il peut diriger la production de protéines virales qui influencent en sa faveur le milieu où il se trouve. Il peut donc être l'objet d'une évolution par sélection naturelle. Mais en revanche il ne peut mettre à exécution son programme génétique et se reproduire que dans un milieu déjà capable d'effectuer les opérations de métabolisme [. . .] On ne peut donc considérer un virus comme un organisme". Mais, page 325, Jacob écrit ceci: "Pour le biologiste, le vivant ne commence qu'avec ce qui a pu constituer un programme génétique. Pour lui, c'est seulement du jour où il donne prise à la sélection naturelle qu'un objet mérite le nom d'organisme". On voit que, d'après ces deux critères, le virus "mérite" bien le nom d'organisme! Maintenant, en ce qui concerne le deuxième "point de rupture", que Jacob place à l'émergence "de la pensée et du langage", on peut se demander tout d'abord si ces deux éléments sont bien apparus de manière concomitante. Bien sûr, tout dépend de la définition que l'on donne aux mots utilisés: le plus grand reproche que l'on puisse faire à des philosophies "confusionnistes" comme celles de Whitehead ou de Teilhard est peut-être de donner à un mot comme "conscience" un sens tellement extensif qu'il perd toute signification réelle et toute valeur opératoire. A un moindre degré, on rencontre des extensions abusives analogues pour les termes "pensée" ou "langage" (ou, encore, "instinct", "intuition" – et "volonté" chez Schopenhauer). Nous ne pouvons entrer ici dans le détail des positions souvent opposées (et parfois chez un même spécialiste) que l'on trouve sur ce problème des limites de la "pensée" et du "langage" et de leur ordre d'apparition. Au surplus, il n'est nullement certain que des faits nouveaux ne doivent pas encore être découverts, qui nous forceront à modifier une fois de plus des conceptions ayant dû, déjà plus d'une fois, faire place à des phénomènes auparavant inconnus qui ne rentraient pas dans les cadres existants: réflexes conditionnés et "deuxième système de signalisation", systèmes sémiotiques des abeilles, des dauphins ou des corbeaux, "alphabet" et signes de ponctuation du code génétique. Jakobson (1970, p. 526–530) a résumé les données qui marquent la "similitude de structure" entre celui-ci et le langage humain, les sous-unités du code génétique correspondant aux phonèmes (et qui comme eux, "en eux-mêmes sont dépourvus de sens, mais servent à constituer les unités significatives minimales", donc, il y a, dans les deux cas, "double articulation"); présence, dans l'un et l'autre code, d'oppositions binaires de traits distinctifs, d'une syntaxe, de la dépendance vis-à-vis du contexte, de phénomènes de synonymie – et même, peut-être, d'une "analogie sur le plan moléculaire avec ce qu'est le dialogue pour le langage". Si, quelque jour, les tentatives jusqu'à présent infructueuses pour réaliser une communication avec des êtres extra-terrestres doués d'une certaine forme de langage devaient réussir (cf. Freudenthal, 1960) – ce qui nous paraît, quant à nous, fort probable – ne serions-nous pas conduits à une "revision déchirante" de notre conception du "langage"? Le même Jakobson, après avoir énuméré les caractères distinctifs qui opposent le langage humain aux diverses formes de communication animale, présente des conclusions nuancées: sans doute, écrit-il, "Le passage de la 'zoosémiotique' à la parole humaine est un gigantesque saut qualitatif"; pourtant "L'opposition traditionnelle entre langage humain et communication animale envisagée comme une opposition entre phénomènes culturels et phénomènes naturels résulte d'une simplification grossièrement exagérée" (1970, p. 521–522) et "Dans leurs diverses opérations, le langage et les autres moyens de communication humaine présentent un grand nombre d'analogies instructives avec le transfert d'informations entre créatures appartenant à d'autres espèces vivantes" (p. 523). De son côté, le psychologue Piaget recherche, comme le linguiste Jakobson, les traits distinctifs qui opposent langage humain et communication animale: il trouve le dialogue, la "composition libre d'éléments", la présence de signes arbitraires, conventionnels (1970, p. 610): pour lui, les "indices imitatifs" du langage animal "correspondent à des schèmes sensori-moteurs, innés ou acquis, mais non pas encore à une conceptualisation"; l'intelligence, "sous sa forme sensori-motrice", précède le langage, mais le fonctionnement de celui-ci "reste lié à des intelligences individuelles [. . .] dont le schématisme sensori-moteur engendre déjà une multitude de significations (schèmes spatiotemporels, objets permanents, causalité, etc.) fournissant la sub-structure des sémantiques verbales" (idem, p. 611); "les sources des structures logico-mathématiques sont à chercher à un niveau plus profond que le langage, au niveau de la coordination générale des actions" qui est celui de l'intelligence sensori-motrice" (p. 612–613). Finalement, "il semble bien que l'intelligence précède le langage et conditionne son acquisition, mais avec actions en retour au sein de processus où l'inné et l'acquis sont tous deux dépassés par un mécanisme plus général d'équilibration progressive" (p. 53).

Jacob mettait en parallèle la dichotomie vie/pensée & langage et celle mémoire de l'hérédité/mémoire du cerveau. Piaget propose un parallélisme plus compliqué, fondé sur la distinction de trois formes principales de mémoire "liée au seul comportement" (en dehors de la "mémoire sens du biologiste": "conservation, durant la vie de l'individu, de tout ce qui est acquis" – c'est encore autre chose, d'ailleurs, que la "mémoire de l'hérédité" de Jacob), et de deux niveaux à l'intérieur du deuxième système de signalisation (p. 309–311 et 609–610); au surplus, l'on pourrait "distinguer au moins dix niveaux de transition entre la récognition des indices (forme de mémoire liée aux conduites innées et aux habitudes sans se confondre avec elles tout en s'appuyant sur elles) et les conduites de reconstitution puis d'évocation" (p. 278). Il est sans doute permis de présenter cette double argumentation sous forme du tableau schématique suivant:

Forme de mémoire	Système de signalisation	Niveau d'émergence
Mémoire "au sens du biologiste"	–	–
Mémoire psychologique (liée au comportement):		
au sens large – récognition	ler système: indices	invertébrés inférieurs
forme intermédiaire-reconstitution, imitation	2e système, ler niveau: symboles, signifiants motivés	abeilles, oiseaux, enfant*
au sens strict – rappel ou évocation	2e système, 2e niveau: signes arbitraires	anthropoïdes, homme à partir de 1½ – 2 ans*

* Piaget semble faire une distinction entre l'apparition de la mémoire d'évocation, qui aurait lieu entre un an et demi et deux ans (p. 311) et celle des deux niveaux d'"instruments sémiotiques" qui, chez l'enfant normal, apparaissent à peu près en même temps, "sauf en général le dessin" (p. 609). D'autre part, la "fonction sémiotique" (2e système de signalisation) est indiquée p. 609 comme apparaissant "chez certains Primates supérieurs et chez l'homme (à partir de la seconde année)", alors que, p. 310–311, il est écrit que "l'évocation semble liée à la fonction sémiotique (images mentales, en tant que symboles représentatifs, et langage), et n'est donc possible que chez l'homme à partir de 1½ – 2 ans et sans doute chez les anthropoïdes". Il paraît y avoir ici quelque hésitation dans la pensée du maître genevois.

Prenons maintenant un autre problème de "coupure", d'ailleurs assez proche du précédent: la limite entre "animal" et "homme". Elle s'est déplacée dans le temps, au fur et à mesure des progrès de la préhistoire. Au temps de Gabriel de Mortillet (1883), le néandertalien – dont il croit qu'il n'avait pas la parole – n'est guère qu'un singe nu; pour Boule (1921), on peut admettre le pithécanthrope dans la série des "hommes fossiles", comme intermédiaire entre les singes et les néandertaliens, mais il souligne encore pour ceux-ci les caractères simiens" de l'homme de la Chapelle-aux-Saints. Teilhard (pour qui, soit dit en passant, "la Pensée" – avec un grand P – est exclusivement humaine: voir, par ex., les p. 169, 178–179 de l'éd. 1970 du *Phénomène humain,* écrit en 1938–40 mais revisé en 1948) hésite sur la place à donner au pithécanthrope et au sinanthrope: p. 190 ils sont "préhominiens"; p. 193 il veut bien admettre qu'ils "étaient déjà [. . .] des êtres intelligents" (donc, pour lui, des "hommes"); au 3e § de la p. 194 ils deviennent des "hominiens", mais les australopithèques représentent encore une "famille d'Anthropomorphes [. . .] où toute une série de caractères hominoïdes apparaissent disséminés sur un fond encore nettement simien" (p. 184, note). En 1962, au symposium de la Wenner Gren foundation, deux schémas de classification différents (Simpson, 1964, p. 29–30; Leakey, idem, p. 48–49) s'opposaient sur certains points (classification des pongidés) mais se rencontraient pour former une famille d'hominidés comprenant deux genres: *Australopithecus* et *Homo.* En 1967, Le Gros Clark discutait les origines des *Australopithecines,* notait que l'on avait suggéré de les inclure dans le genre *Homo* et avançait comme une possibilité d'accueillir le ramapithèque dans les hominidés. Le débat est loin d'être terminé: il a été relancé cette année par de nouvelles découvertes au Kenya, mais aujourd'hui on tend à rebaptiser l'"australopithèque" en "australanthrope" (Chavaillon, 1971, p. 78–79) et on le classe parmi les hominiens (Coppens, 1971, p. 80); l'on donne à *Homo faber* une antiquité de cinq millions d'années (idem, p. 83); s'il était confirmé que le ramapithèque utilisait des outils, il faudrait reculer à quelque trois fois ce chiffre. Coppens écrit en outre que "L'homme, au sens le plus large, c'est-à-dire au sens des Hominidés" a une évolution autonome depuis environ 40 millions d'années. On est bien loin des 50 ou 100 000 ans du schéma de Teilhard (p. 190) ! Ne quittons pas tout de suite le domaine de la biologie. Il est encore usuel de distinguer dans celui-ci deux grands sous-domaines: celui de la botanique et celui de la zoologie (si toutefois l'on ne sépare pas de celle-ci l'anthropologie, aboutissant à une classification ternaire, comme le fait, par ex., Mme Laming-Emperaire, 1971, p. 30). Cette division est consacrée, sur le plan organisationnel, par la présence de chaires ou de départements spécialisés dans les universités, de périodiques primaires et secondaires, de sociétés savantes de "botanique" ou de "zoologie". Bien. Mais, dans une toute récente "Introduction à la classification des animaux", Lerwill (1971, p. 39) reconnaît que "la distinction entre les règnes animal et végétal aux niveaux 'inférieurs' de la vie n'est cependant pas précise", que le groupe des Phytomastigina "contient des organismes qui possèdent à la fois des caractéristiques animales et végétales" et (p. 40) qu'il "form un lien entre les règnes animal et végétal, et est parfois inclus dans le règne végétal". En outre, la coupure primitive du monde vivant – en mettant à part les virus – n'est plus, depuis plusieurs décennies, entre "animaux" et "plantes", mais bien entre organismes procaryotes et eucaryotes – mais elle ne s'est pas encore reflétée dans le découpage des "disciplines", encore moins dans les dictionnaires et les manuels d'enseignement. Si nous nous tournons maintenant vers les sciences de la nature inorganique, une difficulté classificatoire se présente lorsqu'il s'agit de déterminer la place à attribuer à ce que Cournot appelait la série "cosmologique et historique", Cailleux (1951) les sciences "à localisation" opposées à celles "sans localisation", ou encore à ce que Piaget (1970, p. 560) dénomme les "branches latérales du tronc commun": astronomie et géologie. En 1873, Engels plaçait l'"histoire du système solaire, de la terre" entre la chimie organique et la "nature organique", et notait dans un autre passage de *Dialectique de la nature* que la "physiographie" devait occuper une place intermédiaire entre la chimie organique et les "diverses formes de la vie elle-même" (voir les références dans Grolier, 1956, p. 134). Plus tard, en 1877–78, il devait, dans l'*Anti-Dühring,* revenir à la vieille conception comtienne plaçant l'astronomie avant la physique. Aujourd'hui, aux Etats-Unis, les astronomes ont inclus leur association scientifique au sein de l'American institute of physics, mais, au plan international, les géodésiens et géophysiciens, les astronomes, comme d'ailleurs les géologues, et les géographes ont des organisations distinctes de celle des physiciens. Austin (1969, p. 83–84), après avoir mentionné la structure à deux branches de Feibleman (1954) essaie de rétablir un ordre linéaire:

éléments – "nuages gazeux" – étoiles – planètes –
cristaux – paracristaux – "entités vivantes"

Glangeaud (1955) avait – comme Piaget – adopté l'idée d'un "tronc commun" et de branches divergentes, mais hésité à plusieurs reprises sur l'ordre de ces branches et leurs ramifications (voir la discussion que nous avons donnée, Grolier 1965, p. 179–180). Cependant, ses diverses propositions aboutissaient à placer la "série géonomique et astronomique" après la "série biologique", ce qui semble contraire aux idées les plus courantes actuellement, tout au moins chez les biologistes, et à classer les objets astronomiques suivant une disposition – globe terrestre, planètes, étoiles, galaxies – géocentrique, qui paraît contredire au principe fondamental de classification des objets de l'univers en une série (plus ou moins) conforme à celle de l'évolution. On serait sans doute tenté, après tout, de s'en tenir à une classification par disciplines: physico-chimie, cristallographie, astronomie, sciences de la terre, biologie. Mais la minéralogie et la pétrographie, traditionnellement classées dans les sciences de la terre, peuvent-elles y être maintenues, maintenant que l'on étudie les minéraux et les roches de la Lune – et que, peut-être bientôt, on étendra ces études aux associations minérales d'autres planètes? Et, en tout état de cause, un tel schéma aboutit à rompre la liaison naturelle chimie – cristallographie – biologie. On ne voit donc pas d'ordre linéaire "idéal".

La situation s'est considérablement compliquée avec le développement de champs de recherche interdisciplinaires, "mission oriented", comme l'océanologie, qui regroupe l'étude de l'océan comme milieu physique et comme milieu biologique avec celle des moyens d'exploration sous-marine et de l'exploitation des richesses océaniques (minérales comme biologiques), ou les "sciences et techniques aérospatiales", vaste complexe où se retrouvent astrophysique, étude des roches lunaires (en attendant celles d'autres planètes), techniques de l'aéronautique et de l'astronautique et des "automobiles lunaires", etc. (cf. Pinder et Rousselot, 1971). Nous ne proposerons pas ici de solution. Il n'y en a peut-être pas qui soit pleinement satisfaisante.

La géographie pose des problèmes analogues. D'un côté, elle se relie manifestement à la géologie, et d'ailleurs aussi avec la géodésie et la géophysique, pour former un groupe de "sciences de la terre" dans l'ensemble des sciences naturelles, dont la géographie physique, la géomorphologie, la climatologie ne peuvent se laisser dissocier. D'un autre côté, la phytogéographie et la zoogéographie font partie des sciences biologiques, et la géographie humaine, avec ses diverses branches: géographie économique, géopolitique, géographie des religions, géographie linguistique, etc., appartient aux sciences humaines . . . Mme Teslenko (1967, p. 566) n'avait rien trouvé de mieux que de distinguer deux géographies, l'une dans les sciences naturelles, l'autre dans les sciences sociales, soutenant qu'"Une 'géographie unitaire' ne peut avoir de place dans la classification". Solution qui n'en est pas une . . . Aujourd'hui, le "champ d'études interdisciplinaires", *mission oriented,* des "sciences de l'environnement", si à la mode, vient ajouter son lot de complications, avec la prise en considération de questions technologiques (pollution), juridiques, démographiques, politiques, etc.

C'est pourtant dans les "sciences sociales et humaines" – pour prendre l'expression utilisée par l'enquête sur les "Tendances principales de la recherche" de l'UNESCO – que les difficultés de classification sont les plus grandes. Après avoir cherché à réaliser une classification des sciences sociales plus ou moins rationnelle (travaux de Miss Kyle, 1956–63), le Comité international pour la documentation des sciences sociales (devenu cette année Comité international pour l'information et la documentation) a abandonné cette tâche, et s'est rabattu sur la constitution de "thésaurus", listes de termes plus ou moins reliés les uns aux autres par un système de (pseudo-) "champs sémantiques" et de renvois. Ayant suscité une série de rapports sur le même problème en 1964 (publiés dans le vol. 16, numéro 4, de la *Revue internationale des sciences sociales*), l'UNESCO décida de suivre une répartition proposée par Piaget en "quatre grands ensembles": sciences nomothétiques de l'homme, sciences historiques de l'homme, sciences juridiques, disciplines philosophiques (Piaget, 1964, p. 600–602; 1970, p. 2–8). Cette classification est très contestable, et d'ailleurs Piaget lui-même en indique fort honnêtement les limites. L'UNESCO s'était trouvée auparavant aux prises avec ce problème de classification des sciences de l'homme, quand il s'était agi de favoriser un regroupement des associations disciplinaires en un ou plusieurs Conseil(s) parallèle(s) au Conseil international des unions scientifiques, qui ne s'occupait que des sciences "exactes et naturelles". Après quelques discussions, l'opinion prévalut de créer deux Conseils, et non un seul: celui des "sciences sociales" et celui "de la philosophie et des sciences humaines", mais cette décision fut prise plus en vertu de traditions universitaires anciennes, qui rangeaient à part "humanités" et "sciences économiques et sociales", que de vues théoriques précises, et la répartition des "disciplines" entre les deux Conseils est, en fait, arbitraire: elle reflète plus ou moins le clivage qui s'était antérieurement opéré dans certains pays: en France entre les spécialités representées à l'Académie "des inscriptions et belles-lettres" et celles de l'Académie "des sciences morales et politiques"; aux Etats-Unis entre les champs d'action respectifs de l'American council of learned societies ("humanistes") et du Social science research council. Les limites actuelles du groupe des sciences de l'homme et celles de leurs diverses branches sont le produit de circonstances historiques; elles ne sont, certainement, pas définitives. Dans plusieurs pays, la psychologie est maintenant considérée davantage comme une science naturelle que comme une science purement humaine: tel est le cas en Union soviétique et, en France, la Commission de psychologie du Centre national de la recherche scientifique a été transférée, en 1966, de la section "sciences humaines" de cet organisme à la section "sciences naturelles". Mais il est permis de se demander si la sociologie ne suivra pas un jour le même chemin, au fur et à mesure que pourraient être dégagées des lois de développement communes aux associations végétales, aux sociétés animales et aux sociétés humaines – sans compter les "sociétés" que nous rencontrerons peut-être, ou même sans doute, quand les voyages dans l'espace sidéral auront pu nous mettre en contact avec des organisations d'êtres vivants extra-terrestres. Et si devait se réaliser la prédiction de Lévi-Strauss (1958, p. 329) suivant laquelle "L'anthropologie sociale, la science économique et la linguistique s'associeront un jour pour fonder une discipline commune qui sera la science de la communication", cette science devrait alors déborder très largement les frontières des sciences *humaines,* puisqu'aussi bien le problème des communications débutent dès la cellule et concernent l'ensemble des organismes vivants et de leurs artefacts (cf. Morin, 1971). Qui plus est, elle rencontrerait la théorie de l'information, théorie physique s'appliquant à l'ensemble de l'univers (néguentropie). En tout cas, une science de l'information (*information science,* en anglais; *informatika* en russe) est en voie de constitution (Grolier, 1971, a et b), mais ses frontières avec d'autres domaines – sémiologie, informatique (*computer science*) ne sont pas encore précisées.

La sémiologie elle-même, dont on peut tracer les origines lointaines jusqu'à Locke et Leibniz, mais qui n'est guère réellement constituée que depuis quelques années en discipline "autonome", avec la fondation d'une revue spécialisé (*Semiotica*) et d'une association internationale, ne sait pas encore très bien où s'arrêtent ses limites: elle devrait normalement englober tous les systèmes de communication animale, la linguistique (humaine), la science de la littérature, et sans doute même s'étendre à l'analyse des mythes, des contes populaires (voir les idées présentées par Piaget sur les "symbolismes à la deuxième puissance", 1970, p. 573, 615–617, et plus anciennement celles de Bergson sur la "fonction fabulatrice" dans les sociétés humaines – plus, naturellement, les travaux de Propp, de Jung, de Lévi-Strauss . . .). Elle rencontre là les sciences religieuses et la philosophie, dans la mesure où l'on peut reprendre ce qu'il y a de valable dans la *Chiffrelesen* de Jaspers, ou accepter la définition d'Ortega y Gasset (1970, p. 208) sur la philosophie comme "système d'attitudes fondamentales d'interprétation". Nous n'examinerons pas le problème des rapports et des limites entre philosophie et religion: qu'il suffise d'indiquer les distinctions et les rapprochements posés par Hegel (*Leçons sur l'histoire de la philosophie* – voir dans l'éd. Gallimard 1970 p. 28, 40, 70–73, 104–105, 169, 186–187, 218–234), par Jaspers (1965, p. 143–144) ou bien l'identification partielle de Paul Bourde (1939, p. 55–63): les système "dits religieux" ou "dits philosophiques" ont été "catalogués par l'usage" – "parmi les religions ceux qui deviennent populaires et parmi les philosophies ceux qui restent une étude de lettrés"). La même idée avait d'ailleurs été émise par Schopenhauer (*Le Monde comme volonté*, trad. Burdeau-Roos, 1966, p. 856–863). Mais cette question justifierait une étude spéciale, qui chercherait à dégager les "traits distinctifs" communs aux philosophies et aux religions et ceux qui les opposent, et qui pourrait aussi examiner ce qu'il y a de valable dans des thèses comme celle d'Eliade (1957: éd. française, 1965: cf. p. 170–181) sur les formes de la sacralité dans le monde moderne désacralisé.

Nous retrouvons des cas de délimitation difficile et de rapprochement ou de connexion en bien d'autres parties du vaste domaine des sciences de l'homme, ainsi qu'aux frontières entre celles-ci et les sciences de la nature. A titre d'exemple supplémentaire du premier type, on peut mentionner les liaisons entre la science politique, la science administrative et la "science de l'organisation" (*management science*). Ces trois disciplines, fort jeunes toutes les trois et qui se trouvaient plus ou moins réunies au 18e siècle et dans la première moitié du 19e siècle dans la *Kameralistik* d'origine allemande (la première chaire avait été créée à Halle en 1727) ont bien des points communs, ne serait-ce que pour ce qui concerne les processus de décision (*decision making*). Après s'être différenciées, progressivement, depuis la fin du siècle passé, elles tendent de nouveau à se rapprocher (Grolier, 1969a), et la "praxéologie" d'origine polonaise (Slucki, 1926) voudrait même parfois aboutir à les réunir avec l'économie, dans le cadre d'une théorie unifiée de "l'action efficace" (Brus *et al.* 1970, p. 341, 350–351 et notes p. 421–422; Piaget, 1970, p. 593).

Comme exemples du second type, nous prendrons d'une part les rapports entre logique, grammaire et sémantique, et mathématiques; d'autre part le complexe recherche opérationnelle, cybernétique, informatique. Nous avons évoqué ci-dessus (p. 34) la phrase de Piaget sur la logique "cas particulier d'algèbre générale". Pourtant, même s'il écrit aussi (1970, p. 581) que la logique formelle "peut être située au point de départ des mathématiques", il marque bien par ailleurs ses liaisons avec d'autres domaines: systèmes de normes (p. 571, 584–585), théorie de l'argumentation (p. 582) et, surtout, psychologie (p. 583–584, 324–326) – écrivant, notamment, qu'il est "exclu de dissocier la logique des sciences de l'homme, puisque la logique du logicien constitue un prolongement formalisé et largement enrichi de celle du sujet en ses opérations effectives" (p. 48) et qu'elle "appartient à la fois aux sciences exactes et naturelles et à celles de l'homme et assure une connexion entre elles toutes qui échappe aux classifications linéaires" (p. 49), le "problème de l'épistémologie de la logique" faisant "le pont entre cette discipline et la psychologie génétique" (p. 583). Nous avons cité ailleurs (Grolier, 1956, p. 313–314) d'autres positions (Carnap, Hilbert, Russell) et assez longuement analysé les discussions sur le "parallélisme logico-grammatical" (Grolier, 1965, p. 41–65). Ce dernier sujet a été repris et partiellement renouvelé, tout récemment, par les travaux de Chomsky et les discussions qu'ils ont provoquées sur les universaux du langage, comme sur les rapports entre langages formalisés et langage naturel. Les linguistes ont eu de nouveau tendance à relier logique et linguistique, et Jakobson écrit (1970, p. 668 – écrit de la "théorie formelle des grammaires" que les *phrases* qu'elle considère sont "des constructions de l'esprit, inspirées certes par les langages naturels, mais largement autonomes par rapport à ces derniers" et que "La théorie des grammaires appartient donc à un ensemble de théories purement mathématiques pour lesquelles des phénomènes humains sont essentiellement une source d'inspiration"). Bar-Hillel dans ses "Remarks on Carnap's Logical syntax of language" (écrites en 1954, publiées en 1963; cf. 1970, p. 121) accepte les vues de son maître sur "l'unité fondamentale de la syntaxe et de la logique" et a consacré lui-même plusieurs contributions à la "sémantique logique" (1954, in 1964, chap. 2; 1969, in 1970, chap. 15; etc.).

Cybernétique, recherche opérationnelle, informatique: trois domaines en apparence nettement distincts, séparés. Pourtant (dans des notes inédites) Gérard Cordonnier a pu récemment démontrer que le *contenu* des écrits publiés sous ces différentes étiquettes était, dans une large mesure, identique. Tout en marquant par ailleurs la connexion très forte qu'il y a entre R.O. et "science de l'organisation", il faut reconnaître qu'il s'agit là d'un ensemble de recherches en fait fort étroitement liées. Mais où les situer? Nombre de spécialistes ont été d'avis d'en faire des branches des mathématiques (mathématiques appliquées). Boudon (1970, p. 660–661), ayant défini la cybernétique comme l'étude des "systèmes à auto-regulation définis par des ensembles de variables liées par des relations de causalité circulaire" et marqué que ces systèmes sont "plus évolués que les systèmes mécaniques – caractérisés par une causalité unidirectionnelle", constate que "la logique des systèmes considérés par la cybernétique est trop simple pour représenter certains systèmes, notamment les systèmes capables d'apprentissage" et se réfère à la "théorie des systèmes généraux" (*general systems theory*) qui, elle, prend en compte ces "systèmes supérieurs". Piaget (1970, p. 47) considère que la cybernétique constitue "un chaînon essentiel dans le passage de la physique à la biologie" et qu'elle se situe "on ne sait pas exactement où", "entre les sciences de la nature et les sciences de l'homme"; ailleurs (p. 614) il écrit d'elle qu'elle est "une synthèse des théories de l'information ou communication et du guidage ou régulation", et que c'est "sur ce double terrain que des relations plus naturelles qu'une assimilation pure et simple peuvent se constituer entre la linguistique et la logique". Jacob (1970, p. 17) constate que la biologie moderne "correspond à un nouvel âge du mécanisme", que la notion de programme avec laquelle elle travaille "représente un modèle emprunté aux calculatrices électroniques" – mais marque aussi la différence entre ces deux sortes de programmes (p. 17–18); rappelant les écrits de Wiener, il note que "Animal et machine, chacun des systèmes devient alors un modèle pour l'autre" (p. 271–273) et (p. 285) que "tout invite à assimiler la logique de l'hérédité à celle d'une calculatrice. Rarement modèle imposé par une époque aura trouvé application plus fidèle". C'est, "de toute évidence le modèle d'une usine chimique en miniature qui décrit au mieux la cellule bactérienne" (p. 290), et cette usine est "entièrement automatique", grâce à ses "circuits de régulation" (p. 302–304). Et tous ceux qui ont tant soit peu étudié la programmation des ordinateurs ne peuvent manquer d'être frappés, en lisant la description donnée par Jacob de la "lecture du message" chiffré par la chaîne nucléique (p. 298), de la complète analogie des deux méthodes. Cependant, écrit encore Jacob (p. 291), "Si l'on veut regarder la cellule bactérienne comme une usine, il faut [. . .] la considérer comme une usine d'un type particulier. [. . .] Si l'usine produit, la cellule se reproduit".

Mais, tout ceci étant dit, est-il vrai que, comme le soutient Piaget, la cybernétique (et l'informatique qui lui est liée) seraient placées *entre* physique et biologie? Ne s'agirait-il pas plutôt d'une branche d'une "physique générale", s'appliquant aussi bien aux êtres vivants qu'à certains artefacts humains et (probablement) à d'autres systèmes naturels? C'était la thèse qu'exposait naguère A. Ducrocq, dans un livre sans doute prématuré (1960) où, considérant la cybernétique "comme la science exhaustive d'une classe d'effets: aux des effets permettant de *gouverner* un système" (p. 2) et la rapprochant de la recherche opérationnelle comme de la théorie des graphes, il voulait constituer une "intellectique" qui engloberait "les structures classiques de la physique" en prolongeant celle-ci "avec toutes les opérations concernant asservissements, transmission et manipulation des informations", et où la biologie "trouvera sa place [. . .] dès l'instant où l'on définira la classe (*E*) des ensembles organisés possédant deux propriétés fondamentales: 1) aptitude dans un milieu donné *M* d'effectuer toutes les opérations pour la constitution d'autres ensembles (*E*) de cette même classe; 2) autolutte contre le hasard" (p. 4–5). Ce programme ambitieux, et peut-être utopique, est loin d'être réalisé, mais paraît être l'ambition actuelle de la "théorie des systèmes généraux" (cf. Klir, 1969). L'avenir dira s'il s'agit d'une voie sans issue ou de l'ébauche d'une unification réelle de l'ensemble des sciences.

Pour terminer cet examen, "à vol d'oiseau" d'une série de problèmes taxinomiques, nous voudrions revenir à la division et à l'ordre des sciences sociales. Piaget (1970, p. 560) croit pouvoir écrire que "l'on chercherait en vain" dans les sciences humaines un "ordre analogue" à celui, de généralité décroissante et de complexité croissante, qu'il aperçoit (suivant Auguste Comte) dans les sciences de la nature; "que jusqu'ici personne n'en a proposé de pareil: on ne voit guère, par exemple, de raison de placer la linguistique avant l'économie ou l'inverse". Ce n'est pas l'avis, dans le même volume, de Jakobson: pour lui (p. 506), " En raison de leur logique interne, les sciences de l'homme s'ordonnent en une série fondamentalement analogue à celle qui relie et articule des [? pour les ?] sciences naturelles". Il ne précise pas, cependant, l'ordre de cette série; dans son exposé il traite d'abord de la linguistique (concurremment avec les autres branches de la sémiotique), puis de l'économie, de l'anthropologie sociale et de la sociologie, qu'il considère comme "deux branches d'une seule et même discipline" (p. 512). "Il est évident, écrit-il, que le langage est un élément constitutif de la culture, mais, par rapport à l'ensemble des phénomènes culturels, son rôle est celui d'une infrastructure, d'un substrat et d'un véhicule universel" (p. 513). Pourtant, un peu plus loin (p. 515), il semble suggérer un ordre inverse: "Trois sciences appartenant à un ensemble s'englobent l'une l'autre et représentent trois degrés de généralisation croissante: (1) l'étude de la communication de messages verbaux, ou linguistique; (2) l'étude de la communication de messages quelconques, ou sémiotique (y compris la communication de messages verbaux); (3) l'étude de la communication, ou anthropologie sociale et économique (y compris la communication de messages)". On peut toutefois se demander si l'économie se réduit vraiment à des phénomènes de communication: ce que Jakobson dit à ce sujet (p. 513–514) n'est pas entièrement convaincant.

On connaît, d'autre part, le schéma marxiste: infrastructure technico-économique/ superstructure idéologique, schéma qui, comme l'écrit Piaget (p. 616) a soulevé "un nombre considérable de questions quant à la nature et au fonctionnement des divers types possibles de productions idéologiques". La classification des "superstructures" en général n'est d'ailleurs pas chose facile (Grolier, 1956, p. 136, 320–323). Marx n'a jamais eu le temps d'élaborer les indications qu'il donnait dans la lettre à Annenkov de 1846, puis dans l'introduction des *Grundrisse* de 1857, et qu'il publiait en 1859 dans la préface de la *Critique de l'économie politique,* oú il insère l'art entre la religion et la philosophie, suivant d'ailleurs l'analyse qu'il avait faite, deux ans auparavant, de l'art grec qui "suppose la mythologie grecque, c'est-à-dire la nature et les lois sociales élaborées par l'imagination populaire d'une manière non encore consciente mais artistique" *(Fondements de la critique de l'économie politique,* éd. française 1967, p. 41). Engels ne fera guère que paraphraser son ami, modifiant cependant l'ordre des "superstructures", en plaçant l'art (et la littérature) en dernier, et la philosophie avant la religion: dans l'*Anti-Dühring* de 1877, le *Feuerbach* de 1886, les lettres à Conrad Schmidt, en 1890, et à Starkenburg en 1894. Dans la première de ces lettres, toutefois, il rattache la science à la religion et à la philosophie, au moins par son évolution (voir Marx et Engels, *Etudes philosophiques,* 1951, p. 133) et dans la deuxième il inclut dans les "rapports économiques" *toute la technique* de la production et des transports" (idem, p. 136) – alors que Marx en 1857 s'était fixé l'objectif de "préciser les limites" de la "dialectique des concepts de force productive (moyens de production) et de rapport de production", dialectique qui, écrivait-il, "ne doit pas supprimer les différences existant dans la réalité" (*Fondements* . . ., p. 39). On trouve, de nos jours, chez Zvorikine (1964, p. 635–645) une classification se réclamant du marxisme, qui place la "philosophie marxiste" en tête des sciences sociales, et ajoute à celles-ci, et, après elles, un conglomérat de "science de l'homme": anthropologie, linguistique, psychologie, psychologie sociale, pédagogie, médecine sociale, puis les "sciences historiques" qui sont "très étroitement liées à la philosophie marxiste". On notera que Lévi-Strauss (1967, p. 125) donne un ordre sériel des "différents plans entre lesquels évolue toute vie sociale" qui ressemble beaucoup aux prototypes marxiens: "depuis les formes de l'activité techno-économique jusqu'aux systèmes de représentations, en passant par les échanges économiques, les structures politiques et familiales, les manifestations esthétiques, les pratiques rituelles et les croyances religieuses". On peut rapprocher de cette série lévi-straussienne celle que donne Storer (1966, p. 55) de quatre "major social systems":

économique, politique, familial, religieux, "as distinct exchange-systems". Ici, les "manifestations esthétiques" manquent, mais un peu plus loin Storer (p. 91–93) place les scientifiques dans un ensemble de "creative professions", avec les ingénieurs, écrivains et artistes; il est permis de se demander s'il n'y a pas ici quelque lointaine réminiscence des "arts productifs" aristotéliciens. Schopenhauer (*Le Monde comme volonté* . . ., trad. Burdeau-Ross, 1966, p. 1139) avait, quant à lui, proposé de reconnaître "la parenté de la philosophie et des beaux-arts": "la poésie est à la philosophie ce que l'expérience est à la science empirique" (*idem*, p. 1164).

Les divisions traditionnelles ont été mises en cause par Johan Galtung (1965, p. 4–6 et 9–10; 1968, p. 195–197 et 199–201). Il conteste l'utilisation, pour ces divisions, de "critères étranges, tels que événements contemporains ou du passé, sociétés sans écriture ou avec écriture" servant à distinguer la sociologie de l'histoire et de l'anthropologie, ou de l'archéologie "qui essaie de traiter de manière systématique le passé des sociétés sans écriture". "Ces distinctions, écrit-il, devront faire place à d'autres, plus fondamentales". A propos des "recherches sur la paix", il estime que les dichotomies nomothétique/idéographique, diachronique/synchronique, sans écriture/avec écriture n'ont fait qu'"entraver la formulation et la mise à l'épreuve de propositions générales sur le comportement humain en matière de conflit – et d'intégration". Et il propose une classification de "sciences orientées", d'après les "valeurs" sur lesquelles elles sont axées. Mme Laming-Emperaire, tout en donnant un schéma (1971, p. 30) des disciplines intéressant l'anthropologie qui distingue une série "tournée vers le passé" (géologie, paléobotanique, paléontologie animale et humaine, préhistoire) et une série "tournée vers le présent" (géographie, botanique, zoologie et anthropologie physique, ethnologie), prend soin d'indiquer dans une note que "cette dichotomie correspond à une simplification". Mais elle conserve une autre dichotomie récusée par Galtung, en reliant préhistoire et ethnologie pour en faire le domaine de l'anthropologie, en tant que "science de sociétés primitives" – c'est-à-dire sans écriture.

Il serait intéressant de confronter les diverses classifications, autres que celle de Piaget, qui avaient été proposées par les experts consultés par l'UNESCO au début de son enquête sur les "tendances de la recherche en sciences sociales et humaines": Daya Krishna (1964, p. 555–566), Lévi-Strauss (d°, p. 579–597), Romero (d°, p. 617–633) et les remarques de Lazarsfeld: ce dernier récuse à la fois la distinction nomothétique/idéographique acceptée par Piaget (p. 575), et le "faux problème" du rôle spécial des "valeurs" en sciences sociales – qui sert de base à la classification de Krishna (p. 577). Comme l'avait fait Galtung, il exprime ses doutes sur le fait que les divisions existantes entre disciplines correspondent à la réalité, en particulier pour ce qui est de la division entre anthropologie et sociologie (p. 575–576) que Lévi-Strauss essaie cependant de justifier (p. 587). Mais une telle discussion, de même que l'examen de l'essai de synthèse tenté par Hochfeld (p. 519–535) nous entraînerait trop loin.

Si nous voulions offrir une conclusion à l'examen qui précède, nous dirions sans doute que, actuellement, beaucoup de problèmes relatifs à une classification rationnelle des sciences restent encore à résoudre – principalement, mais non exclusivement, dans les sciences de l'homme ou aux confins de celles-ci, des sciences biologiques et des autres sciences de la nature.

2 Les systèmes d'organisation des connaissances utilisés en documentation et en analyse de l'information

21 L'évolution des classifications documentaires

Comme nous l'avons vu au § 037, l'histoire des classifications documentaires a déjà été faite plusieurs fois. Nous en avons nous-même tenté naguère un résumé (Grolier, 1956, p. 144–169), qui d'ailleurs devrait maintenant être revisé, compte tenu des nombreux matériaux apportés par Šamurin – matériaux qu'il faut cependant soumettre à la critique, étant donné l'optique assez partiale du bibliographe soviétique (voir notre compte rendu de son livre, 1969). D'une manière générale, les historiens de la classification bibliographique et bibliothéconomique, depuis Brunet (1865), ont presque tous été eux mêmes des constructeurs de classification, et ont apprécié les efforts de leurs prédécesseurs en fonction de leur propre projet: c'est encore le cas de Ranganathan (1965), pour qui tout le mouvement taxinomique documentaire depuis un siècle se décrit par référence à ses idées et contributions personnelles.

Les rapports entre classifications documentaires et système des sciences, classification "des savoirs", ont été jusqu'ici, à la vérité, assez peu étroits. La plupart des classificateurs, bibliographes, bibliothécaires ou documentalistes ont soutenu l'autonomie de leurs systèmes par rapport à la classification scientifique, leur caractère purement pratique, voire conventionnel (Grolier, 1956, p. 7–8). D'autres – principalement Hulme (cf. Sayers, 1967, p. 48–49; Vickery, 1958, p. 132–133) et Farrell (Grolier, 1956, p. 8–10) – ont soutenu que la classification documentaire devait partir des documents eux-mêmes (le *literary warrant* de Hulme) et chercher à les regrouper par affinités, ressemblances, selon une procédure "inductive", sans avoir à considérer la coïncidence ou non-coincidence de ces groupements avec ceux des classifications "philosophiques": ce point de vue rejoint, en fait, le précédent. Plusieurs auteurs d'histoires des classifications documentaires ont voulu voir dans certaines de celles-ci l'influence de tel ou tel système philosophique de classification des savoirs: si Šamurin (1967, p. 96) a cru trouver l'influence hégélienne dans les classifications bibliothéconomiques allemandes du 19e siècle, son argumentation n'est guère convaincante. Il est vrai que quelques auteurs de classifications pratiques (bibliothéconomiques ou bibliographiques) ont en même temps élaboré des systèmes philosophiques – ou l'inverse: Richardson et Bliss étant des exemples du premier cas, Leibniz et Palermo (1854) du deuxième. Mais, en général, ils ont alors renoncé à appliquer en pratique leur idées théoriques: les schémas du bibliothécaire Leibniz n'ont presque rien de commun avec ceux de Leibniz philosophe et savant; quant à Richardson, c'est à juste titre que Broadfield (1946, p. 47) a pu dire que loin d'appliquer en pratique sa théorie, il s'en évadait (cf. aussi Šamurin, 1967, p. 258).

Bien plus que par l'évolution du système des sciences – qui sans doute les a influencées, mais d'une manière diffuse, et en général avec un long retard – les classifications documentaires ont été conditionnées par l'évolution sociale, l'influence des conceptions idéologiques, voire politiques, régnantes. Nous n'essaierons pas ici d'approfondir leur histoire: nous nous contenterons de marquer quelques traits majeurs de leur lente transformation.

211 Jusqu'à la Révolution française

On ne sait presque rien de précis sur les classifications de bibliothèques de l'antiquité, et l'histoire réelle de la taxinomie documentaire en Occident commence au moyen âge, pour lequel des catalogues, en assez grand nombre, ont été conservés et ont fait l'objet d'études détaillées (une de nos étudiantes de Tours en a donné un bon résumé: Jourry, 1971). Les classifications chinoises ont une histoire qui remonte plus haut (ler siècle avant notre ère), mais elles n'ont exercé aucune influence en Europe et nous les passerons sous silence (voir Grolier, 1956, p. 146; Tsien, 1952; Pélissier, 1969), de même que les classifications arabes (Tarrazi, 1947–48; Eche, 1967).

D'une manière générale, les bibliothèques monastiques étaient classées systématiquement, selon un schéma plus ou moins normalisé, débutant par la Bible et ses commentaires, puis continuant par les Pères de l'Eglise; les oeuvres profanes se trouvaient à part, souvent dans l'ordre – *trivium* et *quadrivium* – adopté pour les études. Les bibliothèques des universités, quand celles-ci furent constituées, adoptèrent pour leur classement l'ordre des facultés: faculté des arts, de théologie, de droit, de médecine. Au début du 15e siècle, la bibliothèque privée de Jean, duc de Berry, présente un classement en cinq grandes divisions: théologie, sciences et arts, belles-lettres, histoire, divers, qui est déjà très proche de celui qui se constituera au 17e siècle sous le nom de "système des libraires de Paris". Ni la Renaissance, ni l'âge classique n'innovent dans ce domaine: il y a une très grande continuité chez les bibliothécaires comme chez les bibliographes, qui s'en tiennent à l'ordre traditionnel de l'enseignement universitaire, mettant seulement parfois la théologie à la fin de leurs séries plutôt qu'au commencement et, en ce cas, débutant soit par la philosophie (au sens large qu'elle avait alors) – c'est la solution adoptée par Devaris pour la bibliothèque du cardinal Ridolfi peu avant 1550, soit par ce que nous appellerions aujourd'hui la philologie: ainsi à l'Escorial à la fin du 16e siècle. En Allemagne, Draud inaugure dans ses grandes bibliographies (1611–1625) le système "alphabético-classé" – par grandes divisions méthodiques, puis par ordre alphabétique de sujets traités – qui aura dans ce pays une grande fortune, jusqu'à nos jours; en Angleterre, Maunsell groupe les notices des livres imprimés en Angleterre en trois grandes classes: théologie, puis – approximativement – *trivium* et le *quadrivium,* enfin tout le reste, et adopte à l'intérieur de chacune la formule du catalogue dictionnaire (auteurs et sujets mélangés dans l'ordre alphabétique) qui est encore la formule la plus usuelle dans tous les pays anglo-saxons. Quant à la France, elle perfectionne les "catalogues raisonnés", méthodiques: Naudé en fait le premier la théorie (1627); les libraires et les bibliothécaires parisiens élaborent une sorte de "norme" qui va s'imposer dans tout le pays jusqu'au milieu de 19e siècle et étendre son influence bien au-delà: jusqu'en Russie, en Angleterre (British Museum), aux Etats-Unis . . . avec cinq grandes classes: théologie, jurisprudence, sciences et arts, belles-lettres, histoire. Dans le détail, les classifications se perfectionnent beaucoup cependant: les bibliothécaires de la Bibliothèque royale subdivisent parfois à l'extrême (2191 divisions pour le seul droit civil dans le cadre établi par Malin à partir de 1753), le vieux schéma de Nicolas Clément (1675: il est toujours en vigueur à notre Bibliothèque, devenue nationale, presque trois siècles après!).

En 1751. l'*Encyclopédie* de Diderot et d'Alembert imprime un plan laissé par l'abbé Girard dans ses papiers, qui doit dater de 1748: il cherche à "rationaliser" l'ordre du système des libraires de Paris en le basant sur les besoins de l'être humain – le premier de ceux-ci étant la croyance religieuse (cf. Šamurin, 1967, p. 203–204).

A la fin de cette période, en 1793, trois Allemands, Schütz, Hufeland et Ersch élaborent une classification qui se veut elle aussi rationnelle: elle commence – comme l'avait fait Arias Montanus à l'Escorial deux siècles plus tôt – par la philologie (qui est alors en plein développement en Allemagne), continue par les trois facultés de théologie, jurisprudence et médecine, et cherche à mettre dans un ordre logique – dichotomique d'un bout à l'autre – les matières de la faculté des arts. L'influence de cette classification sera grande en Allemagne tout au long du siècle suivant.

On peut noter aussi, au crédit des 17e et 18e siècles, l'émergence de la technique du "parallélisme" dans les subdivisions des différents sujets: l'établissement d'un ordre plus ou moins fixe pour les divisions géographiques, celles de l'histoire par pays en histoire religieuse, politique, etc.: on trouve déjà cette méthode dans le catalogue de Clément à la Bibliothèque royale en 1675, et elle est érigée en système par Francke à Dresde (1740–48; cf. Grolier, 1956, p. 34–36; Šamurin, 1964, p. 204–210).

212 Un siècle de développement: 1796–1894

Une nouvelle période s'ouvre avec la discussion entre Ameilhon et Camus à l'Institut de France en 1796. Tous deux, comme l'avait fait Girard, veulent introduire dans les classifications bibliographiques un "ordre naturel", qu'il conçoivent d'ailleurs différemment (Grolier, 1956, p. 152–154; Šamurin, 1964, p. 214–221; Dufresne, 1962). Jusque vers 1803, se succèdent des plans plus ou moins novateurs: – ce mouvement de réforme est inséparable de la grandiose tentative des révolutionnaires français pour doter leur pays d'une organisation nationale de la documentation, complément indispensable, à leurs yeux, d'un système national d'instruction publique et de recherche scientifique. Mais ce mouvement avorte, et l'on en revient très vite à la "norme" des 17e–18e siècles, à laquelle Brunet donne, dans les éditions successives de son *Manuel du libraire* (1810–1865) sa forme définitive, *achevée*, aux deux sens du terme. La Bibliothèque royale de Paris, dont les acquisitions révolutionnaires – et les progrès de la production littéraire – ont doublé le volume, cherche à digérer cette masse énorme (pour l'époque) de documents; le plan de refonte Van Praet-Magnin n'est pas appliqué et, sous le Second Empire, intervient une double décision: reclasser l'histoire et la médecine suivant un nouveau schéma, qui comporte une intéressante novation dans la symbolisation des divisions et subdivisions (Grolier, 1956, p. 31); abandonner le détail des classifications anciennes pour les classes non en cours de revision. Cette oeuvre de démantèlement de ce qui jusque là avait été la plus remarquable classification systématique du monde occidental sera achevée par Léopold Delisle à partir de 1878 (idem, p. 162–163).

On avait auparavant beaucoup discuté sur la refonte des cadres du classement de la Bibliothèque royale: on avait même songé à adapter à son usage la classification des sciences d'Ampère. Parmi les auteurs de ces projets de réforme, dont (malheureusement) aucun ne devait être pris en considération, deux méritent une mention particulières: Merlin et Albert. Le premier, à partir de 1840, élabore un système qui, sans doute, doit beaucoup à Ampère, lui est inférieur sur un certain nombre de points (place donnée à la philosophie et à la théologie, division des sciences naturelles en analytiques et descriptives) mais représente le résultat d'un effort conséquent pour baser une classification pratique sur la classification des sciences sans doute la plus avancée de son temps; Merlin l'applique au catalogue Sylvestre de Sacy (1842–1847), puis, pendant trois ans, à la table systématique de la bibliographie nationale française; il le présente, en 1853, à la célèbre Convention de New York où, pour la première fois, se réunissent les bibliothécaires pour discuter de leurs méthodes au plan international. Le deuxième est plus influencé par la philosophie des Lumières et même – peut-être – par les idées révolutionnaires courantes sour la monarchie de Juillet: ses *Recherches sur les principes fondamentaux de la classification bibliographique* paraissent un an avant la révolution de 1848, et sont un témoignage remarquable du mouvement qui – on l'a rappelé au tout récent symposium de l'UNESCO sur "Science et culture" – substitue au 19e siècle l'homocentrisme au théocentrisme (voir Šamurin, 1967, p. 78–80 – qui, cependant, ne s'est pas aperçu du fait (ou n'a pas voulu reconnaître) que le schéma d'Albert, dans ses grandes lignes, était plus proche des idées de Marx que bien des classifications "marxistes" élaborées en Union soviétique au cours des trois dernières décennies . . .). En 1855, l'Allemand Pilz, dans un manuel pour les libraires, donnera une classification qui est une sorte de synthèse, ou de compromis, entre Merlin et Albert (Šamurin, 1967, p. 118–119).

Pendant toute la période que nous étudions en ce moment, c'est probablement l'Allemagne – où les bibliothèques "savantes" se développement rapidement, en parallèle avec l'essor remarquable des universités, qui s'était d'abord manifesté à Göttingen dès le 18e siècle, mais datait surtout de la fondation par Humboldt de l'université de Berlin (1810) – qui voit surgir le plus grand nombre de nouveaux "systèmes bibliographiques". Les trois plus intéressants sont ceux de Lehmann et Petersen pour la nouvelle bibliothèque publique de Hambourg (1840), d'Andreas Schleiermacher à la Hofbibliothek de Darmstadt (élaborée de 1811 à 1830, publiée en 1847) et – bien plus tard – de Hartwig pour la bibliothèque universitaire de

Halle (conçue en 1879, publiée en 1888). Le plan Lehmann-Petersen se rapproche assez de celui de Merlin; celui de Schleiermacher peut être considéré comme un des nombreux remaniements du schéma Schütz-Hufeland-Ersch, proche notamment de celui donné par Ersch lui-même en 1812; quant à Hartwig, il ne fait guère que remanier sur certain points Schleiermacher, et l'on ne sait trop pourquoi sa renommée à finalement éclipsé celle de ses devanciers. Ce qui est le plus remarquable, chez ceux-ci, c'est le progrès que l'on peut observer de la méthode du "parallélisme", dont nous avons vu, à la fin du § 211, la première systématisation chez Francke, développement qui est aussi manifeste, d'ailleurs, dans la deuxième version (1839) de la classification du Belge Namur (cf. Šamurin, 1967, p. 53–54, 117–118). Schleiermacher va cependant plus loin, et c'est à lui qu'il faut faire remonter la transformation de la technique du parallélisme en celle des tables de "subdivisions communes", qui avait été indépendamment utilisé à Kazan par Fojgt (1834) et qui sera reprise au cours de la dernière décennie du siècle par l'Italien Bonazzi, l'Américain Cutter et les Belges Otlet et La Fontaine (idem, p. 128–129, 179–180, 238–240, 251–256, 268–269). Chez Schleiermacher, on trouve encore une autre innovation formelle, dont on fait en général crédit – à tort – à Dewey: celle d'un index alphabétique détaillé des tables systématiques de la classification (idem, p. 130–131); et il faut encore reconnaître que, avec leurs quelque 12 500 rubriques, celles-ci étaient les plus détaillées qui aient jusqu'alors été jamais publiées.

La décennie qui précède immédiatement la révolution de 1848 en Europe est, certes, l'une des plus intéressantes de l'histoire de la classification. On y trouve aussi ce qui est sans doute le plus remarquable exemple d'une classification "pratique" – bibliothéconomique – créée par un très grand savant, d'ailleurs aussi philosophe: rien moins que le fondateur de l'embryologie moderne, connu en Occident sous le nom de von Baer, et en Russie sous celui de Bér. Revenu à Saint Pétersbourg en 1834, après son long séjour à Koenigsberg, Baer prend, en plus de ses tâches scientifiques, celle de réorganiser et diriger la section étrangère à la bibliothèque de l'Académie des sciences. Il y introduit une classification tout à fait remarquable pour l'époque, publiée en 1838 (une deuxième édition suit en 1841), qui est certainement, avec celles de Merlin et d'Albert, la meilleure de toutes les classifications du 19e siècle du point de vue de la systématisation des sujets. Elle doit sans doute beaucoup aux classifications allemandes, que son auteur avait pu connaître durant les vingt ans qu'il avait passés en Allemagne: notamment, on retrouve chez elle le groupe des *Kameralwissenschaften* déjà pris en compte par Schleiermacher et, à sa suite, par Lehmann-Petersen; le plan, par ailleurs, est moins méthodique que ceux des deux Français, mais par rapport à Merlin – qu'il précède de quatre ans comme date d'édition, présente l'avantage de replacer la théologie et la philosophie à la fin des sciences humaines (comme le fera neuf ans plus tard Albert) et non plus en tête de la classification; par contre, la place donnée à la médecine, aux arts militaires et naval et à l'architecture – tout en queue – est moins heureuse que chez Merlin; de même, celui-ci est-il mieux inspiré en rangeant la linguistique dans les sciences humaines, alors que Baer – suivant ses modèles allemands – la met au début, avant les sciences naturelles (comparer les deux schémas dans Šamurin, 1967, p. 80–81 pour Merlin, p. 185–188 pour Baer).

Cependant, ce ne sont ni Albert, ni Merlin, ni Baer – ni même Lehmann-Petersen ou Schleiermacher – qui ont le plus marqué l'évolution de la classification documentaire de cette époque, malgré leurs mérites, mais bien un obscur petit étudiant d'un obscur "collège" des Etats-Unis: Melvil Dewey. Au départ, en 1876, quand après quatre ans de travail celui-ci publie un opuscule assez insignifiant de 42 pages, sans même le nom de son auteur (qui n'a pas encore alors 25 ans) sur la page de titre, et contenant mille divisions (contre les 12 500 de Schleiermacher!) – rangées selon un ordre à peu près entièrement pris à la classification de Harris, publiée six ans auparavant, et dont la symbolisation elle-même était certainement inspirés de celle-ci: numérotation continue de 000 à 999, alors que Harris avait adopté 101 divisions, de 1 à 101 – nul n'aurait pu prévoir l'immense succès mondial qu'allait rencontrer la *Decimal classification* (qui ne s'appelait pas d'ailleurs ainsi, mais, de manière très neutre "A classification and subject index for cataloguing and arranging the books and pamphlets of a library"). En fait, ce succès fut assez lent à venir: les mille exemplaires de la première édition mirent presque dix ans à s'écouler. Mais, de 1885 à 1894, quatre nouvelles éditions se succèdent régulièrement tous les trois ans, et la deuxième est déjà huit fois plus copieuse que la première: c'est aussi la première à affirmer son caractère "décimal"; de 500 exemplaires pour les éditions 2 et 3, le tirage passe à 1000, puis 2 000. Soixante-dix ans plus tard, en 1965, il paraît que la 16e édition (1958) s'est vendue à 30 000 exemplaires dont 30% à l'étranger dans 90 pays différents . . . Une 18e édition doit paraître cette année même. Cette diffusion, que l'on peut bien qualifier de remarquable, n'est due à aucune innovation véritablement originale: même la notation "décimale" des subdivisions de rang 4 et supérieur, introduite en 1885, utilisait un principe appliqué par Ampère en 1834 (Grolier, 1956, p. 24, 128–129) – Ampère s'étant bien gardé, d'ailleurs, d'identifier son procédé avec des "fractions décimales" (voir la critique de cette expression fautive dans Dobrowolski, 1964, p. 118). Elle n'est pas due non plus à des mérites intrinsèques de la systématisation des connaissances adoptée par Dewey: se rattachant, par Harris et aussi par Battezzati (1871), à celle de Brunet (bien plus qu'à un "Bacon inversé"), assez proche aussi – comme l'a remarqué Šamurin (1967, p. 112, 639) du schéma général d'Ersch (1812), cette systématisation restait très inférieure à celle de la plupart des systèmes européens du 19e siècle. Le "parallélisme" assez poussé des divisions de forme, géographiques, etc., n'allait pas au delà de ce que bien des classificateurs avaient élaboré, depuis Clément; quant à son index alphabétique, auquel Dewey attachait tant d'orgueil, les 20 pages qu'il occupait dans la première édition représentaient bien peu de choses auprès des 340 pages de celui de Schleiermacher. Si la "classification Dewey" a réussi, c'est avant tout à des causes sociales qu'il faut l'attribuer: avec sa simplicité (apparente), son caractère d'instrument résolument *pratique*, elle correspondait aux besoins des bibliothèques publiques de masse qui, justement à cette époque (et en partie, il faut le reconnaître, comme suite à l'action de pionnier, en ce domaine, de Dewey lui-même), se développaient considérablement dans les deux pays alors les plus industrialisés: Etats-Unis et Angleterre – pour répondre, d'ailleurs, à la demande de main-d'oeuvre plus qualifiée de la part de leur industrie et de leurs administrations. Quand Cutter, après douze ans d'"incubation", se décida à publier en 1891–93 la "sixième expansion" de sa classification (la septième, seule complète, ne devait commencer à paraître qu'en 1898 et allait rester inachevée après la mort de son auteur), il était trop tard: les jeux étaient faits, et la "Dewey" avait conquis dans les bibliothèques publiques américaines le quasi-monopole qu'elle y possède encore. Le schéma de l'*Expansive classification* doit beaucoup à celui d'Edwards à Manchester (1859), et n'a rien de particulièrement original ni scientifique, malgré l'insistance que Cutter – contrairement à Dewey – mettait à le rattacher à "l'idée d'évolution". Du point de vue formel, par contre, Cutter doit être considéré avec l'Italien Bonazzi, dont la classification est éditée en 1890, comme l'un des deux auteurs qui ont fait progresser la technique des "subdivisions communes" introduite par Schleiermacher: à cet égard, ils ont tous deux montré la voie que suivra peu après l'Institut international de bibliographie de Bruxelles. D'autre part, c'est le schéma de base de Cutter qui sera repris, au début de la période suivante, par la Bibliothèque de Congrès.

213 Depuis 1895

C'est un peu une ironie de l'histoire que l'année même de la mort d'Engels et de la découverte de Roentgen, 1895 – qui, nous l'avons vu (p. 18) marque un tournant très important et le début d'une révolution scientifique – soit celle où, sous l'impulsion de Paul Otlet et d'Henri La Fontaine, ait été décidée la fondation de l'Institut international de bibliographie (actuellement Fédération internationale de documentation) et, simultanément, l'adoption de la classification de Dewey comme base des travaux d'un Répertoire bibliographique universel. L'idée, en effet, d'établir une classification internationale comme norme permanente, quasi immuable dans ses fondements (voir les citations données dans Grolier, 1956, p. 7 et dans Šamurin, 1967, p. 685), en une période où les sciences – pratiquement *toutes* les sciences – se trouvaient bouleversées et renouvelées, ne pouvait plus mal tomber, alors que ces fondements étaient ceux d'une classification basée sur une tradition datant de presque dix siècles, le schéma deweyen continuant directement, en dernière analyse, le classement utilisé dans les bibliothèques médiévales depuis le 9e siècle. On comprend rétrospectivement la sévérité du jugement de Paul Meyer, critiquant (en privé) le projet otlétien, et qualifiant de "grotesque" ce qu'il appelait la "prétention d'immobiliser les sciences dans un cadre fixe, dans une sorte d'immense *trivium* et *quadrivium*" (Grolier, 1956, p. 177). On comprend aussi que, de nos jours, l'American institute of physics, ayant "testé" à grands frais la possibilité d'utiliser la Classification décimale universelle dans un système automatisé de recherche documentaire pour la physique, l'ait finalement rejetée, à cause, a-t-il été expliqué (A.I.P., 1969, p. XXI), "d'un accident historique: le fait qu'elle ait été développée au même moment où se déroulait la révolution fondamentale en physique représentée par la mécanique quantique, et indépendamment de celle-ci". Ceci étant dit, on peut bien reconnaître que la publication, en 1896, de l'article d'Otlet "Sur la structure des nombres classificateurs" marque une date fort importante de l'histoire de la classification documentaire. C'est là, en effet, qu'il explique pour la première fois les modifications introduites dans la symbolisation de la "Dewey"; si certaines d'entre elles (subdivisions de forme, de lieu, analytiques . . .) ne font que reprendre des techniques déjà utilisées antérieurement, il semble bien qu'il y ait par contre une nouveauté majeure: le "déterminant", appelé plus tard signe de relation, qui représente le premier procédé pour introduire dans ce domaine l'intersection et non plus l'inclusion de classes (Otlet ne connaissant probablement pas les travaux de Dyer qui, quelque neuf ans auparavant – 1887 – avait imaginé des procédés analogues pour une langue artificielle). A partir de 1895 commencèrent à paraître des tables pour diverses classes; en 1897 est publiée, en français et en allemand, une table abrégée; en 1898 un *Manuel* d'"exposé et règles"; de 1899 à 1905 la première édition des tables complètes en français en 35 fascicules, refondue cette dernière année en un seul volume et réimprimée en 1907. La 2e édition complète, également en français, sort en 1927–33; elle est suivie d'une "3e édition internationale", cette fois en allemand (1934–53) – la seule actuellement disponible dans son intégralité, car la 4e édition internationale, en anglais, commencée en 1943, est encore assez loin d'être complète: en 1968, on prévoyait d'éditer toutes les parties encore manquantes en une centaine de fascicules d'ici 1970; en 1971, 47 sont parus; l'édition russe dite complète ne comporte pas les classes philosophie et religion (1 et 2): ainsi amputée, elle a eu deux éditions depuis 1963 (la deuxième en 1969–71). On notera que l'usage de la C.D.U. a été rendu obligatoire en Union soviétique – pour les sciences et les techniques – à partir de 1963. Quoi qu'il en soit, la C.D.U. est actuellement la seule classification encyclopédique qui, mal ou bien, serve de norme internationale – comme Brunet l'avait fait (approximativement) pour le milieu du 19e siècle. Jusqu'à 1962, le Comité central de classification de la F.I.D., qui veille au développement centralisé de la C.D.U., avait voulu préserver la concordance de celle-ci avec les nouvelles éditions de la "Dewey" américaine; à partir de cette date, les deux classifications ont évolué séparément. L'année même où paraissait la première édition abrégée de la C.D.U. débutaient à Washington les travaux préparatoires pour ce qui allait être, pendant trois quarts de siècle, la plus considérable entreprise de classification documentaire dans le monde à côté de celle de Bruxelles: l'établissement d'une nouvelle classification pour la Library of Congress, entreprise qui n'est pas encore terminée, puisque la publication de la classe K (droit) n'est toujours pas achevée, soixante dix ans après que les premiers fascicules de la "L.C." aient commencé de paraître. Nous l'avons dit, le schéma de base adopté était celui de Cutter, d'ailleurs considérablement remanié pour tenir compte des besoins propres de la bibliothèque. Ni du point de vue de sa systématique, ni de celui de sa structure formelle, la L.C. n'apportait d'innovation marquante. Au fond, le travail – énorme – des bibliothécaires du Congrès pour classer, aussi minutieusement que possible, leurs collections, n'est que la reprise – avec des moyens évidemment beaucoup plus importants – de celui accompli par les bibliothécaires du roi de France au 18e siècle, puis par ceux de Napoléon III à la Bibliothèque alors impériale. Il n'est pas jusqu'au procédé de symbolisation qui n'évoque celui adopté à Paris vers 1852 – avec quelques perfectionnements qui s'imposaient (Grolier, 1956, p. 205). Cette classification est en train de concurrencer sérieusement la "D.C." aux Etats-Unis et au Canada, grâce en grande partie au fait que les fiches imprimées de la Bibliothèque du Congrès portent toutes les indices de classification L.C., alors qu'un assez grand nombre d'entre elles n'indiquent pas les indices D.C. – bien que la révision de "Dewey" ait été prise en charge, depuis la 16e édition (1958), par un bureau *ad hoc* de la L.C. Un grand nombre de bibliothèques d'Amérique du Nord, au cours des dix dernières années surtout, ont été "reclassifiées" selon la L.C. Mais celle-ci ne s'est guère répandue dans les autres parties du monde (Birket-Smith, 1970).

C'est trois ans avant la date que nous avons choisie comme *terminus a quo* de la période traitée dans ce paragraphe que l'Anglais James Duff Brown avait commencé ses travaux de classificateur, mais c'est seulement en 1906 que paraît la première édition de sa *Subject classification.* Malgré une troisième édition en 1939, il s'agit ici – contrairement à la D.C., la C.D.U. ou L.C. – d'une classification moribonde, sinon morte (Sayers, 1967, p. 176–179), mais qui présente une certaine importance historique. En effet, Brown a été d'une part le premier en Europe occidentale à tenter à nouveau – quelque soixante ans après Baer, Merlin et Albert – de conformer une classification bibliothéconomique à l'état du système des sciences admis de son temps. Il est vrai qu'il avait été précédé dans cette voie par les auteurs, démocrates ou révolutionnaires, des diverses bibliographies de livres recommandés aux autodidactes publiées en Russie entre 1882 et 1906 (Šamurin, 1967, p. 472–477 et 735–738), mais celles-ci étaient restées à peu près ignorées en dehors de l'Empire tsariste: c'est seulement en 1922 que l'ouvrage de l'un de ces auteurs (Roubakine), *Introduction à la psychologie bibliologique* fit davantage connaître leurs idées. D'autre part, du point de vue formel, la "table de catégories" de la S.C. peut être considérée comme une autre étape, après les "subdivisions analytiques" et le "déterminant" d'Otlet, pour introduire dans la classification documentaire la prise en considération des points de vue et de l'intersection de classes.

A partir de 1902, Bliss commence à travailler à une nouvelle classification, à la bibliothèque du College of the City of New York, où il fera toute sa carrière. Il en publie le schéma en 1910, mais l'édition abrégée sortira seulement en 1935, et l'édition complète en 1940–53 – après deux volumes théorique sur la classification des sciences (1929) et la classification bibliothéconomique (1933). La "Bibliographic Classification" – B.C. – n'obtiendra, malgré ses qualités, aucun succès dans son pays d'origine; elle a, par contre, trouvé quelques adeptes en Grande-Bretagne, où se publie un *Bliss Classification bulletin.* Comme Brown, Bliss veut conformer son système de classification pratique à ce qu'il dénomme le "consensus scientifique", qui correspond à peu près au système des sciences néo-positiviste (Sayers, 1967, p. 216, relève la correspondance entre le plan d'ensemble de la B.C. et le schéma donné par Russell dans *Icarus* en 1926: sciences physiques, biologiques, anthropologiques). Par ailleurs, il reprend à la C.D.U. et à la S.C. la technique des "tables auxiliaires" – mais non pas le "déterminant" otlétien. Il est, en définitive, plus important par la critique qu'il a faite, dans ses deux volumes théoriques, des principes de la classification (théorique comme pratique), que par son propre système.

Revenons un peu en arrière: le début de la période que nous étudions ici voit le développement de la bibliographie scientifique et les premiers efforts sérieux pour son internationalisation, avec l'*International conference on a catalogue of scien-*

tific literature tenue à Londres en 1896 sur invitation de la Royal Society. Otlet vient y défendre la C.D.U., mais ne rencontre aucun écho, et une Commission de la Royal Society fait adopter en 1898 un plan de classification complètement différent de celui de Bruxelles (voir Šamurin, 1967, p. 259–261, 683–684). Cette scission entre classifications des répertoires bibliographiques scientifiques et classifications bibliothéconomiques ou bibliographiques générales, encyclopédiques, va s'approfondir dans les décennies suivantes: la très grande majorité des journaux de résumés scientifiques publiés aujourd'hui ont des systèmes de classification particuliers, *ad hoc*, et seule une minorité d'entre eux (ne comprenant pas, d'ailleurs, les plus utilisés), se donne la peine d'ajouter les indices C.D.U. – au surplus négligés par la majorité des usagers, sauf dans les pays de l'Est où l'Etat en impose l'emploi, avec plus ou moins de succès. Même en U.R.S.S., le VINITI – organisme central d'information scientifique et technique – a ses propres classifications (Šamurin, 1967, p. 584–585 et 763); le Centre national de la recherche scientifique français utilise pour son *Bulletin signalétique* un système individuel, d'ailleurs médiocre.

C'est aussi dans les dernières années du 19e siècle que se produit l'essor des bibliothèques et centres de documentation spécialisés, dont le développement s'accélère alors de manière spectaculaire, bien que leurs obscurs débuts soient à rattacher à la création des premiers établissements d'enseignement supérieur technique, déjà aux 17e et 18e siècles: Académie royale d'architecture à Paris, 1671; Ecole d'artillerie de Douai, 1689: Université technique de Prague, 1717; Ecole des ponts et chaussées de Paris, vers 1744–50 (officialisée en 1775) et toute la série des Ecoles des mines, dont la fondation s'échelonne entre 1716 à Ostrava et 1778 à Paris (on peut aussi noter, comme origines plus lointaines encore de ces collections documentaires spécialisées, celles des observatoires et des jardins botaniques: voir ci-dessus, p. 29). Dans leur majorité, ces centres n'adoptent pas non plus les classifications des bibliothèques encyclopédiques, et élaborent pour leur usage propre des systèmes de classification spécialisés; certains, cependant, emploient la C.D.U., dont ils contribueront à développer les parties techniques.

Les années 1920–1940 voient se développer les idées nouvelles dont nous avons vu le point de départ dans le "déterminant" d'Otlet et les "Categorical tables" de Brown. Il apparaît de plus en plus clairement que le principe "hiérarchique" de classification par inclusions successives ne peut rendre compte de la complexité des relations entre les sujets, tels qu'ils apparaissent dans la littérature. De 1919 à 1929, Eppelsheimer élabore à Mayence une méthode de catalogage par sujets – il évite le mot de classification – basée sur ce qu'il appelle des "clefs" qui, sans doute, doivent beaucoup aux tables de subdivisions communes de Cutter, Bonazzi ou Otlet, – et à Brown pour la "clef générale" *(Allgemeine Schlüssel)* – mais où apparaît l'idée d'appliquer systématiquement à tous les sujets une sorte de classification par points de vue: dans les rubriques 89, 90–93 et 94 de la "Allgemeine Schlüssel", qui sont respectivement pour "relations du concept de groupe avec . . .", "sortes et genres" de l'objet désigné par ce concept, et "questions particulières (problèmes)" concernant celui-ci (Šamurin, 1967, p. 370–376; Brall, 1968;

Stevenson, 1971). De 1927 à 1931, et indépendamment d'Eppelsheimer pendant les deux premières années, Trebst à Dresde travaille à un "catalogue analytique" d'un type très voisin de celui de Mayence, avec cette différence qu'Eppelsheimer regroupait ses *Gruppenbegriffe* par grandes disciplines (médecine, géographie, etc.: 25 en tout, dans l'ordre alphabétique de leurs dénominations en allemand), alors que Trebst rejette une telle division, et veut s'en tenir uniquement aux "choses", aux "objets" (*Sachen, Gegenstände*): tout ce qui concerne, par exemple, la démocratie, doit apparaître dans le catalogue en un seul endroit, une seule rubrique, ensuite divisée par points de vue. Ce qu'il dit du travail d'"analyse chimique" des éléments qui composent le sujet complexe du document (Šamurin, 1967, p. 376–382) fait penser aux pages analogues qu' écrira, quelque vingt ans plus tard, Ranganathan. Mais, contrairement à celui-ci, Trebst refuse absolument toute autre symbolisation que celles du langage naturel, pour les *Sachbegriffe,* et des cartes-guides de différentes couleurs dans le catalogue, pour les points de vue: une symbolisation comme celle de la D.C. ou de la C.D.U., selon lui, n'est qu'une "camisole de force" qui fixe les sujets dans un cadre rigide, alors que le catalogue doit être absolument flexible, pour suivre l'évolution des connaissances.

Bien loin de Mayence, et même de Dresde, et encore une fois de manière indépendante, on retrouve des idées allant tout à fait dans la même direction que celles des deux auteurs allemands: chez Gejnc à Léningrad, cette fois non plus dans une bibliothèque encyclopédique, mais dans un service de documentation spécialisé, à l'Institut d'agronomie. De 1926 à 1930, Gejnc y met au point un système de catalogage "structuré" à deux "niveaux" principaux: celui des "problèmes" et celui des "aspects" – auxquels s'ajoute un "troisième niveau": rubriques détaillées, et éventuellement sous-rubriques (Šamurin, 1967, p. 514–517 et 746).

Nous avons abordé quelque peu en détail les conceptions d'Eppelsheimer, Trebst et Gejnc, afin de montrer combien les problèmes de l'analyse des sujets complexes, de la prise en compte des points de vue différents sous lesquels ils peuvent être étudiés, de la flexibilité nécessaire dans l'organisation conceptuelle adoptée, étaient "dans l'air" en Europe dans les années 20–30. Les premières publications de nos trois auteurs sont respectivement de 1929, 1931 et 1930 – trois ans avant que Ranganathan n'édite à Madras sa *Colon Classification,* élaborée à partir de 1926, et dont la "découverte" essentielle – celle des "facettes" – est ainsi, en fait, une re-découverte (comme il arrive si souvent en science et en technique). Les seules différences sont que, contrairement aux deux Allemands et au Soviétique, qui appliquaient une classification par points de vue à des catalogues, Ranganathan construisait une classification systématique, destinée à être appliquée aux documents eux-mêmes (comme celle de Dewey), et qu'il ne partageait en aucune façon la répugnance de ses trois devanciers vis-à-vis d'une symbolisation indépendante de celle du langage naturel. Sur ce point, par contre, il se rencontrait avec Otlet: en fait, la C.C., au départ, emprunte beaucoup à la C.D.U., du point de vue formel. Le signe qui donne son nom à sa classification – le : – est celui là même qu'Otlet avait adopté pour son "déterminant". Ranganathan l'emploie, il est vrai, pour un autre usage: pour séparer, dans la symbolisation attachée à un groupe de sujets faisant partie d'un même domaine, les différents points de vue. On peut dire aussi que, par rapport à la C.D.U., la C.C. possède une supériorité, du fait que la structure "combinatoire" n'est plus ici – si l'on nous permet cette image – une "superstructure", ajoutée à un schéma de base purement énumératif (celui de Dewey), mais une "infrastructure", prévue dès le départ, un cadre uniforme pour tous les domaines, comme l'était celui d'Eppelsheimer. A partir de 1950, la C.C. évolue dans un autre sens: celui de l'adoption de "catégories" générales, au nombre de cinq (personnalité, matière, énergie, espace et temps, PMEST); sa 4e édition (1952) est remaniée pour tenir compte de cette nouvelle théorie et la réimpression 1963 de la 6e édition attribue à chaque catégorie un signe connectif distinct (entre 1952 et 1960, espace et temps partageaient le même signe). Ce n'est pas le lieu ici de discuter en détail la C.C.; nous nous permettons de renvoyer le lecteur à plusieurs de nos exposés antérieurs (Grolier, 1956, p. 211–213; 1962, p. 47–60; 1965, p. 154–176; 1970, p. 115–120). Qu'il suffise de noter qu'ici encore, la méthode ranganathanienne n'était nullement une innovation absolue: dès 1911, l'Allemand J. Kaiser avait publié une méthode d'indexation qu'il avait appliquée aux Etats-Unis d'abord, puis dans des entreprises anglaises, et qui distinguait deux catégories générales, "concretes" et "processes", correspondant, *groso modo,* à la personnalité et à l'énergie de Ranganathan (voir Sayers, 1967, p. 141; Holmstrom, 1953, p. 46–48 et 199–201). Vers 1947, l'Anglais Holmstrom perfectionna la méthode de Kaiser en introduisant une série plus complète de catégories, chacune indiquée par un symbole particulier (1953, p. 71–81). Enfin, en 1948, l'Autrichien émigré en Australie Charles K. Bliss avait, pour un système de "sémantographie" imaginé par lui (langage auxiliaire universel de caractère pictographique), introduit un jeu de catégories presque identique à celui qu'allait adopter Ranganathan deux ans plus tard (Grolier, 1965, p. 174) – à ceci près que Bliss ne distinguait pas entre "personnalité" et "matière". Il semble bien d'ailleurs que Ranganathan n'ait connu aucun de ses prédécesseurs – du moins il ne fait allusion à leur travaux dans aucun de ses écrits.

En 1942–43, trois ingénieurs travaillant à Paris créent chacun de leur côté une classification de type "combinatoire": les Français Cordonnier et Desaubliaux et le Polonais Dobrowolski. Le premier et le troisième apportent des solutions originales au problème de la symbolisation: Cordonnier avec un système "syllabique" basé sur une théorie mathématique aboutissant à préconiser la base 3 comme la plus économique (ce qui, curieusement, correspond au résultat auquel aboutit, par de tout autres voies, Geza Révész: cf. son ouvrage *Die Trias,* Münich, 1956); Dobrowolski avec un système "à indices brefs" – qu'il exposera ensuite de façon détaillée et théorique dans son ouvrage sur la construction des systèmes de classification (éd. en polonais, 1956; éd. française, 1964, p. 151–216). Nous n'insisterons pas non plus ici sur ces travaux, malgré leur très vif intérêt historique et théorique, renvoyant également le lecteur à ce que nous en avons dit ailleurs (Grolier, 1956, p. 230–242; 1962, p. 67–69). La méthode Cordonnier n'a d'ailleurs jamais été appliquée de façon satisfaisante (c'est-à-dire en utilisant toute ses possibilités combinatoires): les quelques centres de documentation français qui l'on adoptée l'on en général adaptée à des schémas "énumératifs" des plus traditionnels (voir notre critique de l'un d'entre eux, 1962, p. 70–72). Plus récement, Cordonnier s'est tourné vers des recherches d'analyse componentielle des termes scientifiques, en vue d'aboutir à une codification "rationnelle" des concepts (1965, p. 121–130).

Nous avons vu que, depuis 1896, un problème était apparu, d'abord d'une manière encore confuse, indistincte (avec le "déterminant" d'Otlet), puis plus nettement, celui des "relations" entre les termes d'un langage d'indexation, ou d'une classification, et de leurs différents types. Ici, l'on peut dire que se fait pour la première fois sentir dans le domaine pratique de la documentation l'influence – indirecte, il est vrai – des recherches logiques et philosophiques sur le "calcul des relations" qui, commencées par Leibniz (cf. Yost, 1954, p. 21–33), reprises après presque deux siècles d'interruption par De Morgan, dans les années 1840, puis par Peirce, étaient apparues enfin au premier plan de l'actualité scientifique avec le *Formulaire* de Peano et le troisième volume des *Vorlesungen* de Schröder, tous deux publiés en 1895 (cette coïncidence de date presque parfaite avec le travail d'Otlet n'est sans doute pas fortuite, bien qu'Otlet, pour autant que nous sachions, ne se soit jamais intéressé spécialement à la logique symbolique). On trouvera un résumé d'ensemble dans notre étude de 1965 (p. 187–210) et l'on pourra consulter les comptes rendus du Symposium international sur les "facteurs relationnels en classification" tenu à l'université de Maryland en juin 1966 (Perreault, ed., 1967). Nous nous contenterons de donner ci-dessous un tableau des principaux auteurs qui ont apporté une contribution plus ou moins importante à l'étude de ces questions, de 1926 à 1968, avec l'indication des pages de notre rapport UNESCO de 1962 qui les concernent (classés dans un ordre chronologique approximatif):

Auteur	Pays	Date de départ des recherches	Date de première publication*	Pages du rapport UNESCO
Tchakhotine	Russie/France	?	1938	66–67
Selye	Autriche/Canada	1926	1956	105–108
C.B.C.C.**	Etats-Unis	1946	?	109–110
Pagès	France	1948	1955	73–79
Farradane	Royaume Uni	?	1950	98–99
Ceccato	Italie	1949	1951	155, 228–229
Perry *et al.*	Etats-Unis	?	1954	112–126
Čerenin	U.R.S.S.	1954	1959	145–147
Newman	Etats-Unis	?	1956	128–134
Kervégant	France	?	1957	22–23
Gardin	France	1955	1958	86–93
Braffort & Leroy	France	?	1959	83–86
Perreault	Etats-Unis	?	1965	–
Austin	Royaume Uni	1967	1968	–

* Première publication se rapportant expressément à l'analyse des relations
** Chemical-biological coordination center; supprimé en 1957

Il convient cependant d'apporter ici une précision: beaucoup de ces travaux – en fait la moitié exactement de ceux cités au tableau ci-dessus, soit 7 sur 14 (Tchakhotine, Ceccato, Perry, Čerenin, Newman, Gardin, Braffort-Leroy) – n'ont pas eu en vue des *classifications*, mais des *langages* empruntant leurs termes aux langages naturels. Parmi les autres, deux d'entre eux (Krevégant et Perreault) ont eu pour but d'améliorer la C.D.U. en précisant les relations qui sont encore indifférenciées dans le système actuel; Selye, le C.B.C.C. et Pagès ont élaboré des systèmes de classification spécialisés; seuls Farradane et Austin ont comme objectif l'établissement d'une nouvelle classification encyclopédique.

Ceci montre une tendance très nette, et qui s'est encore accentuée au cours des trois ou quatre dernières années, à s'écarter des classifications systématiques et à leur préférer des organisations sémantiques plus souples, mieux susceptibles d'évoluer rapidement avec les modifications des conceptions scientifiques: à cet égard, les idées de Trebst sont partagées par la majorité des constructeurs de ce que l'on tend à appeler les langages documentaires ou langages d'indexation (en anglais: *information retrieval languages, indexing languages*). L'on peut voir dans ce mouvement un effet de l'accélération de la recherche scientifique et technique, et de la prise de conscience du caractère essentiellement transitoire des classifications scientifiques, qui n'évoluent pas tant par accroissement et extension, que par remaniements et refontes (Angell, 1965, p. 110; Selye, 1966, p. 16–18; Ember, 1966, p. 47–48). Nul n'a tenté de faire une statistique des nouveaux "langages documentaires", afin de voir quelle est la proportion représentée par ceux d'entre eux qui retiennent le type systématique, encore le plus courant sans doute il y a une vingtaine ou une trentaine d'années, mais nombreux sont les exemples du passage d'une classification systématique à un langage de "descripteurs" tirés du langage naturel et rangés alphabétiquement – avec souvent l'indication des liaisons sémantiques entre les termes, sous une forme ou sous une autre: renvois d'un terme à d'autres termes représentant une classe d'inclusion ou au contraire une classe incluse, ou à des termes "apparentés"; schémas fléchés; listes de catégories ou de "facettes" ... on aura reconnu, dans cette description, les différents types de thésaurus, auxquels on peut ajouter les listes de termes classés par "champs sémantiques" (qui, le plus souvent, ne méritent d'ailleurs pas cette appellation). Par contre, nous ne connaissons point d'exemple de changement dans le sens inverse. Cette mode – ses détracteurs pourraient dire: cette épidémie, car elle s'est répandue comme s'il s'agissait d'une sorte de virus – des "thésaurus" ne date guère que d'une douzaine d'années, bien que les premières suggestions pour l'emploi de ce genre d'instruments soient antérieures (Luhn, 1951; Bernier, 1957). Elle a été "lancée" en fait par le *Thesaurus* de l'ASTIA (Armed services technical information agency, devenue depuis DDC, Defense documentation center), dont la première édition est de 1960, et par celui de l'American institute of chemical engineers (1961); elle s'est propagée en Europe occidentale avec trois ans de retard (*Descriptor list* de l'Atomkernenergie Dokumentation de l'Institut Gmelin, Francfort, 1963) et a atteint l'U.R.S.S. en 1965, date à laquelle était édité le plan d'un "dictionnaire de descripteurs" pour l'architecture et le bâtiment (des études sur la méthode avaient été publiées dès

1962: Mikhailov *et al.,* 1970, p. 381). Bien sûr, derrière ces organisations lexicales, il y a une sorte de classification cachée, honteuse, qui n'ose pas dire son nom: les systèmes de renvois (BT, *broader term;* NT, *narrower term;* RT, *related term*) correspondent à des inclusions de classes, pour les deux premières catégories, et à des rapports de "voisinage" plus ou moins vagues et imprécis pour la troisième, et les schémas fléchés supposent une classification sous-jacente, si grossière et approximative soit-elle. Mais il s'agit d'une classification toujours partielle, incomplète, et généralement des plus médiocre. Toujours est-il que nous sommes tous peu ou prou actuellement soumis à cette mode; dont le centre créateur n'est pas à Paris, mais à Washington, D.C. Ce n'est d'ailleurs qu'un des effets de la position d'économie dominante des Etats-Unis. Les Anglais ont essayé de résister à la contagion, et de préserver leurs classifications "à facettes", plus ou moins lointainement inspirées de Ranganathan, mais dans la toute récente liste de thésaurus disponibles à la bibliothèque de l'Aslib de Londres (Walkley et Hay, 1971), on compte néanmoins 14 thésaurus d'origine anglaises, (contre 35 américains, 6 provenant d'organismes internationaux, 2 du Canada, et 1 pour chacun des 5 autres pays: Allemagne fédérale, France, Italie, Pays-Bas, Suisse; la liste d'ailleurs n'a nullement la prétention de donner une bibliographie complète, et ne comprend notamment aucun thésaurus des pays de l'Est) – dont un, il est vrai, s'intitule *Thesaurofacet* et veut combiner en un seul outil une classification à facettes et un thésaurus "classique" (Aitchison *et al.,* 1970).

Cette pénétration massive de "lexiques documentaires" anglo-saxons ne laisse pas de présenter quelques dangers. Compte tenu des difficultés considérables – pour ne pas dire insurmontables – qui se présentent quand on veut produire des thésaurus ou, plus modestement, des "listes de descripteurs" multilingues sur une base alphabétique, difficultés qui ont bien été mises en lumière, l'an dernier, par le mémoire d'un de nos étudiants de l'I.N.T.D. portant sur la "Liste commune de descripteurs" de l'O.C.D.E., (Skandalis, 1970), on peut se demander s'il ne résultera pas ce mouvement une prédominance quasi-exclusive de la langue anglaise comme base des langages documentaires spécialisés. Il se peut qu'un tel résultat soit inéluctable, mais peut-être aussi le recours (ou le retour) à une systématisation basée sur un "langage classificatoire" indépendant des diverses langues naturelles permettrait-il d'y échapper. On sait qu'une solution de ce type a été adoptée pour rédiger des terminologies internationales en démographie; cet exemple pourrait et devrait être médité.

Une autre voie pourrait sans doute être préconisée: celle de l'adoption généralisée à l'échelle internationale d'une classification encyclopédique "universelle", et ici la C.D.U. se présente comme candidate: c'est peut-être d'ailleurs pour éviter de voir leur langue par trop défavorisée dans l'organisation mondiale de la documentation que les Soviétiques ont si fortement appuyé les efforts de la F.I.D. pour assurer à la C.D.U. une position dominante. Deux questions se posent cependant à ce sujet: (1) est-il possible et/ou souhaitable d'avoir au plan international *un* langage de communication scientifico-technique basé sur *une* classification encyclopédique des connaissances? ; (2) la C.D.U., telle qu'elle se présente actuellement ou éventuellement remaniée plus ou moins profondément, est-elle appropriée à un tel usage? Otlet répondait oui aux deux questions. Au 17e siècle, Descartes, Leibniz et Wilkins (Vickery, 1953) répondaient oui à la première. Ranganathan raconte (1967, p. 26) qu'en 1950 Chester Barnard, alors président de la fondation Rockefeller escomptait de lui qu'il pose les bases d'une classification pouvant servir à la communication internationale, mais sa conclusion fut négative. Nous avions nous-même fait un certain nombre de recherches, à la suite d'ailleurs de Cordonnier, dans cette voie, que nous ne croyons pas une impasse (1956, p. 84–96), sans nous dissimuler, au surplus les énormes difficultés d'une telle entreprise. Par contre, dans le rapport final qu'il a préparé pour "l'étude de faisabilité" sur la réalisation d'un système mondial d'information scientifique UNISIST, Gardin a pris position à plusieurs reprises contre une telle idée (1971a, § 4.2.3.B, p. 53 de l'éd. française, 45–46 de l'éd. anglaise; 5.1.4.C, p. 79–80 fr., 68–69 angl.; 5.2.1.C–D, p. 85 fr., 73 angl.; 6.2.4.I.A–B, p. 110–112 fr., 95–96 angl.) et a préconisé d'une part une "liste de vedettes-matières générale" pour servir à la "catégorisation des sujets" à un "niveau très général" – sans d'ailleurs prendre position pour aucun des systèmes possibles (voir § 6.2.4.I.A), d'autre part des "thesaurus 'intermédiaires' " *ad hoc* dans chaque spécialité, construits dans un "cadre formel commun" (§ 6.2.4.I.B): l'on peut voir là une allusion au Syntol, conçu justement par lui comme un tel "cadre formel", et aux "lexiques intermédiaires" dont plusieurs de ses collaborateurs ont fourni un modèle (assez peu utilisable pour le moment) dans le domaine de l'information scientifique. Dans un texte antérieur, Gardin (1969, p. 21) s'était également prononcé contre "the interesting fiction of a universal semantic language", en basant ce jugement dédaigneux sur un argument "relativiste": "the way the empirical world is divided up is clearly not the same for the astronomer, for the archaeologist or for the physicist".

La deuxième question pose le problème de l'adéquation de la C.D.U. aux conditions actuelles de la systématisation de la documentation internationale. Sa discussion sort de notre propos présent; il nous suffira de rappeler les nombreux projets de refonte générale qui ont été présentés soit assez anciennement (Màday, 1913, Šamurin, 1937; Donker Duyvis, 1933 et 1948), soit plus récemment (Wåhlin, 1968 – voir Grolier, 1970, p. 103–106), soit enfin tout dernièrement par Dahlberg (1970; trad. anglaise, 1971) sur la base d'une classification des sciences proposée par von Weizsäcker (1969). Il y a fort peu de chances, semble-t-il, pour que cette dernière proposition rencontre auprès du C.C.C. (Comité central de classification) de la F.I.D. plus de succès que les précédents.

En Angleterre, où deux critiques acerbes ont jugé, voici dix ans, que la C.D.U. ne pouvait être amendée (Kyle, 1961; Vickery, 1961), le Classification research group a publié en 1968 et réédité en 1969 les études conduites sous son égide pour une nouvelle classification encyclopédique internationale, sur contrat de l'OTAN. Nous ne croyons pas que les résultats de ces travaux, si intéressants qu'ils soient, permettent à l'heure actuelle beaucoup d'optimisme (Grolier, 1970, p. 110–115).

De leur côté, les Soviétiques – après de très longs débats et de fort nombreuses propositions pour établir une "classification marxiste", dont l'historique qu'en donne Šamurin, 1967, p. 518–525, semble d'ailleurs quelque peu sujet à caution: plusieurs des projets des années 1926–36 qu'il traite assez cavalièrement (notamment ceux de Somov, 1926–33, Novosadskij, 1927–36, et Pisarev, 1936) paraissant avoir été plus valables que les travaux postérieurs – ont publié, de 1960 à 1968, une "Classification bibliothéconomique-bibliographique" (B.B.K.), qui se présente plus ou moins comme un compromis entre C.D.U. et L.C. du point de vue formel, et comme un reflet de l'état actuel de la société soviétique, du point de vue de son contenu. Malgré que les tables en soient en cours de traduction en allemand (dans la D.D.R.), il est peu probable que ce système soit appelé à une diffusion importante en dehors des frontières de son pays d'origine (cf. Grolier, 1970, p. 106–110).

22 Quelques problèmes actuels des classifications documentaires

221 La théorie des classifications documentaires

Comme nous l'avons vu (p. 49), on peut faire remonter la théorie de la classification bibliographique et bibliothéconomique à Naudé (1627); Francke au milieu du 18e siècle et encore Ebert en 1820 ne vont guère au delà des principes mis en lumière par le grand médecin, politologue et bibliothécaire parisien. En 1829, Schrettinger fait un pas en avant (cf. Šamurin, 1967, p. 101–105) et Schleiermacher un autre encore, dans l'introduction à son système de classification, en 1847 (idem, p. 120–121): c'est à lui notamment qu'il faut faire crédit de la première formulation claire du principe de la "caution bibliographique" *(literary warrant)*, plus de soixante ans avant Hulme. Il faut ensuite sauter un demi-siècle pour trouver un nouvel essai en vue de fonder la classification pratique des documents sur une théorie, avec les conférences de Richardson à la New York state library school association en 1900–1901 (notamment la deuxième conférence, "The classification of books", p. 23–44 de l'éd. 1912). Le travail de Richardson, encore sommaire, fut repris par Sayers, dans ses *Canons of classification* (1915) puis dans ses conférences à l'Ecole de bibliothécaires de l'université de Londres en 1920, éditées dans son *Manual of classification* (1926; 4e éd., 1967). Les étapes suivantes furent le deuxième volume de la grande trilogie de Bliss (1933) et les *Prolegomena to library classification* de Ranganathan (1937; 3e éd., complètement refondue, 1967). Nous avons nous-même essayé d'apporter quelques compléments à cet édifice (Grolier, 1956, 1965). Il convient pourtant de ne pas dissimuler le fait que, aujourd'hui encore, une véritable théorie au sens scientifique du mot n'existe pas dans ce domaine: il n'y en a guère que des morceaux, le plus important étant sans doute la théorie de la notation (symbolisation), à laquelle Dobrowolski a apporté une contribution majeure dans son ouvrage cité p. 68.

En 1874, Jevons avait déclaré que la classification des livres par sujets était une "absurdité logique" (Sayers, 1967, p. 38–40); cette assertion – fausse – prise au sérieux par les bibliothécaires anglais devait les écarter, pendant près d'un quart de siècle, de la classification systématique (idem, p. 133–134). En France, la condamnation sans appel des catalogues méthodiques par Delisle (1897), à partir d'arguments d'ailleurs valables en eux-mêmes, mais dont il tirait des conclusions abusives, exerça d'une influence plus néfaste encore, dont on peut discerner les traces jusqu'à nos jours.

L'élément qui ressort sans doute avec le plus de netteté de toutes les discussions récentes sur la classification documentaire peut être brièvement décrit par un mot: *relativisme.* Un des premiers à attirer l'attention sur ce point a été Fairthorne: dans un article de 1947 (réimpr., 1961, p. 3), traitant des conceptions erronées sur la classification, il en mentionnait deux qu'il vaut la peine de rappeler:

– "que la classification devrait être le modèle terrestre de quelque système éternel de tout le savoir passé, présent et futur. En fait, c'est un langage décrivant certains aspects des choses telles qu'elles sont";

– "qu'il soit possible de représenter les différentes relations entre les choses par une carte ou un modèle, disons, par une série de symboles disposés linéairement".

En 1960, Fairthorne écrivait (réimpr. 1961, p. 189) qu'une classification documentaire "ne classe pas 'les connaissances', ni même 'l'état actuel des connaissances' – elle groupe tels ou tels aspects du contenu des documents existants qu'il est utile de distinguer pour répondre aux demandes courantes (des lecteurs)". Plus récemment (1969, p. 74), il a repris la même idée: "aucun système documentaire ne peut traiter les connaissances en tant que telles. Le seul savoir qui rentre dans sa compétence est celui qui concerne le discours (enregistré) – non celui dont traite ce discours". Un peu plus loin dans le même texte (p. 78–80), discutant la notion de "sujet *(aboutness)* d'un discours, il écrit notamment que "Détecter le 'sujet' d'un document en fonction de ce qui intéresse le lecteur dépend nécessairement de quelle sorte de lecteur et de quelle sorte d'intérêt il s'agit". A vrai dire, il retrouve là une théorie qui est à la base de la "psychologie bibliologique" de Roubakine (1922: voir ci-dessus, p. 65) qui, au surplus, correspond aux idées spontanément exprimées par les scientifiques (cf. par ex. Selye, 1966, p. 29) et qui a inspiré aussi en Allemagne dès avant la première Guerre mondiale les travaux de Walter Hofmann (en particulier les catalogues établis par lui à Leipzig en fonction, non d'un système de classification "standard", mais des intérêts de diverses catégories de lecteurs – cf. Grolier, 1956, p. 218–219) et, aux Etats-Unis comme en Angleterre, les partisans du *reader interest arrangement* – qui ignoraient d'ailleurs tout de leurs prédécesseurs russes et allemands (cf. Sayers, 1967, p. 310–312).

Toute une littérature s'est développée depuis une douzaine d'années, surtout aux Etats-Unis, autour de la notion de *relevance*, terme que l'on traduit généralement en français par pertinence, encore que plusieurs auteurs anglo-saxons établissent une différence entre "relevance" et "pertinence". Cette question se rattache de près à celle que nous venons d'examiner: la "convenance" d'un document en réponse à une question portant sur un sujet donné n'est pas, comme le croyaient naïvement il y a quelque temps encore la plupart des chercheurs étudiant le rendement des systèmes documentaires et voulant évaluer leurs performances, une propriété des documents eux-mêmes, mais une fonction complexe où interviennent, en premier lieu, les points de vue subjectifs selon lesquels différents usagers considèrent l'utilité de ces documents par rapport à leurs propres intérêts (Cuadra,

1967; Saracevic, 1970; cf. Fairthorne, 1969, p. 81–83). Le problème de la pertinence se relie, d'autre part, à celui de la définition des termes d'indexation: celle-ci, comme l'a fort bien expliqué Selye (1966, p. 85–86) est elle-même très relative. Assurément, Fairthorne a raison de montrer qu'il ne suffit pas d'étiqueter un document comme "traitant des lapins" (1969, p. 84), étant donné qu'un "discours ayant le lapin comme centre doit traiter de la 'caractéristique lapin' [*rabbitness*] en liaison avec d'autres notions": le lecteur de ce document peut de son côté s'intéresser aux lapins en tant qu'animaux comestibles, nuisibles, ou familiers [*pets*], au système de reproduction d'*oryctolagus cunniculus,* ou aux pattes de lapins en tant qu'organes locomoteurs, ou comme amulettes . . . "En résumé, il n'existe pas un concept du 'caractère lapin', avec l'étiquette 'Lapin' autour de son cou. Le 'caractère lapin' peut faire partie de nombreux concepts, suivant son environnement local [*topical neighbourhood*] dans le discours dont il s'agit". Shera (1951, p. 84–86) avait déjà présenté il y a vingt ans un raisonnement analogue, non à propos du lapin, mais concernant les différents concepts de l'arbre; il en tirait des conclusions "pragmatiques" dont nous avons ailleurs critiqué le caractère extrême, en même temps qu'un certain nombre de conceptions ultra-relativistes du même genre (Grolier, 1956, p. 302–306). Il est, en effet, facile de passer de la constatation, après tout assez banale, qu'un même objet peut être considéré de bien des points de vue différents, à une position, disons métaphysique, niant jusqu'à la simple possibilité d'un "ordre naturel" des connaissances, contestant par là même toute tentative pour établir une classification encyclopédique et aboutissant finalement à une sorte d'indifférentisme taxinomique: n'importe quelle classification, si arbitraire soit-elle, fera toujours bien l'affaire. C'est à une conception de ce genre qu'a semblé souvent se rallier J.-C. Gardin (1962, vol. A. p. iii; 1964a, p. 14–15; 1964b, p. 88–89). Un de ses textes plus récents à ce sujet est assez net (1969, p. 22): "By now it is clear that a single documentary language is inconceivable, in spite of the persistent confusion referred to earlier [il s'agit de l'argumentation que nous avons rappelée ci-dessus, p. 72] : the very diversity of documentary languages is, on the contrary, the reflection of the multiplicity of points of view on the empirical world, not so much for ideological reasons [. . .] as for scientific reasons". Il n'est nullement certain qu'adhérer à une telle théorie soit le meilleur moyen pour arriver à bâtir un langage documentaire (même spécialisé) satisfaisant: en fait, les déficiences assez graves des lexiques du Syntol sur le plan paradigmatique, que nous avons longuement examinées (Grolier, 1965, p. 123–157) semblent bien avoir été parmi les causes les plus importantes des "silences" et des "bruits" observés lors du test effectué sur ce système par F. Lévy (1964, p. 184, 202, 208–209).

Un problème théorique proche du précédent, qui a soulevé beaucoup de controverses ces derniers temps, particulièrement en liaison avec les recherches sur la classification automatique, est celui de la *distance* entre les sujets. Fairthorne lui avait consacré dès 1953 quelques pages qui demeurent à lire (réimpr. 1961, p. 127–132): il avait indiqué que "les divers systèmes locaux apparaissent comme déformés à différents classificateurs, du fait que les distances entre les sujets seront différentes" (p. 130). Il est revenu sur ce point en critiquant (1969, p. 87) un article de Mayne (1968) qui, dit-il, s'attache comme beaucoup d'autres à l'idée d'une "carte des connaissances": "Dans quelle mesure s'agit il d'autre chose qu'une métaphore? Dans des situations particulièrement stables, il peut exister une topologie locale et synchronique mais, en général, les voisinages entre les sujets ne sont pas invariants si l'on change de point de vue. Ainsi, ils ne peuvent satisfaire aux conditions requises même pour des espaces topologiques. La plupart des 'cartes' de sujets sont des cartes des programmes d'enseignement dans un domaine particulier, et leur métrique est celle des salles de classe plutôt que celle des classes". On retombe d'ailleurs, avec cette discussion, sur un relativisme très proche de celui de Gardin: comme ce dernier balayait du revers de la main "l'intéressante fiction d'un langage sémantique universel, sur le plan spéculatif (Platon, Descartes, Leibniz et beaucoup d'autres)", Fairthorne, un peu avant le passage que nous venons de citer, écrit que "les fantômes de Wilkins et de Leibniz ne sont pas complètement ensevelis – le charme de la factorisation unique en un groupe d'atomes sémantiques éternels agit trop fortement sur les nouveaux venus formés par les disciplines à base mathématique – mais les lieux qu'ils fréquentent se sont déplacés vers les marges les plus naives et les moins lettrées de la recherche documentaire". Ces dénonciations sonnent bien, en effet, et il est certes plaisant de voir Gardin ou Fairthorne se moquer des illusions de ces vieilles lunes, de ces ridicules utopistes – Descartes, Leibniz . . . Pourtant, chaque époque a bien dû – *nolens, volens* – construire ce "langage documentaire unique" que Gardin déclare "inconcevable", pour les usages où il est irremplaçable; et, pour s'orienter dans la masse des connaissances accumulées, l'on ne peut se passer d'une "carte" du savoir, si médiocre soit-elle, et quelles que soient les distorsions que, comme chacun sait, implique obligatoirement toute projection. Mais la discussion approfondie de ce que, quant à nous, nous considérons comme une illustration typique de l'état d'esprit qui consiste à "to throw out the baby with the bath water", nous entraînerait trop loin. Assez nombreuses ont été, depuis deux décennies, les tentatives pour élargir la théorie des classifications documentaires, et l'unifier avec celle des méthodes d'indexation alphabétique. La première en date, peu réussie, a été sans doute celle de Jonker (1958): elle était trop manifestement destinée à faire ressortir les avantages attribués par son auteur aux "Uniterms" prônés par la firme qui, à cette époque, l'employait. Parmi les essais plus récents, et plus sérieux, on peut retenir ceux de Coyaud (multigraphié, 1964; imprimé, 1966), Gardin (1966) et Vickery (1971). Aucun, d'ailleurs, n'apparaît comme entièrement satisfaisant. Pour le moment, la seule chose qui ressort avec clarté est que tous les "systèmes d'organisation intellectuelle de l'information" – pour reprendre l'expression qui sert de titre à la série d'exposés publiés sous la direction de Mme Artandi à l'université Rutgers (1964–66) – ont une base commune, et que cette base est la classification. Ce point est reconnu aussi bien par Lancaster (1966, p. 53): "l'indexation est essentiellement un processus de classification", que par Fairthorne (1969, p. 77): "l'indexation est une activité extrêmement complexe, inséparable de la classification – c'est-à-dire de la structure – du champ discursif qu'elle concerne [. . .]". Gardin a proposé de restreindre le terme "classification" en vue de lui faire correspondre uniquement une certaine catégorie de "lexiques documentaires" présentant seulement des "correspondances implicites" entre "langage naturel" et "langage documentaire" sous forme de listes "organisées" (1966, p. 174); cette acception, dont il dit lui-même qu'elle est conventionnelle, ne s'est pas imposée. Nous ne pensons pas non plus que soit destinée à s'imposer la terminologie, fort différente et presque inverse, de Vickery (1971, p. 73) qui propose de définir le "thésaurus" comme une collection (littéralement: une représentation, *display*) des termes d'un langage documentaire "montrant les relations sémantiques entre ces termes", soit sous forme de tables systèmatiques, soit sous forme de schémas, comme les schémas fléchés (mis à la mode par l'Euratom et le Bureau d'organisation Van Dijk), en dénommant dès lors "thésaurus relationnels" les "listes communément appelées thésaurus" qui "combinent la fonction d'un thésaurus [. . .] avec celle d'un dictionnaire".

Artandi (1970, p. 151), quant à elle, dénomme "thesauri" n'importe quel type de "liste" de termes d'un langage d'indexation. Et Fairthorne (1969, p. 94) note que le mot thésaurus "by now is battered almost senseless", ce qu'on pourrait sans doute paraphraser en français en disant qu'à force d'être appliqué à n'importe quoi il a perdu tout sens précis. On notera que le manuel de Mikhailov *et al.*, qui a beaucoup développé dans sa deuxième édition (1968; trad. allemande, 1970, p. 373–391), le chapitre consacré à la description générale des thésaurus par rapport à la première édition (en russe, 1965, p. 265–270), après avoir confronté 13 définitions différentes de ce terme (p. 381–384), finit par décider de l'employer pour désigner un "thésaurus normatif" comportant les termes admis pour l'indexation, (descripteurs)et les mots clefs qui peuvent leur servir de synonymes dans le langage documentaire considéré, les descripteurs étant "systématisés selon leur sens" (?), et "les relations sémantiques existant entre eux étant exprimées de manière explicite" (p. 391). Cette acception se rapproche passablement de celle adoptée par Gardin dans son article de 1966 (p. 175), mais elle est plus restrictive, et au surplus le même ouvrage soviétique propose (p. 244–245) une typologie des langages documentaires complètement différente de celle de Gardin, résumant ensuite cette dernière (p. 245–247) avec une terminologie d'ailleurs sérieusement modifiée, qui en change le caractère. . . Aux 14 définitions du "thésaurus" du manuel de Mikhailov *et al.* (y compris la sienne) s'en sont ajoutées d'autres depuis: notamment celle de Soergel (1969, p. 42–43). On voit, par ces quelques exemples – et il serait facile d'en ajouter d'autres – que nous sommes encore loin d'avoir, en taxinomie documentaire, satisfait à cette première condition préalable d'un théorie satisfaisante: une typologie correcte, permettant une nomenclature unifiée!

Il faut dire que d'autres difficultés, sans doute plus graves, s'opposent encore au progrès de la taxinomie documentaire théorique. Nous n'en prendrons que trois, à titre d'exemples. La première est la tentation de présenter, sous le déguisement d'une théorie générale, la simple rationalisation *a posteriori* d'une pratique préexistante. De ce type, entre autres, sont les exposés "théoriques" de Ranganathan pour son propre système (1937, 3e éd. 1967), de Perreault pour la C.D.U. (1969), de Soergel pour les thésaurus (1969).

La deuxième est la propension à construire une théorie à partir, non d'un examen complet de l'ensemble des faits à synthétiser – suivant la vieille règle cartésienne des dénombrements complets – mais d'une partie seulement de ces faits dans l'optique d'une discipline particulière. C'est ainsi que Coyaud a considéré les "langages documentaires" sous leur aspect purement linguistique (et encore en négligeant complètement l'étude de la "deuxième articulation" et presque entièrement l'aspect sémantique); que Farradane a voulu donner de la classification une théorie à base uniquement psychologique, et qu'un certain "psychologisme" est apparent aussi chez Fairthorne (voir la formule donnée p. 11 de son rapport pour la présente conférence "Identification of the topics discussed in physics, say, is not a part of the science of physics but of physicists, with respect to their habits and habitats of discourse"). Cette dernière tendance va de pair avec l'adhésion – consciente ou non – à des thèses philosophiques (épistémologiques) empiriocriticistes, voire berkeléyennes (dans le rapport de Mme Datta et de Farradane pour cette conférence, p. 2, la formule "The external world remains [. . .] a construct of the mind").

La troisième difficulté (proche de la deuxième) tient à l'absence de recherches dans certaines directions, qui rend pour le moment toute théorie générale impossible: ainsi, l'inexistence de ce que nous appellerions volontiers la "sociologie des classifications documentaires", si cette dénomination ne risquait d'apparaître trop ambitieuse ou même prétentieuse. Peut-être d'ailleurs le terme de "psychosociologie" conviendrait-il mieux ici, surtout si on l'entend comme le faisait Plekhanov, pour qui les idéologies reflétaient la psychologie de l'homme social (1913; trad. française 1948, p. 71). Or, dans de très nombreux cas, on peut constater que les classifications documentaires reflètent elles-même l'idéologie de leur milieu et de leur temps, peut-être même plus clairement que les classifications philosophiques, car elles ont en général (tout au moins pour celles qui réussissent et s'imposent) un caractère moins individuel, plus collectif. Šamurin, qui a – le premier ou l'un des premiers – essayé d'envisager de ce point de vue l'histoire de la classification documentaire, a cependant mal réussi son projet, influencé par un "marxisme" appliqué mécaniquement, sans véritable connaissance de l'histoire, et de plus soumis lui-même aux conditions de son époque qui lui interdisaient de voir dans telle particularité de la B.B.K. ou du projet de classification de la Bibliothèque nationale de Pékin non l'application d'une méthode scientifique, mais la conséquence de l'*idéologie* alors dominante (1967, p. 430–431, 529, 536, 538, 546–548, 560). Ce travail serait à reprendre, et à étendre d'ailleurs aux classifications et, plus généralement, aux langages documentaires spécialisés, que Šamurin n'avait pas traités (la diffusion, de caractère épidémique, du modèle "thésaurus", pourrait sans doute être rattachée à certaines caractéristiques "psychosociales" des milieux de la R. & D. actuelle: parcellisation de la recherche, relativisme pessimiste, cloisonnement entre disciplines, malgré les tentatives, en réalité peu efficaces, d'inter-disciplinarité, etc).

222 La construction des langages documentaires

Pendant toute une période, qui va à peu près de 1915 à 1967, on a cherché à établir des normes que devraient suivre les classificateurs, pour construire leurs systèmes de manière rationnelle. Cette tendance, qui trouve son couronnement (et, espérons-le, son achèvement) avec l'édifice complexe ranganathanien de "lois",

”règles”, ”principes” et ”postulats” (1967, p. 39–43, 113–323), est en régression assez nette partout ailleurs que dans l'école de Ranganathan, comme en témoigne la dernière édition du manuel de Sayers, l'un des auteurs pourtant qui avaient été à l'origine de cette recherche des ”Canons of library classification” (1967, p. 96: ”The dramatic and radical changes in the approach to library classification since 1950 [. . .] have made it both difficult and unwise to lay down definite criteria for a book classification and to state emphatically that the points made should be rigidly observed or taken as advice or standards of permanent value”).

Nous ne parlerons ici ni de lois, ni de règles, ni même de principes et, s'il y est question de postulats, ce sera plutôt, le cas échéant, pour dénoncer ceux que, consciemment ou non, les constructeurs de systèmes ont mis à la base de ceux-ci sans en apercevoir le caractère relatif . . . Nous voudrions simplement passer en revue – fort cursivement – un certain nombre de *problèmes,* qui nous semblent à la fois être parmi les plus actuels de la pratique taxinomique en documentation et parmi ceux qui peuvent présenter quelque intérêt, disons ”philosophique”.

Le premier d'entre eux concerne la *base* même sur laquelle doit être construite un langage documentaire: doit-il partir de *disciplines,* ou bien d'*objets* d'étude et/ou de problèmes? Cette question a été, et est encore, posée aussi en d'autres termes: par exemple, un tel langage doit-il s'attacher en premier lieu aux ”concrets” (*concretes* en anglais, selon une terminologie qui vient de Kaiser) ou aux ”aspects”? Une classification documentaire doit-elle être ”multivoque” ou ”univoque” (terminologie de Gardin: multivoque si le même ”objet” peut se présenter plusieurs fois, selon le point de vue suivant lequel il est étudié; univoque pour les systèmes *one place*, ou chaque ”objet” n'a qu'une seule place dans le ”lexique” du système; cf. Gardin, 1964b, p. 90)? En effet, en règle générale, une classification par disciplines, comme la DC de Dewey et sa fille la CDU, la CC de Ranganathan, sera multivoque; une classification par ”objets” tendra, par contre, à être univoque. Il faut dire, d'ailleurs, que le nombre des LD (nous emploierons ci-après ce sigle généralement admis pour langage documentaire) multivoques est jusqu'à présent beaucoup plus grand que celui des LD réellement univoques: parmi les auteurs de ceux-ci on ne trouve guère que Kaiser, Trebst et Brown *(Subject classification)* ainsi que plus récemment Selye *(Symbolic shorthand system).* Le Syntol de Gardin, lui-même, est en fait résolument multivoque (cf. Grolier, 1965, p. 103–104, 117; sur Brown et Trebst, voir Šamurin, 1967, p. 333, 380). Les désavantages de la ”multivocité”, pour ce qui concerne la C.D.U., ont bien été mis en lumière par Mills (1964, p. 34, 95–96) qui remarque que l'on n'y trouve aucune place où mettre des sujets tels que eau, pétrole, Etats-Unis, enfant ou métal envisagés globalement, sous l'ensemble des points de vue selon lesquels ils peuvent être étudiés.

A cette même question se rattache celle de la définition des ”classes principales” ou classes de base, classes initiales (dans la terminologie de Dobrowolski: 1964, p. 11). En fait, aucune classification n'a réussi jusqu'à présent à expliciter sur des bases rationnelles, et non pas empiriques ou purement pratiques, le choix de ses classes initiales. Les deux seuls essais que nous connaissions en ce sens, ceux de Mrs. Tomlinson et de D. Austin pour le C.R.G. (1969, p. 73–80, 81–103) ne peuvent être considérés comme des réussites (voir nos critiques sur un rapport d'ensemble publié en 1968 par le C.R.G., 1970, p. 110–115). Ranganathan, qui refuse de prendre comme base les "objets", qu'il appelle, selon une terminologie fort contestable, "idées isolées" (pour lui, enfant, or, Inde, sont des "idées isolées", qui peuvent être pris comme composants de "sujets" mais non traitées elles-mêmes comme des "sujets": voir 1967, p. 83), n'a jamais pu présenter une justification cohérente du choix de ses "classes de base" ou "classes principales"; il a varié considérablement dans ses divers écrits à ce sujet (cf. Grolier, 1970, p. 120–121). Dans son principal ouvrage "théorique" (1967, p. 84, 86), il indique que les "classes de base" correspondent aux "sujets de base": ceux-ci doivent être "postulés", en "se conformant aux idées conventionnelles courantes chez les hommes d'étude" et au fait qu'ils "figurent dans les programmes scolaires actuels"! On rapprochera cet aveu d'impuissance de la remarque ci-dessus citée de Fairthorne (p. 76) sur la "carte des connaissances" en tant que carte des programmes d'enseignement.

La plupart du temps, les classes initiales des classifications documentaires encyclopédiques ont plus ou moins correspondu aux grandes "disciplines" issues de l'évolution (fort lente) des "facultés" des universités médiévales. Qu'une telle base soit désormais périmée, on en est en général convaincu (voir la discussion intitulée "L'esclavage des disciplines" à laquelle ont participé en mars 1971 cinq éminents universitaires français, dans la revue *Sciences,* n° 72, mai–juin 1971, p. 5–16, spécialement p. 9–14; on lira aussi avec intérêt les remarques sur les disciplines en tant que systèmes de contrôle de la production du discours présentées par Michel Foucault dans la leçon inaugurale de son cours au Collège de France "L'ordre du discours", 1971, p. 15–17 de la trad. anglaise). D'ailleurs, aujourd'hui, on ne peut plus confondre "spécialité" et "discipline": comme le dit Michel Alliot (discussion citée ci-dessus, p. 13–14), "les spécialités associent généralement plusieurs disciplines et [. . .] cet agencement évolue". A. Moles (1969, p. 503) écrit de son côté que "la symbiose suscitée par les problèmes d'application semble devoir provoquer une dissolution des frontières de la 'spécialité' encore si bien marquées il y a trente ans; on tend vers un champ continu de connaissances où non seulement les noms des sciences s'anastomosent [. . .] mais où les rapport établis entre sciences jusqu'alors éloignées [. . .] disposent le savoir [. . .] en une structure multidimensionnelle de 'noyaux de connaissances' ou de techniques mentales interconnectées par des liaisons multiples". Et Ember, dans ses commentaires sur le système documentaire de Selye, "discipline-oriented" à l'origine, signale que se sont développés "de manière presque totalement indépendante, dans les régions périphériques du système", des "project or mission-oriented subject territories": ce changement, cependant, "from the discipline oriented approach to the problem-centered approach, did not basically modify but rather extended the design, creating several 'off-shore islands' " – ces "problem islands", au cours des années "grew faster than the central disciplinary mainland" (1966, p. 47). Mais il soutient qu'un système spécialisé dont "les classes principales ne sont pas rangées en hiérarchies taxonomiques étanches" peut s'adapter à une telle situation (idem, p. 48–49). Il n'est pas sûr du tout qu'un système encyclopédique fondé sur une division du savoir en classes ”mutuellement exclusives” (comme c'était l'ambition de presque tous les classificateurs jusqu'à ce jour) et divisées hiérarchiquement soit aussi facilement adaptable: il y a même de très bonnes raisons pour soutenir l'opinion contraire. Fairthorne écrivait déjà en 1947 (1961, p. 9) que ”probablement il ne peut y avoir de divisions logiquement pures des activités humaines”, et Licklider (1965, p. 73–74) lui fait écho: ”les connaissances ne semblent pas se prêter de manière naturelle à une analyse hiérarchique [il en résulte que] les systèmes documentaires conçus sur une base hiérarchique se transforment en systèmes de quasi-treillis [*latticelike*] par des catégorisations non exclusives et des renvois”. On lira avec intérêt les remarques présentées par Lenk (1971, p. 38–39) sur la critique faite par Popper et par Wittgenstein de l'”essentialisme” qui est à l'origine de toutes les théories faisant des concepts génériques des entités déterminées de manière absolue.

Il n'en est pas moins vrai qu'une classification documentaire qui serait fondée uniquement sur les sujets d'étude, et ne comporterait aucune place pour les “disciplines” traditionnellement constituées ne serait pas viable: en effet, ces ”disciplines” *existent* – on publie des revues, des traités, des encyclopédies, des manuels qui leur sont consacrés, des études sur leur histoire, leur méthodologie, leur épistémologie, etc., et il faut bien trouver un indice de classification qui permettent de ranger de tels ouvrages.

Une dernière question se rattache à l'ensemble de celles que nous venons d'évoquer: ce que Ranganathan appelé le problème des ”faisceaux de sujets”, c'est-à-dire des recherches que l'on pourrait appeler composites, du type de l'océanologie (au sens large: étude des océans aux points de vue physique, chimique, biologique, comme à celui de leur exploration et de leur exploitation), les sciences de l'espace, celles de l'environnement, de la défense, etc. (1967, p. 496–498). Nous avons brièvement critiqué la ”solution”, qui selon nous n'est qu'une pseudo-solution, proposée par lui (Grolier, 1970, p. 116–117).

Un deuxième ensemble de problèmes s'ordonne autour de la controverse qui a opposé les partisans de classifications documentaires ”énumératives” – prévoyant dans leurs tables une place définie pour l'ensemble des sujets existants – et ceux des classifications ”analytico-synthétiques”, fournissant au classificateur des cadres, des réserves de rubriques qu'il est (plus ou moins) libre de combiner *ad libitum* pour ”fabriquer” des indices s'appliquant à des sujets nouveaux, ou à des sujets déjà connus, mais complexes. Une opposition parallèle existe entre les systèmes ”pré-coordonnés”, où les liaisons entre les éléments ”simples” (laissons de côté la question, fort difficile, de la définition d'un élément simple . . .) pour former des sujets composites sont déjà symbolisées, d'une manière ou d'une autre, par le classificateur au moment de l'entrée des documents dans le système (de leur indexation), et les systèmes ”post-coordonnés”, dans lesquels ces liaisons sont établies seulement au moment de la ”sortie” des documents (de la recherche en réponse à une question). Cette opposition, cependant, ne recouvre pas entièrement la précédente: un système ”analytico-synthétique”, comme prétend l'être la Colon Classification, peut être pré-coordonné, et l'on peut inversement utiliser un système à base énumérative (comme la C.D.U.) en pratiquant une certaine dose de post-coordination. Mais la discussion de cette question, plutôt technique – encore qu'elle ait certains aspects "épistémologiques" – nous entraînerait trop loin. Toute une littérature s'est développée, assez fastidieuse par ses répétitions, autour des systèmes "analytico-synthétiques" et des méthodes que Ranganathan a préconisées pour les construire: "facettes" et "catégories". Le premier de ces termes a été défini de plusieurs façons: Ranganathan lui-même écrit (1967, p. 88) qu'il s'agit d'un "terme générique utilisé pour signifier n'importe quel composant [. . .] d'un sujet composite"; malheureusement, la définition qu'il donne d'un sujet composite, fondée sur les notions de "sujet de base" et d' "idée isolée" est inutilisable, étant donné que ces deux notions ne sont pas elles-mêmes clairement et nettement définies (idem, p. 83–84). Fairthorne (1969, p. 89) définit les facettes comme étant les "principales catégories appropriées à une activité particulière, dont les combinaisons permettent de synthétiser les sujets spécialisés que l'on recherche" *(the specialist subject-matter sought).* C'est encore assez vague . . . Mills (1964, p. 35) écrit que l'on doit, en établissant les tables d'une classification, respecter "le vieux principe logique selon lequel une classe ne doit être divisée que selon un seul principe à la fois, les divisions effectuées selon ce principe devant être épuisées avant de passer à un autre. En d'autres termes, la division doit être opérée strictement en facettes ou catégories [. . .] ". Ici, "facette ou catégorie" s'applique à une série de classes dérivées ayant la même base de division, suivant la terminologie – bien plus précise – de Dobrowolski (1964, p. 11–12). Mills n'a pas de peine à prouver que la C.D.U. ne se conforme pas toujours à ce principe (1964, p. 35–36); Coyaud a d'autre part montré que dans une des "classifications à facettes" du C.R.G. de Londres (celle de Foskett pour l'hygiène et la sécurité du travail), une même "facette" pouvait contenir une série unique de divisions se référant à des caractéristiques différentes (ce qu'il appelle d'ailleurs autrement: utilisant "plusieurs sortes de relations analytiques, 1966, p. 52). Gardin, quant à lui, a assimilé "facettes" et "classes fonctionnelles" (1965a, p. 161–165; 1966, p. 181), en établissant un parallèle entre le procédé des facettes, qui pour lui est une organisation syntactique, et les relations binaires du Syntol. Cette assimilation est sans doute abusive, ou du moins ne vaut que pour certains types de "facettes" des classifications ainsi dénommées. Le terme "facette" reçoit parfois une telle extension qu'il devient "quasi-synonyme" de "mot-clef", ou de "descripteur": ainsi dans un récent article de Thompson *et al.* (1971, p. 64) où il est écrit que pour préparer une liste de manuels sur "l'utilisation des mathématiques ou de la statistique dans l'administration des affaires "la question doit être au préalable décomposée en facettes qui vont être soumises les unes après les autres à la machine sous forme d'expressions booléennes". Or, nous voyons que ces "facettes" sont tout bonnement les termes "textbook", "management", "statistical method", etc.!

Pour ce qui est des "catégories" proprement dites, la seule classification actuellement opérationnelle qui les utilise est la C.C. ranganathanienne; nous ne reviendrons pas sur les critiques que nous lui avons faites sur ce point ailleurs (voir les références ci-dessus p. 68). Mrs. Tomlinson et D. Austin ont proposé pour la "nouvelle classification encyclopédique" étudiée par le C.R.G. des "jeux" de catégories d'ailleurs différents: entités concrètes, "mentefacts", propriétés, activités, temps, lieu, pour la première; attributs, "termes relatifs", "termes de position", propriétés, activités, entités, pour le deuxième (1969). Farradane avait déjà préfiguré cette dernière série avec un schéma un peu plus simple, limité à 4 catégories: propriétés, "abstractions", activités, entités (1967, p. 300–303). Gardin (1964b, p. 53–58) avait proposé un autre "jeu": prédicats, entités, états, actions. Nous avons discuté celui-ci, et quelques autres, dans un chapitre de notre travail de 1965 (p. 154–176) qui demanderait à être actualisé et, sans doute, rendu plus sévère pour toutes ces tentatives qui, disons-le très carrément, ne nous semblent mener, pour le moment, qu'à des impasses, d'une manière générale, et – très souvent – à des "coupages de cheveux en quatre" dont on a, par exemple, un joli exemple avec un passage de Ranganathan (1967, p. 409) où un virus passe de la catégorie "personnalité" à celle de la "matière – propriétés" selon qu'il est considéré en lui-même ou comme agent infectieux . . . On aura d'autres exemples de ce que nous ne pouvons considérer que comme des artifices quand Ranganathan (idem, p. 401) définit la catégorie "personnalité" comme catégorie "résiduelle", quand il veut justifier l'ordre de sa série PMEST (idem, p. 412) en assurant qu'elle suit un ordre d'abstraction croissante, ou encore quand il explique les raisons du transfert des "propriétés" de la catégorie "énergie" à celle de la "matière" (idem, p. 400–401; cf. Grolier, 1970, p. 119). Il convient de noter que l'utilisation des "catégories" chez Ranganathan n'est pas la même que celle proposée par Farradane ou le C.R.G.: chez ceux-ci, tout le vocabulaire du LD est *d'abord* divisé selon les catégories proposées, alors que dans la Colon classification, les "catégories" servent à ordonner les divisions des diverses "classes principales". Dans le Syntol, on a un troisième mode d'emploi: les "catégories" ont servi à "étiqueter", pour ainsi dire, les termes du lexique (sauf ceux de sociologie, pour lesquels cette méthode a été abandonnée), en vue simplement d'orienter les relations syntagmatiques.

Les idées gardiniennes sur l'analogie entre classifications à facettes et langages documentaires à syntaxe libre nous mènent, par une transition commode, au troisième groupe de problèmes que nous désirons évoquer: celui qui tourne autour de la distinction, dans les LD, de deux axes: paradigmatique ou "vertical", d'une part, syntagmatique ou "horizontal", d'autre part, correspondant à deux séries tenues pour distinctes de relations à exprimer entre les termes du LD. Pour employer – peut-être abusivement – un langage plus "philosophique", on pourrait dire qu'il s'agit d'une opposition entre "forme" et "contenu" des LD (pour une justification de cette assimilation, voir Lalande, 1962, article "forme", sens B 1°). L'attention de Gardin et de son école s'est beaucoup plus portée vers l'aspect formel des LD que vers celui de leur contenu, ou de leur substance (au sens helmslevien): on a même pu qualifier le Syntol – non sans raison – de "structure désincarnée" (Gardin, 1965b, p. 29), ce que son auteur principal accepte d'ailleurs en disant que son but n'a été que de fournir un cadre formel *(formal framework)* pour l'enregistrement et la recherche de données diversement organisées; ailleurs (idem, p. 94) il écrit que le Syntol "fournit seulement le format et les opérations [nécessaires] pour traiter des représentations et des classifications préalablement établies". Bien moins encore que les philosophes – qui, à ce qu'il nous semble, n'ont pas encore totalement réussi à s'accorder sur le "sens" (l'emploi) du mot relation (voir par ex. les discussions aux articles "relation" et "inhérence" dans Lalande, 1962, p. 911–913 et 514–515) – les taxinomistes de la documentation ne sont encore parvenus à une théorie claire et cohérente des relations qu'ils manient, sans trop savoir apparemment de quoi il s'agit en fait. Ils ont déjà des opinions (au sens platonicien: en tant qu'opposé à la pensée rationnelle) fort diverses quant à leur nombre: Licklider (1965, p. 82) soutient qu'il y a "au moins quelques milliers de relations significatives entre les choses", ajoutant il est vrai qu'il "apparaît comme concevable, ou peut-être vraisemblable, que les myriades de relations diverses que l'on peut observer soient composées de quelques douzaines – peut-être une centaine – de relations atomiques, et que leur grande variété apparaisse quand ces atomes sont combinés de diverses façons, rappelant les combinaisons de la chimie organique". "S'il en était réellement ainsi, conclut-il, l'analyse relationnelle s'avérerait presque certainement comme une technique efficace à utiliser dans la représentation et l'organisation des connaissances". Ceccato (1967, p. 200) estime que "l'homme, aujourd'hui [. . .] a élaboré un millier environ de catégories, dont [. . .] quelques deux cents sont utilisées comme corrélateurs de la pensée". Pagès (1967, p. 353–354) définit les relations comme "prédicats à plusieurs arguments (deux ou plus)" et estime que les relations "peuvent être comptées par milliers"; toutefois, pratiquement, le "CODOC" (l'analyse codée) de Pagès n'en grammaticalise que quatre: réunion, intersection, thématisation et subordination, exprimant les autres par des procédés "morphologiques" (idem, p. 359–361; Grolier, 1962, p. 74–79). Les auteurs du Syntol ont commencé par un système de 4 relations principales complété par un jeu de 7 "opérateurs" (Gardin, 1964b; p. 48–51, 61–66), mais ceux-ci n'ont en fait jamais été employés pratiquement, car le "système central" des quatre relations est apparu, au cours des tests sur ordinateur, comme provoquant déjà des effets indésirables dans la recherche documentaire (Lévy, idem, p. 196–200). Ultérieurement, Bely *et al.* ont "laissé tomber" une des quatre relations du "système central" et n'en ont donc retenu que trois: coordinative (ou comparative), associative, consécutive (1970, p. 8–11); seules ces deux dernières sont d'ailleurs utilisés dans le "réseau notionnel" (idem, p. 84). On pourra remarquer que ces deux relations, associative et consécutive, correspondent de fort près aux deux catégories de codifications du Symbolic shorthand system de Selye: les codifications "statiques", qui répondent à la question "Qu'est-ce? " et "représentent l'analyse d'une structure déjà existante", et les codifications "dynamiques", qui "indiquent toujours qu'une cause produit un effet sur un certain objectif" (Selye, 1966, p. 18–19). F. Lévy a donné (1967) un commentaire du système relationnel du Syntol le comparant à quelques autres "jeux" de relations. Farradane a suivi le chemin inverse: parti d'un système à 4 relations en 1950 (Šamurin, 1967, p. 434), il est passé en 1955 à un schéma à 9 relations, dont les dénominations ont d'ailleurs quelque peu varié depuis (idem, p. 435; Farradane, 1961, p. 130–134, 1967, p. 305–309). Il serait fastidieux d'énumérer tous les autres systèmes proposés depuis deux décennies; ou en trouvera un certain nombre, classés d'une manière assez contestable, dans Soergel (1967). Il est plus intéressant de mentionner deux ou trois des questions qui ont été le plus débattues dans la littérature récente concernant ces relations. Tout d'abord, le caractère très relatif de la distinction entre relations "logiques" et "sémantiques" (Gardin, 1965b, p. 32–33) et le fait que, en toute rigueur, "la seule information formelle [à fournir par l'indication d'une relation] concerne le fait de savoir si deux termes donnés sont liés ou non, tout élément additionnel étant sémantique au sens large du terme" (idem, citant Taube, p. 91); remarque dans le même sens d'Yngve, p. 90). Fairthorne (1969, p. 91) estime que, "pour retrouver les sujets d'un discours, plutôt que pour la taxonomie pure du 'savoir' [. . .] les relations à employer non développées *(crude)* mais fondamentales sont la co-occurrence et l'inclusion": à partir de ce minimum, ajoute-t-il, "chacun peut raffiner et composer des relations syntagmatiques et paradigmatiques selon ses besoins ou ses goûts". On reconnaît bien là l'empirisme anglo-saxon . . . Un autre débat, qui a provoqué beaucoup d'ardentes discussion il y a une douzaine ou une dizaine d'années, mais qui semble en voie d'apaisement, concerne la forme d'expression des relations. Gardin (1965b, p. 26–27) a apporté des arguments qui nous semblent décisifs contre la technique des "rôles", fort à la mode aux Etats-Unis à la fin des années 50 et au début des années 60, et qu'il avait lui-même utilisée dans son code pour l'iconographie vers 1958. Il justifie, de manière assez convaincante, sa préférence actuelle pour l'utilisation de relations sous forme binaire ou dyadique (idem, p. 29–32) et répond par avance à une objection présentée par Fairthorne (1969, p. 91) concernant les relations ternaires (triadiques, triangulaires) du type "entre" (sur ces relations, et la possibilité de réduire toutes les propositions qui ne sont pas du type sujet-prédicat, et qui ne concernent pas des nombres, à des propositions contenant deux termes et une relation, voir Russell, 1937, p. 95, 207–214).

Fairthorne (1969, p. 89) a mis en garde contre "le danger de confondre les relations d'un sujet [avec d'autres] et [les moyens d'expression de] la syntaxe de la langue naturelle dans laquelle la classification et ses règles sont exprimées". Cette observation est juste, sans doute, et quelques systèmes de relations pour LD n'ont pas évité l'erreur que signale Fairthorne (cf. les remarques de Coyaud, sur le GRISAL d'Euratom et le SM1 soviétique, 1966, p. 107, 110–111), mais il y aurait cependant profit à étudier les procédés d'expression des relations dans les langues naturelles, de manière comparative, en vue de discerner ce que Weinreich appelait les "universaux du langage" à cet égard (cf. son travail de 1961, revisé en 1963, notamment p. 120, 127, 130–134). Se restreignant aux langues romanes, Pottier (thèse de 1955, éditée en 1962, p. 115–121) a cru pouvoir établir pour ces langues une "classification des relations" ternaire: coordination, subordination et "aspect fonctionnel", celui-ci se divisant en deux, statique et dynamique. Nous pensons que la ressemblance assez frappante entre ce schéma et celui du Syntol n'est nullement fortuite.

En tout état de cause, une remarque semble s'imposer: plus un "jeu" de relations est complexe, plus le risque d'interprétations divergentes, menant à des "synonymies" entre syntagmes sémantiquement équivalents, devient grand (Grolier, 1965,

p. 206–208). A cet égard, le jeu fort complexe d'opérateurs auquel est arrivé Austin au C.R.G. (1969, p. 104): 8 opérateurs, dont 4 ayant chacun 7 spécifications, nous semble devoir mener, dans la pratique, à de très nombreuses inconséquences d'indexation.

Pour les systèmes de recherche documentaire mécanisés, qu'ils utilisent des cartes perforées à superposition visuelle, ou l'ordinateur, ou bien encore quelqu'autre procédé, l'ordre sériel des éléments qui composent un indice de classification complexe n'a pas grande importance, mais pour des systèmes "manuels" ou "visuels": fichiers "conventionnels", catalogues imprimés, classement systématique des documents sur les rayons d'une bibliothèque ou dans des dossiers, il est indispensable de fixer un ordre linéaire uniforme; d'où la recherche de règles établissant un ordre de "priorité" ou, pourrait-on dire, de préséance pour la constitution des expressions complexes du langage classificatoire. Ce point a particulièrement retenu l'attention de Ranganathan et de ses disciples plus ou moins orthodoxes: le système des "catégories" est, en fait, une sorte de classification des "parties du discours" en vue de fonder des règles de priorité, comme du reste la catégorisation des termes du Syntol a pour but d'établir des règles "d'orientation" des syntagmes de ce LD (Gardin, 1964b, p. 58–60). Ces dernières sont arbitraires; par contre, Ranganathan a cherché à donner une base "logique" aux "principes pour l'ordre de succession des facettes" qu'il a donnés (1967a, p. 425–428; b, p. 404). Il a cru avoir découvert en 1962 le principe de base *(wall-picture principle)* (1967a, p. 101); en fait, sous une forme moins générale, il est vrai, ce principe avait été appliqué par Kaiser dès le début de ce siècle (priorité du "concrete" sur le "process") et repris par Holmstrom (1953, p. 47–48, 71–72). Selye, pour son Symbolic shorthand system, avait également adopté un "order of precedence" préétabli et formalisé, à la fois pour les classes principales du système et pour l'ordre des syntagmes ("target" précédant les divers "agents": 1966, p. 16).

Une quatrième série de problèmes peut être groupée sous la tête de chapitre *symbolisation* (notation) des LD. Quand on sait toute l'importance qu'a eu, pour le développement de domaines comme la logique, la mathématique, la musique ou même la danse (sans parler de la chimie), l'adoption de procédés de notation plus rigoureux et plus adaptés aux fins recherchées, on ne saurait s'étonner de voir la place occupée par les discussions sur la symbolisation dans la littérature de la taxinomie documentaire, ni de constater l'influence de certaines "inventions" en ce domaine. La Decimal classification de Dewey n'aurait sûrement pas obtenu le succès mondial qui a été le sien si son auteur n'avait pas réalisé un progrès fort important au point de vue de la notation: l'affectation des indices de classification, non plus à des documents individuels, ou aux lieux de placement de ceux-ci (numéros de rayonnages, d'étagères, etc.), mais aux classes elles-mêmes. Dewey a lui-même indiqué (1920, p. 152) le rôle joué par cette innovation dans la diffusion de son système aux Etats-Unis. Les innovations de la C.D.U. en matière de symbolisation ont été également déterminantes dans son adoption par beaucoup d'organismes de documentation. En revanche, si un très grand nombre de ceux-ci se sont détournés récemment de toute forme de classification systématique pour adhérer à la formule du thésaurus, c'est en très grande partie à cause du fait que le manque d' "hospitalité" (pour employer un terme ranganathanien) de la notation de la plupart des classifications dont ils se servaient vis-à-vis des notions nouvelles, et encore bien plus le manque de flexibilité de ces notations par rapport aux remaniements dans l'ordre des sujets imposés par les "révolutions scientifiques" et/ou technologiques de ces dernières décennies, représentaient une gêne considérable, alors que l'hospitalité et la flexibilité de systèmes basés sur les termes de la langue naturelle (dues à "l'arbitraire du signe" – si relative que soit cette notion) paraissaient devoir imposer la formule dite "alphabétique", malgré tous les inconvénients que celle-ci peut présenter par ailleurs. Enfin, si la Colon classification ne s'est pas imposée en dehors de son pays d'origine, il faut en voir une des raisons dans la complication sans cesse croissante de sa symbolisation, qui a rebuté beaucoup d'adeptes potentiels, lesquels ont dû trouver beaucoup plus simple et pratique d'affecter à un article sur "oil transportation and refining in Germany" les mots clefs *oil, transportation, refining, Germany* que de le représenter par l'indice C.C. correspondant: (Neelameghan et Gopinath, 1970, p. 350: XX, D74-9(F8,555)-1-z-(7).55 . . .

Nous nous contenterons de signaler quelques-uns des points qui pourraient figurer dans une discussion plus approfondie du symbolisme des LD;

– le choix d'une notation "expressive" (dite encore structurale, ou hiérarchique) ou, au contraire "non structurale", c'est-à-dire dans laquelle les relations d'inclusion ne sont pas rendues évidentes par la forme des indices (Sayers, 1967, p. 66–69; Ranganathan, 1967, p. 272–279);

– la question de la brièveté des indices; très peu de systèmes ont eu la préoccupation de conformer leur notation à la loi qui régit la symbolisation des langues naturelles: la relation entre la longueur des mots et leur fréquence (Hörmann, 1971, p. 86–87). On peut citer comme exceptions Dobrowolski (ci-dessus, p. 62) et Selye (1966, p. 25). Ranganathan (1967, p. 275) a traité ce problème d'une façon tout à fait superficielle;

– le problème – très lié au précédent – de la "répartition de la notation", c'est-à-dire de l'affectation à chaque domaine d'une réserve de notation proportionnelle à son importance présente et à son développement futur prévisible. Bliss (1933, p. 76–79 et 100–102) a été le premier à attirer l'attention sur ce point, sur lequel nous avons présenté quelques études statistiques (1956, p. 38–62, 159–169; 1970, p. 107–108) qui ont montré combien la plupart des classifications existantes surestimaient des disciplines "statiques", comme la théologie, et sous-estimaient l'ensemble des disciplines qui ne sont montrées "dynamiques" en Europe depuis trois siècles (sciences physiques et naturelles et leurs applications techniques, plus récemment sciences sociales); on est surpris, à cet égard, de voir le plus récent projet de refonte de la C.D.U. (Dahlberg, 1970; trad. anglaise, 1971) proposer un schéma où les sciences et les techniques, correspondant à l'indices de nos statistiques, n'occupent que 55% de la capacité de la notation, contre 61% dans le projet Donker Duyvis de 1948 et 65% pour la B.B.K. soviétique, proportions déjà faibles par rapport au "poids" de cet ensemble dans la R. & D. actuelle: entre 90 et 98%

du budget total R. & D., pour les pays développés (cf. Trist, 1970, p. 879–880, 903–904, 906, 969);

– la discussion sur la prononçabilité des indices, que nous avons nous-même alimentée naguère avec diverses propositions, qui seraient d'ailleurs à revoir à la lumière des travaux plus récents sur la redondance du langage naturel et sur les combinaisons de consonnes (Saporta, 1955; Hörmann, 1971, p. 105–106); les arguments de Ranganathan contre ce genre de recherches peuvent être négligés (1967, p. 229);

– les études – fort discutables – sur la "notation rétroactive" (Sayers, 1967, p. 77–78) et, d'une manière plus générale, sur les "signes connectifs" . . .

– la valeur "symbolique" de la notation: on notera, à cet égard, les remarques de Selye (1966, p. 23, 25, 27–28) sur la "valeur mnémonique" des notations qu'il a adoptées; peut-être aussi y aurait-il quelque enseignement à tirer des études (qui connaissent actuellement un regain de faveur, après avoir été longtemps décriées) sur le symbolisme des sons (Hörmann, 1971, chap. 12, p. 214–229; Guiraud, 1967, p. 65–92 et 125–141; Delamain, 1968).

Une cinquième catégorie de controverses porte sur le choix entre LD proches des langages naturels, acceptant notamment – avec un certain degré de "normalisation" – les mots de ces langages comme termes d'indexation, et LD "classificatoires", dans lesquels les termes reçoivent des codifications destinées à les ordonner dans un ordre (plus ou moins) systématique. Nous en avons déjà touché un mot à propos de la symbolisation (ci-dessus, p. 87), pour marquer l'un des avantages les plus marquants, et les plus souvent soulignés, des premiers. Il convient de mentionner, par contre – et cela en opposition à une énorme quantité d'articles qui ont fait récemment valoir, avec de très faibles arguments, la supériorité prétendument écrasante des LD se contentant d'éliminer quelques inconvénients des langages naturels pour la recherche documentaire (homonymies, synonymies, ambiguïtés, etc.), qui seraient soit-disant les seuls réellement adaptés aux capacités des ordinateurs – les qualités, trop souvent négligées, des "langages classificatoires" systématiques. On rappellera, à cet égard, ce qu'écrit très justement Fairthorne (1969, p. 93): les "vedettes de sujets" (nous supposons qu'il inclut sous cette dénomination les thésaurus qui en sont la forme plus "moderne") "sont journalistiques, ce qui veut dire qu'elles ne peuvent éviter de refléter les modes verbales, au lieu d'exprimer les idées, rendues confuses ou masquées par les mots en vogue. Elles sont aussi affectées par l'esprit de clocher, étant liées à un langage naturel particulier. Même dans des domaines spécialisés, les vedettes de sujets d'une certaine langue ne correspondent pas toujours à celles d'une autre langue". Pour les lexiques du Syntol, l'équipe de Gardin, tout en introduisant des structures hiérarchiques (arborescentes), a cependant gardé les termes d'une langue naturelle – le français – comme descripteurs rangés dans ces structures. Cette solution "intermédiaire" n'a pas produit que des effets heureux: trop souvent, les fabricants des lexiques se sont laissés influencer par la terminologie qu'ils trouvaient disponible et, au lieu de classer des concepts, ont fini par énumérer des mots; parfois, ils ont voulu donner à l'un ou l'autre de ceux-ci une signification "conventionnelle", différente de celle du langage courant, et cette pratique risque fort d'amener des confusions. Selon nous, il eut mieux valu carrément introduire une "notation" systématique, de bout en bout. Un autre point a été bien mis en lumière pas Selye: des LD systématiques sont bien plus appropriés pour la recherche exhaustive portant sur un domaine d'étude dans ses relations complexes avec d'autres sujets; ils possèdent bien davantage cette qualité de "browsability" (peut-être peut-on traduire par "bouquinabilité"!) que le spécialiste apprécie tout particulièrement quand il doit préparer une mise au point ou une monographie et qu'il cherche alors "non seulement des données, mais aussi des idées et des théories pouvant lui apporter des suggestions pour son propre travail" (1966, p. 32). "Le système – il pense à un LD systématique comme le SSS – 'pense' pour nous. Il attire notre attention sur des sujets voisins qui doivent aussi être pris en considération et qui pourraient autrement échapper à notre esprit" (idem, p. 36). La plupart des auteurs qui ont, sans discernement, "joined the band wagon" et adopté d'enthousiasme la mode des thésaurus ont visé beaucoup trop exclusivement les recherches documentaires portant sur une question de détail, sur un point précis, et n'ont pas tenu compte de l'insuffisance manifeste de ce type d'instruments quand il s'agit de prendre une vue d'ensemble d'un champ de recherche plus large. Ils ont par ailleurs sous-estimé l'intérêt des classifications systématiques pour les non-spécialistes, qui doivent cependant s'orienter dans des domaines qui leur sont moins familiers et, bien entendu, ont complètement négligé le fait qu'il faut des classifications encyclopédiques pour répondre aux besoins des lecteurs (étudiants, "grand public", sans compter les élèves des écoles primaires et secondaires) qui recherchent une culture générale. A cet égard, tout en contestant la thèse de ceux qui préconisent l'utilisation de la C.D.U. comme "switch language" (traduisons: langage de commutation), on ne peut qu'être d'accord avec eux sur un point: une classification systématique encyclopédique normalisés a d'immenses avantages pour tous ces besoins (cf. Lloyd, 1969). Mrs. Artandi (1970, p. 153) a raison, d'autre part, d'attirer l'attention sur certaines recherches comme celle de Bower (1969) qui a prouvé expérimentalement l'influence favorable d'une organisation structurale de listes de mots présentés à des sujets à qui l'on demande de mémoriser ces listes: le "rappel" des termes est deux à trois fois plus élevé si ceux-ci sont ordonnée hiérarchiquement et si tout l'ensemble des mots à retenir est présenté à la fois, au lieu que chaque terme soit énoncé séparément.

Enfin, le sixième et dernier ensemble de questions concernant la construction des langages documentaires comprend les relations entre LD encyclopédiques et LD spécialisés. Nous n'insisterons pas sur ce point, le rapport de Dobrowolski au présent Colloque fournissant une bonne base de discussion (il reprend des travaux antérieurs du même auteur: 1964, p. 219–244; 1965, p. 65–80). Nous nous contenterons d'attirer l'attention sur les indications intéressantes que donne Selye (1966, p. 28–29, 59–60, 86–87) au sujet des avantages d'un système adapté aux besoins particuliers d'un groupe déterminé de chercheurs. Le problème de savoir comment concilier cette sorte de "focalisation" nécessaire pour desservir de manière satisfaisante des usagers spécialisés avec les besoins différents, de caractère semi-encyclopédique ou encyclopédique tout court, que nous venons d'évoquer ci-dessus, est loin d'être simple. Aucune des solutions que l'on a préconisées jusqu'ici – utilisation généralisée d'une classification encyclopédique, même dans les centres de documentation centrés sur une discipline ou une "mission"; préparation de concordances; recherche de méthodes, plus ou moins automatiques, pour assurer la "compatibilité" entre systèmes différents (Neville, 1970); mise en oeuvre d'un "macrothésaurus"; utilisation d'un "cadre formel" type Syntol – ne paraît entièrement satisfaisante. De nouvelles recherches seraient donc nécessaires. Elles pourraient, sans doute, s'inspirer du programme tracé dans le rapport Dobrowolski. Mais il convient cependant de signaler un certain nombre de difficultés:

– le nombre de "secteurs scientifiques autonomes" au sens de Dobrowolski est sans doute beaucoup plus grand que celui qu'il avait indiqué en 1964 (2 à 300: p. 231) et repris en 1965 (p. 69). Nous avons suggéré (1970, p. 116) qu'il pourrait y en avoir cinq à dix fois plus. Mais tous n'ont pas la même importance; tous, d'autre part, n'ont pas créé des systèmes documentaires "autonomes" dont on puisse confronter les rubriques;

– il ne conviendrait pas de limiter la recherche proposée par Dobrowolski aux secteurs *scientifiques:* une classification encyclopédique ne peut se limiter à la science; elle est obligée de faire place à l'ensemble des *activités* humaines – même si elle se propose comme but d'ordonner celles-ci en conformité à des principes scientifiques;

– le "postulat" de Dobrowolski (1971, p. 5): "que chaque secteur spécialisé possède un 'noyau' composé des sujets essentiels à la spécialité donnée et *propres à elle seule*" (souligné par nous), postulat qui se trouvait déjà, sous une forme moins catégorique il est vrai, dans ses travaux antérieurs (1964, p. 222; 1965, p. 67), ne nous semble pas toujours vérifié en pratique. Encore moins croyons-nous exacte la proportion moyenne d'environ 50% de sujets fondamentaux avancée par Dobrowolski (1964, p. 223; 1965, p. 71). Si nous prenons une classification spécialisée récente comme la *London classification of business studies* (Vernon et Lang, 1970), il nous sera impossible de trouver *une seule* rubrique "propre" à ce "secteur" – qui pourtant répond à la définition de Dobrowolski selon les critères qu'il propose. Même la classe A, management, est commune à l'administration des affaires et à celle des services publics . . . La classe AZ, l'entreprise, est commune aux "business studies" et à la science économique. Etc. . . ;

– s'il s'agit non seulement d'aligner des rubriques de sujets, mais encore de les *ordonner* d'une manière significative, on se trouvera aux prises avec une autre difficulté: dans un "secteur" donné, il peut exister *plusieurs* schémas de classification des sujets "fondamentaux" de ce secteur, selon les écoles, les spécialisations à l'intérieur du domaine et – dans les sciences humaines beaucoup plus sans doute que dans les sciences de la nature – selon les cultures. Gérard et Bourgeois (1937) avaient déjà signalé ce problème pour la chimie; pour la philosophie, il a inspiré tout un projet de l'Unesco ("Travaux préparatoire à une encyclopédie philosphique comparative", 1971). Ceci ne veut pas dire qu'un travail sur des bases inspirées de celle suggérée par Dobrowolski ne serait pas utile: loin de là. Mais, pour être profitable, il devrait être précédé d'une étude préparatoire portant sur les relations réellement existantes entre les *domaines* scientifiques, techniques, artistiques, etc. (Grolier, 1970, p. 124–125).

223 Les moyens techniques de mise en oeuvre des langages documentaires

Sans entrer dans le détail, nous voudrions signaler ici l'influence qu'a exercée sur la conception même des LD l'évolution des techniques utilisées pour la réalisation matérielle des systèmes de recherche documentaire. Il est certain que l'adoption généralisée des catalogues sur fiches au cours du dernier quart du 19e siècle, se substituant aux catalogues imprimés qui avaient dominé jusque là, a favorisé le recours à des LD "atomisés" utilisant les "vedettes alphabétiques de sujets". Les systèmes d'indexation coordonnée si à la mode depuis les années 1950 correspodaient à un certain état de la technique alors disponible (fiches à superposition visuelle, fiches "uniterms" de Taube, fichiers "inversés" utilisés avec les machines à cartes perforées et les ordinateurs de deuxième génération, etc.). On en sera facilement convaincu en voyant la place qu'occupe la description de ces moyens matériels dans, par exemple, Costello (1966, p. 117–153) ou dans n'importe quel manuel de la décennie 60 sur les systèmes de recherche documentaire. Les facilités offertes par les ordinateurs de la troisième génération ont favorisé des systèmes utilisant les langages naturels, avec des artifices variés (troncature, etc.), permettant, comme l'écrit Fairthorne (1969, p. 98) "to play unprincipled word games on a computer" à des gens pour qui "the symbol is believed to be the sense". On a aussi tenté de "jouer avec l'ordinateur" en utilisant des classifications "traditionnelles" qui n'avaient pas été construites en vue de cet usage. L'ordinateur étant un instrument très "versalile" (pour employer l'excellente expression anglaise, plus parlante que sa paraphrase française "qui se plie à tout"), on a pu "démontrer" la possibilité d'emploi d'une classification comme la C.D.U. dans des systèmes automatisés (Freeman et Atherton, 1968, et les références citées par Fairthorne, 1969, p. 96; Caless *et al.*, 1970; Schneider et Koch, 1970; comptes rendus du 2e Séminaire F.I.D. sur la C.D.U. et les systèmes mécanisés, Francfort, ler-5 juin 1970, à paraître). Mais, dès 1958, Fairthorne pouvait écrire (1961, p. 144) que "mécaniser directement des classifications bibliothéconomiques traditionnelles équivaut à vouloir construire des locomotives qui marchent avec des jambes". Et, plus récemment, Mikhailov *et al.* ont présenté des arguments (d'ailleurs non exempts de contradictions) d'où il ressort que l'utilisation de la C.D.U. pour des systèmes mécanisés est sans doute possible, mais qu'un tel système "partiellement à facettes" (selon leur terminologie: cf. 1967, p. 309) présentait à cet égard des inconvénients (idem, p. 311, 389–390). Mills (1964, p. 83) écrit pour sa part que "UDC is basically for pre-coordinate indexes, usually in card or book form [. . .] mechanized devices for selecting from a UDC index have little to offer". Le Patent office des Etats-Unis, qui a créé un "Computer aided classification (CAC) planning committee" en 1971, semble avoir été mieux inspiré que la F.I.D.; une note sur les travaux de cette commission (F.I.D. news bulletin, 21: 26) nous informe que "it is not contemplated that the existing classification schedule used by the Patent Office will be the final framework of the computer-aided classification system.

Rather, an attempt will be made to use computers to create classification structures based upon the intrinsic information in patent text". On peut cependant noter que, à la British steel corporation, a été développé depuis 1967 un système de diffusion sélective de documents (SDI) mécanisé sur la base de la C.D.U., apparemment avec succès (McCash et Carmichael, 1970).

La 3e Conférence internationale de Cranfield sur les systèmes documentaires mécanisés (20–23 juillet 1971) avait mis à son ordre du jour les relations entre les systèmes mécanisés "généraux" (du type Medlars: traitant les documents d'un vaste domaine) et ceux plus étroitement spécialisés. Sans doute la discussion a-t-elle pu dégager des idées intéressantes: attendons-en les comptes rendus.

Une remarque, pour finir, sur ce sujet qui aurait certes mérité de plus amples développements: Selye (1966, p. 36–37), discutant les projets de mécanisation de son Symbolic shorthand system, exprimait quelque appréhension, craignant de perdre, dans un système semi-automatisé, les possibilités de "browsing" que son système manuel actuel lui procure. On peut cependant noter que les facilités offertes par les ordinateurs pour publier des catalogues imprimés constamment tenus à jour peuvent provoquer un renouveau d'intérêt vis-à-vis des classifications systématiques: en fait, il apparaît bien que le "classified catalog" est l'objet, depuis quelques années, de travaux – surtout américains – reprenant, avec de nouveaux moyens, la longue tradition qui va du moyen âge aux grands catalogues méthodiques de la Bibliothèque royale, puis impériale, de Paris. Par ailleurs, quand seront plus généralement disponibles les écrans de visualisation comme moyen de communication entre l'utilisateur d'un système automatisé (ou, plutôt, semi-automatisé) de documentation et l'ordinateur, il sera possible de présenter à cet usager des listes systématiques de références, sans trop de difficultés: ce progrès technique devrait, lui aussi, contribuer à remettre en honneur ces classifications aujourd'hui si décriées.

224 Vieillissement et mise à jour des LD; la question de la "reclassification"

Tout langage documentaire est destiné à vieillir – de plus en plus rapidement, étant donné le phénomène de "l'accélération de l'histoire" naguère mis en lumière par Daniel Halévy – puis à mourir. En fait, la plupart des LD actuellement connus étaient déjà fort vieillis au moment même de leur naissance, par suite du décalage (allant de quelques années à au moins un demi-siècle, par exemple pour la classification de Dewey, qui était basée sur un état de la science et de l'idéologie datant du premier quart du 19e siècle, sinon du 17e . . .) entre les systèmes "pratiques" et le système des sciences, tel qu'il est conçu par les scientifiques eux-mêmes. Certes, l'on a connu des normes fort durables de classification: le système chinois des "4 classes" est resté en vigueur presque inaltéré de 618 à 1918 (à peu près); l'ordre de base des bibliothèques médiévales n'a pas changé depuis les premiers catalogues manuscrits que nous ayons conservés (Staffelsee, 812) jusqu'à la Renaissance; le "système des libraires de Paris" a duré de 1650 environ (catalogue de Bouilliau) à 1865 (dernière édition du Brunet); la C.D.U. s'est maintenue de 1895 à nos jours . . . Mais il n'est pas d'exemple qu'une classification documentaire ait été immortelle, et l'on observera, de la liste ci-dessus, que leur durée de vie tend – contrairement à celle des être humains – à raccourcir. Dobrowolski (1964, p. 245–247) estime à 25 ou 30 ans le délai après lequel une classification spécialisée a trop vieilli pour pouvoir être seulement revisée et doit donc être mise au rebut; il écrit qu'on "pourrait adopter comme principe le mot d'ordre: à chaque génération, classification nouvelle". Et il propose de clore, au moment de ce changement, l'ancien fonds documentaire, et d'ouvrir une nouvelle série, totalement indépendante de la première. Une suggestion analogue a été faite, cette fois pour une classification encyclopédique, par Otlet (1933): traitant de la refonte générale de la C.D.U., qu'il envisageait comme inéluctable, il pensait que l'on "pourrait aussi décider de la version en usage pour telle période et l'application normale [. . .] de cette version aux publications parues pendant cette période".

Une généralisation de cette méthode a été proposée par Lund et Taube (1937), après Adolf Meyer (1921–23): la classification "par périodes". Un premier projet en ce sens avait d'ailleurs été suggéré par Parent (1801). Dans la version de Lund et Taube, il s'agirait de classer la littérature concernant chaque "époque" selon un schéma inspiré du système des sciences à cette époque. Nous ne croyons pas qu'il y ait eu d'applications pratiques de ces propositions, que nous avons assez longuement discutées ailleurs, ainsi que quelques autres (Grolier, 1956, p. 62–65, 187, 353–354).

Ranganathan (1967, p. 136–139) a exposé ce qu'il a appelé la "loi d'osmose": selon lui, il faudrait changer de classification quand celle jusqu'alors en usage est périmée, et reprendre dans le "nouveau fonds" les documents de l'ancien fonds qui ont été prêtés (ce prêt étant considéré comme la preuve de leur utilité actuelle); en cinq ans de ce processus, tous les documents encore "vivants" seraient reclassifiés selon le nouveau système, et le reste ("dead books") pourrait être laissé à son sommeil éternel dans le tombeau de l'ancien fonds . . . Ce procédé non plus ne semble pas avoir été appliqué par beaucoup de bibliothèques, et ne l'a été, presque certainement, par aucun centre de documentation spécialisé.

Une variante du procédé d'osmose a été suggérée par Angell (1965, p. 108–111): quand une classe du système aurait trop vieilli par rapport aux conceptions contemporaines de la discipline correspondante, elle serait close, et une nouvelle classe serait ouverte, soit contiguë à l'ancienne (par exemple dans le cas des "nouvelles mathématiques"), soit ailleurs (par exemple dans le cas de la psychologie: l' "ancienne" psychologie à caractère philosophique restant avec la philosophie, la "nouvelle" psychologie – expérimentale, etc. – allant dans les sciences biologiques).

Il faut reconnaître franchement que, jusqu'à présent, *aucune* technique réellement satisfaisante pour résoudre ce problème du vieillissement des classifications documentaires n'a été trouvée. En pratique, il y a des bibliothèques qui "s'en tirent" en conservant leurs anciens systèmes, plus ou moins démantelés (par suppression des divisions vieillies: c'est le cas de la Bibliothèque nationale de Paris), et d'autres qui "reclassifient" laborieusement leurs collections, ce qui ne va pas sans soulever de considérables difficultés, du point du vue "management" comme du point de vue financier (Perreault, 1968).

Les "classifications alphabétiques" – listes de vedettes de sujets et autres thésaurus – sont aussi atteintes par le vieillissement, contrairement à ce que prétendent la plupart de leurs partisans: Dobrowolski (1964, p. 247–251) a insisté à juste titre sur ce point, trop souvent négligé.

Une question proche des précédentes est celle des divergences qui se manifestent dans l'application d'une même classification en des pays éloignés, quand il n'est pas prévu de mécanisme central pour surveiller son évolution: c'est un phénomène analogue à celui de la dérive génétique, ou encore à la différenciation progressive d'une même langue naturelle quand elle est parlée par des populations diverses et s'il n'y a pas d'organe "législateur" central (du type de l'Académie française ou de l'Académie espagnole – et encore, même dans ce cas, les efforts "puristes" sont-ils rarement couronnés de succès). On en trouvera un exemple tout récent dans la divergence du thésaurus du Cedocar (Centre de documentation de l'armement) français par rapport à son modèle américain du DOD (Renaud, 1971, p. 8); d'autres cas sont classiques: divergence des classifications inspirées de Dewey en Chine et au Japon, de celles "calquées tout d'abord sur la C.D.U." en U.R.S.S. Bruin (1971) a discuté le cas, assez exemplaire, des difficultés rencontrées pour harmoniser les pratiques suivies à la British national bibliography et à la Library of Congress dans l'application d'une *même* classification: celle de Dewey.

Pour fonder la pratique de la "reclassification" sur des bases théoriques plus solides, il nous semblerait opportun d'étudier de plus près les différentes formes que prend l'évolution des "disciplines" scientifiques: des indications éparses sur ce sujet se trouvent un peu partout, mais n'ont pas (à notre connaissance) fait l'objet d'un ressemblement systématique (cf., par ex., Selye, 1966, p. 16–18; Ostwald, 1929, p. 62–63 – sur le changement de contenu de l'astronomie depuis Babylone). Nous avions cru en 1956 (p. 62) pouvoir distinguer des domaines "dynamiques" (essentiellement les sciences et les techniques) et des domaines "statiques" (art, religion, linguistique, littérature, philosophie) . . . Nous ne pensons plus que cette dichotomie assez sommaire soit justifiée: il est tout à fait évident maintenant que la linguistique, que nous rangions dans les domaines statiques, a subi un profond renouvellement au cours de ces dernières années (sans doute, exactement, depuis Chomsky, 1957). Quant à la religion, c'est seulement une partie des études qui lui sont consacrées (la théologie dogmatique) qui est restée statique, mais il tend à se constituer une "science des religions" globale, qui, elle, est loin d'être statique: on n'en voudra pour preuve que la constitution récente d'une "socio-phénoménologie religieuse" (Roger Mehl), d'une "oecuménologie", d'une mythologie structuraliste (Lévi-Strauss), etc. Une dernière observation sur ce sujet – qui aurait certainement mérité du plus longs développements: suivant le mot profond de Karl Marx (1857; éd. française 1967, p. 35) "L'anatomie de l'homme donne la clé de l'anatomie du singe" – dans un processus évolutif, seule la connaissance de l'état le plus récent permet de comprendre vraiment les états précédents. Cela est aussi vrai du processus évolutif de la connaissance lui-même, et c'est – au fond – ce qui condamne des propositions comme celle de Lund et Taube, et restreint considérablement la portée de la remarque faite par Fairthorne (1960; recueil de 1961,

p. 199) selon laquelle "Toute classification pour la bibliothèque classique d'Alexandrie qui utiliserait notre connaissance actuelle, disons des éléments physiques (autrefois 'chimiques') ne serait pas seulement inutilisable: elle serait aussi stupide". Sans doute – mais seule notre connaissance de l'évolution ultérieure de la géométrie, de la mécanique, ou de la religion, nous permet-elle de mettre à leur juste place les résultats acquis dans ces domaines aux temps des Ptolémée.

225 L'évaluation des langages documentaires

Depuis une douzaine d'années – exactement depuis le début des "expériences de Cranfield" (les tests appliqués sous la direction de Cleverdon à Cranfield, en Angleterre, pour comparer "objectivement" les performances de divers langages documentaires) – l'évaluation des LD est à la mode. Ce sujet occupe à lui seul 158 pages du recueil de Saracevic (1970, p. 537–694), soit environ 1/5 de l'ensemble. Mais, pour l'essentiel, toute la littérature produite à ce jour, et qui doit déjà compter plusieurs centaines de références, sinon plus d'un millier, tourne autour d'un unique problème: étant donné un LD, mesurer deux paramètres supposés correspondre à son "efficacité" – le facteur de rappel *(recall)* et le facteur de précision *(precision)*, naguère appelé facteur de pertinence *(relevance)*, ayant respectivement pour but de mesurer les "blancs" ou le "silence" du système (la proportion de documents pertinents existant dans la collection et non retrouvés – par différence d'ailleurs, puisque le "recall" donne en fait le pourcentage de documents pertinents existant dans la collection et effectivement fournis au demandeur), et le "bruit" de ce système (la proportion de documents non pertinents mélangés, dans la réponse à la question, aux documents pertinents, également obtenue par différence). Nous avons déjà évoqué (p. 74–75) les problèmes que pose la définition même du "document pertinent", dont les premiers chercheurs ne semblaient pas d'être rendus compte; nous n'y reviendrons pas. Mais, quels que soient les raffinements statistiques et/ou mathématiques que l'on puisse apporter à cette méthode de mesure (voir notamment Pollock, 1968), il n'est nullement certain qu'elle suffise à fonder des jugements valables sur les performances réelles des LD. Elle ne permettra jamais, en effet, d'évaluer autre chose que la capacité du système documentaire à répondre, de manière plus ou moins satisfaisante, à des *questions* précises, portant sur des sujets nettement délimités. Or, il s'en faut de beaucoup que ce soit là le seul (ni même peut-être le plus important) usage auquel doivent être adaptés les LD: nous avons effleuré ce problème p. 89. Pour Selye, par exemple (1966, p. 38), "le meilleure preuve d'efficacité du système (son SSS) est donnée par les revues critiques et les livres auxquels il a donné naissance". Et ce n'est certes pas le fait de répondre plus ou moins exactement à des "questions" particularisées qui a permis au SSS d'obtenir ce résultat: c'est bien plutôt la faculté qu'il donne au chercheur de prendre une *vue d'ensemble* d'un sujet ou même d'un domaine, dans toutes ses connexions . . .

3 En guise de conclusion

Nous voici arrivé au terme de cette exploration – qui, malgré sa longuer déjà excessive, n'est guère qu'une vue cavalière, ou à vol d'oiseau, d'un vaste domaine en fait peu exploré systématiquement – où nous avons cherché à éclaircir, pour autant que ce soit possible, les rapports entre une science – la taxinomie théorique – et une technique, celle des langages documentaires.

Il ne saurait y avoir, à pareille tentative, de conclusion à proprement parler. L'on ne trouvera donc, en ce troisième volet d'un tryptique quelque peu artificiel, qu'une ébauche ou même seulement un canevas, esquissant quelques-uns des thèmes de réflexion qu'une telle confrontation peut suggérer.

Une observation liminaire nous paraît s'imposer: la technique a, ici, plus à recevoir de la science qu'elle ne peut lui apporter. C'est, peut-être, parce qu'elle est largement demeurée jusqu'ici au stade des recettes, des procédés empiriques (éventuellement "justifiés" par de pseudo-théories) de l'artisan, et n'a pas atteint ce stade de " synthèse de la science et de la technique" que Moles (1969, p. 500) considère – après Bachelard – comme une caractéristique de notre époque et qu'il dénomme, construisant un néologisme assez discutable, "phénoménotechnique". Les instruments dont se sert le documentaliste pour explorer le contenu informationnel des documents demeurent rudimentaires, fort traditionnels même (et peut-être surtout) quand, affublés de noms à la mode, on les croit "dernier cri". C'est aussi parce que ces documentalistes, si même ils se sont éventuellement rebaptisés informatistes ou informateurs scientifiques, ont jusqu'à présent manifesté un grand dédain pour les bases scientifiques de leurs travaux. L' "unité de valeur" qui a le plus rebuté nos étudiants de l'Institut national des techniques de la documentation a été, l'an dernier, celle que nous avions voulu instituer sur les "bases théoriques" de leur futur métier, tant et si bien que nous l'avons démantelée. On rencontre bien souvent, sous la plume des "gens du métier", des formules qui témoignent, soit de ce dédain, soit d'une certaine naiveté. Comment expliquer autrement les phrases, pourtant signées d'un "ingénieur-documentaliste" membre du personnel de la plus grande société mondiale d'informatique: "Appelons *concept élémentaire* un *mot* isolé d'une langue naturelle quelconque. Un *concept* sera la réunion de concepts élémentaires, reliés entre eux par un certain nombre *d'opérateurs logiques*" (Le Dû, 1971, p. 29 – souligné par l'auteur)? Mais peut-être cette assimilation simpliste du mot et du concept serait-elle établie en fonction du fait que la société en question a tout intérêt à diffuser un certain ensemble de programmes qui permettent " . . . la 'recherche' des résumés, sans intervention préalable d'aucune codification ou traduction en mots-clés" (idem, p. 28)? Intérêt, dans ce cas, mal compris, ou du moins à courte vue . . . N'est-ce pas aussi par méconnaissance des bases théoriques des LD que, sur la suggestion du directeur de la documentation d'un très important groupe industriel français, une pléiade d'éminents spécialistes se sont attaqués au problème de la constitution d'un "macrothésaurus", supposé devoir contribuer à la "compatibilité" des "microthésaurus" spécialisés, en pensant au départ pouvoir le résoudre par la simple réunion de "macrotermes" (non définis de manière tant soit peu précise, et envisagés simplement comme les termes "de base" ou "essentiels" de chaque discipline) proposés par chaque spécialiste et mis bout à bout? Et comment faut-il qualifier tel essai d'utilisation des "cercles d'Euler" pour le "graphic display of concept relationships" dans le domaine de la terminologie, qui ne donne qu'un informe magma, dans un rapport présenté par un non moins éminent fonctionnaire d'Euratom (Rolling, 1965, p. 307)? Cet exemple, malheureusement, n'est pas isolé, et l'on trouve en maints endroits des représentations graphiques presque tout aussi contestables (Ellingham, 1948; Kedrov, 1956, p. 105; Licklider, 1965, p. 83–84; Moles, 1969, p. 503).

Il arrive cependant, parfois, que les réflexions des praticiens débouchent de manière intéressante sur des problèmes théoriques. Ainsi, après avoir pendant deux ou trois décennies élaboré des nomenclatures de produits industriels, d'une part, d'activités économiques d'autre part, sans nullement se soucier de leur donner une justification "scientifique" quelconque, les administrateurs de divers bureaux intéressés dans plusieurs ministères français (Institut national de la statistique et des études économiques, chargé de la Commission interministérielle des nomenclatures; Bureau interarmées des codifications de matériel; Commission centrale des marchés, ayant établi une nomenclature de ces marchés selon leur objet; etc.) ont-ils récemment constitué un groupe d'études dont plusieurs membres ont commencé à réfléchir sérieusement sur la question théorique de la liaison entre une classification d'*objets* (et de services) et une classification d'*activités.*

Ces rapprochements sont encore trop rares; on pourrait souhaiter qu'ils se fissent plus fréquents et plus systématiques. N'en citons que quelques exemples possibles:

– réexamen du "principe de collocation" posé pour les classifications documentaires par Bliss (après Ampère), mais contesté par Pagès (voir Grolier, 1956, p. 302, 305; Sayers, 1967, p. 99, 215), à la lumière d'une étude portant à la fois sur les liaisons organisationnelles entre "disciplines" et champs d'activités et sur leurs liaisons "conceptuelles";

– étude critique des idées de Fairthorne (1971, p. 10–16) sur la "dérivation d'un espace-temps bibliographique" – c'est-à-dire, en termes moins alambiqués, l'étude des "chaînes de citations", en les confrontant aux expériences menées par Price et Schiminovich (1968), puis par ce dernier seul (1971) sur la classification automatique au moyen d'un "algorithme de découverte des structures bibliographiques" et aussi aux recherches menées récemment aux Etats-Unis sur la genèse de certaines innovations pour découvrir les cheminements convergents de pensée qui les ont amenées (I.I.T. research institute, 1968);

– confrontation entre les théories (psychologiques et logiques) sur la formation des concepts et des liaisons entre concepts, et les "présuppositions" qui se trouvent (plus ou moins consciemment) à la base du choix des "descripteurs" et de l'organisation des relations entre ceux-ci dans les LD;

– parallélisme (ou manque de parallélisme) qu'il peut y avoir entre certaines théories de caractère épistémologique sur l'arbitraire des classifications (nous pensons, par exemple, à la théorie proposée par Mandelbrot, 1955, sur la "relation de Willis") et la propension, qui paraît évidente à l'heure actuelle, à poser comme purement arbitraires les structures paradigmatiques des LD (nous pensons, ici, à l'attitude de Gardin);

– liaison qu'il peut y avoir entre le mouvement de pensée qui, depuis le début de ce siècle, paraît s'être si fortement manifesté en philosophie des sciences et en philosophie tout court, pour "dépasser" le rationalisme, poser en face de la prétention – jugée abusive – de la science à englober l'ensemble du savoir les revendications d' "autres voies" de la connaissance, et les tendances analogues que l'on peut reconnaître dans les LD.

Il ne nous paraît pas douteux, cependant, que de tels échanges ce ne soit finalement la pratique documentaire qui doive retirer le plus de profit. Une "science de l'information", telle que les documentalistes les plus "progressistes" veulent la constituer pour asseoir sur des bases plus solides leur *praxis*, sera fort difficile à distinguer de la *science de la science*, si l'on admet, avec beaucoup des partisans de celle-ci, que la science constitue, certes, un système social (Storer, 1966), mais que ce système est essentiellement un système de communication (c'est, à peu près, la thèse de Ziman, 1968, et celle qui ressort de presque tous les écrits récents de Merton). S'il en est ainsi, l'épistémologie des sciences formera l'une des composantes fondamentales de cette *architectonique du savoir* à laquelle devront se conformer les langages documentaires – classificatoires – à venir.

23 septembre 1971 (revisé, 22 novembre 1971)

Additions 1974

§ 034, p. 23 – Le manuscrit du "Dictionnaire étymologique et sémantique de la langue latine" d'Abel Juret, que je viens de publier en microfiches, permettra désormais de discuter – pour le confirmer, éventuellement pour l'infirmer, ou (plus probablement) pour le modifier – le "modèle structurel" auquel ses recherches avaient abouti. Le grand intérêt de ce modèle est qu'il présente, contrairement à d'autres tentatives (Hallig-Wartburg, Buck, et – naturellement – Roget et ses émules) une structure *immanente* au langage étudié, et non pas l'application à celui-ci d'une classification plus ou moins "logique", mais basée sur des critères extérieurs au phénomène linguistique, et que, par ailleurs, contrairement aux recherches sur les "champs sémantiques", il s'agit là de structurer *l'ensemble* d'un lexique, et non pas quelqu'une de ses parties, arbitrairement isolée.

§ 123, p. 45–46 – Depuis que j'ai écrit ces lignes, les exposés de Maurice Leroy et (surtout) de Roman Stopa au colloque international organisé les 28–30 octobre 1971 par l'Académie des Lincei à Rome sur "L'origine de l'homme" ont apporté des éléments intéressants sur les rapports entre langage animal et langage humain: voir, dans le Quaderno N. 182 des Lincei, *Atti del colloquio internazionale sul tema L'origine dell'uomo* (Roma, 1973), p. 277–315. – D'autre part, les recherches des neuropsychologues sur les sujets "split-brain" (aux hémisphères cérébraux déconnectés), que j'ignorais à l'époque, paraissent avoir donné un argument de grand poids à la théorie postulant l'indépendance (relative) du langage et de la pensée (voir l'article du Dr. François Michel. "Cerveau double et cerveau dédoublé", *Le Monde*, 21 août 1974, p. 9).

§ 213, p. 65 – Une nouvelle édition, complètement refondue et appliquant les principes de la "facettisation", de la classification de Bliss doit paraître en 1975 sous la direction de J. Mills (Northern Polytechnic de Londres).

Références citées

AITCHISON, Jean *et al.*

1970 *Thesaurofacet.* Whetstone (Leics.), English electric company. (voir également Aitchison, Jean. The thesaurofacet: a multipurpose retrieval language tool. *The Journal of documentation,* 26 (3) sept. 1970, 187–203.

ALLARD, M. *et al.*

1963 *Analyse conceptuelle du Coran sur cartes perforées.* Paris, La Haye, Mouton.

ALLIOT, Michel *et al.*

1971 L'esclavage des disciplines. *Sciences* (72) mai-juin 1971, 4–16.

AMERICAN INSTITUTE OF PHYSICS. Information division

1969 *A program for a national information system for physics.* Technical report AIP-ID 69 R.

ANGELL, Richard S.

1965 On the future of the Library of Congress classification. *In* Atherton, Pauline (ed.) *Classification research;* proceedings of the second international study conference . . . 14th to 18th September 1964. Copenhagen, Munksgaard, p. 101–112.

ARTANDI, Susan

1970 Document description and representation. *In* Cuadra, Carlos A. (ed.) *Annual review of information science and technology,* vol 5. Chicago, Encyclopaedia britannica, p. 143–167.

ARTZ, Frederick B.

1966 *The development of technical education in France, 1500–1850.* Cambridge, Mass., The M.I.T. Press.

AUGER, Pierre

1961 *Tendances actuelles de la recherche scientifique.* Paris, Unesco.

AUSTIN, Derek

1969 The theory of integrative levels reconsidered as the basis of a general classification. *In* Classification research group, *Classification and information control.* London, The Library association.

BAR-HILLEL, Yehoshua

1964 *Language and information:* selected essays on their theory and application. Reading, Mass., Addison-Wesley publishing company.

1970 *Aspects of language:* essays and lectures . . . Jerusalem, The Magnes press.

BELY, N. *et al.*

1970 *Procédures d'analyse sémantique appliquées à la documentation scientifique.* Paris, Gauthier-Villars.

BEN-DAVID, Joseph

1971 *The scientist's role in society:* a comparative study. Englewood-Cliffs, Prentice-Hall.

BENVENISTE, Emile

1958 Catégories de pensée et catégories de langue. *Les Etudes philosophiques,* n.s., 13 (4) oct.-déc., 419–429.

BERNIER, Charles L.

1957 Correlative indexes, II (et III, avec Karl F. Heumann). *American documentation,* 8 (1 et 3), 47–50, 211–220.

BIRKET-SMITH, Kjeld

1970 *Local applicability of the Library of Congress classification.* FID/CR report No. 10.

BITSAKIS, E. J.

1971 Lénine, le matérialisme dialectique et la physique contemporaine, II. *La Pensée,* (158) Juillet-août 1971, 78–96

BLISS, Henry Evelyn

1929 *The organization of knowledge and the system of the sciences.* New York, Holt.

1933 *The organization of knowledge in libraries.* New York: H.W. Wilson Co. (2e éd., 1939).

BOCHENSKI, I. M.

1967 *La philosophie contemporaine en Europe.* Paris, Payot.

BOUDON, Raymond

1970 Modèles et méthodes mathématiques. Chap. VIII de Unesco, *Tendances principales de la recherche dans les sciences sociales et humaines;* première partie: sciences sociales. Paris, Mouton/Unesco, p. 629–685.

BOURDE, Paul

1939 *Essais sur la révolution et la religion.* Paris, Hartmann.

BOWER, Gordon H. *et al.*

1969 Hierarchical retrieval schemes in recall of categorical word lists. *Journal of verbal learning and verbal behavior,* 8 (3) June 1969, 323–343.

BRALL, Arthur

1968 *Anwendung und Abhandlung der Sachkatalogisierungsmethode Eppelsheimers an deutschen Bibliotheken.* Köln, Greven Verlag.

BRÉHIER, Emile

1937 *La philosophie du moyen âge.* Rééd. 1971. Paris, Albin Michel.

BROADFIELD, A.

1946 *The philosophy of classification.* London, Grafton.

BRUIN, Joyce E.

1971 The practice of classification: a study towards standardization. *Journal of librarianship* 3 (1) January 1971, 60–71.

BRUNET, Jacques Charles

1865 *Manuel du libraire,* 5e éd. Paris, Didot, vol. 6, col. V ss.

BRUS, W. *et al.*

1970 La science économique. Chap. IV de Unesco, *Tendances principales de la recherche dans les sciences sociales et humaines;* première partie: sciences sociales. Paris, Mouton/Unesco, p. 340–426.

CAILLEUX, André

1951 Lieux, époques, espace, temps et classification des sciences. *Revue générale des sciences,* 58 (9–10), 297 ss.

CALESS, T. W. *et al.*

1970 *Strategies for manipulating Universal decimal classification relationships for computer retrieval.* Washington, D.C., George Washington university. AD 717 212.

CARVELL, H. T. et Jan SVARTVIK

1969 *Computational experiments in grammatical classification.* The Hague, Mouton.

CECCATO, Silvio

1967 Concepts for a new systematics. *Information storage and retrieval,* 3 (4) Dec. 1967, 193–214.

CHATILLON, Jean

1968 La culture de l'école de Saint-Victor au 12e siècle. *In* Gandillac, Maurice de, et Edouard Jeanueau (éds.) *Entretiens sur la renaissance du 12e siècle.* Paris, Mouton, p. 147–160.

CHAVAILLON, Jean

1971 Recherches en Ethiopie. *Sciences et avenir,* no spécial: La vie préhistorique, p. 78–79.

CHOMSKY, Noam

1957 *Syntactic structures.* The Hague, Mouton (trad. française: Paris Editions du Seuil, 1969).

COLE, A. J. (ed.)

1968 *Numerical taxonomy.* London, Academic press.

COMTE, Auguste

1822 Plan des travaux scientifiques nécessaires pour réorganiser la société. Réimprimé dans le *Système de politique positive,* Paris, l'auteur, 1854, tome IV.

COPPENS, Yves

1971 Les australopithèques réhabilités. *Sciences et avenir,* no spécial; La vie préhistorique, p. 80–83.

COSTELLO, John C., Jr.

1966 *Coordinate indexing.* New Brunswick, New Jersey, Rutgers – the State university.

COYAUD, Maurice

1964 *Introduction à l'étude des langages documentaires;* Paris, C.N.R.S., multigraphié; nouv. éd. (imprimée), Paris, Klincksieck, 1966.

CUADRA, Carlos A.

1967 *Experimental studies of relevance judgments.* Santa Monica, Calif., Systems development corp., 3 vol. TM-3520/001 à 003.

DAHLBERG, Ingetraut

1970 Möglichkeiten einer Neugestaltung der DK. *Nachrichten für Dokumentation,* 21 (4) August 1970, 143–51; trad. anglaise: Possibilities for a new universal decimal classification. *Journal of documentation,* 27 (1) March 1971, 18–36.

DATTA, Suman et Jason E. L. FARRADANE

1971 *A psychological basis for general classification.* Rapport multigraphié pour la Conférence d'Ottawa.

DAUDIN, H.

1926 *Les classes zoologiques et l'idée de série animale en France à l'époque de Lamarck et de Cuvier (1790–1830).* Paris, Alcan, 2 vol.

DE BIE, Pierre

1970 La récherche orientée. Chap. IX de Unesco, *Tendances principales de la recherche dans les sciences sociales et humaines;* première partie: sciences sociales. Paris, Mouton/Unesco, p. 686–764.

DELAMAIN, Maurice

1968 *Plaidoyer pour les mots:* un essai de phonétique expressive. Paris, Stock.

DELISLE, Léopold

1897 Introduction au *Catalogue général des livres imprimés de la Bibliothèque nationale,* Auteurs, tome I. Paris, Imprimerie nationale, p. I–LXXXII.

DENIFLE, Heinrich
1885 *Die Entstehung der Universitäten des Mittelalters bis 1400.* Berlin.

DEWEY, Melvil
1920 Decimal classification beginnings. *The Library journal,* 45, p. 151 ss. (Cité d'après Šamurin, 1967, p. 666).

DIETERLEN, Germaine
1952 Classification des végétaux chez les Dogons. *Journal de la Société des africanistes,* 22, 115–158.

DOBROWOLSKI, Zygmunt
1964 *Etude sur la construction des systèmes de classification.* Paris, Gauthier-Villars (traduction de l'original polonais: *Budowa klasyfikacji,* Warszawa, 1956).
1965 Analysis of classification systems. *In* Atherton, Pauline (ed.) *Classification research.* Copenhagen, Munksgaard, p. 65–80.
1971 *Secteurs scientifiques autonomes et leur rôle dans la classification des sciences.* Rapport pour le Colloque d'Ottawa.

DONKER DUYVIS, F.
1933 On the future complete revision of decimal classification. Institut international de bibliographie, *XIIe Conférence de bibliographie,* Bruxelles, 1933, Rapports, p. 7–17.
1948 Codification and recodification of classification. F.I.D., *Restricted conference,* The Hague, 1948, Committee on comparative classification, document F48–63.

DURAND (DE GROS), J. P.
1899 *Aperçus de taxinomie générale.* Paris, Alcan.

DUCROCQ, Albert
1960 *Logique générale des systèmes et des effets.* Paris, Dunod.

DUFRESNE, Hélène
1962 *Le bibliothécaire Hubert Pascal Ameilhon (1730–1811).* Paris.

DURKHEIM, Emile et Marcel MAUSS
1901 De quelques formes primitives de classification. *L'Année sociologique,* 6 (1901–2), 1–72.

ECHE, Youssef
1967 *Les bibliothèques arabes publiques et semi-publiques en Mésopotamie, en Syrie et en Egypte au moyen âge.* Damas, Institut français.

EIJKMAN, P. H.
1910 *L'internationalisme médical.* La Haye.
1911 *L'internationalisme scientifique.* La Haye.

ÉLIADE, Mircea

1957 *Das Heilige und das Profane.* Hamburg, Rowohlt. – Trad. française: *Le sacré et le profane.* Paris, Gallimard, 1965.

ÉLISSEEFF, V.

1965 Possibilités du scalogramme dans l'étude des bronzes chinois archaïques. *Comptes rendus du séminaire sur les modèles mathématiques dans les sciences sociales,* année 1964–65. Paris, Ecole pratique des hautes études, 6e section.

ELLINGHAM, H. J. T.

1948 A chart illustrating some of the relations between the branches of natural science and technology. Publié dans Royal society (London), *Scientific information conference,* Report and papers submitted; réimprimé dans Holmstrom, J. Edwin, *Records and research in engineering and industrial science,* London, Chapman and Hall, 3e éd., 1956, en face de la p. 84.

EMBER, George

1966 Some operational aspects of the system. Chap. VII de Selye, Hans, *Symbolic shorthand system.* New Brunswick, N.J., Rutgers – the state university, p. 45–49.

ENGELS, Friedrich

1925 *Dialektik der Natur.* Moscou. Ecrit en 1873–86. Deux traductions françaises: par P. Naville, Paris, Rivière, 1950; par E. Bottigelli, Paris, Editions sociales, 1952, toutes deux sous le titre *Dialectique de la nature.*

FAIRTHORNE, Robert A.

1961 *Towards information retrieval.* London, Butterworths (regroupe des articles et rapports publiés de 1947 à 1960).

1969 Content analysis, specification, and control. *In* Cuadra, Carlos A. (ed.) *Annual review of information science and technology,* vol. 4, Chicago, Encyclopaedia britannica, p. 73–109.

1971 *Temporal structure in bibliographic classification.* Rapport pour le Colloque d'Ottawa.

FARRADANE, J.

1967 Concept organization for information retrieval. *Information storage and retrieval,* 3 (4) Dec. 1967, 297–311.

FEIBLEMAN, James K.

1954 Theory of integrative levels. *British journal for the philosophy of science,* 5 (17) May 1954, 59–66.

FLINT, Robert

1904 *Philosophy as scientia scientiarum.* Edinburgh, Blackwood.

FOUCAULT, Michel

1971 *L'ordre du discours.* Paris, Gallimard. Trad. anglaise dans *Social science information,* 10 (2)

FREEMAN, Robert R. et Pauline ATHERTON

1968 *Audacious* – an experiment with an on-line, interactive reference retrieval system using the Universal decimal classification as the index language in the field of nuclear science. American institute of physics, New York, UDC project. AIP/UDC-7; PB 178 374.

FREUDENTHAL, H.

1960 *Lincos:* design of a language for cosmic intercourse, vol. 1. Amsterdam, North Holland.

GALTUNG, Johan

1965 Peace research. Document UNESCO/SS/41/3.244.1/h/10 – multigraphié. Imprimé dans: *The social sciences, problems and orientations . . .* Selected studies. Paris, Mouton/Unesco, 1968, p. 192–208.

GARDIN, Jean-Claude

1962 *Le Syntol* (Syntagmatic organization language). 3. Exemple de lexique: A – Introduction et champ commun. Paris, Association Marc Bloch. Multigraphié; imprimé ensuite à Bruxelles, Presses académiques européennes, 1964.

1964a *Le Syntol . . .* vol. 3 D: Sociologie. Paris, Association Marc Bloch. Multigraphié.

1964b Aspects théoriques du Syntol. Chap. 3 de Cros, R.-C., J.-C. Gardin et F. Lévy, *L'automatisation des recherches documentaires:* un modèle général "le Syntol". Paris, Gauthier-Villars, p. 39–129. Ce chapitre est une revision du vol. 1 de l'éd. multigraphieé de 1962 (ci-dessus), lequel avait été traduit en anglais par la National Science foundation.

1965a Free classifications and faceted classifications: Their exploitation with computers. In Atherton, Pauline (ed.) *Classification research.* Copenhagen, Munksgaard, p. 161–168. Traduction anglaise d'un rapport multigraphié de juin 1964, Com/23/1964/J.-C.G. Paris, C.N.R.S., Section d'automatique documentaire.

1965b *Syntol.* New Brunswick, N.J., Rutgers – the state university.

1966 Eléments d'un modèle pour la description des lexiques documentaires. *Bulletin des bibliothèques de France,* 11 (5) Mai 1966, 171–182.

1969 Semantic analysis procedures in the sciences of man. *Social science information,* 8 (1) Feb. 1969, 17–42. Trad. anglaise d'un texte paru ensuite en français dans Pouillon, Jean (éd.) *Mélanges Lévi-Strauss,* Paris, Mouton, 1971.

GARDIN, Jean-Claude (suite)

1971 *Unisist;* étude sur la réalisation d'un système mondial d'information scientifique. Paris, Unesco. En anglais: *Unisist:* study report on the feasibility of a world science information system.

GÉRARD, Jean et Pierre BOURGEOIS

1937 Les dépôts de documents à la Maison de la chimie. *Congrès mondial de la documentation universelle, Paris, 1937, Texte des communications,* p. 135–141.

GLANGEAUD, Louis

1955 Classification scalaire des sciences de la matière et de leurs méthodes. *Revue générale des sciences*, 62 (5–6) mai 1955.

GOBLOT, Edmond

1922 *Le système des sciences.* Paris, Colin.

GOLDMANN, Lucien

1967 *Introduction à la philosophie de Kant.* Paris, Gallimard. Réédition d'une traduction, d'abord publiée en 1948 (Paris, Presses universitaires de France) du texte allemand publié en 1945 à Zürich.

GREGG, John Richard

1954 *The language of taxinomy.* New York, Columbia university press.

GROLIER, Eric de

1956 *Théorie et pratique des classifications documentaires.* Paris, Union française des organismes de documentation. Multigraphié. Texte refondu d'un mémoire dactylographié pour le diplôme de l'Institut national des techniques de documentation: *Etudes sur le problème de la classification documentaire,* 1952.

1962 *Etude sur les catégories générales applicables aux classifications et codifications documentaires.* Paris, Unesco. Texte revu d'un rapport multigraphié en 1960. Traduction anglaise: *A study of general categories applicable to classification and coding in documentation.* Paris, Unesco, 1962.

1965 *On the theoretical basis of information retrieval systems.* Final report. AFOSR contract AF61 (052)–505. Paris, Sept. 1965. Multigraphié.

1969a Document no. 20 pour le Colloque du C.I.S.S. sur "University achievment and management needs", Fontainebleau, avril 1969; Multigraphié.

1969b *Draft of an outline of a comparative study on international scientific NGOs in the field of social sciences.* International social sciences council, Paris. Document ISSC/AR/64–69 – Introduction, Sept. 1969.

GROLIER, Eric de (suite)

1969c L'histoire des classifications: une interprétation soviétique. *Bulletin des bibliothèques de France;* Compte rendu de Šamurin, 1964–1967 14 (8) août 1969 * 652–657

1971a *Formation et recherche en science de l'information.* Colloque... de Tours, 11–12 juin 1971. Rapport préliminaire.

1971b Note pour la préparation d'un Colloque international sur la communication et l'information. Note interne C.I.S.S., dactylographiée.

GUSDORF, Georges

1966 *Les sciences humaines et la pensée occidentale. I; De l'histoire des sciences à l'histoire de la pensée.* Paris, Payot.

1967 *Idem.* II.: *Les origines des sciences humaines.* Paris, Payot.

HALLIG, Rudolf et W. von WARTBURG

1952 Begriffssystem als Grundlage für die Lexikographie. *Abhandlungen der deutschen Akademie der Wissenschaft zu Berlin, Klasse für Sprache, Literatur und Kunst,* 1952, no 4.

HEGEL, Georg Wilhelm Friedrich

1970 *Leçons sur l'histoire de la philosophie.* Paris, Gallimard. Trad. par J. Gibelin des *Vorlesungen über die Geschichte der Philosophie,* professées en 1829–1830.

HOCHFELD, Julian

1964 Introduction. *Revue internationale des sciences sociales,* 16(4) 519–535.

HOLMSTROM, J. Edwin

1953 *Facts, files and action.* Part two: filing, indexing and circulation. London, Chapman & Hall.

HÖRMANN, Hans

1971 *Psycholinguistics:* an introduction to research and theory. Berlin, Springer-Verlag. Trad. de *Psychologie der Sprache,* même éd., 1967.

HULME, E. Wyndham

1922 *Statistical bibliography in relation to the growth of modern civilization.* London, Grafton.

HUXLEY, Julian

1967 *Science et synthèse.* Dans le recueil du même titre, textes des exposés et débats d'un colloque international Unesco pour le 10e anniversaire de la mort d'Einstein et de Teilhard de Chardin. Paris, Gallimard, p. 67–83.

ILLINOIS INSTITUTE OF TECHNOLOGY RESEARCH INSTITUTE
1968 *Technology in retrospect and critical events in science.* Contract NSF–G535. December 15, 1968. Multigraphié.

IRSAY, S. d'
1933–35 *Histoire des universités françaises et étrangères.* Paris, Picard, 2 vol.

JACOB, François
1970 *La logique du vivant.* Paris, Gallimard.

JAKOBSON, Roman
1970 La linguistique. Chap. VI de Unesco, *Tendances principales de la recherche dans les sciences sociales et humaines;* première partie: sciences sociales. Paris, Mouton/Unesco. p. 504–556.

JASPERS, Karl
1932 *Philosophie.* 2e éd., 1948, Berlin, Springer.
1949 *Ursprung und Ziel der Geschichte.* Zürich, Artemis. Trad. française: *Origine et sens de l'histoire,* Paris, Plon, 1955.
1964 *Kleine Schule des philosophischen Denkens.* München, Piper Verlag. Trad. française: *Initiation à la méthode philosophique.* Paris, Payot, 1970.
1965 *Introduction à la philosophie.* Paris, Plon.

JONKER, F.
1958 The descriptive continuum: a generalized theory of indexing. Preprints of papers for the *International conference on scientific information.* Washington, National academy of sciences, area 6, p. 21–41.

JOURRY, Maryse
1971 *Les sources de l'histoire des bibliothèques médiévales.* Mémoire I.U.T. de Tours. Dactylographié.

JURET, Abel
1960 *Les idées et les mots;* essai de philosophie linguistique. Paris, Vrin. Edition partielle du Dictionnaire étymologique et sémantique de la langue latine, publié sur microfiches en 1974 en reproduction du manuscrit original (Microfiches Eric de Grolier).

KARADY, Victor
1971 Projet d'étude sur la genèse historique de la recherche scientifique en lettres et dans les sciences humaines en France, *Information sur les sciences sociales,* 10 (3), Juin 1971, p. 103–15.

KEDROV, Bonifatij Mikhajlovič
1947 *Engel's i estestvoznanie.* Moskva.
1961–65 *Klassifikacija nauk.* Moskva, 2 vol.

1956 La classification des sciences. *Recherches soviétiques, philosophie,* cahier 1, p. 83–111. Trad. de son article: O klassifikacij nauk. *Voprosy filosofii,* 1955, 2, 49–68. Présenté d'abord comme rapport au Congrès international de philosophie des sciences, Zürich, 1954.

KENDALL, Henry W. et Wolfgang K. H. PANOFSKY

1971 The structure of the proton and the neutron. *Scientific American,* 224 (6) June 1971, 61–77.

KLIR, George J.

1969 *An approach to general systems theory.* New York, Van Nostrand Reinhold company.

KRISHNA, Daya

1964 De la distinction entre les sciences naturelles, les sciences sociales et les sciences humaines. *Revue internationale des sciences sociales,* 16 (4) 555–566.

KUHN, T.

1962 *The structure of scientific revolutions.* Chicago University press.

KYLE, Barbara

1961 The UDC: present position and future developments. *Unesco bulletin for libraries,* 15, March–April 1961, 53–64.

LALANDE, André (éd.)

1962 *Vocabulaire technique et critique de la philosophie;* 9e éd. Paris, Presses universitaires de France (publié d'abord dans le Bulletin de la Société française de philosophie, 1902–1923).

LAMING-EMPERAIRE, Annette

1971 Primitifs d'hier et d'aujour'hui. *Sciences et avenir,* no spécial: La vie préhistorique, p. 26–34.

LANCASTER, Frederick W.

1966 Seminar panel discussion. *In* Selye, Hans, *Symbolic shorthand system.* New Brunswick, N.J., Rutgers – the state university, p. 50–69.

LATIL, Pierre de

1971 Le 3e âge de la physique. *Sciences et avenir* (295) Sept. 1971, 708–715.

LEAKEY, L. S. B.

1964 East African fossil hominoidea and the classification within the super-family. In Washburn, Sherwood L. (ed.) *Classification and human evolution.* London, Methuen & Co., p. 32–49.

LE DÛ, Jacques
1971 Utilisation d'un ordinateur pour la diffusion sélective et la recherche rétrospective de la documentation chimique. *Documentaliste,* no spécial: L'automatisation documentaire en France et dans les pays de langue française, p. 28–38.

LEFER, Pierre
1971 *La mesure des activités scientifiques dans les sciences sociales et humaines.* Paris, Unesco (document CSR-S-1, septembre 1971).

LE GROS CLARK, Wilfrid E.
1967 *Man-apes or ape-men?* The story of discoveries in Africa. New York, Holt, Rinehart and Winston.

LENIN, Vladimir Iliič
1951 *Matérialisme et empiriocriticisme.* Paris, Editions sociales. Trad. française de l'original russe, paru en 1908.

LENK, Hans
1971 Prolegomena zur Wissenschaftstheorie der Planung. *In* Deutsche Forschungsgemeinschaft, *Kolloquium über Forschungsplanung,* Wiesbaden, Franz Steiner Verlag, p. 13–41.

LERWILL, C. J.
1971 *An introduction to the classification of animals.* London, Constable.

LÉVI-STRAUSS, Claude
1958 *Anthropologie structurale.* Paris, Plon.
1964 Critères scientifiques dans les disciplines sociales et humaines. *Revue internationale des sciences sociales, 16 (4),* 579–97.
1967 L'anthropologie sociale. *Sciences* (47) janv.–fév. 1967, 115–28.

LÉVY, Francis
1964 Expérimentation sur calculateur. Chap. 5 de Cros, R.-C., J.-C. Gardin, F. Lévy, *L'automatisation des recherches documentaires.* Paris, Gauthier-Villars, p. 167–209.
1967 On the relative nature of relational factors in classifications. *Information storage and retrieval,* 3 (4) Dec. 1967, 315–325.

LICKLIDER, J. C. R.
1965 *Libraries of the future.* Cambridge, Mass., M.I.T. press.

LLOYD, G. A.
1969 The UDC in its international aspects. *Aslib proceedings.* 21 (5) May 1969, 204–208.

LOTKA, A. J.
1926 The frequency distribution of scientific productivity. *Journal of the Washington Academy of sciences,* 16, 317–323.

LUHN, Hans Peter

1951 A new method of recording and searching information. Daté 10 sept. 1951. Appendix 1 de *The IBM electronic information searching system.* New York, IBM.

LUND, J. et Mortimer TAUBE

1937 A non-expansive classification system. *Library quarterly,* 7, July 1937, 373–394.

LYONS, John

1968 *Introduction to theoretical linguistics.* Cambridge university press. Trad. française: *Linguistique générale.* Paris, Larousse, 1970.

McCASH, W. H. et J. J. CARMICHAEL

1970 UDC user profiles as developed for a computer-based SDI service in the iron and steel industry. *Journal of documentation,* 26 (4) Dec. 1970, 295–312.

MADAY, Stefan von

1913 Verbesserte Dezimaleinteilung. *Zentralblatt für Bibliothekswesen,* 30, 161–180.

MANDELBROT, Benoît

1955 On the language of taxonomy. *In* Cherry, Colin (ed.) *Information theory;* third London Symposium. London, Butterworths scientific publications, p. 135–145.

MARIETAN, Joseph

1901 *Problème de la classification des sciences d'Aristote à saint Thomas.* Paris, Alcan (thèse de lettres, Fribourg).

MARX, Karl

1857 *Grundrisse der Kritik der politischen Ökonomie.* Ecrit en 1857–58 (l'Introduction est de 1857); publié à Moscou, 1939. Trad. française: *Fondements de la critique de l'économie politique.* Paris, Editions Anthropos, 1967.

1951 Karl Marx, Friedrich Engels. *Etudes philosophiques;* nouv. éd. Paris, Editions sociales.

MATORÉ, Georges

1953 *La méthode en lexicologie;* domaine français. Paris, Didier.

MAYNE, A. J.

1968 Some modern approaches to the classification of knowledge. *Classification society bulletin,* 1 (4) 13–17.

MIKHAJLOV, A. J. *et al.*

1968 *Osnovy informatiki.* Moskva, Nauka. Trad. allemande: *Grundlagen der wissenschaftlichen Dokumentation und Information.* Köln, Westdeutscher Verlag, 1970, 2 vol.

MEYER, Adolf

1921–23 Probleme des Realkatalogs. *Zentralblatt für Bibliothekswesen,* 38 (9–10) 227–238; 39 (9) 388–399; 40 (5) 208–213. – Der Realkatalog *Idem,* 40 (8–10), 412–424.

MILLS, Jack

1964 *The universal decimal classification.* New Brunswick, N.J., Rutgers – the state university.

MOLES, Abraham A. et André NOIRAY

1969 La pensée technique. In Noiray, André (éd.) *La philosophie.* Paris, Centre d'étude et de promotion de la lecture, p. 496–525.

MONOD, Jacques

1970 *Le hasard et la nécessité.* Paris, Editions du Seuil.

MONTFAUCON, Bernard de

1739 *Bibliotheca bibliothecarum manuscriptorum nova.* Paris, 2 vol.

MORGAN, Convy Lloyd

1923 *Emergent evolution.*

MORIN, Edgar

1971 *Communication et communications.* Document Unesco SHC./WS/ 187, 28 mai 1971. Multigraphié (non signé).

MOUNIER, Emmanuel

1962 *Introduction aux existentialismes.* Paris, Gallimard (réédition de l'ouvrage paru sous de même titre. Paris, Denoël, 1947).

The NATIONAL COMPUTING CENTRE

1968 *Commodity coding:* its effect on data recording and transfer. Manchester (il s'agit d'un rapport préparé par Scientific Control Systems, d'octobre 1967 à avril 1968, et financé par le N.C.C.)

NAUDÉ, Gabriel

1627 *Advis pour dresser une bibliothèque.* Paris, Targa.

NEELAMEGHAN, A. et M. A. GOPINATH

1970 Transport economics: depth classification version of CC. *Library science,* 7 (4) Dec. 1970, 314–352.

NEVILLE, H. H.

1970 Feasibility study of a scheme for reconciling thesauri covering a common subject. *Journal of documentation,* 26 (4) Dec. 1970, 313–336.

OHMAN, Suzanne

1953 Theories of the "linguistic field". *Word,* 9, 123–134.

ORNSTEIN, Martha

1913 *The role of scientific societies in the seventeenth century.* New York.

ORTEGA Y GASSET, José

1970 *L'évolution de la théorie déductive.* Paris, Gallimard.

OSSOWSKI, S. et OSSOWSKA, M.

1936 The science of science. *Organon.* 1, 15–20.

OSTWALD, Wilhelm

1929 *Die Pyramide der Wissenschaften.* Stuttgart, Cotta.

OTLET, Paul

1933 Sur le développement ultérieur de la classification décimale. *XIIe Conférence de l'Institut international de bibliographie, Bruxelles, Rapports,* p. 1–6.

PAGÈS, Robert

1967 Relational aspects of conceptualization in message analysis. *Information storage and retrieval,* 3 (4) Dec. 1967, 351–372.

PAULSEN, Friedrich

1919–21 *Geschichte des gelehrten Unterrichts auf den deutschen Schulen und Universitäten.* 3. Aufl. Leipzig, 2 vol.

PEIGNOT, Gabriel

1802 *Dictionnaire raisonné de bibliologie.* Paris, 2 vol.

PÉLISSIER, Roger

1969 *Les bibliothèques en Chine pendant la première moitié du XXe siècle.* Paris, Ecole pratique des hautes études, 2 vol.

PERREAULT, Jean M.

1967 On the articulation of surrogates. *Information storage and retrieval,* 3 (4) Dec. 1967, 177–189.

1968 *Reclassification: rationale and problems;* ed. by Jean M. Perreault. College Park, Md., University of Maryland.

1969 *Towards a theory for U.D.C.* London, Bingley.

PETZHOLDT, Julius

1860 *Chronologische Übersicht von bibliographischen Systemen.* Dresden, Schönfeld.

PIAGET, Jean

1964 Classification des disciplines et connexions interdisciplinaires. *Revue internationale des sciences sociales,* 16 (4), 598–616.

1970 La situation des sciences de l'homme dans le système des sciences. Introduction de Unesco, *Tendances principales de la recherche dans les sciences sociales et humaines;* première partie: sciences sociales. Paris, Mouton/Unesco, p. 1–65; La psychologie, chap. III du même ouvrage, p. 274–339; Problèmes généraux de la recherche interdisciplinaire et mécanismes communs, chap. VII, idem, p. 559–628.

PINDER, Richard et Robert ROUSSELOT

1971 Techniques spatiales. *La Banque des mots,* 1, 103 ss.

PLEKHANOV, G.

1949 *L'art et la vie sociale* (recueil de divers articles). Paris, Editions sociales.

POLLOCK, S. M.

1968 Measures for the comparison of information retrieval systems. *American documentation,* 19 (4) Oct. 1968, 387–397.

POTTIER, Bernard

1962 *Systématique des éléments de relation;* étude de morphosyntaxe structurale romane. Paris, Klincksieck (thèsè de doctorat, écrite en 1955).

PRICE, Nancy et Samuel SCHIMINOVICH

1968 A clustering experiment: first steps towards a computer-generated classification scheme. *Information storage and retrieval,* 4 (3) August 1968, 271–280.

RANGANATHAN, Shiyali Ramamrita

1965 Library classification through a century. *In* Atherton, Pauline (ed.) *Classification research.* Copenhagen, Munksgaard, p. 15–35.

1967 *Prolegomena to library classification.* Bombay, Asia publishing house, 3e éd., refondue; 1ère éd. en 1937; 2e en 1957.

RÉVÉSZ, Geza

1957 *Die Trias.* Bayerische Akademie der Wissenschaften, Philosophisch-historische Klasse, Sitzungsberichte, Jahrg. 1956, Heft 10. München.

REY, Abel

1933 *La science dans l'antiquité.* Paris, Renaissance du livre.

RICHARDSON, Ernest Cushing

1901 *Classification, theoretical and practical.* Reprinted with additions. New York, Chas. Scribner's sons, 1912.

ROGET, Peter Mark

1852 *Thesaurus of English words and phrases.* London.

ROLLING, L.

1965 The role of graphic display of concept relationships in indexing and retrieval vocabularies. *In* Atherton, Pauline (ed.) *Classification research.* Copenhagen, Munksgaard, p. 295–325 (voir en particulier la discussion, p. 321–325).

ROMERO, José Luis

1964 La particularité de l'objet dans le contexte socio-culturel. *Revue internationale des sciences sociales.* 16 (4), 617–633.

RUSSELL, Bertrand

1937 *The principles of mathematics.* London, George Allen & Unwin. 2e éd. (1ère éd., 1903).

ŠAMURIN, E. I.

1964–67 *Geschichte der bibliothekarisch-bibliographischen Klassifikation.* Leipzig, VEB Bibliographisches Institut, 2 vol. Trad., avec un complément par Mme Teslenko, de *Očerki po istorii bibliotečno-bibliografičeskoj klassifikacii,* Moskva, 1955–1959.

SAPORTA, S.

1955 Frequency of consonant clusters. *Language,* 31 (1), 25–30.

SARACEVIC, Tefko

1970 The notion of "relevance" in information science. *In* Saracevic, Tefko (ed.) *Introduction to information science.* New York, R. R. Bowker, p. 111–151.

SAYERS, W. C. Berwick

1967 *A manual of classification for librarians.* London, André Deutsch. 4e éd. refondue par Arthur Malby. 1ère éd. en 1926.

SCHIMINOVICH, S.

1971 Automatic classification and retrieval of documents by means of a bibliographic pattern discovery algorithm. *Information storage and retrieval,* 6 (6) May 1971, 417–435.

SCHNEIDER, Karl et Karl Heinz KOCH

1970 *The use of the UDC in the production of mechanized indexes.* Berlin, Beuth Vertrieb.

SCHOPENHAUER, Arthur

1966 *Le monde comme volonté et comme représentation.* Presses universitaires de France. Trad. A. Burdeau, revue, de *Die Welt als Wille und Vorstellung,* Leipzig, 1819

SELYE, Hans

1966 *Symbolic shorthand system.* New Brunswick, N.J., Rutgers – the state university.

SHERA, Jesse H. et M. EGAN

1951 *Bibliographic organization.* Chicago university press.

SHIELDS, Charles W.

1888–89 *Philosophia ultima,* or science of the sciences. New York, 2 vol.

SIMPSON, George Gaylord

1945 The principles of classification and a classification of mammals. *Bulletin of the American museum of natural history,* 85, i-xvi, 1–350.

1961 *Principles of animal taxonomy.* New York, Columbia university press.

1964 The meaning of taxonomic statements. *In* Washburn, Sherwood L. (ed.) *Classification and human evolution.* London, Methuen & Co., p. 1–31.

SKANDALIS, G.

1970 Mémoire inédit de l'Institut national des techniques de la documentation: *Etude critique de la "Liste commune de descripteurs" de l'O.C.D.E.*

SLUCKI, E.

1926 *Ein Beitrag zur formal-praxeologischen Grundlegung der Oekonomik.* Kiev.

SOERGEL, Dagobert

1967 Some remarks on information languages, their analysis and comparison. *Information storage and retrieval,* 3 (4), Dec. 1967, 219–291.

1969 *Klassifikationssysteme und Thesauri.* Frankfurt am Main, Deutsche Gesellschaft für Dokumentation

SOKAL, R. R.

1966 Numerical taxonomy. *Scientific American,* 215, 106–116.

SOUÈGES, René

1938–51 *Embryogénie et classification.* Paris, Hermann, 4 vol.

SPARCK JONES, Karen

1971 *Automatic keyword classification for information retrieval.* London, Butterworths.

STEVENSON, Gordon

1971 The Eppelsheimer subject catalog. *Library resources & technical services,* 15 (3), Summer 1971, p. 309–328.

STORER, Norman W.

1966 *The social system of science*. New York, Holt, Rinehart and Winston.

TARRAZI, Philippe de

1947–48 *Histoire des bibliothèques arabes dans l'univers.* Beirut.

TEILHARD de CHARDIN, Pierre

1970 *Le phénomène humain.* Paris, Editions du Seuil. Réédition: 1ère éd. en 1955; rédigé en 1938–40, remanié en 1947–48.

TESLENKO, O. P.

1967 Die Klassifikation in der UdSSR von 1958 bis 1965. Chap. 13 de Šamurin, E. I. *Geschichte der bibliothekarisch-bibliographischen Klassifikation.* Leipzig, vol. II, p. 555–592.

THOMPSON, G. K. *et al.*

1971 Isis – un petit guide à l'intention des bibliothécaires et des documentalistes. *Documentaliste,* no spécial: L'automatisation documentaire en France et dans les pays d'expression française, p. 62–69.

THUILLIER, Pierre

1971 Comment se constituent les théories scientifiques. *La Recherche,* (13), Juin 1971, p. 537–554.

TOMLINSON, Helen

1969 Problems arising from first GCS papers. *In* Classification research group, *Classification and information control.* London, The Library association, p. 73–80.

TRIER, Jost

1931 *Der deutsche Wortschatz im Sinnbezirk des Verstandes.* Heidelberg, Carl Winter.

TRIST, Eric

1970 Organisation et financement de la recherche. Chap. XI de Unesco, *Tendances principales de la recherche en sciences sociales et humaines,* première partie: sciences sociales, Paris, Unesco/Mouton, p. 825–980.

TSIEN, Tsuen Hsuin

1952 A history of bibliographic classification in China. *The Library quarterly,* 22 (4), 307–324.

UNESCO

1971 Travaux préparatoires à une encyclopédie philosophique comparative. Document interne.

VERNON, K. D. C. et Valerie LANG

1970 *The London classification of business studies.* London, Graduate school of business studies.

VICKERY, B. C.

1953 The significance of John Wilkins in the history of bibliographic classification. *Libri,* 2, 326–343.

1958 *Classification and indexing in science.* London, Butterworths scientific publications. Dans la 2e éd., 1959, le passage cité est p. 175–176.

1961 The Universal decimal classification and technical information indexing. *Unesco bulletin for libraries,* 15, 126–138.

1971 Structure and function in retrieval languages. *Journal of documentation,* 27 (2) June 1971, 69–82.

VOILQUIN, Jean (éd.)

1964 *Les penseurs grecs avant Socrate.* Paris, Garnier-Flammarion.

WÅHLIN, Ejnar

1968 *US: universal system of classification.* Stockholm, l'auteur.

WALKLEY, Janet et Barbara HAY

1971 An annotated list of thesauri held in the Aslib library. *Aslib proceedings,* 23 (6) June 1971, 292–300.

WEINREICH, Uriel

1962 Lexicographic definition in descriptive semantics. *In* Householder, F. W., Jr., and Sol Saporta (eds.) *Problems in lexicography.* Baltimore, Publications of Indiana research center in anthropology, folklore and linguistics, 21, p. 25–43.

WUNDT, Wilhelm

1889 Ueber die Eintheilung der Wissenschaften. *Philosophische Studien,* 5, 1–55.

YOST, R. M., Jr.

1954 *Leibniz and philosophical analysis.* Berkeley, University of California press.

ZIMAN, J.

1968 *Public knowledge;* an essay on the social dimension of science. Cambridge university press.

ZVORIKINE, A.

1964 Les sciences sociales en URSS: situation et tendances. *Revue internationale des sciences sociales,* 16 (4), 634–649.

Ganesh Bhattacharyya (Bangalore) and S.R. Ranganathan (Bangalore)

From Knowledge Classification to Library Classification

[The main landmarks in the development of Philosophers' Classification Schemes are, 1) The Socio-centric Scheme of the Vedic Period, 2) The Utilitarian Scheme of the Greek period, 3) the University-centred Scholastic Classification of Medieval-Europe, 4) the Psychology-centred Classification of Bacon, 5) the Dichotomic Scheme of Kant, 6) the Triadic Scheme of Hegel, and 7) the Serial Classification of Hobbes, Comte, Ampère, and Spencer, inserting each applied science after the pure science forming its foundation. The interest of philosophers in classification was largely speculative and theoretical; and they were not carried into details. On the contrary, the schemes of Library Classification designed during the last one century are intended to arrange documents in a helpful "Leakage-free, Noise-free" sequence; therefore, the Library Classification schemes have been worked out to minute details, so as to classify even micro subjects. Another distinguishing feature of Library Classification is the representation of each subject by a distinctive ordinal number to enable the mechanisation of arrangement of subjects in the preferred sequence. In the light of actual need, the notational system has led Library Classification successively from an enumerative scheme with a pure decimal fraction notation as in DC, through the slightly faceted scheme as in UDC, to the rigidly faceted, almost freely faceted, and the freely faceted versions of CC. In the present, Library Classification has come to be based on a Theory of Classification. From the static theories of the *Classification: theoretical and practical,* the *Manual of library classification,* and the *Organisation of knowledge in libraries,* we have now reached the dynamic theory given in the *Prolegomena to library classification.* This Theory is applicable to any Scheme for Classification.]

0 SELECT TERMINOLOGY

To minimise refraction in communication, a select terminology is given in this section. In this paper, each term defined in this section, is to be interpreted according to the definition given. One novel feature, that might be felt by some, concerns the terms 'Idea' and 'Knowledge'. A fuller terminology with illustrations will be found in the *Prolegomena to library classification* (6).

01 Memory

A compartment, so to speak, of the mind, in which percepts, concepts and ideas – whether got through sensory organs, or intellect or intuition, that is, sensory, intellectual or transintellectual – are deposited and stored, and from which any assortment of them can be taken out for forming new ideas.

* The School of S.R. Ranganathan uses capital letters to indicate special concepts.

011 *Pure Percept*

A meaningful impression, produced by any entity through a single primary sense and deposited in the memory.

012 *Compound Percept*

The impression, deposited in the memory, as a result of the association of two or more pure percepts formed simultaneously or in quick succession.

013 *Concept*

The formation, deposited in memory, as a result of the association of percepts -- pure as well as compound -- already deposited in memory.

014 *Intellectual Idea*

The product of thinking, reflecting, imagining, etc got by the intellect by integrating, with the aid of logic, an assortment from the memory.

015 *Transintellectual Idea*

Idea apprehended directly by intuition, unmediated by the sensory organs or the intellect.

02 Knowledge

Universe of ideas -- past, present and future.

03 Subject

An organised body of ideas, whose extension and intension are likely to fall coherently within the field of interest and comfortably within the intellectual competence and the field of inevitable specialisation of a normal person.

031 *Work*

The exposition of a subject in a language, or pictures, or symbols, or in any other mode, to make communication possible.

04 Universe of Subjects

The totality of subjects -- past, present, and future.

05 Document

Record of work on paper or other material, fit for easy physical handling, transport across space, and preservation through time.

06 Universe of Documents

Totality of documents -- past, present, and future.

07 Universe of Discourse

The Universe of Ideas taken at one time for discussion or for classification as the case may be.

1 KNOWLEDGE CLASSIFICATION

10 Schemes by Philosophers

For nearly over 40 or 50 centuries, thought has been devoted to Knowledge Classification --that is, the classification of the Universe of Ideas. This fell into the province of the philosopher. The philosopher's interest was in grouping ideas according to their degree of filiation. He also attempted to arrange the resulting groups in the helpful sequence preferred by him. There are still extant some schemes of Knowledge Classification, of the older time.

101 *Two Kinds of Schemes*

These schemes of Knowledge Classification appear to fall into two sets:

1 Pure-Application-Sequence-Free Scheme.- These do not disclose any attempt to put the application of a discipline immediately after the discipline forming its foundation. Except for the scheme of Thomas Hobbes, all these had occurred mostly before the nineteenth century.

2 Pure-Application-Sequence Schemes.- These schemes were developed mostly in the nineteenth century. But as stated already, even in the seventeenth century Hobbes had put forward a scheme of this kind.

2 PURE-APPLICATION-SEQUENCE-FREE SCHEME

21 Vedic Classification of India

The earliest of the schemes of Knowledge Classification known is the Vedic one. The Vedic Scheme divided the Universe of Ideas into four groups and arranged them as follows:

1 *Dharma.*- The dominant subjects included in this group are what are now denoted by the terms 'Law', 'Theology', 'Ethics' and 'Sociology'. This group was taken to comprehend the subjects involved in the maintenance of society as a coherent organism.

2 *Artha.*- The dominant subjects included in this group are what are now denoted by the terms 'History', 'Political Science', 'Economics', and the 'Applied Sciences', forming the basis of economic wellbeing. This group was taken to comprehend the subjects leading to social wellbeing.

3 *Kama.*- The dominant subjects included in this group are what are now denoted by the terms 'Literature', 'Fine Arts', and 'Pure Sciences'. This group was taken to comprehend the subjects arising out of the fulfilment of the involuntary creative urge in man, and/or for the enjoyment of the result of such creation. A work of this group has generally an elevating effect on its readers.

4 *Moksha.*- The dominant subjects included in this group are what are now denoted by the terms 'Philosophy' and 'Spiritual Experience'. The former seeks to understand the universe by intellect and thereby get over the miseries of life, and attain Peace. Spiritual Experience, on the other hand, apprehends the totality of the universe, at once in it entirity, through intuition, and leads to Peace. Fine Arts and Literature lead to an approximation to Peace.

211 *Sequence of the Groups*

It can be seen that this sequence of the four groups corresponds to the different stage through which a society or an individual tends to move. The first prerequisite is coherent existence. Then further progress requires production of commodities, and their equitable distribution. A few persons do not get self-satisfaction merely by the fulfilment of these two factors. They want, in addition, the joy of creative work. Their personality and creative work help in the progress of the society. Last comes the state of bliss attained by a few through spiritual realisation.

212 *Character of the Filiation*

To put it schematically, Exist ⟶ Grow towards full Happiness ⟶ Seek creative Joy ⟶ Attain ultimate Delight.

213 *Beneficence of "Moksha" Group*

The transcendental quality of those belonging to the *"Moksha"* group, is their power to lift the other groups gradually along the line of the filiatory sequence, each in his own way, and in his own measure. This beneficence of those in the *"Moksha"* group, may not however be easily perceived by those in the other groups -- particularly in the first and the second groups.

Vedic Classification was socio-centred.

22 Greek Classification

Greek Classification may be taken to have developed from the time of Plato onwards. Aristotle's scheme of Knowledge Classification may be taken as representative of the Greek Schemes. His scheme divided the Universe of Ideas into the following successive groups:

1 Theoretical philosophy.- The dominant subjects included in this group are "Logic", "Metaphysics", "Mathematics" and "Physics".

2 Practical philosophy.- The dominant subjects included in this group are "Ethics", "Political science" and "Economics".

3 Productive arts.- The dominant subjects included in this group are "Applied Sciences" including "Useful Arts".

Greek Classification was utility-centred.

23 Scholastic Classification of Mediaeval Europe

Scholastic Classification may be taken to have developed after Greek Classification. This scheme of Knowledge Classification may be taken as based on the practice of the mediaeval universities. This scheme divided the Universe of Ideas into the following successive groups:

1 Trivium.- The dominant subjects included in this group are "Linguistics", "Dialectics" and "Rhetoric".

2 Quadrivium.- The dominant subjects included in this group are "Arithmetic", "Geometry", "Astronomy", and "Music".

3 Third group.- The dominant subjects included in this group are "Theology", "Metaphysics", and "Ethics".

Scholastic Classification was university-centred.

24 Baconian Classification

In his *Advancement of learning,* (1605), Francis Bacon (1561–1626) has divided the Universe of Ideas into the following 3 successive groups:

1 History -- emanation from Memory.- The dominant subjects included in this group are "Geography" and "History".

2 Poesy -- emanation from Imagination.- The dominant subjects included in this group are "Fine Arts" and "Literature".

3 Philosophy -- emanation from Reason.- This group included all the other subjects.

Baconian Classification was psychology-centred.

25 Kant's Classification

In his *Critique of pure reason* (1781), Immanuel Kant (1724–1804) has given the following dichotomic scheme of Knowledge Classification. This scheme is apparently based on the primitive human instinct. To say of anything, either "This is A," or "This is not A"; in other words, the attempt to exhaust all the entities under consideration, within two groups.

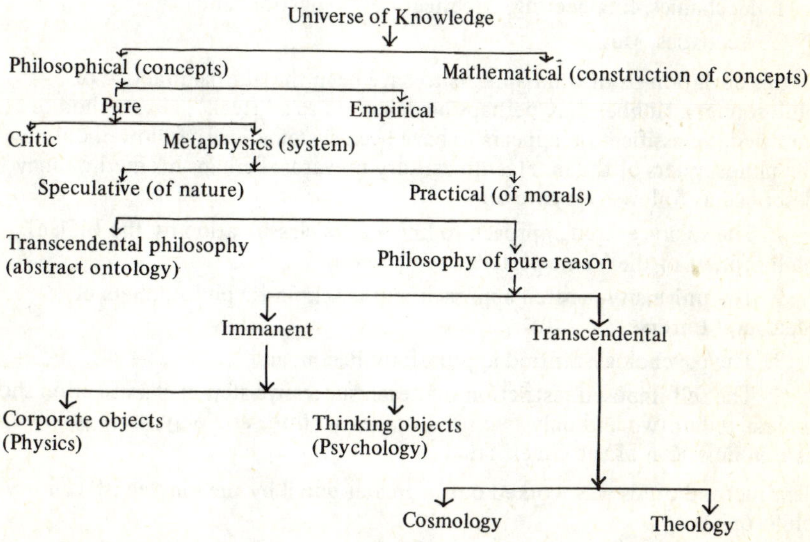

26 Hegel's Classification

In his *Logic* (1812), Georg Wilhelm Friedrich Hegel (1770–1831), has given the following triadic scheme of knowledge classification:

- Universe of Knowledge
  - Logic (Hegelian sense)
    - Ontology
    - Theology
    - Epistemology
  - Natural Sciences
    - Mechanics
    - Physics
    - Biological Sciences
  - Mental Sciences
    - Anthropology
    - Philosophy, Ethics and Law
    - Art, Religion and Philosophy

This scheme is apparently influenced by the tradition of Greek Triad.

3 PURE-APPLICATION-SEQUENCE SCHEME

31 Hobbes's Classification

Even during the period, when philosophers do not appear to have thought of putting a pure discipline before its applications, in their respective schemes of Knowledge Classification, in his *Leviathan* (1651), Thomas Hobbes (1588–1679) had divided knowledge into groups such as follows:

1 Mechanics, Engineering, Architecture, Navigation; and

2 Acoustics, Music.

In this deviation from what appears to have been the normal practice of philosophers, Hobbes may perhaps be described as a "freak". His method of knowledge classification appears to have been systematically followed only by the philosophers of the nineteenth century onwards. Perhaps his method may be described as follows: He gave up

1 The socio-centred approach to knowledge classification by the Indian philosophers of the Vedic period;

2 The university-centred approach by the scholastic philosophers of Mediaeval Europe;

3 The psychology-centred approach by Bacon; and

4 The self-imposed restriction that each successive step in classification should include either two and only two subdivisions, or three and only three subdivisions, as if in imitation of the Greek Triad.

This method of his was worked out in greater detail by the nineteenth century philosophers.

32 Comte's Classification

In his *Cours de philosophie positive* (1830), Auguste Comte (1798–1857) arranged the subjects in the following sequence: Mathematics, Astronomy, Physics, Chemistry, Biology and Social Physics. In respect of this sequence, Comte's claim is that each subject is virtually an application of the preceding one. The name "Serial Classification" is used to denote this kind of Knowledge Classification.

33 Ampère's Classification

In his *Essai sur philosophie de science ou exposition analytique d'une classification naturelle de toutes les connaissances* (1834–1843), André Marie Ampère (1775–1836) appears to have worked out the system of serial Classification more thoroughly. He has named the subjects in his scheme of knowledge classification, in a clear way. A sample of his Serial Classification is given below:

> Physics, Engineering; Geology, Mining; Botany, Agriculture; Zoology, Animal Husbandry, Medicine.

34 Spencer's Classification

In his *Classification of sciences* (1864) Herbert Spencer (1820–1903) gave the following as his version of Serial Classification:

> Logic, Mathematics, Mechanics, Physics, Chemistry, Astronomy, Geology, Biology, Psychology, and Sociology.

We cannot regard this sequence to be as helpful as that of Ampère. Indeed, Ampère's scheme should be taken to be the best Serial Scheme, given by the philosophers of the nineteenth century.

35 Theory of Classification

Herbert Spencer appears to have been the first philosopher to have attempted a Theory of Classification to be used as guideline in designing a scheme of Knowledge Classification.

4 KNOWLEDGE CLASSIFICATION *vs* LIBRARY CLASSIFICATION

41 Interest of the Philosopher in Classification

All along, the philosopher appears to have been deeply interested in knowing the mutual relation between ideas, and also in their sequence. This he tried to examine on the basis of some assumed principle or another. This work of the philosopher naturally resulted in Knowledge Classification, perhaps, for its own sake, as it were, or at best for mental satisfaction. He does not seem to have had in view the application of any detailed scheme for classification for arranging documents in a helpful sequence as an aid to the reader of documents in his easy, leakage-free, noise-free selection of the relevant documents satisfying his interest at the moment. In other words, the interest of the philosopher in classification was, in the main, probably, speculative and theoretical, though there might have been some use of it in arranging broad subjects in some form of helpful sequence or another. The philosopher does not seem to have carried out his classification of the Universe of Ideas to a sufficient degree of details, to be of much help to the librarian.

42 Parallel Numbering of Bookcases and Shelf-Planks, and of Books

In course of time, many of the major subjects were subdiyided. But even these divisions were only of the "major" variety. Perhaps, by the fifteenth century, their number had become too large to place the books on each of the different subjects consistently together. The first device to be naturally improvised for this purpose, was to number each of the Bookcases and each of its Shelf-Planks. Each subject was assigned to one or more bookcases and each of its divisions to one of more of its shelf-planks, depending upon the number of books in them. This has been called "Absolute Arrangement" of books. No doubt, some extra space would have been left in each bookcase and shelf-plank to receive newly added books. This would have proved adequate during the few centuries, when book-production had been at a low rate.

43 Press Mark *vs* Subject Numbering

But by the ninteenth century, the Fifth Law of Library Sciences (A library is a growing organism) had begun to assert itself, and the rate of growth of the number of books added to a library became very great. Therefore, "Absolute Arrangement" of books had to give place to "Relative Arrangement". In this arrangement, in course of time books will move on from one shelf plank to another and from one bookcase to another. Therefore, the number of the bookcases and of the shelf-planks -- the "Press Mark" as it was called -- proved futile and it was given up. On the contrary, each subject was given a number of its own.

44 Second Implication of the Fifth Law

The second implication of the Fifth Law is the continued subdivision of subject to an ever increasingly minute degree. These minute subdivisions of a major subject had also to be arranged helpfully according to their own mutual filiation and given distinct numbers of their own so as to lay bare and represent the filiation. It is here that the distinctive quality of Library Classification, as against Knowledge Classification of philosophers, becomes clearly visible. This problem became acute in the later half of the ninteenth century. To meet the situation, the right man appeared at the right time with the right-scheme for classification. It was the Decimal Classification (= DC) (1876) of Melvil Dewey.

441 Inverted Baconian Sequence

Generally speaking, the sequence of the Main Subjects in DC may be described to be Inverted Baconian. The Universal Decimal Classification (=UDC) (1899) adopted the Main Subjects of DC as its own Main Subjects in the self-same sequence. Thus the sequence of the Main Subjects of UDC also was Inverted Baconians.

45 Serial Sequence: Colon Classification

Generally speaking, the Colon Classification (1933) of S R Ranganathan arranged its Main Subjects in the Natural Sciences in the Serial Sequence putting each Applied Science immediately after the dominant Pure Science used by it. Further, it arranged the Pure Sciences themselves in the sequence of Increasing Concreteness. In its second edition (1939), CC introduced, after the most concrete of the Serial Sequence of the Natural Sequences, the new Main Subject "Spiritual Experience and Mysticism" as the concrete subject *par excellence.* Here everything and the universe as a whole are directly apprehended by intuition unmediated by the intellect or the primary senses. Intellectual apprehension gives only a partial view, and therefore, not fully concrete. Viewed from another angle, "Spritual Experience and Mysticism" represents what is natural *par excellence.* After it, CC arranges the "Humanities" and "Social Sciences" in a decreasing sequence of naturalness and ends with "Law" the most artifical subject -- the furthest remove from naturalness. Further, among these subjects also, the pure-applied sequence has been followed, wherever feasible.

5 DECIMAL CLASSIFICATION

50 Decimal Integer Notation

Beginning from the middle of the sixteenth century to the middle of the nineteenth century, the term 'Decimal' has been used to describe some of the schemes for classification. There the term 'Decimal' meant only an integral number expressed in the scale of ten. In the integral decimal notation any digit added at the end of the integer will not make it a subdivision of the host integer. But, convenience requires that the number for the subdivision of a subject should itself be a subdivision of the number for the host subject. Therefore, the decimal integer notation was boldly given up by DC.

51 Pure Decimal Fraction Notation

The designer of DC (1876) retained the Decimal Scale, but ingeniously made all his numbers Pure Decimal Fractions. Since all the numbers are pure decimal fractions, he omitted the insertion of the decimal point at the beginning of each of the DC numbers. This he should have done by the subconscious prompting of the Law of Parsimony. A word of warning is due here. He inserted a dot after the first three digits, as a punctuation mark – a dummy without any meaning. But this should not be mistaken for a decimal point. It was put in only to help the physiology of the eye which makes three digits an optimum number capable of being taken up within a sweep of the eye.

511 *Absence of Evidence of Class Numbers in Knowledge Classification by Philosophers*

The bold use of the decimal fraction notation was beyond doubt the first supreme step taken in Library Classification. This was in sharp contrast to the contribution of philosophers to Knowledge Classification. They do not appear to have felt obliged to carry out the subdivision of any subject to an ever-increasing degree of minuteness, to arrange them helpfully, and to maintain the arrangement consistently. As a result of the latter, they do not appear to have felt the need for representing each subject -- major or minor, or tiny -- by a system of Ordinal Numbers, now commonly called "Class Numbers".

52 Inadequacy of the Pure Decimal Fraction Notation

Within about thirty years of its invention, the pure decimal fraction notation was overpowered by the Fifth Law of Library Science. For, it brought about the division of a subject after any digit in its class number, along different lines simultaneously and not merely along the original line of march of the decimal fraction digits of the host original compound subject. This quality of a subject disclosed by the Fifth Law of Library Science laid bare the inadequacy of the pure decimal fraction notation of DC. Its packing successive digits in the class number along one and the same line could not meet the new situation brought about by this new quality of a subject.

6 THEORY OF CLASSIFICATION

601 *No Theory Used by Philosophers is Known*

So long as we confine ourselves to the classification of broad subjects, it is possible to manage the work with any help furnished by general experience and reinforced, if at all, even by a dash of help from the sub-conscious. Perhaps, this is the reason for our not having been able to find till now any substantial and consistent Theory of Classification from which guiding principles might have been drawn by the philosophers in their work on Knowledge Classification.

602 *Need for Theory of Library Classification*

On the contrary, while extending classification to minute subjects, dependence upon general experience helped by a dash from the sub-conscious proves to be insufficient. It becomes necessary to have a theory of classification to give us guiding principles to meet different situations. As this need arises only in Library Classification, we shall call it Theory of Library Classification. During the first 60 years of the practice of Library Classification, only broad subjects figured. For, only they had literary warrant. The powerful Decimal Fraction Notation of DC (1876) the veneering of the DC core with some facets by UDC (1899) and the changing of the foundation of classification from the Enumerative one to the Faceted one by CC (1933), were all done without the aid of any objectively stated Theory of Library Classification. This was so in spite of the subjects of smaller extension and greater intension gradually gaining literary warrant. But their pressure on the design of Library Classification had not increased sufficiently to demand a full-blown Theory of Library Classification. An extra flair here or there, and any possible little help coming from the subconscious, proved sufficient to meet the situation. But this hand to mouth adjustment could not be sufficient for long. The rate of proliferation in the Universe of Subjects went on increasing. This soon brought a situation quite beyond the capacity of hand to mouth existence. This situation made a regular Theory of Library Classification an absolute necessity.

603 *Dynamic Theory of Library Classification*

As the Universe of Subjects is becoming increasingly dynamic and its proliferations happen in unanticipatable ways, the Theory of Library Classification has itself to be equally dynamic. However, the first three attempts at the Theory of Library Classification – the *Classification: theoretical and practical* (12), the *Manual of library classification* (15), and the *Organisation of knowledge in libraries* (1) – were largely static and descriptive of what had been already done, with, if at all, a slight element of dynamism. The fourth attempt was made in 1937, in the *Prolegomena to library classification* (4). Even this was largely static though it contained a little more of dynamism than the earlier ones. But, a fairly dynamic theory appeared in 1967 in the third edition of the *Prolegomena.* The Normative Principles formulated in this theory fall into the following groups:

1 General laws of the basic process of thinking;

2 The Five Laws of Library Science, published in 1931 and forming the basis for the development of every discipline in Library Science, including Library Classification.

3 Postulates, Canons, and Principles to give guidelines in the work of Library Classification in the Idea Plane;

4 Canons to give guidelines in the work of Library Classification in the Verbal Plane; and

5 Canons to give guidelines in the work of Library Classification in the Notational Plane.

604 *Verbal Plane*

Naturally, work in the Verbal Plane cannot fall solely within the sphere of the Library Profession. The only function of the profession is to stress that there should be a continuing effort to establish a homonym-free, and synonym-free technical terminology for each discipline.

605 *Separation of Work in the Idea and the Notational Plane*

The Idea Plane has unrestricted freedom in the development of the Universe of Subjects. But, the Notational Plane is unable to have it. At the same time, the result on the two Planes of Work should register absolutely, if Library Classification is to be of the best use possible. One effect of the inevitable restriction in the development of the Notational Plane has been to inhibit the work in the Idea Plane and to prevent it from spreading its wings progressively towards its own fulness in the measure of the successive development and proliferation in the Universe of Subjects. The advantage of separating the work in the two Planes was brought out by the *Prolegomena* in 1957 (5). It was further stated that the work in the Idea Plane was paramount; and that it was the duty of the Notational Plane to improve its capacity and versatility so as to implement in full measure the findings of the Idea Plane. The paramountcy of the Idea Plane had been stated even in DC as early as 1876. But, some removal of the restrictions in the Notational Plane disabling it from respecting the paramountcy of the Idea Plane, was first attempted in 1899 by UDC. To help the progressive removal of this disability of the Notational Plane, the Theory of Library Classification is being developed actively in the Notational Plane also since 1950. This work is actively occupying the DRTC Team (Documentation Research and Training Centre, Bangalore) since 1965. Its results are being published at intervals either in the quarterly *Library science with a slant to documentation* (1964–) or in the *Proceedings of the DRTC Annual Seminar* (1963–). CC attempts to implement the findings of the dynamic Theory of Classification as much as practicable.

We shall devote the next two sections to the Idea Plane and the Notational Plane respectively. Any Guiding Principles by the philosophers in their Knowledge Classification, are applicable to the work of Library Classification in the Idea Plane. But, we have not been able to get a full objective statement of all the Guiding Principles used by philosophers.

7 IDEA PLANE

71 General Guiding Principles of Philosophical Origin

The *Prolegomena,* 3rd ed., 1967 (8) assumes the following Laws of Philosophical origin:

1 Laws of Interpretation;
2 Law of Impartiality;
3 Law of Symmetry;
4 Law of Parsimony; and
5 Law of Local Variation.

The help of these Laws has to be invoked only when resolution of a conflict among the later-mentioned Normative Principles needs their aid.

72 The Five Laws of Library Science

The *Prolegomena,* 3rd ed., 1967 (7) assumes also the following Five Laws of Library Science.

1 Books (documents) are for use;
2 Every reader his book (document);
3 Every book (document) its reader;
4 Save the time of the reader; and
5 A library is a growing organism.

The basic book containing these Laws is the *Five laws of library science* (1931) of S R Ranganathan.

The help of the Five Laws of Library Science has to be invoked only when a resolution of a conflict among the Postulates, Canons, and Principles guiding the work in the Idea Plane in the area of Library Classification needs their aid.

721 *Ekavakyata*

The Five Laws of Library Science have the qualities of *ekavakyata* (all in one statement). All the theories and practices of library science are lying in a potential form in them. They thus unify them. Teaching of the different branches of library science get intimately unified if they are related to the Five Laws, wherever possible.

722 *Quality of Five Laws*

In the words of W C Berwick Sayers, the *Five laws of library science* (1931), is "a work of great simplicity which conceals depths and yet reveals what may be called the spiritual but intensely practical springs of his activity" (14).

723 *Light form Shri Aurobindo's "Mother"*

Some of the passages in the *Mother* (1928) by the late Shri Aurobindo, perhaps, the greatest of the realised souls of our own times, lend themselves to be adapted to this context;

Annotation.– The *Mother* is the Supreme Energy Principle of the Ultimate (Thesists would use the term God to denote the Ultimate). Maheswari, Mahakali, Mahalakshmi, and Mahasaraswati are the names by which the Supreme Energy Principle is denoted in relation to its different functions. As adapted to Library Service, Science, and Work, the passages are stated below:

1 The role of Maheswari.– Laying down the large lines of the world-forces in so far as they are applicable to Library Science, Service, and Work;

2 The role of Mahakali.– Driving their energy and impetus;

3 The role of Mahalakshmi.– Discovering their rhythms and measures;

4 The role of Mahasaraswati.– Directing the details of the organisation and execution, the relation of parts and the effective combination of forces and the unfailing exactitude of result and fulfilment; and for this purpose:

1 Giving the intimate and precise knowledge, the subtlety and patience, the accuracy of intuitive mind, and conscious hand and discerning eye of the perfect worker;

2 Making strong, the tireless, the careful and efficient builder, organiser, administrator, technician, artisan, and classifier;

3 Leading to action, laborious and minute and often seems to our impatience slow and interminable, but it is persistent, integral and flawless;

4 Making the will put into work scrupulous, unsleeping, and indefatigable;

5 Noting and touching every little detail;

6 Leaning over us;

7 Finding out every minute defect, gap, twist or incompleteness;

8 Considering and weighing accurately all that has been done and all that remains still to be done heareafter;

9 Not allowing anything to be treated as too small or apparently trivial, for attention;

10 Allowing nothing to escape attention even if it is unpallable, disguised or latent;

11 Moulding and remoulding each part till it has attained its true form;

12 Putting everything in its exact place in the whole so as to fulfil its precise purpose;

13 Arranging and rearranging all things keeping eyes on all needs at once, and on all the ways to meet them;

14 Marking out what is to be chosen and what to be rejected;

15 Determining the right instrument, the right time, the right condition, and the right process;

16 Abhorring carelessness, negligence, indolence, scamped hasty and shuffling work, clumsiness, misfire, false adaptation, misuse of instruments and facilities and undone or half-done things;

17 Securing that nothing is forgotten, misplaced, omitted or left in faulty condition when the work is finished;

18 Securing that all is solid, accurate, complete and admirable; and

19 Finding satisfaction in nothing but perfect perfection.

73 Additional Specific Normative Principles

To implement the directions of the Five Laws, each discipline of Library Science, has its own sets of additional specific Normative Principles to give the necessary 'guidelines' specific to its purpose. We are here concerned only with the discipline of Library Classification. In this discipline, additional specific Normative Principles are necessary for guidance in the work in the Idea Plane, the Verbal Plane, and the Notational Plane respectively.

74 Postulates for Classification

The fundamental set of Normative Principles for work in the Idea Plane, are called "Postulates". As is well known a Postulate cannot be said to be either true or false. They can only be either helpful or unhelpful for the work in hand. The Postulates formulated in the dynamic Theory of Classification, given in the *Prolegomena,* 3rd ed. 1967, (11) are

1 Postulate of Fundamental Categories.– There are five and only five Fundamental Categories -- *viz,* Time, Space, Energy, Matter, and Personality.

11 These terms have no metaphysical significance.

12 The Fundamental Category Matter may manifest itself in a subject either as "Matter Property", or as "Matter Method" and/or as "Matter Material".

13 Every Compound Subject has a Basic Facet as the first Facet, and it is not a manifestation of any one of the five Fundamental Categories.

14 Each of the other components of a Compound Subject is called an "Isolate Facet"; and each "Isolate Facet" is deemed to be a manifestation of one and only one or the other of the Five Fundamental Categories.

2 Postulate of Round for Energy. – The Fundamental Category "Energy" may manifest itself in one and the same subject more than once. The first manifestation is taken to end Round 1 of the manifestation of the three Fundamental Categories "Personality", "Matter", and "Energy". The second manifestation is taken to end Round 2, and so on.

21 Each of any of the three Fundamental Categories "Personality", "Matter", and "Energy" may manifest itself in Round 1, Round 2 etc.

22 Each of the Fundamental Categories "Space" and "Time" will normally manifest itself only in the Last Round.

3 Postulate of Level. – The Fundamental Category "Personality" may manifest itself more than once within one and the same Round; each of these is called a Level within the Round. So also with the Fundamental Category "Matter".

4 Postulate of Facet Sequence. – Each manifestation forming a component of a Compound Subject is called a Facet of the subject. Within a Round, the sequence of the Isolate Facets will be each successive Level of Personality, each successive Level of Matter, and Energy.

741 *Example*

Consider the following Compound Subject:

The prevention of the side effect of the antibiotic muscular injection to a cow with mouth disease in Tropical Asia in the rainy season of 1971.

S N	Name of the Facet	Focus in the Facet
1	Basic Facet	Animal Husbandry
2	Round 1, Level 1, Personality Isolate Facet	Cow
3	Round 1, Level 2, Personality Isolate Facet	Mouth
4	Round 1, Level 1, Matter-Property Isolate Facet	Virus Disease
5	Round 1, Energy Isolate Facet	Treatment
6	Round 2, Level 1, Personality Isolate Facet	Antibiotic

S N	Name of the Facet	Focus in the Facet
7	Round 2, Level 1, Matter-Method Isolate Facet	Muscular Injection
8	Round 2, Level 2, Matter-Property Isolate Facet	Side Effect
9	Round 3, Energy Isolate Facet	Prevention
10	Space Isolate Facet	Tropical Asia
11	Time Isolate Facet	Rainy Season of 1971

742 *Name of the Subject Resulting from the Postulates*

If the terms expressed as well as implied in the name of the subject are analysed and arranged, in accordance with the Postulates for Library Classification, the name of the subject will read as follows:

Animal Husbandry. Cow. Mouth. Virus Disease. Treatment. Antibiotic. Muscular Injection. Side Effect. Prevention. Tropical Asia. Rainy season of 1971.

743 *Resulting Helpful Sequence*

If the names of all the subjects are reduced to this form, their sequence will be helpful to the majority of readers. The aim of Library Classification is to arrange minute subjects -- such as this, and even more minute ones -- in a helpful sequence. It is not known whether Knowledge Classification by philosophers has ever had this as its objective.

75 Canons of Classification: Idea Plane

Prior to the formulation of the Postulates for Classification, Canons of Classification for work in the Idea Plane had been formulated. The number of the Canons, and the names of the books first formulating them are given in the following table.

Book 1: *Classification: Theoretical and practical.* 3rd ed. 1964. (13).
Book 2: *Manual of classification.* 3rd ed. 1962. (16).
Book 3: *Organisation of knowledge in libraries.* 1933. (2).
Book 4: *Prolegomena to library classification.* 3rd ed. 1967. (9).

SN	Canons for	Number of Canons			
		Book 1	Book 2	Book 3	Book 4
1	Characteristic	1	2	2	4
2	Succession of Characteristics	0	1	0	3
3	Array	0	2	2	4
4	Chain	0	2	1	2
5	Filiatory Sequence	0	0	0	2
	Total	1	7	5	15

76 Probable New Canons in Library Classification

Out of the fifteen Canons formulated for work in the Idea Plane in Library Classification, twelve are probably found explicitly or implicitly in the work of philosophers in Knowledge Classification. If at all, the following three Canons are probably the only new ones formulated for use in Library Classification, which had to take within its purview the classification of minute subjects also.

1 Canon of Consistent Succession.– The succession in the scheme of characteristic chosen as relevant for the specific purpose of the classification of the Universe of Subjects should be consistently adhered to. (This canon is the third among the Canons formulated in the *Prolegomena* for the Succession of Characteristics).

2 Canon for Subordinate Classes.– In an array in which the coordinate and the subordinate classes of different orders are arranged in a single sequence, if A_1, A_2, A_3, etc are subclasses of any order whatever of Class A, arising in one or another of the Chains arising from the Class A, the Classes A_1, A_2, A_3, etc should immediately follow the Class A in succession, without being separated from it or among themselves by any other class. (This Canon is the first among the Canons formulated in the *Prolegomena* for Filiatory Sequence).

3 Canon of Coordinate Classes.– In an array in which the coordinate and the subordinate classes of different orders are arranged in a single sequence, if Class A and Class B had originated in one and the same array and had been consecutive in it, they should not be separated from each other by any class other than the classes A_1, A_2, A_3 etc having A as their common immediate universe. (This is the second among the Canons formulated in the *Prolegomena* for Filiatory Sequence).

77 Principles of Helpful Sequence

771 An Omnibus Canon

of the 15 Canons mentioned in Sec. 75, the Canon of Helpful Sequence is as follows:

The sequence of the Classes in an array of classes, and of the ranked isolates in an array of ranked isolates, should be helpful to the purpose of those for whom it is intended. (This Canon is the third among the Canons formulated in the *Prolegomena* for Array).

This canon implies a tall order as it were. Indeed, it is an Omnibus Canon. There can be many ways to satisfy this Canon. For some purposes, there may be only one way. For some other purposes, there may be two or more ways of satisfying the Canon. In the latter case, judgment should be used in the selection of the most paying from among the alternatives in the light of the subject context. And if this does not prove helpful, the help of the Canons, the Five Laws, and the General Laws should be invoked as stated in Sec. 71, and Sec. 72.

722 *Formulation of Principles for Use as Guidelines in Applying the Canon*

As guidelines to help in deciding the truly useful helpful sequence for a specific purpose in a particular subject context, 17 Principles for Helpful Sequence have been formulated in the *Prolegomena,* 3rd ed., 1967.

773 *Principles without Alternatives*

The following seven principles are without alternatives. Provided the subject context needing them is sensed, the classifier will not have any difficulty in choosing the right principle.

1 Principle of Later-in-Time;
2 Principle of Later-in-Evolution;
3 Principle of Away-from-Position;
4 Principle of Increasing Complexity;
5 Principle of Canonical Sequence;
6 Principle of Literary Warrant; and
7 Principle of Alphabetical Sequence. (10)

774 *Alternative Pairs of Principles*

The remaining ten Principles for Helpful sequence fall into the following five pairs of alternatives.

11 Principle of Bottom-Upwards, and
12 Principle of Top-Downwards;
21 Principle of Left to Right, and
22 Principle of Right to Left;
31 Principle of Clockwise Direction, and
32 Principle of Counter-Clockwise Direction;
41 Principle of Periphery to Centre, and
42 Principle of Centre to Periphery;
51 Principle of Increasing Quantity, and
52 Principle of Decreasing Quantity. (10)

One Principle of a pair often gives a direction opposite to that of the second. This is a matter of necessity. Because, two subject contexts of otherwise similar nature will indicate opposite sequences as helpful.

Example 1.

In arranging the Organs of a Tree visible above the surface of the earth, the Principle of Bottom Upwards is easily applicable. For, the tree proliferates into many branches, subbranches, twigs, leaves, flowers and so on. Therefore, there is no means of arranging them uniquely from Top-Downwards. Again, there may be further unanticipatable growth at the top.

On the contrary, in arranging the proliferations of the root beginning from the surface of the earth and forming branches below the surface of the earth, the experience is just the opposite.

775 *Knowledge Classification*

There should have been little likelihood for the philosophers pursuing Knowledge Classification to have found the need for such detailed principles in the Theory of Classification, as thos mentioned in Sec. 773 and Sec. 774. But these form the very essence of the Theory of Library Classification. This is increasingly realised as the intension of the subjects classified becomes deeper and deeper. The most difficult, important, and never-ending pursuit of Library Classification concernes such ultimate Principles. So long as the Universe of Subjects goes on decreasing in extension and increasing in intension, and thereby throwing forth proliferations in unanticipatable directions and ways, the Theory of Library Classification will have to continue to be dynamic in solving the problem arising in the Idea Plane.

8 NOTATIONAL PLANE

81 Limitation of Notational Plane

The development of the Universe of Ideas and the development of the Theory of Library Classification for work in the Idea Plane can be without any restriction whatever. These can keep step with whatever ideas the human mind produces. But work in the Notational Plane is severely restricted. It is a great struggle for the Theory of Library Classification to develop the work in the Notational Plane in the full measure required by the continuing development in the Universe of Subjects. To put it in other words, the versatility of the Notational Plane is negligible when compared with the versatility of the Idea Plane. The delay in the development of Library Classification during the last 100 years is due to this difference in versatility. DC has virtually dropped out very early in the race between the Idea Plane and the Notational Plane (3). It has now declared itself to be unsuitable for use in the classification -- depth classification as we call it -- of micro subjects, that is, subjects of very small extension and of very deep intension usually appearing as articles in learned periodicals. And yet it is these articles that are of vital importance for the progress of industry and research. Without continuous progress in industry and research, the present rate of increase of world population, threatens to bring its own destruction.

82 Contribution of UDC to the Notational Plane

UDC was the first scheme to face this problem. On account of its continuing the DC core of Compound Subjects as the basis for further classification, it is not able to do its work as well as it can. However, it is bravely facing the fight. It is moving forward though at an extraordinarily slow speed.

83 Contribution of CC to the Notational Plane

It was CC that went the whole hog in facing this problem – in narrowing the gulf between the versatility of the Idea Plane and that of the Notational Plane. It has made successive improvements during the last 40 years by its

1 Abandonment of starting with a schedule of hundreds of thousands of Compound Subjects as DC and UDC have done;

2 Starting only with a short schedule of less than one hundred Basic Subjects – non of which was a Compound Subject;

3 Admitting during all these 40 years about one thousand non-compound Main and other Basic Subjects of recent origin;

4 Analysing all Compound Subjects into elemental component facets in accordance with the Postulates mentioned in Sec. 71;

5 Providing only short schedules of Isolates belonging to each of the Rounds and Levels of each of the Fundamental Categories likely to appear in Compound Subjects going with the respective Basic Subjects;

6 Providing a distinctive indicator digit to indicate the Fundamental Category to which the succeeding facet belongs; and

7 Increasing the capacity of its Notational System by means such as

71 Lengthening each array so as to accomodate 1166 coordinate Isolates (as it stands to-day, in the place of the 10 or at most 20 in DC and UDC),

72 Postulating a few "Empty and Emptying Digits" and thereby providing the interpolation of any number of new Main Subjects between any two already enumerated consecutive Main Subjects,

73 Postulating certain "Empty Digits" -- such as, z, 9, and Z -- which enables the breaking up of 1166 isolates in an Array into 66 Sectors,

74 Postulating the concept of Compound Isolates, and

75 Postulating the concept of Special Components for use in the formation of Compound Isolates.

84 Telescoping of Consecutive Arrays and Shortening of Class Number

The Sector Notation mentioned in Sec. 83 (73), enables the "telescoping" of the Isolates in two consecutive Arrays within a Schedule. This enables the shortening of the Class Number of Many subjects. This satisfies the Law of Parsimony. The following Table illustrates it:

Isolate Term	Telescoped Isolate N in CC	Number in UDC omitting the initial digits 582
Cryptogamia	1	2
Thallophyta	2	22
Bryophyta	3	32
Pteridophyta	4	35
Phanerogamia	5	4
Gymnosperm	6	42
Monocotyledon	7	52
Dicotyledon	8	61

Note.– 11 In CC, "1 Cryptogamia" and "5 Phanerogamia" belong to the Array of Order 1, viewed both from the Idea Plane and the Notational Plane.

12 So also it is with "2 Cryptogamia" and "4 Phanerogamia", in UDC.

21 In CC, "2 Thallophyta", "3 Bryophyta", and "4 Pteridophyta" belong to the Array of Order 2 when viewed from the Idea Plane; but they appear to belong to Array of Order 1 when viewed from the Notational Plane. This is the effect of "telescoping" of Array of Order 2 into Array of Order 1. The telescoping has reduced the number of digits from two to one.

22 In UDC, "22 Thallophyta", "32 Bryophyta", and "35 Pteridophyta" belong to the Arreay of Order 2 viewed either from the Idea Plane or from the Notational Plane. Therefore, there is neither telescoping, nor economy in the number of digits.

3 Remarks similar to those in 21 and 22 apply also to the subdivisions of "Phanerogamia".

85 Compound Isolate and Shortening the Length of a Schedule

CC postulates the digit "–" (hyphen) for compounding two Isolates in the same schedule. This results in the shortening of the schedule of all the possible Isolates to be otherwise provided in it.

Example

The following is an extract from the schedule for "Y Sociology" in CC (7th ed.):

Y	SOCIOLOGY
1	*By speciator "Age and Sex"*
	By speciator "Age"
11	Child
12	Youth
13	Adult
14	Old
	By speciator "Sex"
15	Woman
16	Man

Note.– In the above schedule there are 4 isolates derived on the basis of the speciator "Age"; and there are 2 isolates derived on the basis of the speciator "Sex". It is possible to compound a "sex-isolate" with an "age-isolate" to form a Compound Isolate -- such as, "Old woman". In this case, there may be 8 such Compound Isolates. The number for each of the Compound Isolates, can be constructed by using the indication digit "–" (hyphen). For example, the number for "Old woman" will be "15–14".

On the other hand, if all the possible Isolates are enumerated in the schedule it will be as follows:

Social group
- Child
- Youth
- Adult
- Old
- Female
  - Female child
  - Young woman
  - Adult woman
  - Old woman
- Male
  - Male child
  - Young man
  - Adult man
  - Old man

Note.– CC Schedule enumerates only 6 isolates, and yet provides isolate numbers for 14 isolates. This indicates the measure of economy achieved in the length of the schedule in CC, with the aid of the concept of Compound Isolate.

851 *Special Component for Compound Isolate*

CC defines the term "Special Component" as follows:

"A Special Component" is an idea fit to form a Component of a Compound Isolate, but not by itself fit to be an Isolate".

For the purpose of shortening the length of the schedules, wherever necessary, CC gives a list of Special Components to be compounded with any Isolate in a specified schedule. CC postulates the digits "=" (equals) for compounding an Isolate with a Special Component. Here is an example extracted from the schedule for "P Linguistics" in CC (Ed 7):

15	Consonat
151	Glottal
152	Circular
153	Velar
154	Platal
155	Post-dental alveolar
156	Predental
157	Labiodental
158	Bilabial

Special components for Compound Isolates
beginning with 15 and its subdivisions

1	Plosive
2	Lateral
3	Rolled
4	Fricative
5	Semi-vowel
6	Nasal
	Compound Isolates *(illustrative)*
151 = 6	Glottal-nasal
154 = 6	Platal-nasal
158 = 6	Bilabial-nasal

Note.– A schedule of eight isolates and a list of six Special Components -- total number of items 14 -- replace a long schedule of 48 isolates. This indicates the measure of economy achieved in the length of the schedule in CC with the aid of the concept of Special Component.

86 Theory of Classification for the Notational Plane

All these means of increasing the versatility of the Notational Plane are to enable it to implement the findings in the Idea Plane.

87 Relatively Slow Growth

Due to the constriction inherent in the Notational Plane, the Theory in regard to that Plane could not make either adequate or rapid progress. Still, it is not left stagnant. The DRTC Team is constantly looking for means for improving this theory so as to enable the Notational System to increase its versatility and thereby carry out the demands of the Idea Plane. The example in the next section will show the appreciable success of the work of the DRTC Team in this matter. The examples in Sec. 881 and 882 give the UDC and CC Numbers respectively for the following subject reading as follows when facet-analysed and made ready for translation into Class number:

Animal Husbandry. Cow. Mouth. Virus disease.
Treatment. Antibiotic. Muscular injection.
Side-effect. Prevention. Tropical Asia.
Rainy season of 1971.

It may be repeated here that this facet-analysed name of a subject will be of help to any scheme for classification as it is based on a general theory for work in the Idea Plane, quite independent of any particular scheme for Classification.

88 Illustrative Class Number

881 *Universal Decimal Class Number*

UDC Number	Number of digits
636.21:611.31:616.988–085: 615.331:615.032.73: 615.065:616–084(5:213)"1971:Rainy Season"	83

882 *Colon Class Number*

CC Number	Number of digits
KX,311,21;423:6,66;63;4:5 .4=3 'N71-u	35

883 *Relative Obesity of UDC and CC Numbers*

The obesity of the Universal Decimal Class Number is unmistakable when compared with the Colon Class Number. This difference in obesity is discernible not only in the whole numbers but also in most of their respective components.

884 *Colon Classification*

The relative slimness of the Colon Class Number was possible because its present version was designed, after a General Theory of Library Classification for the Notational Plane had advanced sufficiently. There is a continuous progress in this. (*See* Sec. 85).

885 *Universal Decimal Classification*

But, the UDC has not followed the Theory. For, it had adopted the long initial schedule of compound subjects. In fact, this long schedule excludes only the extra subjects derivable by the addition of Form Isolates, Space Isolates, and Time Isolates and analytical divisions in UDC. No doubt, during the last few years, UDC has been slowly changing the initial schedule of Compound Isolates. However, it does not appear to have taken courage, so to speak, to break up each subject in its initial schedule of Compound Subjects into their respective ultimate facets Basic as well as Isolate; in other words, it has not yet thought of de-freezing each of the Facet Numbers, other than the last one, forming components of the Class Number of a Compound Subject.

9 FREELY FACETED CLASSIFICATION

Ignoring Form, Time, and Space Facets, the progress in the Theory of Library Classification has been gradually changing Library Classification from

1 Enumerative Classification with its impenetrable rigidity (DC); through

2 An essentially Enumerative Classification with a dash of elasticity for penetration (UDC); through

3 A rigidly Faceted Classification (Version 1 of CC – 1933); through

4 A slightly less rigidly Faceted Classification (Version 2 of CC – 1952); to

5 A freely Faceted Classification (Version 3 of CC), the press copy of which is nearing completion.

In a freely Faceted Classification,

1 Facets are no longer mistaken to belong to the Basic Subjects;

2 There is no rigid Facet Formula prescribed for all the Compound Subjects going with a specific Basic Subject; and

3 Every Compound Subject, whatever be its Basic Facet, can bring with it as many or as few Isolate Facets, as it prefers.

This is the summary of the latest result obtained from the dynamic Theory of Library Classification *in progress.* The term 'in progress' should be emphasised. It is difficult to prophesy now what new changes will come in Library Classification as the theory on the subject progresses. But there is no doubt that we all wish that the Theory makes significant progress and that the Library Classification increases its efficiency in fulfilling the never-ending exacting demand of the Five Laws of Library Science.

91 BIBLIOGRAPHICAL REFERENCES

(Numbers in brackets refer to the section in the text containing the reference)

1) H.E. BLISS Organisation of knowledge in libraries and the subject approach to books. 1933. (Sec. 603)

2) –.–. 2nd ed. 1939. p. 37–46. (Sec. 75)

3) B.A. CUSTER, Mr Phillips. Library Association record. 62; 1960; 407. (Sec. 81)

4) S.R. RANGANATHAN, Prolegomena to library classification. 1937. (Sec. 603)

5) –. –. 2nd ed. 1957. Chap. 35 and 36. (Sec. 605)

6) –. –. 3rd ed. 1967. Part C. (Sec. 0)

7) –. –. –. –. Chap. DB. (Sec. 72)

8) –. –. –. –. Chap. DC to DG. (Sec. 71)

9) –. –. –. –. Part E. (Sec. 75)

10) –. –. –. –. Part F. (Sec. 773, 774)

12) E.C. RICHARDSON Classification: theoretical and practical. 1901. (Sec. 603)

13) –. –. 3rd ed. 1930. p. 41–2. (Sec. 75)

14) W.C. Berwick SAYERS *In:* S.R. Ranganathan. Prolegomena to library classification. 2nd ed. 1957. p. 19. (Sec. 722)

15) –. Manual of classification for librarians and bibliographers. 1926. (Sec. 603)

16) –. –. 3rd ed. 1962. p. 79–82. (Sec. 75)

Alwin Diemer (Düsseldorf)

L'ordre (classification) universel des savoirs comme problème de philosophie et d'organisation

0. Les discussions, quel qu'en soit le thème, sont bien souvent menacées de deux erreurs: la *perte de vue du thème même de la discussion* d'une part, et d'autre part *l'oubli de l'intention*, du sens qui fut au point de départ de la reflexion. Comme cette situation peut se présenter dans le cadre de ce congrès (qui précisément se définit dans un cadre très large et doit même se présenter comme le début d'une evolution qui doit aboutir à l'édification d'un institut de recherche en classification) il nous semble justifié de choisir ce problème comme thème de notre communication. Mais tout d'abord quelques remarques préliminaires.

0.1 En premier lieu, la *détermination de l'objet* même qui doit être traité ici est déjà un problème: dans le texte anglais du programme il est question de "*Knowledge*", dans le texte français de "*Savoirs*" – donc un pluriel – alors qu'une version allemande parlerait plutôt de "Wissen". Beaucoup d'exposés en outre semblent impliquer une compréhension des "sciences".

Si maintenant nous comparons les termes (ce mot sera désormais employé dans la suite de l'exposé car il est par delà toute querelle sur le mot, le concept et l'idée) nous nous apercevons que leur signification n'est que relativement identique. Dans quelle mesure peut on dire que "Savoirs" est identique à "Knowledge", et "Knowledge" identique à "Wissen". En allemand il faudrait encore ajouter "Erkenntnis". D'autre part, si nous nous limitons aux seuls savoirs scientifiques la situation n'est pas moins problématique, car le mot allemand "Wissenschaft" n'est en aucune manière identique à "science". Qu'en est il des "Geisteswissenschaften" qui sont appelées en anglais "Humanities"? Que faire de "christian science", de la théologie et de tous les systèmes dogmatiques? Qu'en est il de l'astrologie, sans parler du banal savoir pratique de la vie courante comme les recettes de cuisine, en tant que manifestation du "savoir ménager", ni de la philatélie ou des résultats de sport?

0.2 Si nous nous penchons sur le thème de la *classification*, nous constatons qu'il ne pose pas moins de problèmes. Sans doute ce thème est d'actualité, mais la philosophie aussi bien traditionnelle que contemporaine ne pense le problème de la classification que dans le sens d'une classification des "Wissenschaften", que comme un vaste système de déduction, et y voit plus l'occasion d'un jeu d'esprit pour epistemologues que matière à réflexion. La philosophie n'a pas encore trouvé le chemin de l'actualité scientifique contemporaine, en particulier en ce qui concerne le problème d'une politique et d'une organisation des sciences.

C'est à partir d'un tout autre domaine que la problématique de la classification a pris son importance: le domaine de la documentation. Cette problématisation s'est développée à partir de préoccupations essentiellement bibliothécaires: un domaine que les "véritables savants" ont toujours considéré comme secondaire. Les sciences de l'information, la documentation, dans leur forme classique ont développé et appliqué les classifications comme un simple instrument d'ordre. Quand elles sont passées des préoccupations purement bibliothécaires à leur forme moderne de sciences de l'information, la nécessité d'une problématisation fondamentale s'est fait jour: les formes classiques de la classification, comme la classification décimale, furent restructurées, spécifiées; une orientation polyhiérarchique (facettes) remplaça l'orientation monohiérarchique. Mais ces tentatives ne résolurent pas le problème dans sa totalité et il vint s'ajouter le développement des "thesauri" et les deux formes d'ordre et d'orientation se regroupèrent dans les "langages de documentation". Mais justement à l'occasion de ces efforts se pose avec acuité le problème de savoir ce qu'est encore un système de classification, s'il existe un système universel? Et quelle que soit cette universalité (qu'elle soit limitée aux sciences ou qu'elle s'applique à tous les domaines du savoir) reste à savoir ce qu'est un système de classification et ce que l'on peut en attendre. S'il s'impose de discuter sur un plan théorique le problème de l'ordre et de la classification qui n'en est qu'une forme particulière, et le problème de *l'universalité,* nous devrons toutefois nous demander si il est possible de cerner cette problématique en ne la considérant que sous son aspect formel.

0.3 Il nous apparaît en effet nécessaire, à la lumière de recherches théoriques ainsi que d'une pratique de l'organisation dont nous parlerons par la suite, de considérer le problème de l'ordre du savoir sur le plan théorique des recherches sur l'ordre, la *classification* etc. mais également sur le plan de *l'organisation.*

Il nous semble même que l'on puisse regarder la nécessité d'une telle division comme un développement historique dans lequel les quatre phases suivantes se laissent distinguer: (voir Fig. 1, p. 146)

1 phase –Orientation ontologique
= ordre des niveaux de réalité

2 phase –Orientation gnoséologique scientifique
= ordre, classification des sciences

3 phase –Orientation bibliothécaire
= possibilité d'ordre pour la localisation des livres en tant que savoir représenté

4 phase –Orientation informationelle
= schémas d'orientation en vue du gain et de la négociation d'informations (Informèmes) dans un processus informatif.

Il est certain que dans l'ordre ontologique aussi bien que dans l'ordre gnoséologique – dont le fondement est d'avant tout philosophique – (il s'agit d'établir l'ordre immanent à la réalité ou au savoir, le "système naturel" comme on l'appel-

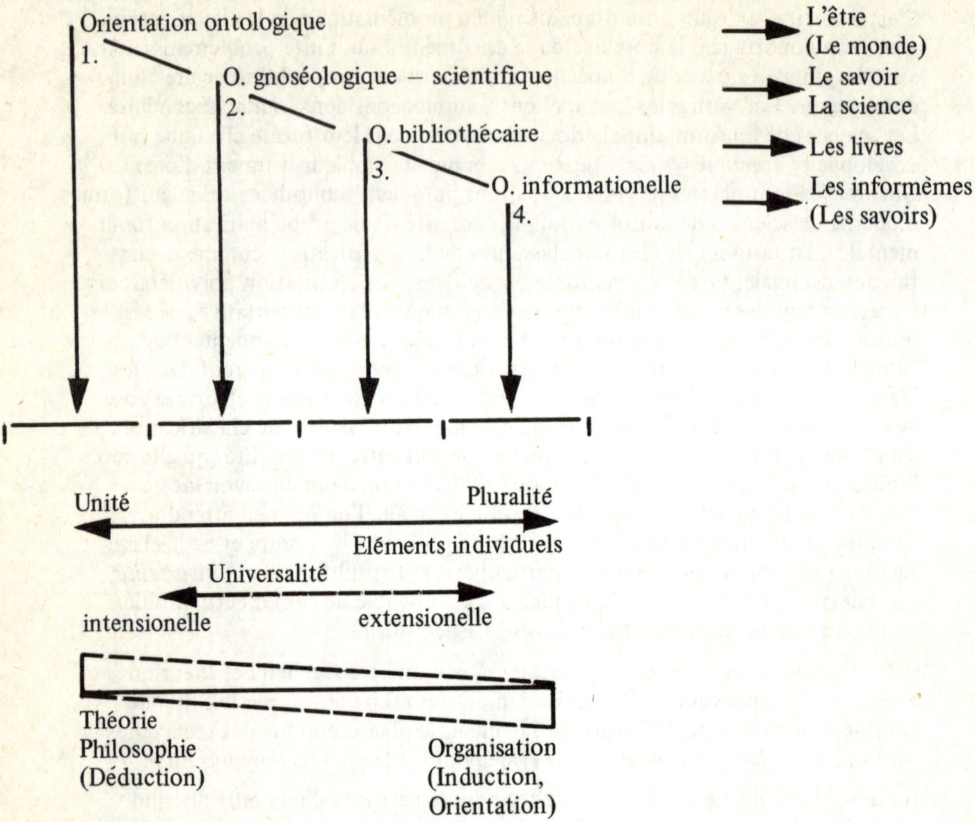

Fig. 1

lera ensuite) certains aspects fonctionnels ont déjà joué un rôle; au XVIIIe siècle même le problème de l'organisation fut déjà ça et là sujet de discussion (heuristique), mais ce n'est qu'actuellement qu'il est nettement thématisé.

Un petit modèle peut illustrer ce que nous voulons dire – ou bien de plus éclairer la situation actuelle du problème de classification. Nous croyons donc que ces quatre phases constituent en même temps les quatre champs où se pose la question de la classification aujourd'hui.

ad 1) Nous avons comme premier champ le champ de la philosophie même, ou plus exactement le champ ontologique. Le problème de constitution de la réalité – de l'être – lié aux problèmes des catégories donne un excellent exemple du lieu ou se pose la question de la classification. La tradition la connait comme l'idée de la stratification. Chaque état appartient à une classe; les classes comme telles sont constituées par l'ontologie materiale.

ad 2) Plus actualisée est la classification dans le cadre de la science. Ici le problème se pose en deux configurations, comme problème intradisciplinaire et comme problème interdisciplinaire. La tradition connaît le problème premièrement comme problème de la classification des sciences. Ici nous avons toutes les questions spéciales sur lesquelles nous aurons encore à revenir. Pour le problème intradisciplinaire on pourrait citer la taxonomie en biologie, la comparaison p.e. dans la littérature comparée etc.

C'est surtout ici que naît la nécessité de la classification comme telle. La *théorie de la classification* peut être conçue comme une théorie autonome dans le cadre d'une théorie de science. Assez sur ce sujet, des détails plus tard.

ad 3) Séparément – à l'égard de l'évolution de la thématisation – des champs traités jusqu'à maitenant s'est développée la classification bibliothécaire. Ici, il fallait – et il faut toujours – d'un côté créer des possibilités pour donner des lieux où poser les livres, et – de l'autre côté – créer des possibilités pour classer les savoirs manifestés par et dans les livres. Personne ne doute qu'ici la classification ait une autre intention. Jusqu'à maitenant la classification était orientée vers les unités que sont les livres. Une classification est parfaite si elle donne à chaque livre une place exacte.

ad 4) Cette tendance est exprimée encore plus fortement dans le quatrième champs, celui de l'information. L'entité constitutive est ici l'informème – comme nous nommons le continu individuel et identique dans tout procès informatoire. Nommons X le livre, A, B, C . . . les savoirs – les contenus –, l'évolution se présente:

1)	$X_{(A)}$	lieu bibliothécaire
2)	$X'_{(A, B, C \ldots)}$	lieu bibliothécaire catalogue des sujets
3)	$A_{(X)}B_{(X)}C_{(X)}$	les informèmes

0.4 C'est autour de ces remarques que nous organiserons notre développement. Elles résultent à la fois d'un travail théorique dans le domaine de la théorie des sciences et de l'information (recherches sur la classification en particulier) et d'expériences pratiques dans la mise en oeuvre d'un système de documentation philosophique (PHILIS). Les considérations théoriques seront au premier plan, et les problèmes de l'organisation qui dépassent le thème immédiat de cette réunion ne seront traités que dans le dernier point où nous indiquerons à titre d'exemple le schéma de PHILIS.

Nous traiterons donc:

1 Le problème du savoir
2 Le problème de l'ordre et de la classification
3 Le problème de l'universalité qui relève à la fois de 1 et de 2 mais dont l'importance exige une discussion particulière
4 Le problème du rapport de l'ordre théorique et de l'organisation.

1. Le problème du savoir se presente sous quatre aspects:

1.1 Le problème de la quantité
1.2 Le problème de la qualité
1.3 Le problème de la présentation
1.4 Le problème de l'interprétation considéré sous l'aspect de l'intersubjectivité de l'interprétation

1.1 Quand on considère le savoir sous l'angle de la quantité, une question se pose: De quoi se compose l'ensemble du savoir? Une première division, apparemment banale mais décisive pour la suite du développement doit être posée: l'ensemble du savoir peut se diviser en deux parties: le savoir scientifique d'une part et d'autre part le savoir autre que le savoir scientifique.

Il est certes très difficile de déterminer ce qui est considéré comme science avant d'avoir réalisé une classification correspondante! C'est là la tache de la philosophie des sciences. Nous prendrons donc un critère provisoire: Nous considérerons comme savoir scientifique l'ensemble du savoir des "sciences universitaires". En conséquence nous appelons science toute discipline qui fait l'objet d'un enseignement et de travaux de recherche dans le cadre d'une université. Mais déjà ici surgissent les problèmes. Faut-il considérer la théologie comme une science? En réalité LA théologie n'existe pas. Il y a d'un côté un ensemble de "Weltanschauungen", de dogmatiques religieuses et de l'autre des sciences, comme les sciences religieuses. C'est pourquoi il nous semble opportun pour une tentative de classification du domaine scientifique d'établir certains domaines non scientifiques qui seront toutefois à prendre en considération lors du classement des sciences.

En réalité bien qu'il s'agisse encore ici d'un "savoir sérieux" pour reprendre la terminologie de la classification des sciences du XVIIIe siècle, nous atteignons le vaste domaine du savoir autre que le savoir scientifique. A ce domaine, dont il nous importe peu de savoir s'il est composé de "Wissen" ou de "Erkenntnis", il faut maintenant ajouter le domaine de la vie courante dont nous ne donnerons que quelques exemples: informations sportives, résultats de football, informations concernant les loisirs, la philatélie, les produits de beauté, les recettes de cuisines etc. ... Il ne faut pas oublier le domaine de l'astrologie pour lequel la définition de ce que nous appelons savoir devient très sensible: le savoir de l'astrologie est considéré comme tel et demandé par une partie de la population, donc il faut le tenir pour un savoir – tout au moins dans le cadre d'une banque d'information.

Pour en terminer avec ce point de la quantité du savoir, il faut signaler que le problème de la totalité du savoir, que d'aucuns étaient en droit d'attendre dans ce paragraphe, sera traité quand nous aborderons le thème de l'universalité du savoir.

1.2 Problème de la qualité du savoir.
Quand nous parlons du savoir, nous entendons généralement par là certaines idées, certaines représentations qui sont exprimées, désignées par certaines expressions et se trouvent le plus souvent imprimées dans des textes (articles, livres, notes).

Ce qui a pour conséquence que la documentation consiste pour la plus grande part dans l'emmagasinement et la négociation de textes.

Cette qualification du savoir ne nous semble pas complète, et dans la perspective d'une saisie plus large et plus détaillée de l'information nous tenons la classification suivante pour nécessaire:

1. Savoir numérique
2. Savoir non numérique
   - 2.1 transmis à l'aide d'un support linguistique généralement textuel
   - 2.2 transmis de manière non textuelle
     - iconographique
     - acoustique
     - divers, gustatif par ex.

La distinction entre 1. et 2. est largement admise et ne nécessite pas d'explication. Elle n'a pas grande importance pour la classification mais elle est très importante quand on la considère sous l'angle de l'organisation. Prenons par exemple la question de savoir si les informations numériques doivent être emmagasinées en fonction de leurs domaines d'application, ou si il doit y avoir un seul centre d'emmagasinement d'informations numériques qui seront interprétées en fonction de chaque domaine particulier; ce sont des raisons d'organisation qui résoudront ce problème.

La distinction entre un savoir textuel et non textuel est très actuelle pour nous. Cette question n'a encore jamais été étudiée de près, pas plus dans le cadre d'une théorie de la connaissance que dans celui d'une théorie des sciences ou de l'information. Le problème du savoir iconographique devrait cependant attirer l'attention des autorités compétentes de la critique d'art: peut-on dire en effet que le savoir relatif à un tableau s'épuise avec sa description? Ou bien doit-on considérer le savoir iconographique comme un savoir sui generis? On peut de la même manière poser la question de l'existence d'un savoir gustatif, question qui est déjà discutée par l'esthétique moderne. Ces remarques peuvent donner l'impression de s'éloigner du thème précis de cette discussion. Or il n'en est rien, car notre propre expérience nous a montré la nécessité de d'abord reconnaitre ce que l'on doit mettre en ordre avant de pouvoir décider quelle sorte d'ordre est le plus adéquat.

1.3 Comment se présente le savoir?

Pour beaucoup de documentalistes la distinction entre un savoir "en soi" donné dans les idées, les concepts etc. . . . et de son expression, de sa dénomination, ne pose aucun problème. On pose ainsi ce "savoir en soi" comme une sorte de monde des idées à la manière platonicienne, donné une fois pour toute, et de l'autre côté le langage, l'expression linguistique qui ne fait que représenter ce monde des idées. Il arrive alors suivant cette théorie que la même idée soit différemment représentée que plusieurs mots expriment la même idée.

C'est sur cet arrière fond théorique, généralement non réfléchi (Vorverständnis) que s'inscrivent la plupart des systèmes de classification, en particulier le "thesaurus". Le principe fondamental d'une telle classification est en effet la réduction des données linguistiques à un certain nombre de "descripteurs" qui seuls expriment l'idée désignée par les différents mots qui ne sont alors que des variantes de ce descripteur unique. La même chose s'applique aux équivalents étrangers qui sont réduits à ce même descripteur.

Aussi nécessaire que puisse paraitre une telle conception pour la réalisation pratique – en particulier pour l'établissement d'un système d'information international – nous la mettons radicalement en doute! Pour notre part nous considérons les mots, plus exactement les termes du langage comme des pions. Chaque pion représentant une idée spécifique. Ainsi nous n'admettons pas que plusieurs pions désignent la même idée (comme si un même concept pouvait avoir des dénominations équivalentes). Si, bien souvent, plusieurs termes semblent se rapporter à la même idée, cette identité ne peut être dite absolue que sur le plan de l'intentionalité, et non en fait. Nous aurons l'occasion de revenir sur ce point quand nous traiterons de l'intersubjectivité de l'interprétation.

Un exemple illustrera notre conception: on distingue en allemand entre "Ross", "Pferd" et "Gaul". La classification habituelle, généralement utilisée dans les sciences de la nature réduit ces trois désignations au seul terme de "Pferd" puisqu'elle admet que ces trois termes désignent la même idée, le même concept. Une telle réduction peut être pratique en biologie, mais une information touchant l'histoire de la culture se doit de distinguer ces trois termes.

Cette remarque s'applique aux différentes langues: ainsi les termes "Wissenschaft" et "science" sont posés comme ayant une signification identique. Or nos recherches sur les rapports entre "science" et "Wissenschaft" nous ont montré qu'il n'en est rien. Sans développer plus avant cette remarque nous indiquerons simplement les difficultés que soulève le simple terme: "Geisteswissenschaft".

Nous pouvons donc résumer notre position de la manière suivante: chaque mot est considéré comme un pion auquel correspond une et une seule idée. Chaque proposition exprime un savoir spécifique. Il faut du reste souligner que c'est le pion seul qui est emmagasiné dans la mémoire électronique, et non l'idée.

Est ce qu'une telle conception tombe sous le coup d'une critique du nominalisme? En aucun cas, et nous repousserons ce reproche dans notre théorie de l'intersubjectivité de l'interprétation-qui est aussi fondamentale pour notre théorie des sciences que pour notre conception de l'information.

Pour terminer ce chapitre signalons quelques autres niveaux sur lesquels le savoir se présente.

Il est possible de distinguer un savoir primaire, qui est le savoir que le demandeur attend, et que l'information doit lui fournir. Nous appelons "Informème" ce savoir fourni conformément à une attente d'information. D'autre part le savoir secondaire est constitué par le savoir des références où l'on peut trouver les informèmes, où il est traité d'un thème actuel, des données bibliographiques par exemple. La documentation courante fournit le plus souvent ce savoir des références et laisse au demandeur le soin de se reporter aux originaux. Mais passons à l'interprétation!

1.4 L'intersubjectivité de l'interprétation.

L'intersubjectivité de l'interprétation s'actualise à trois niveaux:

1 – au niveau immédiat d'une situation locutive et compréhensive qui se manifeste surtout dans le mot pris dans son contexte.
2 – au niveau très stratifié des domaines particuliers comme les disciplines scientifiques
3 – au niveau d'un horizon universel, qui a le caractère d'un absolu intentionnel

1.41 Les données du langage s'actualisent toujours dans le champ d'un "parler" et d'un "comprendre". Que se passe-t-il dans une telle situation? Les mots sont distribués à chaque partenaire comme des pions. Chacun associe avec chaque pion une signification intentionellement identique qu'il thématise (du moins en est il certain subjectivement) dans une perspective historique ("J'ai toujours voulu dire la même chose par cela"). Chaque partenaire transpose cette attitude sur les autres. Chaque protagoniste postule que les autres attribuent à son pion la même signification que lui. Naturellement chacun apporte avec lui un système qu'il postule intentionellement identique avec les autres. Ainsi chacun a son "monde naturel" comme il a son "langage naturel". Monde est à comprendre ici au sens où l'entend la phénoménologie contemporaine c'est à dire comme un horizon d'interprétation, qui se traduit dans un système correspondant.

1.42 A cette situation de départ, à ces mondes propres, viennent se superposer toute une couche de schémas, de modèles différents: schématismes propres à chaque science ou autres.

Les schématismes scientifiques, surtout ceux des sciences exactes sont le résultat d'une critique du monde naturel et du langage naturel trop equivoques et inexacts. Ainsi un horizon et un langage artificiels sont ébauchés, présupposant un schéma d'ordre artificiel, le plus souvent de structure simple.

1.43 Ces deux niveaux sont eux mêmes surmontés par un plan universel. Ce plan universel est lui aussi présupposé – d'une manière intentionellement identique – absolu en tant que condition de possibilité de toute compréhension, effective ou possible.

Il fut un temps où l'on pouvait recourir à un être divin pour assurer ce niveau. On lui attribuait un savoir absolu, et il était le garant du caractère absolu de tout schéma de classification. Mais nous avons perdu cet arrière fond divin sans que la philosophie ni même les sciences de l'information s'en soient aperçu. En ce sens on peut dire que la théorie du pré-savoir (Vorverständnis) absolu se trouve dans la même situation que la physique au moment de son passage de la conception newtonienne à la conception moderne. Newton avait présupposé un Dieu absolu à la base de sa conception de l'espace absolu, pour lequel cet espace faisait fonc-

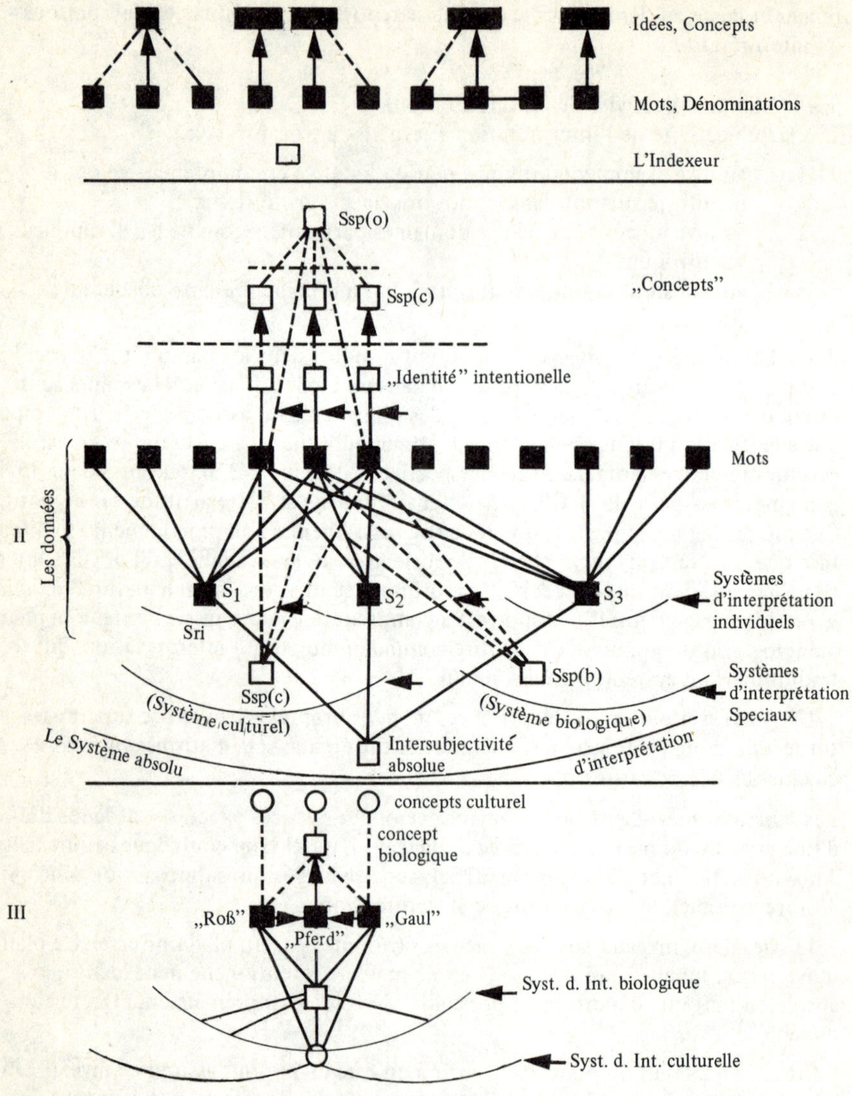
Idées, Concepts
Mots, Dénominations
L'Indexeur
Ssp(o)
„Concepts"
Ssp(c)
„Identité" intentionelle
Mots
II
Les données
S1
S2
S3
Systèmes d'interprétation individuels
Sri
Ssp(c)
Ssp(b)
Systèmes d'interprétation Speciaux
(Système culturel)
(Système biologique)
Le Système absolu
Intersubjectivité absolue
d'interprétation
concepts culturel
concept biologique
III
„Roß"
„Pferd"
„Gaul"
Syst. d. Int. biologique
Syst. d. Int. culturelle

Fig. 2

tion de "sensorium Dei". En conséquence, la "philosophia naturalis" était le savoir de Dieu lui même et comme tel savoir absolu. Savoir absolu aussi bien sur le plan de la valeur que de la totalité de ce savoir. La conception hégélienne de la philosophie comme encyclopédie universelle, qui est le dernier système philosophique, n'a pas d'autres fondements.

Nous avons perdu aujourd'hui tout fondement absolu-Dieu est mort- et nous devons construire un horizon à la fois intentionellement identique et intentionellement absolu, et cela de manière analogue à l'établissement d'une théorie générale et d'une théorie spéciale de la relativité. C'est là une des grandes tâches de la philosophie contemporaine, et c'est dans ce sens que s'inscrivent les remarques précédentes ainsi que nos travaux de philosophie générale.

Cet horizon a sa plénitude matérielle et surtout sa structuration formelle dont nous allons parler dans notre seconde partie.

2. Le problème de l'ordre et de la classification

Dans toute discussion sur l'ordre, il faut distinguer deux problèmes: d'une part le problème d'une théorie de l'ordre en général, et d'autre part l'effort d'établissement d'un schéma universel d'ordre qui remplisse les conditions requises. Il s'agit ici d'une théorie de l'ordre. Les discussions actuelles autour du problème de l'ordre montrent toujours plus nettement que les questions soulevées ne peuvent être résolues que dans le cadre d'une théorie générale de l'ordre. Mais quelques remarques de terminologie s'imposent. Tout d'abord, il nous semble nécessaire de séparer nettement trois plans fondamentaux: le plan de l'ordre, de la division (Einteilung), et de la classification.

2.1 Une définition explicite de l'*ordre* est impossible car elle doit toujours être présupposée comme condition transcendantale de possibilité de l'ordre sous toutes ses formes. L'ordre ne peut être explicité que d'une manière implicite. L'ordre présuppose:

– Un ensemble (fini ou infini) d'éléments susceptibles d'être ordonnés. N'importe quel donné défini peut constituer un élément.
– Les éléments ordonnables doivent être de même type. Cela est surtout valable quand il n'est question que d'objets.
– Les éléments doivent permettre d'établir des relations.
– L'ordre, en tant que résultat d'un "ordonner" dépend toujours d'un principe d'ordre.

2.2 La forme classique de l'ordre est la *division.* Pour la division valent les déterminations idéales suivantes:

– Il faut qu'un principe d'ordre unitaire soit donné.
– La totalité des éléments (donnés ou possibles) doit être totalement saisie.
– Chaque élément doit être ordonné de manière univoque à un domaine.

La conception classique de la division, ébauchée pour la première fois par Aristote, se manifeste dans la conception classique de système telle que, en particulier Kant l'a développée. Pour Kant, un système est un ensemble de connaissances ordonnées suivant des principes. Sur le plan philosophique, cette idée se manifeste en considération de leur système de pensée respectif, dans la "Wissenschaftslehre" de Fichte quant à son aspect formel et dans l'encyclopédie de Hegel quant à son aspect matériel. Ce qui pour Fichte comme pour Hegel est déterminant, c'est la déduction à partir d'un principe premier: la proposition formelle: "Ich bin Ich" de Fichte et l'idée matérielle du "Weltgeist", la proposition spéculative absolue "la substance est sujet", chez Hegel. Bien que ces deux systèmes, dans leur version effective, soient aujourd'hui dépassés, ils n'en conservent pas moins par leur intention et leur développement une grande valeur actuelle; nous y reviendrons par la suite. Qu'il nous soit encore permis d'ajouter que la théorie de la division sert à son tour de base à la théorie classique de la définition. Mais dans ce cas, et beaucoup de critiques modernes l'oublient, il s'agit de la définition de données de fait, et pour cela, c'est la théorie traditionelle d'orientation ontologique qui sert de base (arbre de Porphyre; genus proximum; deferentia specifia).

2.3 Une autre forme particulière de l'ordre est la *classification.* Avant d'étudier ce problème en détail, précisons les termes de "classe" et de "système". L'utilisation de ces termes est très imprécise et bien souvent on les confond, surtout au sujet de la classification où il est question de système de classification (voir Fig. 3, p. 155).

La "définition de Düsseldorf" représente sur ce point une tentative d'éclaircissement terminologique différenciée. Nous entendons par *classe* le groupement d'éléments semblables en raison de leurs propriétés communes. La classe est donc une caractéristique. Elle joue le rôle de concept supérieur par rapport aux éléments qui lui sont sous-ordonnés. Les rapports entre les classes et leurs éléments ainsi que les rapports que les classes entretiennent avec les sous-classes représentent des problèmes particuliers.

Au contraire le *système* est une structure de différents éléments qui entretiennent entre eux des rapports correspondants. La cohésion du système est assurée par un principe systématificateur. Les éléments sont ordonnés au système en fonction de leur place dans la configuration générale. Les classes peuvent bien sûr être des éléments de système.

On peut distinguer de nombreux types de système. Mais dans le contexte actuel, ce sont surtout les systèmes structuraux et génétiques qui retiendront notre attention. Le premier type de système joue le rôle de base structurelle d'un établissement d'un système synchronique; tel que l'exige par exemple l'élaboration d'une taxinomie en biologie moderne. Mais son importance ne saurait être négligée en vue de l'établissement d'un système universel des sciences et du savoir en général. La classification se distingue de la division par le fait qu'elle part en premier lieu des éléments à classifier, et les assemble ensuite dans une structure d'ordre. Des principes de division – aussi vagues soient ils – sont bien sûr présupposés; ceci apparaît clairement dans la classification décimale qui se définit par sa maniabilité et non par sa conformité à l'objet.

Les classifications, au sens actuel du terme'peuvent être considérées comme une mosaïque: on donne des pierres isolées avec lesquelles on élabore une image d'ensemble d'après un plan préconçu. L'utilisation de toutes les pierres est ici une condition sine qua non. Nous croyons qu'une théorie de la classification pourrait être conçue de la manière suivante:

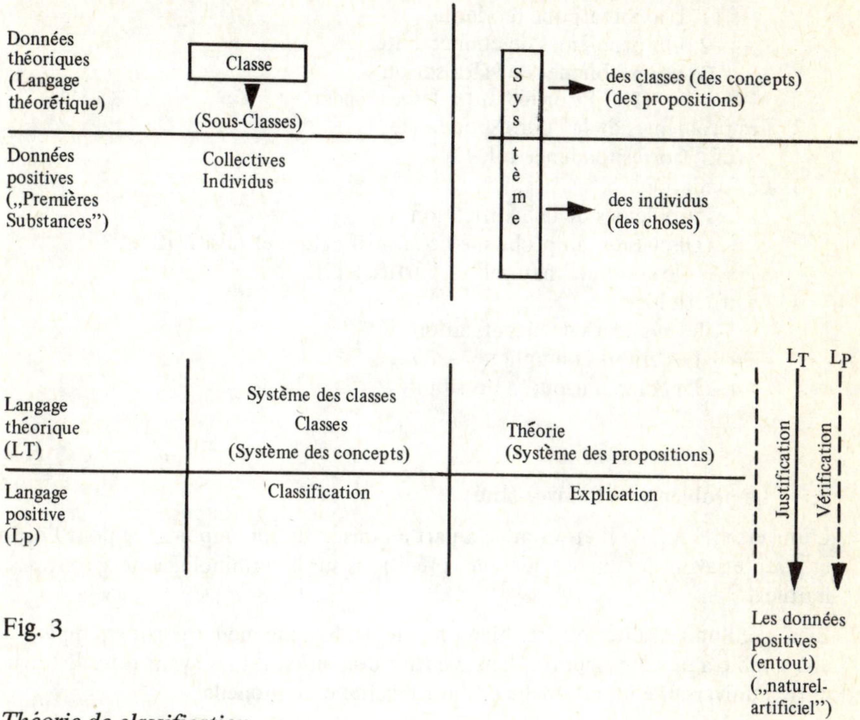

Fig. 3

Théorie de classification

1. Le fondement ontologique
   Problèmes
   1. Le problème de constitution des données
      = le problèmes de l'être
   2. Le système d'interprétation

le système des catégories

le s.d.i. pratique — le s.d.i. théorique

S. de classification

– S. d'explication

2. Théorie de Système

3. Théorie de classification
   comme part d'une théorie des sciences
   (= part des problèmes des données théoriques
   cf. les problèmes de la théorie en sciences)

   1. La logique intérieure
      p.e. la systématification
      Les problèmes d'une axiomatisation
      (1. L'axiomati que moderne
      2. Les problèmes de connectivité
      3. Les problèmes de valorisation
      (La "Priorité" entre les sciences)
   2. Les problèmes de la "correspondance"
      (cf. "Correspondence rules")
   3. La fondation
      Les problèmes de la justification
      (Discussion du problème de la vérification et falsification;
      le système "naturel" – "artificiel")
   4. La critériologie
      des systèmes de classification
      4.1 Les critères (simplicité, . . .)
      4.2 La comparaison; sa possibilité

3. Le problème de l'universalité

L'universalité mérite d'être traitée à part en raison de son importance pour l'ordre et pour le savoir. Ici encore quelques précisions sur la terminologie ne seront pas inutiles.

Bien que l'universalité soit très bien connue, en logique moderne par exemple, elle semble à peu près ignorée dans d'autres domaines. Il faut avant tout distinguer entre l'universalité intentionelle et l'universalité extentionelle.

L'universalité intentionelle est identique à la généralité; c'est l'Universel des classes et des catégories. Et c'est dans ce sens que l'on peut parler d'un problème des universaux.

L'universalité extentionelle par contre est orientée vers la saisie globale de l'ensemble des données aussi bien de fait que possibles. C'est cette universalité qui est au coeur des discussions sur l'élaboration d'un système de classification universel. Nous renvoyons l'historien de la philosophie à la discussion cartésienne du intellegere (intentionel) et du comprehendere (extentionel) de l'infini. C'est évidemment cette universalité extentionelle qui nous intéresse ici. Elle pose le problème des rapports d'une approche de la question sur un plan théorique et sur le plan de l'organisation.

3.1 Il s'agit sur le plan théorique de concevoir un schématisme universel qui permette de saisir tous les savoirs possibles. Un tel schéma devra avoir deux dimensions: une dimension synchronique et une dimension diachronique, dimensions qui résultent du caractère foncièrement historique de tout savoir humain.

Un système universel devrait être systématique en regard des données actuelles, c'est à dire qu'il devrait être structuré de telle manière qu'il puisse saisir la totalité du savoir actuel dans une perspective synchronique. La question de savoir comment ce système devrait être structuré ne sera pas abordée ici.

Mais un système universel doit également avoir une dimension diachronique – génétique. Le savoir, qu'il soit considéré dans sa totalité ou uniquement comme savoir scientifique est toujours défini et structuré historiquement. Ces structurations n'excluent en aucune façon la possibilité de re-structuration, bien au contraire, les "Révolutions scientifiques" (Kuhn) sont un élément vital de la science (voir Fig. 4).

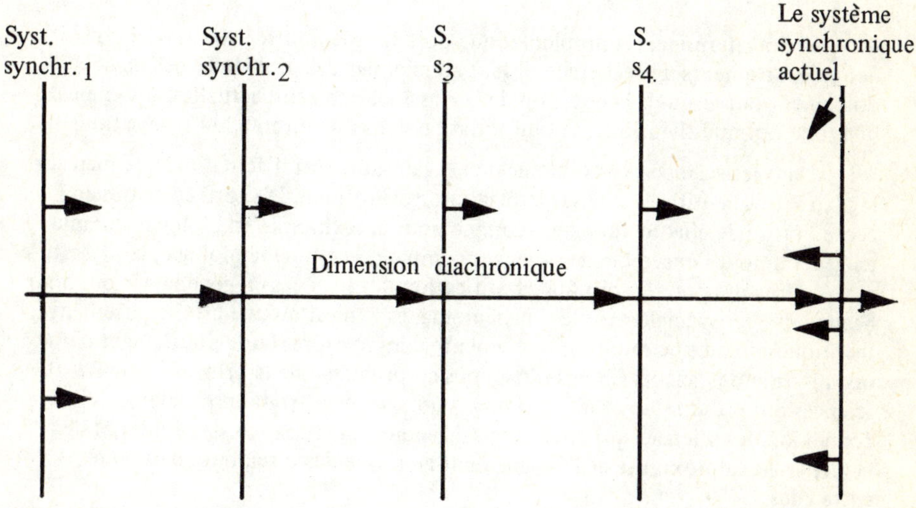

Fig. 4

Cet aspect n'a pas encore été remarqué en ce qui concerne l'ordre et la classification; on ne s'est même pas encore rendu compte à quel point les systèmes de classification actuels sont dépendants du "Geist der Zeit"! On ne devrait pas négliger les profondes remarques de Hegel sur ce sujet. Il ne s'agit pas ici du projet de l'encyclopédie, mais de l'idée hégélienne de l'historicité de tout système. Certes Hegel n'a pas réalisé ce système, mais il a été le premier et peut-être le seul jusqu'à présent qui ait reconnu cette tâche.

3.2 A l'autre pôle de l'universalité théoriquement discutée, nous trouvons l'universalité envisagée sous l'angle de l'organisation. Comme la première discussion vise la systématification et l'unification – donc la totalité – celle-ci s'occupe de la pluralité: Il s'agit de parvenir au détail, rapidement si possible, avec des critères semblables, et ceci toujours en fonction de tout savoir possible. Elle utilise pour cela les moyens les plus variés, et même à l'occasion elle utilise les systèmes de classification. Mais les éléments essentiels sont d'une part les théories d'organisation, comme la théorie du jeu, ou des modèles économiques, et d'autre part, un appareillage correspondant. L'appareil le plus important est bien sûr la calculatrice – qui garantit la rapidité etc.: D'autres appareils tels microfilm ou microfiche peuvent également être utilisés.

3.3 Si l'on compare ces deux aspects de l'universalité il faut se garder de les mettre en opposition, mais bien plutôt les considérer comme nécessairement complémentaires. On peut interpréter cela sous deux aspects: d'un côté comme polarité dans la tension entre l'exigence d'ordre et de saisie totale et la pluralité des données particulières, et de l'autre côte on peut y voir la philosophie comme l'âme de cette entreprise et l'organisation comme le corps.

4. Si pour terminer, la problématique doit être démontrée aux travaux des deux départements de recherche, il faut alors parler des problèmes des deux domaines pour lesquels la question de la classification est d'actualité: les sciences, dans une optique théorique, et l'information dans une optique d'organisation.

4.1 Dans les sciences, le problème de l'organisation est d'actualité aussi bien sur le plan intrascientifique que sur le plan interscientifique. D'abord en ce qui concerne l'ordre de chaque domaine, comme pour la taxinomie en biologie, comme en général pour l'ensemble des sciences comparées comme la biologie, la littérature etc. . . . Ensuite pour établir une classification des sciences en général. Ce qui pour beaucoup – et spécialement les épistémologues – n'est qu'une affaire purement théorique, prend une importance éminente dans le cadre d'une politique et d'une organisation des sciences; nous renvoyons au problème de la priorité entre les sciences qui présuppose avant tout une typologie, et surtout au problème de la connexion des sciences qui rend nécessaire pour l'organisation des études et de la recherche la proximité et l'éloignement ainsi que des disciplines particulières entre elles.

Trois plans thématiques semblent appropriés:

4.1.1 Le premier plan est un plan purement épistémologique. A ce plan on essaie une typologie des sciences en fonction de la constitution de leur objet; on peut distinguer trois types:

- orienté vers le domaine physique
- les sciences du comportement
- les sciences des données culturelles

4.1.2 A l'autre pôle nous trouvons des préoccupations scientographiques qui visent à établir une carte topographique pour l'ensemble des régions du savoir permettant de mettre à jour des points, des regions d'établissement possibles

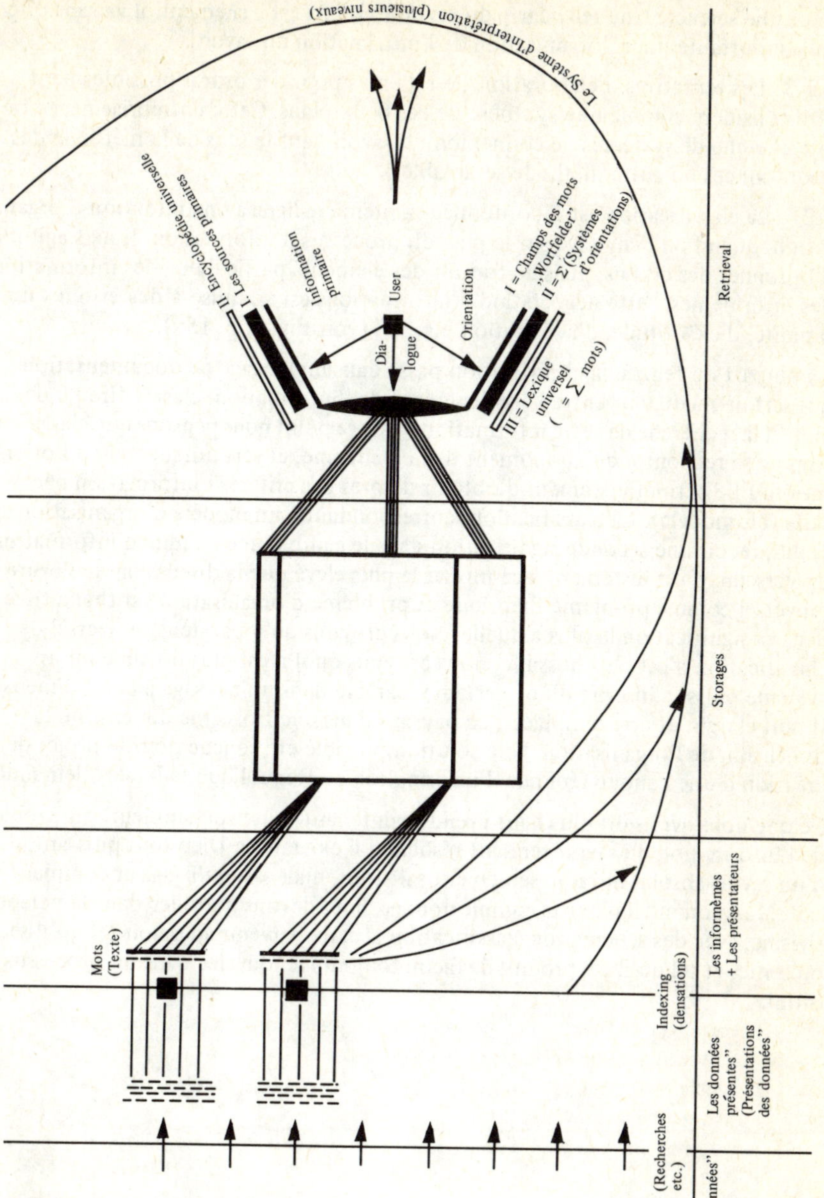
Le Système d'interprétation (plusieurs niveaux)
I. Les sources primaires
II. Encyclopédie universelle
Information primaire
User
Dia-logue
Orientation
I = Champs des mots
„Wortfelder"
II = Σ (Systèmes d'orientations)
III = Lexique universel (= Σ mots)
Mots (Texte)
Retrieval
Storages
Les informèmes + Les présentateurs
Indexing (densations)
Les données présentes" (Présentations des données"
„Les données" (Recherches etc.)

Fig. 5

pour une science. Une telle carte du savoir existant est également, il va sans dire, très importante pour l'organisation de l'information du savoir!

4.1.3 Les tentatives d'élaboration de schémas pour des ordres possibles peut être considéré comme une synthèse de ces deux plans. Cette entreprise nécessite la recherche de systèmes de connexion du savoir dans le sens de la méthode des coordonnées ou autre méthode semblable.

4.2 La classification est subordonnée en première ligne à une intention d'organisation, quand on l'envisage sur le plan du processus d'information. Il ne s'agit plus d'ordonner des classes, mais de fournir des éléments particuliers, des informations, des informèmes; cette négociation d'informations est soumise à des critères de rapidité, d'exactitude, d'adéquation etc. . . . (voir Fig. 5, p. 159)

La plupart du temps, la classification passe dans un langage de documentation qui sert de médiateur entre l'établissement de l'information, c'est à dire l'indexation et la recherche de cette information! Sur ce point nous pensons que classification ne se rencontre qu'au moment de la recherche, et sert uniquement à l'orientation. Elle permet justement d'obtenir d'après des critères l'information nécessaire (cf. modèle). La classification, correspondant à un modèle d'organisation établi, reçoit une seconde signification dans le cadre d'une banque d'information universelle. C'est justement à ce niveau le plus élevé que la discussion de l'ordre universel comme problème théorique et problème d'organisation est thématisée dans sa signification la plus actuelle. Nous croyons qu'un système universel de classification n'est plus possible, nous croyons qu'il n'est plus possible qu'un système puisse s'inscrire d'une certaine manière dans un paysage à deux dimensions. Il faut élargir, et non remplacer, ce paysage dans une troisième dimension, la dimension de l'organisation. Une solution possible été conçue pour le projet de création d'une banque fédérale d'information en République fédérale allemande.

Ce que nous avons dit plus haut prend maintenant toute son ampleur: les problèmes de l'information du savoir seraient résolus si il existait un Dieu tout puissant, doué d'un savoir absolu, qui non seulement sait tout, mais sait qu'il sait et comment il sait. Mais comme il n'existe comme donnée, nous devons travailler dans le perspectivisme, avec des schémas de classification afin de parvenir au savoir tel qu'il se présente, et tel qu'il est produit de façon toujours renouvelée dans un processus infini.

R. Mölgaard-Hansen (Copenhagen)

On the problem of universality in knowledge classification

The philosophers feel that the theme "Conceptual Basis of the Classification of Knowledge" should be discussed in the light of the present state of the problem, mainly because a satisfactory general classification of knowledge and an adequate clarification of the philosophical problems underlying such classification are lacking. Clarification is important from theoretical as well as practical point of view because the general systems of classification of knowledge depend on theories of knowledge. This paper, therefore, contributes a survey of the relevant aspects of the experiences of an electrical engineer and information scientist, obtained through work on the international level within the field of building industry during the 1950s (International Building Classification Committee, IBCC), and the field of general classification during the 1960s (FID/CR Committee on Classification Research). For the sake of limitation the presentation and the references will almost exclusively be concerned with the activities of these two bodies.

Natural *vs.* arbitrary classification systems

Already in the beginning of the 1930s, Bliss (1) made the distinction between "natural" and "arbitrary" classification systems. In his opinion the aim of the speculation of philosophers and the experiments of scientists is continuously to refine schemes which are believed to "explain" natural phenomena, and, therefore, to be true or "natural". An illuminating example is Niels Bohr's scheme of the periodic system of chemical elements, so true in its power to explain a wide range of fundamental causalities in physics and chemistry.

Opposite to this the classification schemes, designed by classificationists, i.e. librarians and documentalists, with the aim to bring scientific literature in a helpful order for the sake of effective retrieval, may be called "arbitrary"; they have a pure, practical purpose, and must be adjusted to a number of special conditions immanent in library techniques. Nevertheless, there is a connection. As stated by Bliss the order of the classes in these systems should always aim at being "consistent with the consensus of scientists of to day". It was ever so, and it still holds true that the scientists who produce the literature and formulate the titles of publications also present their queries to the libraries. Therefore, it is a fact that the collocation of basic subjects in the main structure of universal, bibliographic systems, always was based on the dominant philosophy of the day.

Practical information people have only a vague impression of the contributions of philosophers and scientists to the classification of knowledge. Ranganathan (2) has outlined the history of classification as a philosophical problem much more than two thousand years back, mentioning the Vedic, the Greek, the Medieval, the Baconian, and other systems; it is emphasized that these systems only present main structures of knowledge in accordance with a certain philosophy, and do not go into any details like modern bibliographic systems. A fruitful discussion between documentalists and philosophers directly also presupposes a knowledge of each others terminology, thus Lewis' works, e.g. "Mind and the world order" (3) ought to be studied by the information people.

For the following it is necessary to stress a couple of contributions from special science philosophy to epistemology in general. We know how deeply the development of modern physics, both Einstein's theory of relativity and the quantum mechanics is influencing our conception of the "world order". To our subject it is useful to be reminded by Heisenberg's "Uncertainty Relation" and Bohr's "Principle of Complementarity" (PC-principle) that the analysis of a continuum of phenomena by discrete units (measuring units, concepts) probably always ends in a paradox. The value of the PC-principle is perhaps doubted by many philosophers, but is now generally accepted by physicists. The lesson of this as regards the classification of the universe of knowledge is that the dualism between the subjective and the objective world is real, and must be fundamental in a modern scheme, just as in the human intellect which apparently organizes the chaos of sense impressions in a cosmos of concepts.

When the philosophers aim at a general classification of knowledge, i.e. to present today's "natural" main structure of knowledge, it must be based on a consistent philosophic view concerning the relations between the fundamental categories. It is realized that a great obstacle to this stems from the inevitable specialization which takes place within all scientific disciplines, and makes it more and more difficult to separate and define classes which are mutually exclusive.

The development of universal library classification usually dates back to 1876 when Dewey published his Decimal Classification (DDC). Introducing the decimal notation and the idea of a "relative index" he displayed the main structure of knowledge in 1000 classes. As is wellknown, a number of other systems developed in the course of time along with the practical and theoretical needs of the library world: the Universal Decimal Classification (UDC), the Library of Congress Classification (LC), Bliss' Bibliographic Classification (BBC), and Ranganathan's Colon Classification (CC). The need for comparative research became apparent about 1950 when Ranganathan founded and, until 1962, chaired the FID/CA Committee on Classification Theory; a survey of this work is given by de Grolier (4). In recent years big computer-based projects like the LC centralized cataloging project (MARC) and a world wide information network (UNISIST) have stimulated the need for research. Relative valuation of the systems, DDC and LC, UDC and CC, was carried out and many papers have been published, e.g. by Angell (5), Birket-Smith (6), Mölgaard-Hansen & Rigby (7), Neelameghan (8), Ranganathan (9), and Richmond (10).

One more universal classification system should not be forgotten, especially in these days when the computer so strongly accentuates the importance of special thesauri as retrieval tool. Credit as the first pioneer should perhaps be given to P. Roget who in 1851 published his "Thesaurus of English Words and Phrases", classified in a structure of 1000 classes. If the formulation of a main structure of the universe of knowledge consistent with the consensus of scientists today should be tried out, it could formally be restricted to a "third" edition similar to Dewey's and Roget's 1000 classes. Such a proposal was actually made privately in 1966, including the idea that the new scheme should be published in 1976 on the occasion of the Dewey Centenary. To do this might be a suitable task, if a Research Center for the Classification of Knowledge is to be established.

The consideration of natural versus arbitrary classification systems seems to raise the following questions. Is a collocation of basic subjects as main division of a universal classification possible today on philosophical grounds? How should it look, and in which respects would it differ from those, already well known to us in existing universal systems?

General versus special classification systems

Following the Dorking Conference in 1957, this theme was thoroughly discussed at Elsinore, 1964, and published in the Proceedings, pp. 65–100. Today information retrieval covers a wide spectrum, dealing with publications denominated by single subjects, compound subjects, and complex subjects. Depth classification, therefore, has been a keyword, and the extensive subdivision of the universal or general, hierarchical systems needed, has created difficult problems. The rigidity and lack of flexibility involved in this just make the information specialists doubtful as to the usefulness of macro-hierarchical structures. Nevertheless, the state of main divisions of the general systems, be the order Baconian, inverted Baconian, or inspired by the 19th century philosophy, is still a main concern with philosophical background for all those, who deal with current amendment and adjustment of the systems to modern conditions.

If you want to locate a place in a town, it is most easy to look at a map, Perreault says, but you can also ask your way at each streetcorner, and avoid any map. Now, the computer works so fast that the development today is dominated by the second approach (i.e. to avoid the map) and the interest in classification consequently decreasing. For the use in "coordinative indexing" an increasing number of thesauri, covering special subject fields, is being collected at the Bibliographic Center at Cleveland (English language) and at CIINTE in Warsaw (other languages). Nevertheless, the need for systematic order cannot be ignored, and a most urgent task is to create a general framework of knowledge ("roof classification" or "umbrella classification") to serve as a switching system for large tape collections (databanks).

In the 30-ties, Ranganathan introduced the coordinative method, used in punchcard and computer systems, known in the library world under the name of "Facet-classification". Today, the Colon Classification (CC-system) is a free-faceted, general classification, resting on a consistent set of principles and postulates, forming a true general classification theory. The CC has a main division of canonical classes, disciplines, and fields of activities, but complex subjects are located consistently by combining uniterms (isolates) from different facets according to the citation order of the fundamental categories, PMEST (Personality, Matter, Energy, Space, and Time). Ranganathan (11) postulates that the distinction between general and special schemes is a fallacy, because an Indexing Language like the CC covers equally systematic location of complex subjects on all levels.

The opposite view, the possibility to create a general classification out of special subject field systems, is advocated by Dobrowolski (12) who says "One could imagine that, after having created autonomous classifications for all special scientific fields, one could then reunite them to form a single systematic scheme, having as skeleton the classification of sciences". Within the field of welding Dobrowolski finds that only 25% are core-subjects on welding as such, but 75% are fringe subjects from other parts of science and technology.

The number of special field systems amounts to thousands. Most of these are enumerative (hierarchical), but today many are facet-classifications. An investigation made by the author (13) in 1959 of more than 50 systems on the domain of the building industry showed that each of the 5 most important systems principally reflected one of the well known general systems. Therefore, it is evident that just as the designers of general systems look for guidance from the philosophers, the designers of special systems copy the principles of the universal systems. Another experience related to the problem in question is expressed by Foskett and Sabel, namely that the specialists find the general systems, e.g. UDC, and the most comprehensive thesauri, (e.g. E.J.C. or TEST) useful outside their own subject field, but unsatisfactory as regards the presentation of core-subjects.

Theoretically, it is a main issue whether special systems should be derived from a general classification, based on a consistent theory of classification (Ranganathan), or whether the universal classification should be derived by synthesis of special systems (Dobrowolski).

From a retrieval point of view the British school advocates that a general framework of knowledge is not needed, e.g. as expressed by J. Mills at Elsinore, "There is no reason to think of a general classification as something which must be physically embodied on 5x3 cards. It may conceivably be in the background of a post-coordinate system as a programming instrument", or even possible, as Fairthorne argues, "Although the vague and naive idea of a 'general' classification must be abandoned, we can design attainable and useful compromises". This viewpoint also dominates the British memorandum on classification to the organizers of the UNISIST project.

Since the late 1950s members of the British Classification Research Group (CRG) designed a number of special faceted classifications, based on the Ranganathanian principles. The fundamental categories were accepted, but specified or extended to the socalled "Standard Citation Order" by B. C. Vickery. Nevertheless, when it came to the problem of a general, faceted classification the CRG could not accept, as in the CC-system, a main division of canonical classes. At the Montague Conference in 1963 the CRG introduced the idea of developing a general classification, based exclusively on concept-combinations and the "Theory of integrative levels". Nevertheless, this is a project which apparantly involves many difficulties.

Facet classification and coordinative indexing have during the 1960s greatly. stimulated the research on specification of categories of attributes and relationships, denominated as role-indicators, operators, syntagmata or relators,etc. Perreault (14) has presented, besides his own classification of relators, a survey of such categories, listed by Aristotle, Lull, Kant, Mills, Costello-Wall, E.J.C., W.R.U., Pages, Leroy-Braffort, Kervegant, Gardin and Farradane. (See p. 170 + 171 Appendix 1–2)

Actually, the Ranganathanian analytic-synthetic method of classification was introduced in Europe when a Swedish building committee, 1945–1948, worked out the faceted scheme for building production, named the SfB-system (15). As chairman of IBCC, L.M. Giertz succeeded in 1959 in getting the system internationally authorized by the Conseil International du Bâtiment (CIB), and it is now in use in more than 20 countries. It is daily run on computer in several countries, exemplifying the capacity of a special facet-scheme in handling documents on quantity-, cost-, and operational analyses of the building production.

In practice a faceted classification must be simple, i.e. limited to the core-subjects of a well defined special field of activity, e.g. building production. For the handling of the numerous fringe subjects belonging to the building industry as a whole, the SfB had to be supplemented by a special edition of the UDC, namely the Abridged Building Classification (ABC), published by the IBCC in 12 languages for the CIB abstract exchange between 18 countries. In 1959 IBCC generalized this solution of the problem in the following way (See ref. 13)"Firstly, all of the subject fields may be displayed in a hierarchical scheme consistent with the organization of knowledge established in the consensus of scientists and educationists. Secondly, in a complementary way, each of the subject fields may be detailed displayed in special, faceted classifications". Essential in the formulation of this, socalled "Principle of Complementarity", is that it exclusively deals with a combined use of the analytical, hierarchical method and the synthetic, faceted method of classification. When Ranganathan has realized that "the facets do not belong to the subject", and has based the Colon Classification on a hierarchical scheme of canonical classes, specified by facet combinations, it may be said that the PC-principle in a way is inherent in his system, while the British CRG viewpoint fully denies the principle. It is, therefore, symptomatic that Mills (16) in the latest, national edition, the Construction Index SfB (CI/SfB) has extended the number of facets, and made the system autonomous. The use of UDC is thereby avoided.

The trends toward integration of information networks makes simultaneous use of different filing systems, whole or in part covering the same subject field, very essential. Most of the quiries in daily library work require searches in indexes with quite different filing order. For the sake of rationalization and standardization increase of the compatibility between indexing languages is urgent. In this respect Angell (17) has investigated a number of the leading descriptor-languages. Simple, supplementary use of one or more indexing languages is often applied to cope with the problem, e.g. in the MARC project, ISO standards, the Japanese JICST system, subdivided by UDC, or the Israel Atomic Center's combination of UDC with descriptor indexing. Nevertheless, it is intended in the following to demonstrate that a PC-principle in classification has a more distinct meaning.

Attributional *vs.* relational axes of the universe of knowledge. "The Facet Globe"

Four approaches from classificationists to the problem of a universal classification of knowledge have been presented:

1) Ranganathan's, 2) Dobrowolski's, 3) The CRG's, and 4) The IBCC's. Of these, Dobrowolski suggests a complete hierarchical procedure, and the CRG on the contrary a complete synthetic procedure. Ranganathan and IBCC work from both ends, suggesting a hierarchical structure of basic subjects, complemented with pure synthetic procedures.

In 1959, following the idea of the PC-principle, the possibility of making a classification of classifications was considered, i.e. to display all special systems over the surface of the Globe, applying the geodetic, notational system of longitudes and latitudes which is equally universal with the decimal notation, and much more hospitable. Progress was made during the 1960s, and a model of a "Facet Globe", displaying the relations between the fundamental categories of the universe of knowledge now seems to be possible.

It is a great help to the presentation that G.A. Borden and W.F. Nelson (18) in 1969 published their theoretical considerations toward a viable classification scheme, based on current psychological, philosophical, and rhetorical theories. It appears that L.S. Vygotsky (19) already in 1962

> "compared conceptualization of knowledge with a totality of concepts distributed over the surface of the Globe. The location of each concept is defined by a system of coordinates corresponding to longitude and latitude in geography. Longitudinal ordinates include the objective content of a concept, while latitudinal ordinates represent the acts of thought apprehending content. Their intersection determines all the relationships of a given concept to others – its coordinate, superordinate, and subordinate concepts."

Borden and Nelson hold the view that the structure of knowledge is affected by the structure of the human cognitive system through which it is conceptualized. It seems to be generally acknowledged, that at least two axes continually appear, one accounting for the objective content of a concept (attributional qualities), and one accounting for the acts of thought (relational qualities). Attributes and relations are expressed by very abstract, superordinate terms which are neutral classifiers for conceptual processes. "They are not ideas: they are areas from which ideas are formed". The authors formulate the following conclusions:

1) Knowledge factors according to attributional and relational axes.
2) Contents of both attributional and relational axes are hierarchically organized.
3) A system of superordinate terms identifying content clusters along attributional and relational axes is identifiable.
4) Variance in superordinate terms is minimal and relatively unaffected by time, space, and cultural influences.

Based on a list by Wilson and Arnold (20) of 11 attributes (which includes and specifies the five, fundamental Ranganathanian categories, PMEST), and 5 relationships, e.g. causality, correlation, and genus-species relationships, the authors, moreover, make a first attempt to concretize the idea of Vygotzky by placing the said attributes as longitudinal superordinates, and the relationships as latitudinal superordinates.

Now, the conception that these neutral attributes and relationships are parallel with their own kind but orthogonal to each other is a very elucidating illustration of fundamental, categorical relationships within the universal of knowledge.

Our idea of a "Facet Globe", although similar to Vygotzky's, is less abstract, based on practical documentation work as it is. Here, the orthogonality illustrates the classificatory "Principle of Complementarity". The collocation of classes along longitudinal ordinates is imagined as a main structure of the universe of knowledge, displaying the basic subjects of science (disciplines) and practical fields of activities, etc. Notationally they may be separated in an East and a West section. The collocation of attributional qualities along latitudinal ordinates presents first of all the basic dualism between object and subject, i.e. between the physical concepts, representing the outside world (the southern hemisphere), and the mental constructions, representing the inside world (the northern hemisphere). Thus, from the South Pole, indicating physical force, attributes like space/time, energy, matter, artefacts, make an open ended scheme towards equator, following the theory of integrative levels. Now, from the North Pole, indicating psychical forces, attributes denominating mental activities (mentefacts) could be presented. For the collocation of mental categories it might be proposed to draw on the recherches as regards the classification of human needs, made by the philosophers of needs. Let it be hypothetical if a division, e.g. according to spiritual, psycho-genic, and physiological needs would be feasible. In any case a development on these lines would greatly prevent the criticism of the CRG's use of the theory of integrative levels which concerns physical categories, and does not give equivalent room for mental constructions. (See ref. 10).

In this way an infinite, but nevertheless closed model of the universe of knowledge possibly could be realized. It appears that the subjects covering the practical, human activities will be distributed over the tropical zone, while more and more abstract, scientific and philosophic subjects will be placed in the temporal and arctic zones.

The realization of a modern, universal classification of knowledge, based on philosophic grounds, is important, not only for the world of documentation, but also as a means against the effects on the society as a whole of the atomization of knowledge, caused by the ongoing process of specialization within science and technology. Therefore, promotion of research on this domain is urgent.

Bibliographical References

1) H.E. BLISS, Organisation of knowledge in libraries and the subject approach to books. 1933

2) S.R. RANGANATHAN, Library classification through a Century. (Elsinore Proceedings, Editor: Pauline Atherton. Published by FID/CR Committee on Class. Research and Danish Centre for Documentation. Copenhagen 1965, pp. 15–35)

3) C.I. LEWIS, Mind and the world order; outline of a theory of knowledge. New York 1929 (with corrections 1956)

4) E. de GROLIER, A study of general categories applicable to classification and coding in documentation. Publ. by UNESCO. 1962. 248 p.

5) R.S. ANGELL, On the future of the Library of Congress classification. (Elsinore Proceedings. Copenhagen 1965, pp. 101–112)

6) K. BIRKET-SMITH, Local applicability of the Library of Congress classification. A survey with special reference to Non-Anglo-American libraries. Copenhagen 1970. 69 p. (FID/CR Report Series, No. 10)

7) R. MÖLGAARD-HANSEN & M. RIGBY, Proceedings of the First Seminar on UDC in a mechanized retrieval system. Copenhagen 1969. (FID/CR Report Series, No. 9)

8) A. NEELAMEGHAN, Classificationist and the study of the structure and development of the Universe of Subjects. Copenhagen 1967. 22 p. (FID/CR Report, Series, No. 5)

9) S.R. RANGANATHAN, Choice of scheme for classification. (Library Science with a slant to documentation, *5*, 1968, 1, pp. 1–69)

10) Ph. RICHMOND, Contribution toward a new generalized theory of classification. (Elsinore Proceedings. Copenhagen 1965, pp. 39–64)

11) S.R. RANGANATHAN, General and special classifications. (Elsinore Proceedings. Copenhagen 1965, pp. 81–93)

12) Z. DOBROWOLSKI, Analysis of classification systems. (Elsinore Proceedings. Copenhagen 1965, pp. 65–80; and: FID/CR Report Series, No. 2 (in French))

13) R. MÖLGAARD-HANSEN, A study on building filing systems. (IBCC Report No. 2; in: Recent Developments in Building Classification, IBCC, Reports 1–5, Bouwcentrum, Rotterdam, 1959)

14) J.M. PERREAULT and J.C.G. WESSELING, On the Perreault Schema of Relators and the rules of formation in UDC. Copenhagen 1966, 20+26 p. (FID/CR Report Series, No. 4)

15) SfB building register and its application. Ed. by the SfB, Coordination Committee for the Building Trade, Sweden. Report of June 11th, 1948. 18+72 p. (Prepared by the Secretary, L. M. Giertz)

16) J. MILLS, Progress in documentation. Library classification. (Journal of Documentation, *26*, 2, 1970, pp. 130–160)

17) R.S. ANGELL, Two papers on thesaurus construction:
a) "The language of term relation designations in subject access vocabularies"
b) "The specific-to-general see reference in thesaurus construction".
Copenhagen 1968. 27 p. (FID/CR Report Series, No. 8)

18) G.A. BORDEN and W.F. NELSON' Toward a Viable Classification Scheme: Some Theoretical Considerations .
(American Documentation, Oct. 1969, pp. 298–301)

19) L.S. VYGOTSKY, Thought and Language. John Wiley & Sons, New York, 1962, pp. 112–113.

20) J. WILSON and C.C. ARNOLD, Public Speaking as a Liberal Art. Allyn & Bacon, Boston, 1962, pp. 103–106

Appendix 1

Systematic tabulation of Perreault' Relators

+	ALTERNATION, Relation in General
:	CONJUNCTION, Logical Relation
a	Reciprocal
b	Converse
c	Negative
d	Subsumptive
dd	Type/Kind
ddd	Principle/Manifestation
dde	Genus/Species
ddf	Species/Individuum
de	Whole/Part
ded	Organism/Organ
dee	Composite/Constituent
def	Matrix/Particles
df	Subject/Property
dfd	Substance/Accident
dfe	Possessor/Possession
dff	Accompanance
dffg	Presence; With
dffi	Absence; Without; Lack
e	Determinative
eg	Active
egg	Productive
eggd	Causing
egge	Originating; Source
eggf	Influencing; Environmental pressure; Catalytic
egh	Limitative
eghd	Restrictive
eghe	Orienting; Establishing goals or applications
eghf	Frame of Reference; Point of view (active)
egi	Destructive
egid	Injuring
egie	Suppressing; Eliminating
egif	Curing
eh	Interactive
ehg	Concordant
ehgd	Association; Community; Symbiosis
ehge	Imitation; Simulation
ehgf	Cooperation (conscious)
ehh	Differing
ehhd	Borrowing; Parasitism; Instrumentality; Utilization
ehhe	Barrier; Exclusion
ehhf	Distinction; Withdrawal
ehi	Contrary
ehid	Attack; Aggression
ehie	Competition; Antipathy
ehif	Resistance; Defence
ei	Passive
eig	Produced
eigd	Effected; Product; Independent Variable
eige	Derived; By-Product; Dependent Variable
eigf	Influenced; Catalyzed
eih	Limited
eihd	Restricted
eihe	Applied; Oriented
eihf	Frame of Reference; Point of View (passive)
eii	Destroyed
eiid	Injured
eiie	Suppressed; Eliminated
eiif	Cured
f	Ordinal
fd	Conditional
fdd	State
fddg	Necessary
fddh	Contingent
fddi	Arbitrary
fde	Attitude
fdeg	Favorable
fdeh	Indifferent
fdei	Unfavorable
fdf	Energy
fdfg	Potent; Capable
fdfh	Latent; Virtual
fdfi	Impotent; Incapable
fe	Comparative
fed	Degree
fedg	Superior; More
fedh	Equivalent
fedi	Inferior; Less
fee	Size
feeg	Larger
feeh	Equal in size
feei	Smaller
fef	Duration
fefg	Longer; More Durable
fefh	Equally Durable
fefi	Shorter; Less Durable
feg	Identical
feh	Similar; Analogous
fei	Dissimilar
ff	Positional
ffd	Figurative
ffda	Outside; Transcendent
ffdb	Parallel
ffdc	Inside; Immanent
ffdg	Near
ffdh	Between
ffdi	Far
ffe	Spatial
ffea	Lateral
ffeaa	Right
ffeab	Middle
ffeac	Left
ffeb	Axial
ffeba	Front
ffebb	Center
ffebc	Back
ffec	Vertical
ffeca	Above; Upon
ffecb	Level
ffecc	Below; Under
fff	Temporal
fffa	Prior; Before
fffb	Simultaneous; During
fffc	Posterior; After
ffg	Toward
ffh	At
ffi	Away

Appendix 2

ARISTOTLE

substance
accident
- quantity
- quality
- relation
- place
- time
- position
- state
- action
- passion

LULL [12, 13]

general questions
- possibility
- definition
- materiality
- formality
- size
- quality
- temporality
- locality
- modality
- instrumentality

relative principles
- difference
- concord
- contrariety
- beginning
- middle
- end
- majority
- equality
- minority

KANT [14]

quantity
- unity
- plurality
- totality

quality
- reality
- negation
- limitation

relation
- inherence/ subsistence
- causality/ dependence
- community

modality
- possibility/ impossibility
- existence/ non-existence
- necessity/ contingency

MILLS [15]

- whole thing
- kinds
- parts
- materials
- processes
- properties
- agents
- operations

COSTELLO-WALL [16]

1, application
2, cause
3, matter
4, means
5, medium
6, by-product
7, product
8, research
9, dependent variable
10, design
11, processed (passive)

E.J.C. [17]

1, matter
2, product
3, by-product
4, application
5, environment
6, cause
7, effect
8, major topic
9, passivity; location
10, means
0, bibliographic data

W.R.U. [18]

A/E, categoric intrinsic
I/O, inclusive/ aggregate
Q, affected
U, productive
W, instrumental
X, negative
Y, attributive
Z, simulative

PAGES [19]

a, *order in general*
ab, equal
ac, prior
ad, intermediate
af, first
ag, last
e, *determinative*
eb, cause
ec, influence
ed, source; origin
ef, suppression; injury
eg, frame of reference
i, *concrete*
ib, simultaneous
ic, means
id, barrier
if, aiding
ig, supply; transfer
ij, competition
il, aggression
im, attack
in, resistance
o, *capacity*
ob, high cap.
oc, average cap.
od, low cap.
u, *reciprocity*
ü, *converse*

LEROY-BRAFFORT [20]

action
A, relation to
B, for the purpose
L, location
M, by means of
R, results in

KERVEGANT [21]

:, *relation in general*
→1 →, **appurtenance**
→11 →, inclusion; implication
→12 →, parts; organs
→13 →, constituents
→14 →, properties
→141 →, physical pr.
→142 →, chemical pr.
→143 →, biological pr.
→15 →, aptitudes; predispositions
→2 →, **process**
→21 →, action
→211 →, favorable ac.
→212 →, unfavorable ac.
→2121 →, retardation
→2122 →, inhibition
→2123 →, destruction
→21 ←, interaction
→211 ←, favorable i.a.
→212 ←, unfavorable i.a.
→22 →, operation; product
→3 →, **dependence**
→31 →, causality
→32 →, origin
→33 →, conditions
→31 ←, correlation
→32 ←, association
→33 ←, combination
→4 →, **orientation**
→41 →, aspect
→42 →, application
→43 →, utilization
→5 →, **comparison**
→51 →, resemblance
→511 →, analogy
→512 →, equality; identity
→52 →, non-resemblance
→521 →, difference
→522 →, opposition
→0 →, negation of the relation

GARDIN [22, 23]

- subject
- object
- qualifier
- location
- instrument

SYNTOL words
- predicates
- entities
- states
- actions

SYNTOL relations
- coordinative
- consecutive
- associative
- predicative

FARRADANE [24]

- concurrence
- comparison
- association
- equivalence
- state; dimension
- appurtenance
- distinctness
- reaction
- causation

William R. Shea (Montreal)

The classification of scientific terms as "theoretical" and "observational" in contemporary philosophy of science

The mainstream of the philosophy of science in the second quarter of this century – the so-called "logical empiricist" or "logical positivist" movement – made two basic assumptions. First, it classified the terms occuring in scientific discourse as "theoretical" or "observational", and, secondly, it took for granted that theoretical language is dependent upon observational language and can be eliminated by explicit definition.(1) Lately, these assumptions have been queried, and in the present paper some of the reasons why the theory-observation dichotomy has been criticized will be discussed.

The empirical component of logical positivism is best understood as a development of British empiricism. To the question: "What do theories describe? ", seventeenth-century empiricists would have answered: "They describe the behaviour of concealed entities which are known by means of aids to the senses, such as the microscope, or are inferred from their observable effects". Ordinary language was considered adequate to describe these small entities, which were held to obey the laws of newtonian mechanics. Locke's misgivings about the possibility of gaining actual knowledge of causes that were completely concealed beneath the appearances does not seem to have checked their optimism. Assuming that unobservable entities can be known, increasingly elaborate accounts were produced which purported to describe entities of increasing hiddenness and more and more recondite properties. Their outcome can be seen in the contradictions of aether theories in the nineteenth century.

The collapse of metaphysical mechanism and the bizarre characteristics of modern theoretical physics marked the end of this headlong progress of uncritical realism. By the same token, it compelled philosophers of science to ponder how familiar language can be used to describe hidden entities. The logical positivists met this question "by retreating to the firm basis of the observable, where there was thought to be no doubt as to what we are talking about, what our language means, or how we know our assertions to be true".(2) In other words, the question of *meaning* (What do theories mean?) and the question of *acceptability* (How do we know what theories are true?) were believed to hinge on the question of *reference* (What are theories about?) which, in turn, was taken to raise no difficulty. This methodological assumption was embodied in the basic classification of scientific terms as self-evident "observation terms" and derivative "theoretical terms". This constituted the *empirical* component of logical empiricism.

The logical component consisted in the reliance on mathematical logic for formulating the problems of the philosophy of science, often rechristened "the logic of the sciences" to emphasize its two main characteristics. First, just as formal logic, since Aristotle, deals with the form rather than the content of propositions and statements, so philosophy of science should concern itself with the logical form of scientific statements rather than their content. The task of the philosopher, interpreted in this light, became the representation of scientific expressions in general, leaving to the practising scientist the job of confronting his conclusions with actual scientific procedure. The following important advantages were said to accrue from the adoption of the analogy of formal logic: (a) disengaged from the specific tenets of particular scientific theories, philosophy of science acquired immunity from the vicissitudes of change and the overthrow of current beliefs, and (b) it became possible, in principle at least, to outline the characteristics of all possible explanations, and, by the same stroke, the characteristics of all future explanations. This was one of the most attractive aspects of the logical empiricist programme: it appealed to the administrative machinery of our minds, accustomed to reduce understanding to linear arguments and dichotomous classifications capable of development by deductive logic.

The second characteristic of the "logic of the sciences" was its conception of scientific theories as formal systems, axiomatizable with the aid of the techniques of mathematical logic. The philosopher of science concentrated on perfected or idealized systems, and took for granted that science grew by the automatic incorporation of earlier hypotheses into later theories as special cases applicable in limited domains of experience.

Briefly, the empirical and the logical aspects of logical empiricism can be expressed in the following two tenets: theoretical terms are (a) grounded in self-evident observation terms, and (b) open to exhaustive manipulation through modern mathematical logic.

This unquestioning faith in the truth and obvious meaning of observational predicates was really "a cover for a much more deeprooted belief, stemming from the British empiricist tradition, that there are entities (and their properties) which are directly given in perception and that there are more or less well-defined areas of language which directly describe these".(3) The positivist distinction between "observation" and "theory" is a direct lineal descendant of the distinction between "impressions" and "ideas" formulated in the seventeenth and eighteenth centuries as a criterion of meaning and acceptability of concepts. The "observational-theoretical" dichotomy inherited the function of its predecessor but with three refinements Frist, an attempt was made to avoid the misleading psychological overtones of the word "ideas" by speaking of "terms" and "statements". Secondly, Hume's problem of the origin of such terms was eschewed by justifying the acceptance or rejection of theories on the grounds of their "definability" and "reducibility". Finally, the new distinction was employed to tackle a problem to which the more general classification between ideas and impressions had not been applied. Dealing with theories, namely a network of interrelated concepts rather than with single ideas, the modern classification had to order rival sets of terms and statements. Hume and the early empiricists did not discuss competing sets of ideas, and it is here that the logical empiricists added a new dimension to the classical debate. Assuming that observational statements had direct empirical reference and were immediately decidable, they saw in the overlapping observational vocabularies of different theories a basis for instituting a comparison between them.

In view of the importance given to observational statements in assessing the relative merits of rival theories, it is surprisingly difficult to find in recent works in philosophy of science any detailed explanation of their nature. Accounts of the observational language are usually dependent on circular definitions of observability and its cognates, while the theoretical language is generally defined negatively as consisting of those scientific terms which are not observational. For instance, R. C. Braithwaite writes:

> . . .experience, observation and cognate terms will be used in the widest sense to cover observed facts about material objects or events in them as well as directly known facts about the contents or objects of immediate experience.(4)

Hempel states that

> . . . an *observation sentence* might be construed as a sentence. . . which asserts or denies that a specified object, or group of objects, of macroscopic size has a particular *observable characteristic,* i.e. a characteristic whose presence or absence can, under favorable circumstances, be ascertained by direct observation.(5)

Although Hempel acknowledges that the classification is not always easy to apply, he regards it as unambiguous:

> . . .no precise criterion for distinguishing between experimental laws and theories is available, and none will be proposed here. It nevertheless does not follow that the distinction is spurious because it is vague, any more than it follows that there is no difference between the front and the back of a man's head just because there is no exact line separating the two.(6)

The most explicit statement is probably still to be found in Carnap's classic paper "Testability and Meaning" where he makes the following two points. First, he holds that "observable" must be a primitive of the meta-logic of theory-structure, in terms of which other concepts such as "confirmation" and "testability" are defined. The reason he adduces for this is his conviction that what it is to be "observable" is a question for psychology and the behaviourist theory of language, not philosophy. Next, in approximate replacement of a definition, Carnap describes the concept of observability as follows: a predicate "P" is observable for a person *N* if for some object *b, N* can under suitable circumstances come to a decision to accept or reject 'P*(b)*' with the help of a few observations. He continues:

> This explanation is necessarily vague. There is no sharp line between observable and non-observable predicates because a person will be more or less able to decide a certain sentence quickly, i.e. he will be inclined after a certain period

> of observation to accept the sentence. For the sake of simplicity we will draw here a sharp distinction between observable and non-observable predicates. . . Nevertheless the general philosophical, i.e. methodological question about the nature of meaning and testability will, as we shall see, not be distorted by our over--simplification. Even particular questions as to whether or not a given sentence is confirmable, and whether or not it is testable by a certain person, are affected, as we shall see, at most to a very small degree by the choice of the boundary line for observable predicates.(7)

Carnap presupposed, therefore, that an initial distinction between observables and non-observables on pragmatic grounds leads to no distortion in the subsequent discussion of the relation of theory to observation. But there is no such guarantee. What is overlooked in this pragmatic account of observability is that the pragmatic conditions themselves are theory-laden: they are always formulated against a background of current scientific views and opinions. Furthermore, it is by no means obvious on Carnap's account that any predicate is, in principle, non-observable. Perception depends on training, past experience, and contemporary hypotheses, and there are circumstances in which suitably educated persons can come to a quick decision about any predicate, however apparently theoretical such as "meson" or "positron". The recourse to psychology or physiology to determine what an "observable" is offers no easy way out. Even if these sciences could provide us with a clear answer to the question, we would not escape the problem of a vicious regress. For what is it to be accepted as an "observable" for psychology and physiology, and hence to be an empirical basis for their theories of "observability"?

But the reason why it has been for so long considered necessary to accept the comparatively unproblematic character of the observation predicates raises an interesting psychological point. Could it be, as Mary Hesse suggests, that a half-conscious fear of vicious circularity has inhibited the desire to approach the problem from a different angle? "If we have no firm observational basis on which to stand, how can we begin to examine the meaning and justification of theories which are erected upon observables? "(8) It was good conservative policy, therefore, to assume that there could be explicit definition and hence complete equivalence between every theoretical predicate and a function of observation predicates. Attempts to implement the programme, however, proved that this was illusory. First it was shown that in many existing theories, including quantum theory and even Newtonian mechanics, such exhaustive translations cannot be carried out, yet no scientist would wish to abandon otherwise satisfactory theories on this ground alone. Second, and more important, it was pointed out that explicit definitions of each theoretical predicate in terms of observation predicates would cripple the theory in its proper role of expanding to correlate other and different observation statements.

By this time the "firm basis" of observation language had become shaky, and philosophers had started looking elsewhere for a sure footing. But the breakdown of the theoretical-observational dichotomy was yet to come.

One of the reasons why the crisis was delayed was the possibility of shifting one's attention from the empirical to the logical aspect of the problem, and to concentrate on the deductive character of theories. In the deductivist account, theoretical predicates need not be given explicit definition nor even reduction definition in terms of observable. On the analogy of mathematical logic, the postulates of the theory give the syntax and basic logic of the system, and the question of meaning becomes the problem of determining the extra-logical semantics or interpretation. Campbell and before him Maxwell held that this interpretation is given by a physical model which is intelligible, ultimately in observation terms, independently of the theory and its explanandum, and which therefore contributes to the theoretical predicate an interpretation not derived from any direct connection with its explanandum. More commonly, however, deductivists regarded such physical models as non-essential heuristic devices, and argued that the meaning of theoretical predicates was implicit or contextual.(9) Just as "point" and "straight line" are not explicitly defined in a formalized geometry, but their meaning is fixed by the relation that two and only two points determine a straight line, so it is maintained that a predicate such as "proton" means just that entity which has the relation to other entities of atomic physics which are specified in the postulate system of physics. The difference between physics and formalized geometry lies in the correspondence rules (also called coordinating definitions, semantical rules, epistemic correlations) whereby the formal postulates of physics are translated into observation statements and are given empirical content.

This interpretation is open to severe criticism. First, the analogy between physics and a physically interpreted geometry is imperfect and cannot adequately explain the notion of implicit definition of theoretical predicates. Secondly, the function of a logical model is to specify a set of entities with their predicates such that when the postulate system is interpreted in the domain of these entities, the postulates are true in that domain. When no model is specified for the postulates of a scientific theory, the most that can be said is that the postulate system "implicitly defines" the set of possible models that would realize the postulates. "Thus the "implicit definition" is not an interpretation of the theoretical predicates but a definition of a set of possible models which partially intersect in the observable consequences of the theory. In any interesting case there are likely to be an indefinite number of such models. It is not at all clear that the notion of their "set" is well-defined, and in any non-trivial case some of them are likely to be inconsistent with others. The notion of a set of models therefore cannot add anything to the cognitive content of the theory over and above its already interpreted observable consequences."(10) Hence deductivist accounts cannot provide a solution to the problem of the *meaning* of theoretical predicates in its original form. A further question is whether they can provide a solution to the problem of *confirmation* or *acceptability* of theories.

Carl Hempel's famous paper "Studies in the Logic of Confirmation", which appeared in 1945, was the first major attempt to outline criteria of adequacy for the confirmation of statements related to evidence in deductive systems. Hempel enquires what general requirements ought to be satisfied by a confirmation theory which does justice to accepted forms of theoretical inference. Two are particularly relevant to our discussion.

1) *Special Consequence Condition.* If an observational report E confirms a hypothesis H, then it also confirms every consequence of H.(11) This seems intuitively desirable in order to allow for the familiar type of inference where a theory is advanced on the basis of experimental laws which are entailed by it. For instance, Newton's theory is proposed on the evidence of Galileo's law of freely falling bodies and Kepler's laws of planetary motion which are entailed by it. Further as yet untested consequences are then drawn from the theory, for example the variation in the period of a pendulum as it approaches the equator and the influence of the moon on the tides. These are predicted with assurance on the basis of the confirmation of the theory by its observed consequences. Indeed, it is sometimes the case that confidence in the consequence of the theory is so great that it enables corrections to be made to laws that have already been accepted, as when Newton's theory amended Kepler's third law as received on the observed evidence of the periodic time of the planets.
2) *Converse Consequence Condition.* If an hypothesis H entails an observational report E, then the observational report E confirms any stronger hypothesis of which H is a consequence.(12) This condition seems at first sight to be required to explicate why evidence entailed by a theory confirms the theory. But if it is taken together with 1), we get the following counter-intuitive case: suppose $H \rightarrow E_1 E_2$, then E_1 confirms H by 2), hence E_1 confirms E_2 by 1). But suppose H is equivalent to $E_1.E_2$, then certainly $H \rightarrow E_1.E_2$, but there is no reason to expect E_1 to confirm any arbitrary E_2 which may be conjoined with E_1 to form H, indeed in general it will not do so. This example shows that if 1) and 2) are both accepted, any evidence will confirm any statement whatever by the mere construction of a suitable H by conjunction.

For this reason, Hempel originally rejected 2) and retained 1), thus demanding a stronger relation between hypothesis and evidence than mere entailment if evidence is to confirm the hypothesis. Carnap demonstrated however that this solution gave rise to serious difficulties, for if 1) is accepted and 2) rejected, then the confirmation function cannot be a probability function, for in a probalistic confirmation theory 2) is always satisfied for empirical E.(13) A further consequence is that 1) is then not in general satisfied, for if E_1 is to confirm E_2 probabilistically there must be a stronger relation between E_1 and E_2 than their joint deducibility from some hypothesis H. For these reasons, Hempel has now accepted that condition 1) rather than 2) should be abandoned.(14)

But whether a probabilistic theory is adopted or not, the upshot of the discussion is that the deductive criteria thus far suggested are unsatisfactory. It is hard to see how they could even begin to be relevant to the crucial problem of predictive inferences of the kind illustrated by Newton's theory and the laws it explains. For instance, let us suppose we are given an axiomatic system S from which some observation statements can be derived with an appropriate set of correspondence rules. Since S itself, on the deductive view, contains no observation predicates, no reason can be derived from the statements of S why one observation predicate should occur in a given correspondence rule rather than another. On the one hand, if the conjunction of S and the correspondence rules is merely to explain given observations, then the observation predicates in these rules are determined by the known relations of observation predicates in observation statements. On the other hand, if S is to be used predictively, this involves deriving statements containing new observation predicates which do not appear in the correspondence rules. But S is powerless to determine what additional correspondence rules should be introduced, and consequently no predictions involving some particular new observation predicate rather than some other can be made with confidence from the theory. There are, however, countless examples of scientific inferences in which just such predictions have been made with confidence, for instance how the earth would appear when seen by astronauts landing on the moon.

In spite of its detailed analysis of the logical structures of postulate systems, the deductivist account of theories had shown itself grossly inadequate. A strong reaction was inevitable, and by the end of the fifties the complaint was increasingly voiced that in their concentration on technical problems of logic, the logical empiricist movement has lost contact with real science.

The development of history of science from a hobby pursued by retired scientists to a full-fledged academic discipline revealed that it was naive to read the past, as the positivists did, as the record of great men throwing off the shackles of a dark inheritance and heralding the dawn of scientific objectivity. Many older theories that were allegedly laden with superstition, for instance, medieval mechanics and the phlogiston theory, were found to contain far more than simple-minded error and prejudice. It was discovered that Newton not only framed non-empirical hypotheses but that he was strongly influenced by the alchemical tradition, and that Galileo, the father of "empirical science", neither dropped balls from the Leaning Tower of Pisa nor cared for experiments as much as had hitherto been believed. As closer attention was paid to the framework of theories it became apparent that the theoretical context determined not only the questions that were raised but the terms in which the answers had to be expressed to be judged acceptable. This was a direct challenge to the logical empiricist position that there is an absolute, theory-independent observation language whose terms have the same core of common meaning for all competing theories. According to the new view, the meaning of all scientific terms, whether factual or theoretical is governed by the paradigm or general intellectual stance which underlies them and in which they are imbedded.

This new interpretation has been urged with considerable vigour by Thomas Kuhn(15) and Paul Feyerabend. "The meaning of every term we use", Feyerabend writes, "depends upon the theoretical content in which it occurs. Words do not 'mean' something in isolation: they obtain their meanings by being part of a theoretical context".(16) Whereas the logical empiricists considered theoretical terms as wholly dependent on observational ones, from the new viewpoint the exact reverse is true. In his characteristically forceful language, Feyerabend declares:

> The philosophies we have been discussing so far assumed that observation sentences are meaningful *per se,* that theories which have been separated from observations are not meaningful, and that such theories obtain their interpretation by being connected with some observation language that possesses a stable interpretation. According to the point of view I am advocating, the meaning of observation sentences is determined by the theories with which they are connected. Theories are meaningful independent of observations; observational statements are not meaningful unless they have been connected with theories. . . It is therefore the *observation sentence* that is in need of interpretation and not the theory.(17)

It follows that a basic shift in the theoretical entails a change in what counts as a real problem, a correct method, an acceptable classification, and even a fact, since the meaning of observational terms is determined by the theory in which they occur.

Kuhn and Feyerabend have mercilessly exposed the inadequacies of the usual formulation of the distinction between theory and observation, and they have shown that the concept of "explanation" in science cannot be divorced completely from a consideration of the history of scientific explanations. In their reaction against logical positivism, however, they seem to have embraced an equally extreme view. For all its value and its suggestiveness, the position of these writers has not been formulated in a way that resolves the major problems of philosophy of science. It has only succeeded in making them more glaring. If the meaning of every term depends on its theoretical framework, a change of theory must produce a change of meaning of every term in the theory. But this sweeping solution brings a new series of problems in its wake. For instance, does the mere extension or application of a theory make a difference to the theoretical content and hence to the meaning of the terms involved? Does an alternative axiomatization alter the theoretical content so that the meanings of the expressions axiomatized change with reaxiomatization?

Apart from these technical difficulties, more fundamental objections can be raised. For instance, what is the point of making experiments if they can be interpreted to support any theory? If the meaning of observational terms depends completely on the theoretical framework, how can we ascribe any continuity to the different usages of the same terms in successive theories? Rival hypotheses can no longer be said to contradict one another, for in order for two sentences to be contradictory what is denied by one must be affirmed by the other, and this is meaningless unless they have something in common. If Kuhn and Feyerabend have established the bankruptcy of a philosophy of science based on the allegedly firm foundation of "observations", they have produced an alternative account that severs it from empirical evidence altogether. Their position eventuates in a complete relativism in which it becomes impossible to compare any two scientific theories and to choose between them on any but the most subjective grounds. The positivists may well retort that if observation itself is said to share the uncertainties of theories in any important sense, we cannot avoid being cast adrift in a sea of hypotheses. Without a solid observational basis, we would seem to have no moorings, no plan, and no science.

Any viable account of the observation language must be able to show not only that we can keep afloat in such a sea but that we can make progress through it. Several attempts have been made recently to plot such a course. One of the most promising is Mary Hesse's outline of a self-correcting confirmation theory.(18)

For the sake of brevity and ease of comprehension, Dr. Hesse introduces a model of a world consisting of a finite number *X* individuals described by means of a fixed set of "K" primitive monadic predicates each of which may or may not apply to each individual.

An essential feature of the model is that a predicate may be wrongly ascribed to an individual and that a definite small prior probability can be assigned to the occurence of this error. Since this value is small, the ascription of predicates will for the most part be correct and certain regularities in the relation between predicates will begin to emerge. For instance, it may be the case that when a predicate P has been applied to some individual, Q has usually been found to belong to it as well. Each predicate can be visualized as being a knot in a network of relationships with other predicates, where the strength of the strings between the knots stands for some increasing function of the proportion among individuals in which the predicates have been reported as co-occurent or co-absent. Assuming that there exists a probabilistic confirmation theory which explicates the more elementary kinds of inductive inference, and that prior probabilities can be assigned with the aid of some postulate such as Keynes' Principle of Limited Independent Variety, Bayes' Theorem can be used to calculate the probabilistic confirmation value of any as yet unobserved application of a predicate to a given individual on the basis of existing evidence.

The crucial problem is to detect and correct the erroneous ascription of a predicate. Since all evidence is necessarily in terms of reports that P applies to *b,* and it is assumed that there is a finite probability that such a report may be wrong, a distinction must be drawn between a report and "what really is the case". What must be done in this situation is to try to identify errors on the basis of other reports, some of which may make the correctness of a given application extremely improbable. It is at this point that a charge of circularity could be brought against Dr. Hesse's model. She believes that it can be avoided, however, for while it is true that all predicates could be erroneously applied, it is not the case that all could be known to be wrong at the same time. For instance, alchemical terms used in modern chemistry are no longer predicated of individuals in the same sense as they were in the middle ages, but the transformation of their meaning did not happen for all of them at the same time. What is "contrasted with the *report* something or other is the case is not what *really* is the case, to which we have no access, but rather what *in a given theory* would be the case, which may in a small proportion of cases contradict what has been reported."(19) The fact that this is not as arbitrary as may seem can be illustrated by the following case-history in geology. When Louis Agassiz, fresh from studying glacial action in Switzerland, visited Scotland in 1840, he immediately recognized moraine, roches moutonnées, and glacial striae as distinctive features of the Scottish scenery. In this case, his extensive knowledge of the effects of glaciers and glaciation may be said to have enhanced his perception. But this way of looking at objects became so ingrained that later he began to *see* evidence of glacial action where certainly there had been none, as in the Tijuca hills behind Rio de Janeiro. His reports, however, were only revised when it was shown that they were *inconsistent* with later and more detailed accounts.

It does not follow from this that the theory which contradicts the fewest reports is the theory with the highest confirmation value, for the amount of "correct evidence" according to this theory may be counteracted by the higher prior distribution of some other theories. For instance, if Keynes' Principle is built into the prior distribution, "preference will be given to theories which cluster individuals, that is to say, which allow individuals to be classified in more or less well-defined classes with high similarity between members of a class and a non-member of that class."(20) Without going into further refinements which could be introduced, we can see how this model, however oversimplified, offers an alternative to the positivist observational language.

On the new interpretation, the meaning of descriptive predicates is no longer determined exclusively by direct empirical reference and in a way that is radically different from theoretical predicates. The meaning of both observational and theoretical terms has elements of stimulus-response in situations of easy empirical reference, and also of contextual relation with other predicates which co-occur or are co-absent. It follows that whereas error on the positivist view is taken to be relative to 'what is really the case", on this interpretation, it is considered relative to a theory. This has the interesting consequence of allowing the possibility that the source of error may not merely be lack of attention to what is observed, but can also occur in a sincere and careful report, when ascription of a predicate does not fit in well with the general outline of the theory. "This is to say that the correct application of a predicate depends both on a careful response to a stimulus situation *and* the relations of this predicate to other predicates in this and other individuals."(21) A theory is constituted by a set of such relations. Hence confirmations and falsification will depend not on one allegedly crucial experiment, but on all the reported evidence in other situations, together with the verdicts of different theories on the correctness of these reports. Thus the directness of empirical reference and the immediate decidability which the positivists considered as the characteristics of observation predicates marking them off from theoretical predicates appear illusory.

Whether the approach suggested by Mary Hesse can be shown to be adequate to all types of theoretical inference is too extensive a question to prejudge here, for it stands in need of a confirmation theory based on a language containing the real number continuum, for which many as yet unsolved mathematical problems arise. But perhaps enough has been said to conclude that we are not compelled to adopt a kind of historical relativism and to regard theoretical inference as fundamentally irrational. The problems and the techniques of the logical empiricists can be pushed further than they themselves thought possible. Both the question of meaning and the question of confirmation are more complex and interrelated than had hitherto been imagined, but there is, as yet, no reason to despair of a logic of scientific discoveries.

While Mary Hesse tackles the problem of scientific discourse from the logical angle, Dudley Shapere is mainly concerned with the lessons to be learned from the history of science. He grants that Kuhn and Feyerabend are right in stressing that the framework of contemporary hypotheses, the paradigm or "disciplinary matrix" determines to a large extent what questions can be raised, and what views can be suggested about a given scientific problem. He objects, however, that the "incommensurability" of theories, which is a consequence of the extreme form of their position, distorts the very nature of the scientific enterprise. Shapere argues, for instance, that it makes sense to compare the medieval theory of impetus with Aristotelian mechanics on the one hand, and with the principle of inertia on the other.

> The impetus theory stands in a transitional phase between these two traditions--still in the tradition of Aristotelian physics in its view that impetus is a cause of motion, but heading toward the modern view in making impetus an internal and incorporeal force rather than an external, corporeal one.(22)

The impetus theory encouraged a fresh approach to traditional problems by removing long-standing conceptual barriers. For instance, the Aristotelians rejected outright any suggestion that the earth rotated because the physical assumptions of their system entailed that a strong wind would be set up in the direction opposite to the earth's motion. By affirming that the air would receive an impetus and would be carried around as though nothing had happened, the impetus theory made it possible to entertain the idea that the earth might move. Likewise, by internalizing the cause of motion it shifted attention to a new set of possibilities. Since air could no longer be the force producing motion but merely an impediment, motion in a vacuum ceased to be implausible, and thus the impetus theory cleared the ground for thinking about the idealized case of a body moving in the absence of impeding forces. Furthermore, by treating all cases of motion, whether terrestrial or celestial, natural or constrained, in terms of one kind of cause, namely impetus, it paved the way for a unified account of all motion and provided an alternative approach to the traditional Aristotelian division.

On the Kuhn-Feyerabend view one is practically driven to describe scientific change in revolutionary terms, to speak, for instance, of the overthrow of Aristotelian mechanics by the impetus theory and of the latter by Newtonian science. From the viewpoint Shapere adopts a more balanced description seems possible. The Aristotelian-Scholastic tradition applied, as a template, an intricately connected web of concepts and propositions to the data of perception and everyday experience. In laying down this system of interpretation, it simultaneously set up obstacles or limitations, both by theoretical precept and by suggestion to thinking in certain other ways. The vacuum and the actual infinite, for example, appeared self-contradictory, while the motion of the earth seemed physically impossible. Scientific advance is related to a reassembling of the pattern of our experience, and there is immense resistance to this. "In this sense", writes Shapere,

> . . . we can speak of the Aristotelian view as having involved certain 'presuppositions' specifying (for example) what could and what could not count as an explanation. To this extent Kuhn and Feyerabend have made an important point. But these 'presuppositions' were not mysterious, invisible, behind-the-scene 'paradigms' (Kuhn) or 'high-level background theories' (Feyerabend), but were involved in the straightforward scientific statements themselves, even though there were disagreements about details (and even about fundamentals), and even though the way in which they restricted thought, or the importance of these restrictions, could not be seen so easily.(23)

After all, they were seen by Philoponos, Oresme, Buridan and others who reassembled the facts in a different pattern!

The relativism of the Kuhn-Feyerabend position is not the result of an investigation of actual science and its history but the purely logical consequence of a narrow presupposition about what "meaning" is. These writers hold that if scientific terms do not retain precisely the same meaning over the history of their incorporation into more general theories, then these theories cannot be compared, and the similarities they exhibit must be considered at the best as superficial and at the worst as deceptive and misleading. This claim rests on the assumption that two expressions or sets of expression must either have exactly the same meaning or else must be completely different. The only possibility left open by this rigid dichotomy of "meanings" is that history of science, since it is not a process of development by accumulation, must be a completely noncumulative process of replacement.

The inherent weakness of this position turns out to be its retention of a positivistic concept of "meaning". If anything the revolution isn't radical enough. Kuhn and Feyerabend, in spite of their spirited attack on the positivist view that "theories" are parasitic on "observations", nevertheless approach their problems with that distinction in mind. They have applied the old classification to a new purpose rather than invented new conceptual tools for dealing with old problems.

They have merely inverted the respective roles of the two members of the classical distinction: it is now the "theory" that determines the meaning and acceptability of the "observation", rather than the other way round.

It is much more radical to call the distinction itself in question and to escape from the horns of the dilemma by breaking them. It may well be that several problems that bedevil contemporary philosophy of science are heightened (if not created) by the deficiencies of the classification of scientific language as theoretical or observational. If this is the case, it is no longer the solution that is seen to be problematic but the very way in which the question is framed. For instance, the notorious problem of the ontological status of theoretical entities or the question whether a realistic interpretation of scientific theories can be upheld may be partly generated by an inadequate concept of the working of theories. It is all too easy to view the distinction between observational and theoretical as paralleling a distinction between existent and non-existent. If observation terms are said to have a clear and direct reference to entities that exist while theoretical terms do not, it becomes difficult to know what theories are about. This is not due to any intrinsic opaqueness of the concept of existence, but to the sharpness of the distinction between theory and observation. As long as it was believed that theoretical terms could be exhaustively described via observational ones, theories could be handled as a convenient shorthand. When it became apparent that such a reduction was only possible in part, the extra meaning of theoretical terms was sought in the position they occupied in the context of the system they belonged to. Thus theoretical entities could not exist in the same sense as tables and chairs, and it became a first-class puzzle to know in what sense they could now be said to exist, short of being merely useful fictions.

The reaction of writers such as Kuhn and Feyerabend was to that this indicates that observations are governed by theories which are, in last resort, irrational guesses at what the universe really looks like. But this did not solve the problem of meaning: it merely replaced the positivist thesis of meaning invariance with the doctrine of incommensurable meanings. Dudley Shapere suggests a less rigid interpretation of meanings were they are considered similar or analogous, namely comparable in some respects while differing in others. By taking this path, we can hope to preserve the fact that theories, for instance, Newtonian and relativistic dynamics, are not incommensurable, although they are more fundamentally different than the most usual logical empiricist views make them.

The difficulty in this interpretation lies in the concept of *similarity* or degrees of likeness of meanings. It is only too tempting to distinguish, against the background of a particular theory, between what is and what is not an essential part of the meaning of a term. For instance, it seems obvious (in the light of subsequent developments) that Newton's absolute space and absolute time are "irrelevant features" of his mechanical theory. Yet those very features, for some purposes, may prove to be the very ones that are of central importance in comparing two uses. An absolute distinction between essential and accessory features could only rest on an *a priori* conception of scientific understanding and would reintroduce the fallacy of the theoretical-observational dichotomy. It would thus seem wiser to allow all features of the use of a term to be equally potentially relevant in comparing the usage of the terms in different contexts.

Whether the post-positivistic approach of Mary Hesse, Dudley Shapere, and other radicals can be shown to be both faithful to the history of science and adequate to the logic of theoretical inference remains to be seen. The discussion, which is only in its early stages, will involve a critical examination of the assumption that most scientific inferences rest on analogy, and, if so, in what sense. Whatever the final outcome of the debate, the benefits already accrued from the radical reappraisal of the classification of scientific terms as theoretical and observational make the venture a promising one.

Bibliographical References

* Acknowledgements are gratefully given to the editor of *Dialogue* for permission to include material which appeared in "Beyond Logical Empiricism", *Dialogue* X(1970), 223–242, and to Mr. Maurice Alarie and Mr. Gilles Bergevin for their help and advice.

1) "We assumed that there was a certain rock bottom of knowledge, the knowledge of the immediately given, which was indubitable. Every other kind of knowledge was supposed to be firmly supported by this basis and therefore likewise decidable with certainty." (Rudolf Carnap, "Intellectual Biography" in P.A. Schilpp (ed.), *The Philosophy of Rudolf Carnap.* Lasalle, Ill.: Open Court, 1963, p. 57).

2) Mary B. Hesse, "Positivism and the Logic of Scientific Theories" in P. Achinstein and S. Barker (eds.), *The Legacy of Logical Positivism,* Baltimore: John Hopkins Press, 1969, p. 89.

3) *Ibid.,* p. 103.

4) Richard C. Braithwaite, *Scientific Explanation,* Cambridge: Cambridge University Press, 1959, p. 8.

5) Carl G. Hempel, *Aspects of Scientific Explanation,* New York: The Free Press, 1965, pp. 102–103.

6) Carl G. Hempel, *The Structure of Science,* London: Routledge & Kegan Paul, 1961, p. 83.

7) Rudolf Carnap, "Testability and Meaning" in Herbert Feigl and May Brodbeck (eds.), *Readings in the Philosophy of Science,* New York: Appleton-Century-Crofts, 1953, pp. 63–64.

8) Mary B. Hesse, *art. cit.,* p. 103.

9) For a critical examination of these two interpretations, see Mary B. Hesse *Models and Analogies,* London: Sheed and Ward, 1963.

10) Mary B. Hesse, "Positivism and the Logic of Scientific Theories" in P. Achinstein and S. Barker (eds.), *The Legacy of Logical Positivism,* Baltimore: John Hopkins Press, 1969, pp. 94–95.

11) Carl G. Hempel, *op. cit.,* p. 31.

12) *Ibid.,* p. 32.

13) Rudolf Carnap, *Logical Foundations of Probability,* Chicago: Chicago University Press, 1962, pp. 478–479.

14) Carl G. Hempel, *op. cit.,* pp. 49–50.

Zygmunt Dobrowolski, (Varsovie)

Secteurs scientifiques autonomes et leur rôle dans la classification des sciences

1. La nécessité urgente d'une classification générale des sciences, pour l'organisation rationelle des activités scientifiques et la planification des recherches à l'échelle nationale et internationale est incontestable. Le problème n'est pas nouveau.

Les classifications des sciences ont été créées à travers les siècles par de célèbres philosophes tels que Aristote, Francis Bacon, d'Alembert, Comte . . . Elles sont le reflet de leurs systèmes philosophiques. On peut suivre cette tradition. En se fondant cette fois sur les études théoriques concernant les principes de la science même, on pourrait peut-être établir une classification conforme aux opinions actuelles sur la spécialisation rationelle de la science.

Il serait cependant très difficile d'obtenir un consentement universel à un système basé sur les considérations théoriques.

Nous pensons donc qu'il serait plus facile de résoudre ce problème en commençant par mettre de l'ordre dans l'ensemble des secteurs de la science spécialisés existant. En groupant les secteurs apparentés en classes de niveau de plus en plus élevé, on pourrait obtenir un schéma plus ou moins complet représentant presque toute la science.

Ensuite, il sera peut-être possible – par des perfectionnements successifs – d'aboutir à une classification acceptable à la majorité des intéressés.

2. Il est naturel que l'ensemble des secteurs scientifiques spécialisés ne présente pas une image de développement harmonieux et rationnel de la pensée humaine. Les spécialités scientifiques les plus utiles aux grandes industries et à la défense nationale ont toujours été favorisées aux dépens des sciences abstraites, qui avaient moins de chances à être dotées. Toutefois, étant donné que la classification envisagée doit illustrer les relations existant effectivement entre les sciences, ce sont les secteurs spécialisés, formés sur la base des tendances naturelles de l'homme et de ses besoins vitaux, qui doivent figurer dans l'ensemble assujetti à la classification.

Les champs d'activité des secteurs scientifiques particuliers se recouvrent partiellement, tandis que leurs définitions trop générales ne permettent pas de déterminer leurs extensions avec une précision nécessaire.

Il semble donc indispensable de fixer les conditions que doivent remplir les secteurs scientifiques spécialisés pour figurer dans l'ensemble classifié. Il faut aussi choisir le contenu de cet ensemble de façon à ce que la science entière, dans toute sa diversité, y soit représentée.

3. Un secteur scientifique spécialisé, suffisament développé pour disposer de son organisation propre, de son institut de recherche avec un centre de documentation, de chaires d'enseignement supérieur, de ses associations scientifiques et professionnelles, un secteur auquel sont consacrées des revues spéciales, des ouvrages et manuels peut – dans une certaine mesure – être considéré comme un secteur scientifique "autonome".

En relations constantes avec d'autres secteurs à degré analogue de spécialisation et d'évolution, il occupe une place distincte dans la grande famille des sciences.

La spécialisation des connaissances ne s'arrête pas, bien entendu, à ces secteurs scientifiques qui remplissent les conditions requises pour être considérés comme autonomes. Il importe toutefois d'envisager ici cette étape de spécialisation, car c'est un ensemble d'organismes de ce genre justement qui puisse être considéré comme représentant la science entière, dans toute sa diversité, approximativement au moins.

Sous le terme "secteurs scientifiques" nous comprendrons donc dans la suite les secteurs "autonomes".

4. En analysant la documentation d'un secteur spécialisé on constate qu'elle concerne avant tout les sujets qui sont essentiels au secteur donné et ensuite des sujets appartenant aux autres domaines des connaissances que nous déterminerons "sujets accessoires".

Les mêmes sujets peuvent être "essentiels" à plusieurs secteurs spécialisés, mais ils y figurent alors en liaison avec des problèmes différents. Ainsi, les métaux – sujets "essentiels" dans la Métallurgie au point de vue de leur production – sont "essentiels" également au point de vue de leur utilisation, dans de Génie Civil par exemple.

Quant aux sujets "accessoires", qui expriment les relations du secteur donné avec d'autres secteurs spécialisés, ils confirment seulement le principe de "l'unité de la science". Les sciences naturelles et humaines, par exemple, sont bien représentées dans les sciences techniques et l'inverse est aussi vrai. On peut dire en général qu'un secteur spécialisé présente, dans une certaine mesure, l'univers entier du point de vue de la spécialité scientifique donnée.

On émet une opinion que l'unité de la science rend impossible la construction d'une classification des sciences, les "extensions" des secteurs spécialisés autonomes ne pouvant être délimitées. Ce n'est pas juste, il faut seulement pour le groupement en classes des secteurs "apparentés" ne pas prendre en considération que leurs "noyaux" composés de sujets "essentiels", en faisant abstraction des sujets "accessoires". Le tableau de la classification des sciences établi de cette manière illustrera justement l'unité de la science.

En conclusion, nous admettrons que chaque secteur spécialisé possède un "noyau" composé des sujets essentiels à la spécialité donnée et propres à elle seule, le noyau qui décide de sa place dans la classification des sciences.

5. Pour déterminer ce "noyau" il est nécessaire de construire une classification du secteur donné, car la classification seule permettra de séparer ce qui est "essentiel" dans le secteur scientifique donné de ce qui est "accessoire".

Pour faciliter la discussion, nous appellerons classifications "pures" les classifications des secteurs spécialisés ne contenant que leurs noyaux, en opposition aux classifications "complètes" contenant également les sujets accessoires représentant les relations avec d'autres domaines de la science. Nous proposons donc que la classification des sciences prenne la forme d'une classification des classifications pures des secteurs spécialisés scientifiques autonomes.

6. La classification "complète" d'un secteur spécialisé de la science est aussi nécessaire, avant tout pour cataloguer systématiquement la documentation et normaliser la terminologie. Mais ce n'est pas son seul usage. La classification est très importante pour la planification des recherches. Celui qui effectue des travaux scientifiques, surtout dans les techniques, sait combien il est difficile de s'y retrouver si l'on ne dispose pas d'une classification moderne. La classification est comme une carte topographique des problèmes de chaque spécialité, elle est une sorte de guide, qui fait apparaître même aux moins initiés les thèmes scientifiques de celle-ci. Pour une spécialité jeune, cristallisée depuis peu, comme il en existe tant dans tous les domaines de la science, l'établissement d'une classification propre est le point culminant de l'organisation de collaboration des savants dans les travaux de recherche, dans les rapports internationaux, etc. Lorsque l'on possède une classification moderne appropriée on lui découvre toujours de nouveaux avantages et il devient alors difficile de croire qu'il fût un temps où l'on ne possédait un instrument permettant de mettre de l'ordre dans les travaux aussi variés.

Les intéressés trouveront dans les publications sur la théorie de la classification plus de détails à ce sujet, ainsi que le modèle d'une classification à facettes, destinée aux secteurs scientifiques spécialisés. Nous pensons que l'insuccès des tentatives récentes en vue de créer une classification des sciences, proviennent justement de ce que certains secteurs scientifiques spécialisés n'ont pas respecter le principe, qu'une classification propre indépendante de tout schéma général puisse seule assigner au secteur donné sa place dans la famille des sciences.

7. On peut se demander à quelles spécialités scientifiques incombe la tâche d'élaborer une classification des sciences? Il n'y a pas de doute que c'est la science des sciences, une branche relativement jeune de la philosophie qui doit considérer la classification des sciences comme l'un des points principaux de son programme d'études.

Les mêmes raisons pour lesquelles tout secteur scientifique doit disposer d'une classification propre, sont applicables dans le cas de la science des sciences. Alors on peut avoir des doutes: est-ce qu'il soit juste de procéder à monter une classification des sciences avant d'établir une classification de la science des sciences?

Dans une classification de la science des sciences la classification des sciences jouera le même rôle que la classification des plantes dans la classification de la botanique, elle sera l'une de ses grandes divisions.

Il semble donc que, parallèlement aux travaux de montage d'une classification des sciences, il serait nécessaire de procéder à l'établissement d'une classification de la science des sciences, cette dernière comprenant la classification des sciences comme l'une de ses divisions principales.

Resumé

La caractéristique des secteurs scientifiques autonomes est donnée et leur rôle dans la classification des sciences est décrit en détail. La différence entre une classification "pure" et "complète" étant expliquée, l'auteur propose de baser la classification des sciences sur l'ensemble des classifications "pures" des secteurs scientifiques, en soulignant la grande utilité aussi des classifications "complètes" dans la pratique. Dans le dernier point de son exposé, l'auteur caractérise la classification des sciences comme l'un des problèmes de la science des sciences.

Frederick Suppe (Urbana)

Some philosophical problems in biological speciation and taxonomy [(1)]

Abstract

This paper is concerned with a number of basic philosophical problems in scientific taxonomy and classification which arise in biological taxonomy. Its consideration of these problems, and the conclusions it draws, are general and not restricted to the biological versions of these problems. The main problems dealt with are

(1) What makes a taxonomic system natural?

(2) What definitional forms are appropriate for defining taxa and taxanomic categories?

(3) What is the role and function of taxonomy in scientific theorizing?

The paper considers the empirical properties of a large variety of different kinds of taxonomic systems which have been proposed by various taxonomic schools. From this consideration it draws a number of general conclusions, the most important being that natural taxonomies are factually true or false; taxa in a natural taxonomy will be "natural kinds"; the conditions for taxa membership are empirically true or false generalizations in a natural taxonomy; hence, the appropriate definitional form for taxa (and, to an extent, for taxonomic categories) is an empirical issue; it is possible that there may be different and inequivalent factually true natural taxonomies for a given domain of individuals. Because of this possibility, naturalness and factual truth are not the only criteria by which taxonomic adequacy is to be assessed. In particular, the usefulness of the taxonomy in a wide range of scientific applications also is important. It is argued that numerical taxonomists are correct in insisting on this, but that the criterion they advance for being useful (their "naturalness" condition) is incompatible with what is required of a taxonomy if it is to be useful in a wide range of scientific theorizing. In showing this a general-discussion is given of relationship between taxonomic system and scientific theories, and of the role taxonomy can and should play in scientific theorizing. Other functions of taxonomy in science also are considered. From this, certain recommendations are made concerning the direction biological taxonomy should proceed in.

In this paper I am concerned with a body of problems arising in biological taxonomy: What is a species? How are taxa defined? What is the difference between a natural and an artificial taxonomy? Are taxonomies true or false? What theoretic properties do taxonomic systems possess? My interests in these problems are primarily philosophical, not biological. Although these problems arise most dramatically in biological taxonomy, they are by no means peculiar to biology;

indeed similar problems can be raised for most branches of science as well as for common-garden varieties of knowledge. Although my treatment will proceed in terms of the biological versions of these problems, my investigation has as its aim the establishment of a number of general epistemological and metaphysical results about taxonomic systems – results which will lay the foundation for a general philosophical analysis of taxonomic and classificatory systems in science. As such my interests here are different from most of the recent philosophical treatments of biological speciation and taxonomy which have as their focus such questions as "How is 'species' to be defined in biological taxonomy? ", "What is a natural biological taxonomy? ", and "What sorts of definitions are or ought to be given for biological taxa? "[2] Whereas these discussions attempt to provide philosophical solutions to biological problems, my purpose is to exploit the biological problems in an attempt to provide solutions to some quite general philosophical problems about classification.

I. Key problems in taxonomic theory: An historical survey

By way of providing background for my investigation, I want to begin by characterizing what a taxonomic system is, introducing some key terminology, and indicating more precisely the problems concerning taxonomy and classification I intend to consider. In doing the latter, my approach will be largely historical.

By a *taxonomy* (or *taxonomic system*) I mean a system of categories for classifying individuals on the basis of similarities; these similarities may be morphological, functional, social, or whatever. A *standard taxonomy* for domain $\underline{D}$ is a finite collection of *taxa* (classes of individuals in $\underline{D}$) such that each taxon is assigned to a unique category in a hierarchial ordering of categories and each individual in D belongs to exactly one taxon of each category. More precisely, a standard taxonomy must meet the following conditions: (1) There is a finite serially ordered sequence of *taxonomic categories,* $\underline{C}_1, \ldots, \underline{C}_{\underline{n}}$; (2) each taxonomic category $\underline{C}_{\underline{i}}$ $(1 \leq \underline{i} \leq \underline{n})$ contains $\underline{m}_{\underline{i}}$ *taxa,* $\underline{T}_{\underline{i}, 1}, \underline{T}_{\underline{i},2}, \ldots, \underline{T}_{\underline{i},\underline{m}_{\underline{i}}}$; (3) the $\underline{T}_{\underline{i},1}, \ldots, \underline{T}_{\underline{i},\underline{m}_{\underline{i}}}$ are each collections of individuals in $\underline{D}$ such that each member of $\underline{D}$ is a member of exactly one $\underline{T}_{\underline{i},\underline{j}}$ $(1 \leq \underline{j} \leq \underline{m}_{\underline{i}})$; (4) all individuals in a given $\underline{T}_{\underline{i},\underline{j}}$ $(1 \leq \underline{i} < \underline{n};\ 1 \leq \underline{j} \leq \underline{m}_{\underline{i}})$ must be members of the same $\underline{T}_{\underline{i}+1,\underline{k}}$ $(1 \leq \underline{i} < \underline{n};\ 1 \leq \underline{k} \leq \underline{m}_{\underline{i}+1})$.[3] Most biological taxonomies are standard taxonomies. For example, the usual phylogenetic taxonomies are of this sort, where $\underline{D}$ is the class of living organisms. In such taxonomies, the taxonomic categories $\underline{C}_1, \underline{C}_2, \underline{C}_3, \ldots$ respectively will be the classes of species, genera, families, etc., where, e.g., the category *species* is the set consisting of all those taxa (classes of organisms), which qualify as species. As such individual organisms are members of taxa but not of taxonomic categories, and taxa are members of taxonomic categories whereas the individuals belonging to the taxa are not – e.g., a man is a member of the taxon *homo sapiens* (a species) but is not a member of the category *species,* and the taxon *homo sapiens,* being a species, is a member of the category *species.*

Certain features of standard taxonomies need to be emphasized. A given taxon of whatever taxonomic category will always be a collection of individuale in D. Thus a species is a collection of living organisms, a genus is a collection of living organisms, a family is a collection of living organisms, etc. When we say that the species *Rosa multiflora* is of genus *Rosa*, which in turn is of family *Rosaceae*, we are saying that every living organism which is a member of the taxon *Rosa multiflora* is also a member of the taxon *Rosa* and of the taxon *Rosaceae*; we are *not* saying that the species *Rosa* is a member of the genus *Rosa multiflora.*

Taxa of whatever taxonomic category are collections of individuals grouped on the basis of some "similarity". A given taxon (e.g., species *Rosa multiflora*) is defined by specifying the similarity characteristic of its members, the similarity in virtue of which they are classed together.[4] The hierarchal nature of standard taxonomies simplifies the definition of a given taxon. If an individual a is a member of taxon T_{1,i_1} of category C_1, it follows that a also must be a member of a taxon T_{2,i_2} of category C_2, a member of taxon T_{3,i_3} of category C_3, . . ., and a member of taxon T_{n,i_n} of category C_n (where $1 \leq i_j \leq m_j$ for $j = 1, \ldots, n$). Since $T_{1,i_1}, \ldots, T_{n,i_n}$ each are collections of individuals having certain similarities, to specify a taxon T_{j,i_j} of category C_j all that need be done is to specify that an individual is a member of T_{j,i_j} if and only if it is a member of $T_{j+1,i_{j+1}}, \ldots, T_{n,i_n}$, and has similarity characteristics c which are characteristic of T_{j,i_j} but of no other taxon of category C_j. That is, we need not specify all the similarities in virtue of which individuals are grouped together in T_{j,i_j}, but rather only those not common to all members of $T_{j+1,i_{j+1}}$. Thus to define the taxon Class *Mammalia* (T_{j,i_j}) we need only specify that its members are those living organisms of Subphylum *Vertebrata* ($T_{j+1,i_{j+1}}$), Phylum *Chordata* ($T_{j+2,i_{j+2}}$), and Kingdom *Animalia* (T_{n,i_n}) which have hairy skin, the transverse division of the body cavity by a diaphram, the nonnucleated condition of mature red blood corpuscles, and milk-producing mammary glands used to nurse the young. To define the Subclass *Prototheria* ($T_{j-1,i_{j-1}}$) all that need be said is that it consists of those individuals of Class *Mammalia* (T_{j,i_j}) which are egg-laying.

Depending upon what sorts of similarities are chosen to define taxa, a number of different standard taxonomies may be defined for a given domain, D. For example, one may define plant species on the basis of morphological similarity, genetic similarity, similarity of sexual parts, affinity by common ancestry, size etc. Some choices (e.g., length of body) may result in highly arbitrary taxonomies. Since the recorded beginnings of taxonomy (Aristotle and Theophrastus), the attempt has been made to distinguish such artificial from "natural" taxonomies, and throughout the history of taxonomy much theoretical dispute has centered on the issue of what makes a taxonomy natural. Most accounts accept the intuitive idea that *natural taxonomies are those which classify in accord with the objective reality confronting us in nature,* but differ in what is required to do so. Dispute over this question often has become intertwined with dispute over three other questions: What form of definition should be used in defining the various taxa (e.g., What definitional form is to be given for *Rosa multiflora*?)? How are the taxonomic categories to be defined, (e.g., What is a species?)? Should taxonomic systems be natural? These three subsidiary questions obviously are interconnected with the basic question about natural taxonomies. For if one denies taxonomies should be natural, the choice of definitional form for taxa and taxonomic categories will be restricted only by what makes the taxonomy useful for its intended purposes. And if one maintains that taxonomies ought to be natural, then what is taken as being natural must be reflected in they way one defines taxonomic categories and how one may define the individual taxa. It is a useful oversimplification to view differing theoretical views on taxonomy as constituting differing answers to these four questions[5]).

The earliest recorded theoretical position on the nature of taxonomy is to be found in Aristotle's writings.[6]) According to Aristotle, systems of classes can be formed in a number of ways, but only certain classes are *natural.* Natural classes are those which have a *form* or an essential nature *(essence)*; artificial classes lack a form. This form is immutable and eternal, being the secondary substance of things with being (individual existents, or primary substances). A natural taxonomy is one comprised wholly of natural classes; an artificial taxonomy is one comprised, at least partially, of artificial classes.[7]) Aristotle's notion of natural taxonomies, which he maintains ought to be adopted, strongly influences the account he gives of taxonomic categories. According to him, species are the essences of natural classes which are indivisible in the sense that none of their parts are essences or forms of primary substances; derivatively, the individuals which possess that species essence are of the species. Two individuals are of the same genus if their forms differ only in the quality of their parts, but have the same parts. This characterization of the taxonomic categories largely determines the admissible definitional form for the individual taxa. For the parts of a form or essence are, according to Aristotle, characters. And the form of a species thus is determined by its combination of parts and their qualities – i.e., by the combination of characteristics of specified qualities. The form of a genus will be determined by the combination of parts without regard to quality – i.e., by the combination of characteristics without regard to quality. (E.g., men are of the same species, since their essences have the same parts with the same qualities – e.g., eyes are similar as are noses, etc. – whereas not all fishes are of the same species, but are of the same genus, since their parts are the same, but differ in quality--e.g., fins are located differently, etc.) Thus the appropriate form of definition for a taxon is the specification of those characteristics (with regard to kind and, possibly, quality) which constitute its essence. That is, the definition of the individual's species and genus is to be given by listing a set of necessary and sufficient characteristics which constitute its essence. Insofar as Aristotle deals with the higher taxonomic categories, it is on the basis of parts common to the essences of genera. Thus Aristotle's basic theory of taxonomy is one wherein natural taxonomies classify on the basis of the essences of individuals (on the species level) and their similarities (in the higher taxonomic categories), where essences can be specified in terms of the necessary and sufficient characteristics of the essence or forms of the individuals. These essential characteristics, which collectively constitute the form of an individual, are largely (but apparently not exclusively) morphological.[8)]

Through the Middle and Dark Ages, taxonomy remained essentially Aristotelian, with nothing in the way of significant change. By the 16th century, taxonomists (e.g., Cesalpino, Harvey, and DeGraaff) were beginning to discover more and more morphological characters which were added as "essential" characters to the Aristotelian definitions of essences, and by the 17th century (with the work of Malpighi and Grew), plant anatomy had come into being. The net effect of this increase in "essential" characters for classifying organisms was that the doctrine of essential natures was eroded until the point that Aristotle's doctrine of species as essences was rejected. The influential John Ray (1627–1705) maintained that all parts of the plant should be used in classification, and not just selected essential ones. For Ray, species were natural and immutable, but not determined by essences.

The changes in taxonomic theory begun under Ray find fairly full articulation in Linnaeus' (1707–1778) work. According to Linnaeus, a natural taxonomy is one which serves to teach the "nature" of plants and animals, rather than merely distinguishing one from another; natural taxonomies do so by being based on "the simple symmetry of all the parts." Linnaeus is differing from Aristotle on what it is to be a natural taxonomy, though not as radically as it might seem. For according to him, species are fixed,[9)] being stable and timeless classes of organisms which are to be defined in terms of the symmetry of all parts, including physiological as well as morphological characters. In a word, species are like Aristotle's except that all characters are essential. And the form of definition is exactly the same as Aristotle's– in terms of necessary and sufficient characteristics. Although Linneaus did succeed in developing a very good artificial taxonomy (based on sexual organs) for plants, he never succeeded in developing a natural taxonomy.

And it is no wonder that Linnaeus failed; for if his dictum that natural species are to be defined in terms of *all* characteristics of the organism, where these become necessary and sufficient characteristics for species membership, is taken at face value, there will be far too many species (e.g., males and females, varieties, deformities, etc., will end up as different species). If species are to be defined in terms of necessary and sufficient characteristics, and such proliferation of species is to be avoided, of necessity only certain of an organism's characteristics may be used to define species. Two obvious ways of resolving this dilemma present themselves: reintroduce essentialism, or maintain that species are arbitrary.

The latter approach was taken by Buffon (1707–1788), who initially rejected Linnaeus' fixity of species, but accepted the idea that species are not essences. Noting the variations and diversity of characteristics within a species, he concluded that species are arbitrary and only can be defined arbitrarily. Later he changed his position and maintained that while genera, orders, classes, etc., are arbitrary and unreal, species are real, being non-arbitrarily defined on the basis of interfertile breeding and the production of fertile offspring. Buffon's taxonomic theory thus is that only on the species level are taxonomic categories real, and they are to be defined solely in terms of an interbreeding criterion. There is no place for diagnostic characteristics in defining species. On the higher levels, no taxonomy can be natural: taxonomic categories are to be defined arbitrarily in terms of necessary and sufficient characteristics.

Similar positions, which nonetheless retreat from Buffon's nominalism, were advanced by Adanson (1727–1806) and others. Adanson maintained that natural taxa are based on affinity which is to be measured by taking all characters into consideration, taxa being separated from each other by means of correlated features. This notion was further developed by Vicq-d'Azyr (ca. 1792) who treats a natural class as the result of assembling species which have the greatest number of resemblances to each other where these resemblances are greater than they have to specimens in other classes. Here we find a genuinely different approach to taxonomic theory. As with Linnaeus, a natural taxonomy is based on all characteristics, but it does so in terms of resemblances, allowing speciation in the face of divergence; it is a crude approximation to the modern numerical taxonomic approach of defining taxa via clusters of characteristics which do not provide necessary conditions for inclusion in the species. The upshot of doing things this way is that one can obtain a kind of natural taxonomy which admits intraspecies variation.

Closely related to this approach to handling variations is the neo-Platonist account developed by Louis Daubenton (1716–1800) and Cuvier (1769–1832), wherein classification is on the basis of prototypical organisms – basic plans inherent in nature – of which individual organisms were descendent variations. Here, a natural taxonomy would be one which classified on the basis of these prototypes with the species definition giving the prototypical ancestor's necessary and sufficient conditions. Thus, like Buffon, breeding true (but with variations allowed) is the mark of a species. This is the ancestor of typological classification.

Lamarck (1748–1829) was a staunch opponent of Cuvier's neo-Platonism. In 1812 he had discovered geographical variation among species, which prompted him to develop a theory of evolution. With his theory of evolution, the doctrine of fixity of species (which Linnaeus, Cuvier, et al. had defended) was rejected. For, following de Candolle (ca. 1819), he maintained that physiological (functional) characteristics were useless in classification, the only legitimate basis being comparative morphology. A species, then, is a group of individuals with the same or similar morphological characters, and as these morphological characters change as a result of evolution, species, too, must evolve. Species at best can be construed as sufficiently similar collections of organisms *at the moment,* and such a collection of organisms can only be called a species for the time being. Species, classes, genera, et al., thus are unreal. All taxonomy is artificial. The form of definition in an artificial taxonomy still is the Aristotelian one of necessary and sufficient characteristics. Although Darwin's (1809–1882) evolutionary theory was quite different than Lamarck's, his basic positions on these taxonomic issues were much the same as Lamarck's.

In these 17th, 18th and 19th century positions we find the seeds of most commonly held contemporary theoretical positions on taxonomy. First, there is the school of taxonomy known as *typology*. Its chief tenets are that species are natural groups in virtue of an ideal type of which individual specimens are variants, that the task of taxonomy is to discern the essences of these ideal types, and that these ideal types are to be defined in terms of necessary and sufficient characteristics. While this approach is not widely held today, it is the dominant method in many branches of paleontology.[10]) It also finds widespread employment in the social science.[11]) This position obviously is heir to the tradition associated with Cuvier.

The widespread acceptance of evolution has had the effect of most taxonomists abandoning anything like an Aristotelian essentialism or typology. Some taxonomists admit evolution, but maintain that there are fixed species which can be defined in Aristotelian fashion in terms of necessary and sufficient characteristics; this position, however, has found it virtually impossible to define species accordingly.[12]) Other taxonomists admit evolution and genetic continuity, adhere to Aristotelian definitions, and thereby deny species reality. This position is heir to Lamarck's position. Neither of these positions is particularly happy, however,[13]) and various different approaches have been tried to avoid these problems. One of these stems from Buffon's later position, maintaining that there is some sort of functional characteristic in terms of which natural species are to be defined. In its modern guise, this position, known as the New Systematics, takes as its basic position the idea that natural species represent phylogeny. Various attempts have been made to provide a definition of the species category under this approach. The more important definitions proposed for the species category are: (1) The *Genetic Definition:* "A species is a group of genetically identical individuals"; (2) the *Interfertility Definition:* "A species is a group of interfertile individuals"; (3) the *Biological Definition:* "Species are actual or potentially interbreeding natural populations which are reproductively isolated from other such groups"; (4) the *Evolutionary Definition:* "An evolutionary species is a lineage (an ancestral-descendant sequence of populations) evolving separately from others and with its own unitary role and tendencies." Unfortunately, none of these definitions proves satisfactory.[14]) Buffon's taxonomy was similar in that he attempted to define the species category along the lines of the biological definition; but to do so he had to eliminate the use morphological characteristics in defining taxa names because of the intra species morphological variation which occurs under the biological definition. To avoid this defect, the New Systematics has replaced the Aristotelian form of definition by a disjunctive *polytypic definitional form.* Simply put, this form of definition consists of clusters of sufficient characteristics which are characteristic of each variety or subspecies of a species, these clusters being disjunctively joined together. Unlike the Aristotelian definition it usually will be the case that no member of a species will have all the characteristics used to define a species, but rather will have only some of them. As originally proposed, polytypic definitions were limited to a finite number of clusters disjoined together.[15]) But Hull has shown that such polytypic definitions are inadequate.[16]) Rather taxa are cluster concepts because they are definable only disjunctively and because the number of disjuncts is indefinite.[17]) That ist, they are polytypic definitions with an indefinite number of disjuncts. The reason an indefinite number of disjuncts is required is that:

> "A definition of a species name is sufficient if it distinguishes the population of that species from all other known species. As evolution proceeds or as new species are discovered, the list of defining properties must be changed and expanded. A definition that was once sufficient is no longer sufficient. . . but the definition of the name of a taxon need not be prepared in advance for all possible contingencies. But it must be modified as new contingencies arise.[18])

That is, the definition must have an indefinite number of disjuncts since, as evolution progresses, new varieties of a species will emerge, and their defining clusters must be included as defining disjuncts; since we don't know what these clusters will be, these disjuncts cannot be specified in advance. The advantage of this approach is that a species definition remains fixed throughout the species' evolutionary course; the means consists in only partially specifying the definition; many taxonomists (and also philosophers) find this definitional form unacceptable qua definition because of its incompleteness, and so reject the proposal.

The failure of the New Systematics to provide a phylogenetic criterion of species, what they view as a failure to supply an adequate means for defining taxa under a phylogenetic account of the species category, and the extremely loose correlation between genotypic and phenotypic characteristics, has prompted another group of taxonomists to reject the idea that species definitions should have a phylogenetic basis. More precisely, they accept that species do have a phylogenetic basis, but deny that it can be employed practically to define species.[19]) Instead, they take their lead from Adanson and maintain that in a practical taxonomy species are those classes of organisms which have the greatest affinity to each other, where affinities are determined on the basis of as many characters as possible.[20]) This position, known as numerical taxonomy, takes such a taxonomy as being a natural one in the sense that it is adequate for general use by all scientists; an artificial taxonomy is one designed for restricted use for a special purpose.[21]) On the basis of a number of rather naive and questionable arguments, they maintain that there is a practical incompatibility between being natural in this way and defining species names phylogentically.[22])

In the survey just given, we have seen a number of different answers given to the four questions we asked about taxonomy. It will help if we summarize the various sorts of answers we have seen. For much of what follows in this paper consists in the philosophical evaluation and examination of these answers and their consequences:

1. Should taxonomies be natural?
   a) Yes, they have an objective basis in nature.
   b) No, they are arbitrary groupings made for convenience in achieving some goal.
2. What is the objective basis in nature which makes a taxonomy natural?
   a) Organisms have an essential nature or form, and a natural taxonomy classifies on the basis of common essential forms.
   b) Organisms are variations of one ideal type or another; a natural taxonomy is one which establishes taxa on the basis of these ideal types.
   c) Organisms stand in some phylogenetic relationship to some organisms, and fail to be phylogenetically related to others; a natural taxonomy is one which classifies on the basis of shared phylogenetic relationship.
   d) Organisms are to be classified on the basis of affinity in their characteristics; a natural taxonomy is one which classifies on the basis of affinity in the totality of characteristics of organisms.
3. How are taxa to be defined?
   a) By listing a set of necessary and sufficient characteristics.
   b) By a polytypic finite disjunction of clusters of sufficient characteristics.
   c) By a polytypic disjunction of clusters of sufficient characteristics, where there is an indefinite number of disjuncts.
   d) On the basis of some statistical measure of affinity which proceeds in terms of a large number of characteristics.
4. How are taxonomic categories to be defined?
   a) So as to yield a natural taxonomy.[23)]
   b) So as to be the most useful for the intended purposes of the taxonomy.

In the remainder of this paper, I try to philosophically investigate and evaluate these various answers to these questions. Doing so will lead me to some more general problems about taxonomic systems, their theoretical properties, and their epistemological status. Although my treatment thus stems from the biological considerations just sketched, my investigation will be much more general, and will give some insight into taxonomic systems in general.

II. Natural taxonomies as factual descriptions of nature.

For a while I want to confine my attention to natural taxonomies. Common to answers a)–d) for question 2 is a core of agreement: natural taxonomies have an objective basis in nature which artificial ones lack. What is this objective basis? A plausible answer, which is implicitly contained in several of the answer to question 2, is that natural taxonomies, if correct, are factually true, whereas artificial ones are not factually true or false. Resumably the underlying idea here is that artificial taxonomies are essentially conventional, whereas natural ones are factual descriptions of nature. This suggestion is an attractice one; for if it is correct, then a natural taxonomy apparently is nothing other than a variety of true or false realistic scientific theories, and presumably has the general properties characteristic of scientific theories.

In order to assess this suggestion, it is necessary to know what it is to be factually true. Elsewhere I have developed a philosophical model of factual truth which I want to assume here.[24)] As Strawson says, facts are what statements (when true) state or assert about the world.[25)] As such any detailed account of facts and empirical truth must be in terms of the world and its characteristics. The world is composed of *particulars* which have *intrinsic characteristics*--i.e., properties they possess and relationships which they enter into with other particulars independently of how anybody characterizes, conceptualizes, or conceives of them. There are two sorts of particulars--*simple particulars* and *complex particulars.* A complex particular is formed by the continuous instantiation of an intrinsic relationship by particulars; these instantiating particulars may be either simple or complex. Simple particulars are those particulars which would remain if all intrinsic relationships were eliminated from the world. Depending upon one's ontology, intrinsic characteristics may or may not themselves have characteristics, and these characteristics of characteristics themselves may or may not have characteristics, etc. It should be emphasized that whereas all objects are particulars, there are particulars which are not objects.

Intrinsic characteristics are extensional in the sense that they in principle could be specified in an extensional language having infinite disjunctions and distinct temporal names for each particular in the following way

$$\underline{p}\ \ [\text{an intrinsic property}] = \hat{\underline{x}}\ (\underline{x} = \underline{a} \vee \underline{x} = \underline{b} \vee \ldots)$$

where '$\underline{a}$', '$\underline{b}$', etc. are temporal names for particulars or else characteristics which have intrinsic properties; intrinsic relationships are specified analogously.[26)] It has been shown that the following principles of individuation hold for a worls of the sort just sketched:

Leibniz law: If $\underline{a}$ and $\underline{b}$ are identical particulars or characteristics, then for every intrinsic property $\underline{f}$, $\underline{a}$ is $\underline{f}$ if and only if $\underline{b}$ is $\underline{f}$.

Absoluteness of Identity:
If $\underline{a}$ and $\underline{b}$ are particulars which are the same $\underline{f}$, and $\underline{a}$ is $\underline{g}$, then $\underline{a}$ and $\underline{b}$ are the same $\underline{g}$ (where $\underline{f}$ and $\underline{g}$ are sortal intrinsic characteristics).[27)]

When an empirical predicative proposition of the form 'S is P' is asserted where 'S' is the logical subject and 'P' is the logical predicate of the proposition, its sense determines definite ostensible referents for '$\underline{S}$' and '$\underline{P}$'; the ostensible referent for '$\underline{S}$' is a possible particular or a possible characteristic capable of having characteristics;

the ostensible referent of 'P' is a possible intrinsic property--i.e., an extensional property which particulars or characteristics in the world *could* possess. Analogously, relational propositions determine ostensible subject and predicate referents. A proposition 'S is P' is *empirically true* if and only if (i) the ostensible referent of 'S' is a particular or a characteristic capable of having characteristics which exists in the world; and (ii) the ostensible referent of 'S' has an intrinsic property which has the same extension as does the ostensible referent of the predicate 'P'. 'S is P' is *empirically false* if it is not empirically true but refers to ostensible particulars. Whenever a proposition 'S is P' is empirically true, then it is a fact that the referent of 'S' has an intrinsic property whose extension is identical with that of the predicate 'P'. 'S is P' then is said to be *factually true.* Empirical truth and facts can be defined analogously for relational propositions, and these definitions can be extended in the usual fashion to include compound and general propositions. Put crudely, this analysis says that the empirical truth of 'S is P' depends on whether 'S' refers to a particular in the world possessing the intrinsic property specified by the predicate 'P', and if it does then it is a fact that S is P; Whether that particular has the intrinsic property specified by 'P' is in no way dependent upon how anyone conceptualizes the world.

If natural taxonomies are factually true, how is that feature to be analyzed under this account of facts and empirical truth? Several possibilities present themselves:

(A) Whether an individual belongs to a particular taxon T is a matter of empirical fact; hence, in a correct natural taxonomy, *being* a T is an intrinsic characteristic.

A second possibility is

(B) The definitions given for taxa, if correct, will be empirically true.

This in effect means that taxa definitions will be universal generalizations which qualify as scientific laws. This would make a taxonomic system a kind of scientific theory.

A third possibility is

(C) The definitions of taxonomic categories, if correct, are factually true.

This admits of at least two different interpretations:

(C') Being of ______ taxonomic category is a characteristic of the intrinsic characteristics which determine taxa.

(C") The definition of a taxonomic category, if true, specifies intrinsic characteristics common to each of the taxa it comprehends.

These possible explications of what it is to be a natural taxonomy in the sense of being factually true or false are, of course, not mutually exclusive; indeed it is logically possible that a taxonomy could be factually true in all four of these senses. The next several sections are devoted to the systematic development of these suggestions, and an examination how they relate to the various answers which have been given to the four main theoretical questions about taxonomy given at the end of the previous section. The account of the world, facts, and empirical truth just outlined is assumed throughout the discussion.

III. Is taxa membership a property of individuals?

Interpretation (A) of what it is for a taxonomy to be natural was that belonging to a particular taxon was a matter of empirical fact. More precisely (A) says that if an individual a is correctly classified as belonging to taxon T in a natural taxonomy, then the characteristic is a T (or *being* a T,)[28] is an intrinsic characteristic of a. Differently put, interpretation (A) requires that in a natural taxonomy, membership of an individual in a taxon be in virtue of possessing an intrinsic characteristic common to all members of the taxon which is possessed by no individuals which are not members of the taxon. Thus, for example, under (A) *being a homo sapiens* would be an intrinsic characteristic of men, and nothing correctly could be classified as a *homo sapiens* if he didn't have this characteristic; and the only flowers which correctly could be classified as *rosa multiflora* would be those having the intrinsic characteristic *being a rosa multiflora.* Thus interpretation (A) postulates that *being a rosa multiflora* is just as genuine a property as *being a desk, being red, being a Thompson gazelle, being a dog,* or *being an electron;* and a natural taxonomy will be one in which species membership will be in virtue of a single shared common intrinsic characteristic such as *being a homo sapiens* or *being a rosa multiflora.*

As such, (A) takes extremely seriously a doctrine on the problem of universals attributed to the later Wittgenstein by Renford Bambrough.[29] In the *Philosophical Investigations,* Wittgenstein raises the question what is common to all games. He then considers a number of characteristic features of games such as requiring skill, having a winner, etc. But Wittgenstein observes that none of these are common to *all* games, and concludes that "if you look at them you will not see something that is common to *all* [games], but similarities, relationships, and a whole series of them at that." That is, all we see is "a complicated network of similarities overlapping and criss-crossing: sometimes overall similarities, sometimes similarities of details."[30] Such similarities he calls *family resemblances,* and concludes that while there are no distinguishing characteristics common to *all* games, they all are games in virtue of standing in a family resemblance to each other. Concepts designating family resemblance classes have come to be known as *cluster concepts,* and biological taxa concepts often are alleged to be such cluster concepts. Bambrough interprets Wittgenstein's treatment of family resemblances as not denying that, e.g., games have anything in common, but rather as maintaining that "games have nothing in common except that they are games."[31] The problem of universals can be formulated as the question what all the referents of a general term must have in common, and Bambrough generalizes his interpretation of Wittgenstein's discussion of family resemblance to the following solution to the problem of universals: General terms in a language are non-arbitrary in that the class of individuals they designate is not a closed class; all the referents of a general term must have in common the possession of the characteristic specified by the general term; in addition there may be other characteristics common to all of them, but there need not be; what is required, however, is that there be objective similarities between the members of the class; these objective similarities sometimes will be such that there is only a family resemblance between the members of the class, but in other cases it may consist in the possession of certain *other* characteristics common to the members of the class but to not other individuals.[32])

If taxa are family resemblance classes and taxa names are cluster concepts, as is often suggested, then (A) together with Bambrough's interpretation of Wittgenstein has the following consequences for taxonomy: (1) In a non-arbitrary natural taxonomy, an individual belongs to a taxon T in virtue of possessing a characteristic common to all members of the taxon; (2) the common characteristic will be that of *being* a T; (3) wherever taxa are family resemblance classes, *being* a T will be the only characteristic common to all members of the taxon T; (4) in any natural taxonomy, there will be some objective similarities common to all members of the taxon. Thus in the ordinary language classification scheme, the individuals in the taxon *chair* all possess the common characteristic of *being a chair* and possess no other common characteristics, but do evidende various similarities to each other. And the common characteristic, *being a chair,* is an intrinsic characteristic in the sense of the previous section. Viewed from the perspective of Bambrough's interpretation, (A) stipulates that in a natural or non-arbitrary taxonomy, there is a single objective or intrinsic property characteristic of all and only those individuals which are members of the taxon T; this is the property normally designated by the predicates '*being* a T' or 'is a T', but may be designated by other predicates as well. (A) leaves open the question whether there are any other characteristics common to all members of T. Crudely put, then, (A) amounts to nothing more than the requirement that in a natural taxonomy there must be a single intrinsic property characteristic of all and only those individuals belonging to a given taxon.

Many biological taxonomists are not of the opinion that *being* a T, where T is a taxon, is an intrinsic or objective characteristic of the organisms placed in biological taxon T, [33]) and so would not be inclined to accept (A). Other than a committment to a conventionalist theory of taxonomy, I think much of the resistence to accepting *being* a T as an intrinsic characteristic can be traced to the fact that if *being* a T is an intrinsic characteristic of organisms, it does not have the same status as other intrinsic characteristics of organisms; in particular, it is not a diagnostic characteristic, and is not as readily recognized as diagnostic or other characteristics.[34]) That *being* a T is not as readily recognized as diagnostic characteristics does not, of course, mean that it is not an intrinsic characteristic – any more than the difficulty in recognizing electrons makes *is an electron* not be an intrinsic characteristic (though it certainly makes it ineligible for being a diagnostic characteristic). When this difficulty of recognizing *being* a T is coupled with taxonomy's typical insistence on operational definition, it becomes clear why taxonomists are reluctant to view *being* a T as an intrinsic characteristic. According to operationalism, the descriptive terms a science uses are of two sorts--those which refer to directly observable entities or attributes and those which do not; the former are called *observational terms* and the latter are called *theoretical terms.* The presence or absence of the characteristics referred to by observational terms can be checked readily. Since the presence of characteristics referred to by theoretical terms cannot be so readily checked, the only theoretical terms allowed are those which are fully "operationally" defined in terms of observation terms. In biological taxonomy this becomes the requirement that taxa names (being theoretical terms) must be defined in Aristotelian or polytypic fashion in terms of diagnostic characteristics specified by observational terms. This requirement in effect requires that taxa names be nothing other than abbreviations for the combined occurences of certain diagnostic characteristics--e.g., if Aristotelian definition is required, then the definition becomes

x is a T if and only if x has characteristics $c_1, c_2 . . ., c_n$,

and 'is a T' designates the combined occurence of the characteristics $c_1, . . ., c_n$; thus 'is a T' does not designate any distinct characteristic of the organism. This in turn prompts a denial of (A) and requires embracing a conventionalist position wherein taxa names are nothing other than conventional abbreviations for the combined occurence of certain diagnostic characteristics.

It has been amply demonstrated, however, that the above picture of theoretical terms and how they function in science generally is untenable.[35]) In particular, most theoretical terms in science do not admit of full specification and cannot be reduced to abbreviations for combinations of observational terms. The reasons leading to this conclusion, coupled with the notable lack of success in developing theoretically viable taxonomies in accordance with the dictates of operationalism. suggest there is little basis for hope that taxonomy will be any exception to these results.

If the claim is rejected that taxa names are merely abbreviated descriptions of clusters of diagnostic characteristics, then it becomes plausible to suppose that 'is a T' does refer to or designate a characteristic possessed by some organisms. But, then, *being* a T becomes an intrinsic characteristic of organisms. Once this is accepted, it is plausible to suppose that natural taxonomies are those whose taxa are such that *being* a T is an intrinsic characteristic of all and only those organisms which are in taxon T. But this is nothing other than mantaining (A). Thus, developments in philosophy of science strongly suggest that the operational imperative is nonsense, which in turn suggests that (A) plausibly might hold for taxonomic theory; in any case, the denial of (A) commits one inextricably to a pure conventionalist theory of taxonomy.

The acceptance of (A) suggests a new perspective for viewing taxa definitions. Consider an Aristotelian definition of a taxon- e.g.,

x is a T if and only if x is c_1 & x is c_2 & . . . . & x is c_n, where $c_1, \ldots, c_n$ are diagnostic characteristics. By the rejection of operationalism, 'is a T' does not mean the same thing as 'is c_1 & c_2 & . . .&c_n'; rather, the definition asserts a correlation between the occurence of the characteristic *being* a T and the combined occurence of characteristics $c_1, \ldots, c_n$ in organisms, stating that whenever *being* a T occurs, $c_1, \ldots, c_n$ also jointly occur, and vice versa. This asserted correlation is a factual matter, and so the definition takes on the status of an empirically true or false generalization about correlations of occurrences of characteristics in organisms. Thus, if (A) holds, then (B) – that taxa definitions must be empirically true or false in a natural taxonomy – follows naturally. This in turn indicates that if (A), hence (B), holds, question 3 (How are taxa to be defined in a natural taxonomy?) becomes an empirical issue. If (A) holds, *being* a T is an intrinsic property of organisms, and by (B), taxa definitions must be empirically true or false. If there is no set of characteristics $c_1, \ldots, c_n$ such that the Aristotelian definitional form is factually true, then there can be no empirically true taxonomy which includes T as taxon if Aristotelian definition is insisted on. Similary, whether T can be defined by polytypic definition in an empirically true taxonomic system depends upon what sorts of correlations occur between occurrences of *being* a T and diagnostic characteristics.

These last observations bring us back to Bambrough's interpretation of Wittgenstein. It is often alleged that taxa names are cluster concepts, and that there are no diagnostic characteristics common to and characteristic of all members of a taxon. Whether this is so is a matter of contingent fact. If it is so and (A) is accepted, then on Bambrough's account *being* a T will be a common characteristic of all and only those organisms in T; and new "definitional" forms will be required for taxa membership in an empirically true taxonomy. But regardless what form is required, Bambrough's account provides a means for still providing natural taxonomies with an objective factual basis in such cases, and thereby escaping from the unsatisfying conventionalist account of taxonomic systems. Doing so requires, however, acceptance of both (A) and (B). And the acceptance of (A) and (B) amounts to taking quite seriously the idea that natural taxonomic systems are a species of scientific theory, that they are factually true or false--an idea which must be taken seriously if the operational imperative is rejected for biological taxonomy.

Having seen the rationale underlying both (A) and (B), we can return to a detailed consideration of (A). As stated, (A) is ambiguous since it is not clear whether is a T must be an intrinsic property or relationship. Suppose an individual a is a T and is a T is an intrinsic relationship T. Presumably a will then be a complex particular, and so consists of the continuous instantiation of some intrinsic relationship R by particulars. This relationship R will have to be the T relationship. That this is so follows from the individuation properties of complex particulars--which are such that if the same particulars instantiate two different intrinsic relationships R_1 and R_2, two different particulars result, one from the instantiation of R_1 and the other from the instantiation of R_2.[36] From this it follows that if a is a T and T is an intrinsic relationship, then T must be the particularizing relationship R whose continuous instantiation comprises a. Thus, if a is a flower of species *Rosa multiflora,* under this proposal a would be comprised of particulars (e.g., perhaps cells, organs, or parts of the plant) standing in the *Rosa multiflora* relationship to each other; then it would be an empirical fact that a is *Rosa multiflora,* but it would not be an intrinsic property of a that it is a *Rosa multiflora.* Since is a T is an intrinsic relationship, *is a Rosa* is also an intrinsic relationship; it is obvious that being *Rosa multiflora* and being *Rosa* have different extensions, and so are different intrinsic relationship. As complex particulars are individuated on the basis of the intrinsic relationships which are continuously instantiated, [37]) it is impossible for the same particular to be both a *Rosa multiflora* and a *Rosa.* But under a standard taxonomy, anything which is a *Rosa multiflora* automatically is a *Rosa*; hence *is a T* cannot be an intrinsic relationship, and thus must be an intrinsic property.

Under (A), then, *is a T* must be an intrinsic property. Hence, if a is correctly classified as a T in a natural taxonomy, a must have an intrinsic property whose extension is idential with that of 'is a T'. Thus, if a is a *Rosa multiflora,* it is factually true that a is a *Rosa multiflora,* which is to say that a has an intrinsic property whose extension is identical with that of 'is a *Rosa multiflora'*--just as my desk is a desk since it has an intrinsic property whose extension is identical with that of 'is a desk'. Since every individual of species *Rosa multiflora* is also of genus *Rosa,* it also follows that a has an intrinsic property whose extension is idential with that of 'is a *Rosa'.* To summarize, (A) is to be interpreted viz.:

(A–1) In a natural taxonomy, whether an individual belongs to a particular taxon is a matter of empirical fact in the sense that being of taxon T *(being* a T, is a T) is an intrinsic property the individual does or does not have.

We mentioned above that if interpretation (A–1) is adopted, question 3 (How are taxa to be defined?) becomes primarily an empirical issue. Suppose, for example, that answer a) were correct--that taxa are to be defined in terms of necessary and sufficient conditions, then it would be the case that for appropriate intrinsic characteristics $c_1, \ldots, c_n$,

x is a T if and only if x is c_1 & x is c_2 & ... & x is c_n.

This implies that there is a constant correlation between the presence of the intrinsic property is a T and combined occurrence of the intrinsic characteristics $c_1, \ldots, c_n$. Whether such a correlation of intrinsic characteristics occurs is, of course, a contingent empirical matter, and if no such correlation occurs for any $c_1, \ldots, c_n$, then such a form of definition is impossible. And if such a correlation does occur, there may be more than one such correlation--e.g., in addition is a T being correlated as well with the combined occurence of $b_1, \ldots, b_m$, where $b_1, \ldots, b_m$ are not all the same as $c_1, \ldots, c_n$. Should this eventuality occur, then, more than one natural taxonomy using the same taxa would be possible for D, where the taxonomies differ on what characteristics are to be used to define D; these taxonomies, if natural in the sense of (A–1), would not disagree on which species an individual will be (except in cases of misapplication of definition). Thus there is the possibility that different but equivalent natural taxonomic systems could be adopted for different purposes. There is also the possibility that natural taxonomies could be formed for D which select different taxa; this would result in different and inequivalent taxonomies.

These considerations indicate that whether answer a) to question 3 is acceptable and the question whether more then one characterization is possible for a given taxon will be factual, empirical questions. And their answers will consist in establishing empirical generalizations about characteristic correlations within the taxon. Similarly, answers b)–d) to question 3) are empirical claims about characteristic correlations within the taxon. Actually, a)–d) are more generalized than this suggests. In their typical versions, these answers state that all taxa within the taxonomic system admit of the same form of generalization and, sometimes, that the same kinds of diagnostic characteristics are to be used in all such definitions. Again, these are empirical generalizations which will be empirically true or false. Thus, if (A–1) is accepted, then the issue of definitional form is empirical, not philosophical.

Before going on to consider interpretation (B), interpretation (A–1) should be distinguished sharply from a significantly weakened version of (A) which is satisfied by every plausible taxonomic system, whether artificial or natural:

(A–2) In a taxonomy where taxa are defined in terms of diagnostic characteristics, the specified diagnostic characteristics must be intrinsic characteristics of any individual belonging to a species defined in terms of them.

That is, whether natural or artificial, a taxonomic system must define its taxa and taxonomic categories in terms of the intrinsic characteristics which members of the domain D may or may not possess. That (A–2) is significantly weaker than (A–1)[38] indicates that (A–1) is a non-trivial notion of what it is to be a natural taxonomy.

IV. Taxa definitions as empirically true or false: fixed taxa

In the previous section we saw that interpretation (B) – that in a natural taxonomy the definitions given for taxa, if correct, will be empirically true–follows from the acceptance of (A–1); we will see below that it presupposes (A–1); it of course presupposes (A–2). In this section I want to consider (B) for cases where *fixed taxa* (in the sense that diagnostic characteristics for membership in the taxa are immutable) are assumed.

If (A–1) is assumed and taxa are fixed, two basic sorts of taxonomic systems are possible: essentialist and non-essentialist systems. Essentialism is the doctrine that of the intrinsic properties an individual has, certain of these, the essential characteristics, are necessary for the individual being the thing it is, whereas others, the accidental characteristics, could be different than they are without changing the thing the individual is. To take an example of Aristotle's, man is a rational animal; if an individual is not rational or is not an animal, then it couldn't be a man. Thus *being rational* and *being an animal* are essential characteristics for being a man. Men also either are or are not pug-nosed; but whether an individual is pug-nosed or not is of no relevance to whether it is a man. Thus, *being pug-nosed* is an accidental characteristic for men.

As a taxonomic theory, essentialism maintains that taxa are to be defined in terms of essential characteristics (in the above sense). Two versions of essentialism can be distinguished, which I call *invariance essentialism* and *nomic-essentialism.* Invariance essentialism takes the position that for each kind of individual there are certain intrinsic characteristics possessed by every individual of the kind. Thus, if something is a *homo sapiens,* then there are certain intrinsic characteristics common to all *homo sapiens.* And the taxon *homo sapiens* is to be defined in terms of these essential characteristics. A variation on invariance essentialism, which I call *polytypic invariance essentialism,* allows that for a given kind (taxon) there may not be any single set of essential characteristics, but requires that there be a finite number of sets of sufficient characteristics such that each individual of a given kind will always possess one of these sufficient sets of characteristics.

Both versions of invariance essentialism implicitly assume that (A–1) is met; for they assume that certain characteristics must be possessed to be of a given kind, and so being of that kind must be an intrinsic characteristic. Then invariance essentialisms are nothing other than versions of (A–1) wherein the definitions of taxa are in keeping with answers a), c), oder d) to question 2. By the discussion of (A–1) we thus see that taxa definitions are empirically true of false, depending on whether or not the diagnostic intrinsic characteristics correlate with the intrinsic characteristic is a T.

It should be apparent that invariance essentialism makes a taxonomic system relative to an initial choice of taxa. In effect essentialism assumes that there is a set of intrinsic characteristics (namely, *being* a T) such that every individual in the system's domain possesses one or more of them,[39]) and then views its problem as how to define the taxa corresponding to these intrinsic characteristics (e.g., the taxon T determined by possession of *being* a T). A set of definitions constitutes a solution to its problem, and this solution will be empirically true or false–in exactly the way specified in the above discussion of (A–1). Thus, we obtain the following version of (B):

(B–1) A taxonomy is natural if it is an invariance essentialism, meets condition (A–1), and its taxa definitions, if correct, are factually true empirical generalizations about the members of its taxa.

Notice that, depending upon the initial choice of taxa-determining characteristics, a number of different and inequivalent invariance essentialist taxonomic systems may be possible for a given domain D. Whether more than one is possible will depend upon the kinds of correlations which do obtain between intrinsic characteristics. And if more than one is possible, to argue that one is more natural than another requires a premise to the effect that one is more natural than the other; that is, in this sense of natural, all taxonomies meeting (B–1) are equally natural, and so if any is the more natural, it is in some other sense of 'natural'; most post-Aristotelian and pre-Linnaean taxonomies seem to have been invariance essentialisms.[40])

Whereas invariance essentialism holds that for each kind of individual there are certain intrinsic characteristics possessed by every individual of the kind, nomic essentialism maintains that for each kind of individual there are certain characteristics which *must* be possessed by all individuals of the kind. These are the essential characteristics for taxon membership. Thus, according to nomic essentialism, whether intrinsic characteristics are essential for taxon membership is a function of the laws of nature holding in the world.

An intrinsic characteristic possessed by an individual is an essential characteristic for membership in taxon T under a nomic essentialism if and only if any individual in the taxon has the characteristic in every possible world containing it which is consistent with the laws of nature. Thus, the essential characteristics of an individual in taxon T are intrinsic characteristics which it is *causally necessary* that the individual must possess if it is to remain in T; moreover, the essential characteristics for each taxon must be such that no organism not of taxon T possesses all of them.[41]) Nomic essentialism clearly assumes that (A–1) holds; for it amounts to the requirement that the definitions in a taxonomy stipulate a law-like correlation which holds between taxa membership and the possession of diagnostic characteristics in all causally possible worlds. In effect, then, nomic essentialism is an invariance essentialism on which has been imposed the requirement that the same correlations between taxa membership and diagnostic characteristics will hold in all other worlds compatible with the laws of nature. As such, (B) clearly is assumed in a nomic essentialism, though more stringent truth conditions are imposed on taxa definitions. We obtain the following version of (B):

(B–2) A taxonomy is natural if it is a nomic-essentialism, meets condition (A–1), and its taxa definitions, if correct, are factually true law-like generalizations about the members of its taxa.

Are alternative natural taxonomic systems possible for a given domain under (B–2)? There are two possible ways alternative taxonomic systems could be possible. First, it may be that there is more than one selection of taxa-determining properties which, given appropriate definitional forms for taxa and selection of diagnostic characteristics, will result in true nomic essentialisms. This, of course, will result in inequivalent taxonomic systems. Second, relative to a choise of taxa-determining properties, hence taxa, the correlations between possession of the characteristic *being* a T and diagnostic characteristics may be such that different choices of diagnostic characteristics will result in different but equivalent taxonomic systems for a domain D. Whether such alternative taxonomies are possible for a given domain D depends, of course, on the law-like correlations which obtain, and so is a matter of empirical fact. As such, the possible definitional forms for taxa is a factual matter.

In discussing both versions of essentialism, we have assumed tacitly that species are fixed in the sense that criteria for taxa membership are invariant over time. For essentialism this amounts to the assumption that the essential characteristics are invariant over time. Certain possible versions of evolution are compatible with this assumption – namely any version of evolution in which the variation in intrinsic characteristics of members of a given species is limited to accidental characteristics of individuals. No existing evolutionary theories meet this condition, but, as a matter of empirical fact, some such theory might be true.

Other than essentialist views, for fixed species only two other views seem plausible. First, one may deny that any characteristics are essential, and maintain that all that is required is that taxa definitions be empirically true correlations (possibly polytypic) of intrinsic characteristics with taxa membership. This position is nothing other than invariance essentialism minus the appilation 'essentialism', and differs in no significant way from invariance essentialism. Mentioning it does serve, however, to emphasize how little weight 'essentialism' bears in invariance essentialism. Second, one can maintain what I call an *affinity taxonomy.* According to an affinity taxonomy, taxa are not defined in either Aristotelian or polytypic fashion. Rather, taxa are groupings of individuals who have a mutual affinity. This affinity typically need not be in all characteristics, and allows some divergence in shared characteristics. Typically it proceeds by looking at a large variety of intrinsic characteristics for a group of individuals, then applies a measure of affinity and groups the individuals on the basis of greatest affinity. Numerical taxonomy is a paradigm example of an affinity taxonomy, though nonstatistical ones are conceivable. A number of different versions of affinity taxonomies are possible. Suppose (A–1) holds. Then there are two possibilities. First, the affinity relationship may be such that it is an intrinsic relationship; if so, then we obtain the following version of (B):

(B–3) A taxonomy is natural if it is an affinity taxonomy, meets condition (A–1), and, if correct, it is true that all members of a taxon T stand in an intrinsic "affinity" relationship to each other and also have the intrinsic characteristic *being* a T.

In such a theory, the definition of a taxon T will be that x is a member of taxon T if and only if x is in the largest class of individuals which stand in . . . .intrinsic relationship of affinity to each other. Taxa definitions thus are empirical generalizations and are factually true or false. It is also obvious that, depending on the constitution of the world, more than one taxonomy meeting (B–3) may be possible for a given domain D; relative to a given choice of taxa-determining characteristics, these alternatives all would be equivalent.

Second, (A–1) is met but the affinity relationship is not an intrinsic relationship. At first blush it boggles the mind to conceive of an affinity taxonomy where the affinity relationship is not an intrinsic relationship, but (A–1) is met. Yet with some thought and ingenuity, it emerges that the possibility is highly plausible, though making it so will require somewhat extended discussion. Doing so is worthwhile, however, because it yields what seems to me a highly plausible taxonomic theory.

As a starting point, I want to consider an account of scientific definition presented by Peter Achinstein in his book, *Concepts of Science.*[42]) He observes that in actual scientific practice the definitions given for terms such as 'copper', 'electron', 'oxygen', 'insect', etc., rarely provide necessary and sufficient conditions (i.e., they are non-Aristotelian definitions). Rather, the characteristics listed are cited as being *relevant* either in the sense that their presence tends to count in and of themselves, to some extent, toward classifying something as being, e.g., copper or in the sense that their absence tends to count, to some extent, against classifying something as, e.g., copper. Such properties are said to be *semantically relevant* to being copper. But for a thing to be copper, it need not possess all of the semantically relevant properties, as the semantically relevant properties are neither necessary nor sufficient. For example, having atomic number 29 is semantically relevant to 'copper', and in many contexts is required if something is to be correctly classified as copper, but in other contexts, it is not required. After presenting detailed and careful arguments in support of the above claims, Achinstein concludes that scientific definitions often are definitions meeting the following conditions:

> Suppose that as the term "X" is used by a speaker or group of speakers, $P_1, \ldots, P_n$ are properties semantically relevant . . . for X, though not logically necessary or sufficient; suppose also that they are among the most central ones for X. Then we might have:
> D_2: As the term "X" is normally used by . . ., items (both actual and hypothetical) are correctly classifiable as X's if and only of they have most, or at least many of properties $P_1, \ldots, P_n$ among others.[43])

Achinstein's discussion makes it quite clear that much of actual scientific definition does proceed in terms of D_2, and that this form of definition cannot be assimilated to polytypic definition.[44])

It is clear that Achinstein's "cluster" definitions do require that there be a certain affinity between all the members of a given class – namely that they stand in the relationship to each other of having a goodly number of the characteristics $P_1, \ldots, P_n$ in common. Now it is entirely plausible to suppose that this affinity relationship might not be an intrinsic relationship the members of a taxon enter into. At the same time it may be possible that the assertion 'x is a T if and only if x has a goodly number of the intrinsic characteristics $P_1, \ldots, P_n$' is empirically true or false.[45])
If so, then Achinstein's definitions establish an affinity between members of a taxon which is not an intrinsic relationship of affinity holding between members of the taxon. At the same time it does require that (A–1) and (B) hold. Since (A–1) holds, *being* a T is an intrinsic property; and the semantically relevant $P_1, \ldots, P_n$ must be intrinsic properties (in virtue of (A–2)). Thus the definition that a is a T if and only if a possesses a goodly number of the intrinsic characteristics $P_1, \ldots, P_n$ asserts a factually true or false correlation between the occurence of *being* a T and the joint occurence of a goodly member of $P_1, \ldots, P_n$, wherein not every one of the P_i always must be possessed by a T for the correlation to obtain. Thus we have the following version of (B):

(B–4) A taxonomy is natural if (A–1) is met, taxa are defined in terms of semantically relevant properties $\underline{P}_1, \ldots, \underline{P}_n$ (for a given taxon $\underline{T}$) in such a way that an individual is of taxon $\underline{T}$ if and only if it has most or at least many of the intrinsic characteristics $\underline{P}_1, \ldots, \underline{P}_n$, and this correlation of *being* a T and having most or many of $\underline{P}_1, \ldots, \underline{P}_n$, if correct, is an empirically true generalization.

Since Achinstein's definitional form has received little attention in the taxonomic literature, it is worth making a few comments about its potential applications in taxonomy. First, it can be incorporated into modified versions of polytypic and indefinite polytypic definitions. Taxonomies actually employed in biology, of whatever sort, in effect define the conditions for being *ideal* or *perfect* specimens of a taxon; they do not take into consideration damaged, deformed, etc. specimens. Accordingly, it is possible that there may be specimens of a given species and variety who lack certain of the defining characteristics which show up in the disjunct for the variety in the species' polytypic, or indefinite polytypic definition.46) In such a case, the lack of these characteristics by the specimen need not falsify the definition, require the introduction of a new variety, or any other revision in the taxonomic system. Rather, it's just an organism belonging to the taxon which lacks some of the defining characteristics. But, if polytypic or indefinite polytypic definitions are employed in the system, this is impossible since the definitions are exhaustive exceptionless biconditionals: Strictly speaking it can't be of the species if it lacks any of the conjoined diagnostic characteristics specified in the appropriate disjunct of the definition unless the definition is false. This indicates that, contrary to what is claimed, in actual practice polytypic definitions rarely meet the official criterion wherein to belong to a taxon an organism *must* have *all* of the characteristics occuring in one of the disjuncts; rather, these disjuncts really specify semantically relevant characteristics, of which a goodly number must be present in an organism for it to qualify the organism as belonging to the taxon.47) It might be objected here that deformed specimens in actual practice do not falsify taxa definitions and can be disregarded – perhaps with the explanation that, e.g., it is a crab which has lost a foot. I grant that this is what is done; but the question here is how is this practice legitimate. If taxa are classes of actual organisms and the definition *is* of the form '$\underline{x}$ $\underline{is}$ $\underline{a}$ $\underline{T}$ and only if $\underline{x}$ is $\underline{c}_1$ &. . .& $\underline{x}$ is $\underline{c}_n$' (or the polytypic analogue) and a deformed specimen of taxon $\underline{T}$ lacks, say, $\underline{c}_1$, then the specimen either cannot be of the taxon, or else the definition is false. If the definition is not falsified and the specimen is of taxon $\underline{T}$, it follows that the definition does not have the claimed form. Rather, it is a misleading way of stating, say, that *ceteris paribus* $\underline{x}$ is a $\underline{T}$ if he has all of the characteristics $\underline{c}_1, \ldots, \underline{c}_n$, or most of them, or that if it were normal, it would or some such hedge clause. That is, in actual practice, something like Achinstein' definitional form is used in biological taxonomy.48)

I can see only two ways of avoiding this conclusion. One is to adopt an essentialism (most likely, a nomic one); historically this problem was one of the reasons for adopting an essentialism, the idea being that deformities which did not affect taxa membership were deformities in accidental characteristics. Of course, the difficulty encountered in finding such essential characteristics makes the possibility of such an essentialist taxonomy highly unlikely. The second way is to deny that taxa are collections of organisms, and construe the members in taxa as abstracted replicas of organisms. E.g., one could construe taxa as collections of phenotypes of which actual specimens are imperfect realizations; then a deformed specimen would just be an imperfect specimen of phenotype P, and hence its phenotype explicitly satisfies the definition (having $\underline{c}_1, \ldots, \underline{c}_n$) even though the specimen does not. Thus, unless this latter proposal is accepted,[49]) something like Achinstein's definitional form will be required; and if the proposal is accepted, something like Achinstein's definitional form may be required to specify the conditions for specimens being of a given phenotype.[50])

It is also possible to have affinity taxonomies in which (A–1) is denied. In such taxonomies, taxa will be classes of individuals standing in some relation of affinity to each other. This affinity relationship may or may not be an intrinsic relationship. But since (A–1) is denied, being of a given taxon is not an intrinsic property; hence, either way the taxon definition cannot be factually true. Rather, the definition does nothing more than to establish the convention that taxon name 'T' will be used to refer to the largest class of individuals in domain D which stand in such and such an affinity to each other. (B) clearly does not hold for such taxonomies; and it is quite clear that such taxonomies will be artificial in any reasonable sense of the term. Given appropriate choices of affinity relationships, such artificial taxonomies may be of considerable scientific use (cf. Section VII).

So far in this section I have concentrated my attention on various versions of (B) and the ways they render taxonomies natural. Now I want to consider systematically what happens if (B) doesn't hold; (A–2) is assumed to hold.[51]) First, suppose that (B) doesn't hold for a taxonomic system. There are two ways in which (B) could fail to hold, depending whether is a T is an intrinsic characteristic or not. If it is an intrinsic characteristic, then the failure of (B) must be in virtue of the fact that the correlation between the intrinsic characteristic is a T and the diagnostic intrinsic characteristics specified by the definition fails to be realized--i.e., if the definition is factually false. That is, the taxonomy is a putative natural taxonomy, but fails to actually be one because it asserts false generalization about the world in its taxa definitions. Strictly speaking, this is incompatible with a denial of (B); for (B) says only that the taxa definitions, *if correct,* will be empirically true. Thus, if (B) fails, (A–1) also must fail. Suppose then, that (A–1) doesn't hold. Then (B) could not be met since the definition fails to establish an empirically true or false correlation between diagnostic characteristics and *being* a T. The affinity taxonomies where (A–1) fails (discussed above) constitute a species of this sort of taxonomy and the same considerations raised above with respect to such affinity taxonomies apply equally well here. And the conclusion is the same: If both (B) and (A–1) do not hold for a taxonomy, then the taxonomy is artificial and conventional. The above, of course, is an insufficient discussion of conventional taxonomies; they will be given more extended treatments in Section VII below.

V. Taxa definitions as empirically true or false: variable taxa

In examining (B) so far, I have confined my attention to taxonomic systems with fixed taxa in the sense that the criteria for membership in a taxa are invariant over time. I now want to remove that restriction and consider cases where taxa are not so fixed. The considerations raised at the end of the previous section in effect establish that whenever (B) holds, (A–1) also must hold; since the considerations raised there in no way depended on the assumption that taxa were fixed, the result is completely general. Thus we may assume (A–1) holds whenever (B) holds.

If a taxonomic system has reasons for not assuming fixed taxa, it is because there is reason to suppose that the diagnostic characteristics which could be used to define taxa vary over time; often the motivation underlying this assumption is some sort of evolutionary theory. If (B) holds for a taxonomic system, (A–1) also must. Hence, we obtain the following:

Principle of limited variation:
There is a limit to the variation an organism of taxon T may undergo--namely, it cannot lose the intrinsic characteristic, *being* a T and still remain in the taxon.

This is a double-edged principle. One edge allows unlimited variation in an organism's characteristics; the other edge puts limits on the variation the organism can undergo and still remain in the taxon. It is the task of the taxa definitions to specify the latter edge, by stating the extent of allowable variation.

In our survey of taxonomic history only two main approaches to specifying this limit were found – the ideal-type or topological approach and the phylogenetic approach. I will begin by considering the ideal-type approach.

An *ideal-type-taxonomy* asserts that for each taxon there is an ideal specimen of the taxon, and that all members of the taxon must stand in some relationship to that ideal specimen. Presumably the taxon is to be defined by means of an Aristotelian-form definition of the ideal spcimen and a specification of the relationship which must hold between the ideal specimen and the remaining members of the taxon. E.g., in Cuvier's version of an ideal-type taxonomy, the relationship is one of ancestral descendence. This much is clear about such taxonomies if (B) is to hold: (i) all the members of a taxon T must possess the intrinsic characteristic *being* a T; (ii) the relationship relating the ideal specimen to other members of T must be an intrinsic relationship which holds between the ideal specimen and all and only those individuals which have the intrinsic characteristic *being* a T; (iii) the ideal specimen must be a particular. Should any of (i)-(iii) fail the taxon definition could not be factually true or false, and so (B) would fail.

Depending on the way (iii) is satisfied, different versions of ideal-type taxonomies are obtained. First, (iii) could be satisfied because the ideal specimen is an actual specimen (e.g., a fossil). On this version, which might be called a *non-Platonic-ideal-type taxonomy,* a number of different and inequivalent taxonomies may be obtained for a given D, since there will be different specimens which could be chosen as the ideal specimen for a given taxon. Since it is unrealistic to suppose that a single specimen could be, e.g., ancestor to all members of a taxon, this approach to taxonomy is unlikely to be acceptable for biological organisms. Second, (iii) could be satisfied in virtue of the ideal specimen existing, not as an organism, but as an ideal abstraction which also is a particular. The paradigm of this case is Plato's metaphysics wherein physical chairs existed as well as the *form* chair (the ideal specimen) itself, and all physical chairs stood in a *mimetic* (resemblance) and *methexic* (participatory) relationship to the form chair. The ideal specimen could be an independent existent, as were Plato's *forms,* or it could exist as an intentional particular.[52] Such a position is generally viewed as methaphysically excessive, and is repugnant to most contemporary philosophers. Nonetheless, if such a position, which might be called *Platonic ideal-type taxonomy,* were viable, condition (B) obviously would be met. If the ideal type is an independent existent, presumably only one correct taxonomic system would be possible for a given domain D; under the intentional particular version a number of different taxonomic systems could be correct for a given domain, depending on what ideal specimes were conceived. Thus, for ideal-type taxonomies, we obtain the following version of (B):

(B–5) A taxonomic system is natural if it is an ideal-type taxonomy where each taxon has an ideal specimen with specified intrinsic characteristics, (A–1) holds, and the taxon definition, if correct, specifies a factually true relationship which holds between each member of the taxon and the ideal specimen.

The shortcomings mentioned above for both sorts of ideal-type taxonomies are sufficient to render such taxonomies implausible in virtually every biological application as well as for many other areas of investigation.[53]

If an ideal-type taxonomy fails to meet (i)-(iii), then (B) is not met. Such a taxonomy will be a version of one of the sorts of taxonomies considered in Section III: e.g., if (A–1) holds, it will be one where no matter how much variation there is among members of taxon T, the members of T are similar to the ideal type (an affinity taxonomy); or, if (A–1) fails to hold, one where 'is a T' is conventionally reserved to apply to those individuals who show a certain affinity to the ideal specimen. Some versions of numerical taxonomy seem to be of this latter, conventional, sort.

The second approach to meeting the principle of limited variation in biological taxonomy is that of the phylogenetic approach. Several different possible versions can be distinguished. The first, which might be called *population-desdendence taxonomy,* is quite similar to the ideal-type approach. On this theory, one recursively defines a taxon viz: An initial natural population T_1, consisting of all of the individuals which have the characteristic *being* a T at time t_o, is defined so as to meet (B–1), (B–2), (B–3), or (B–4). Then the taxon T is defined as the class of individuals which either belong to T_1 or else stand in a stated intrinsic relationship to one or more individuals in $\underline{T}_1$. For example, $\underline{T}_1$ might be the class of all *Rosa multiflora* at given time (or evolutionary stage), and then $\underline{T}$ might be the class of all organisms which either are in that initial class or else are descendents of members of that initial class. Buffon's taxonomy (Section I) seems to have been of this sort; also a phylogenetic taxonomy which requires that species be defined in accord with the Evolutionary Definition (cf. Section I) could be of this sort (cf. also, below). The taxa definitions for such taxonomies will be factually true or false. Thus, the following version of (B) is obtained:

(B–6) A taxonomic system is natural if it is a population-descendence taxonomy wherein (A–1) holds, each member of a taxon $\underline{T}$ is either a member of a subclass $\underline{T}_1$ of $\underline{T}$ whose definition meets (B–1), (B–2), (B–3) or (B–4), or else stands in a specified intrinsic relationship to one or more members of $\underline{T}_1$.

It should be noted that a number of different and inequivalent taxonomic systems which meet (B–5) may be possible for a given domain $\underline{D}$, depending upon the choise of $\underline{T}_1$ and the intrinsic (descendence) relationship chosen.

Most phylogenetic taxonomic systems are not population descendence taxonomies. Typically a phylogenetic taxonomy begins by stipulating a criterion for being a taxon of a given taxonomic category – e.g., the Biological Definition of species might be adopted, which requires that species be classes of individuals which are actual or potentially interbreeding natural populations which are reproductively isolated from other such groups. This defines what it is to be a taxon in the taxonomic category *species.* Then the individual taxa must be defined in terms of diagnostic characteristics so as to meet the condition that the class so defined also qualifies as a taxon of the requisite taxonomic category. E.g., a specimen of *Rosa multiflora* must belong to exactly one species taxon. The definition of that taxon must be such that the class of all and only those individuals which satisfy that definition qualifies as a species – e.g., if the Biological definition is adopted, the class defined must be a population of actual or potentially interbreeding natural populations which are reproductively isolated from other groups. For (B) to be satisfied it is clear that the taxa definition if correct, must be factually true, and also that the criterion for a class being a taxon of the specified taxonomic category must be factually true of the class defined by the taxa definition. Thus, the following version of (B) is obtained:

(B–7) A taxonomic system is natural if (A–1) is met, it is a phylogenetic taxonomy, the definition for each taxon $\underline{T}$, if correct, is factually true, and the definition of what it is for a taxon to be of the appropriate taxonomic category $\underline{C}$, if correct, is factually true of the class determined by the definition of taxon $\underline{T}$.

Since the majority of contemporary biological taxonomic systems (those advocated by adherents to the New Systematics) are phylogenetic taxonomies which are supposed to meet (B–7),[54]) some extended discussion of such systems is in order. Depending on the forms of taxa definition and the choice of criterion for being of a taxonomic category, a number of different taxonomies meeting (B–7) could be possible for a given domain D.

The many attempts to produce a taxonomy meeting (B–7) have not been particularly successful. Two different and distinct sorts of difficulties have been experienced. First, it has proved extremely difficult to produce an adequate criterion even for being a taxon of the species category. None of the definitions (Biological, Genetic, etc.) considered in Section I are satisfactory,[55] though recently a polytypic definition has been suggested which may prove adequate.[56] Second, there has been a problem defining the individual taxa in terms of diagnostic characteristics. The problem, simply put, is that all the plausible criteria for being a taxon of a given taxonomic category allow extensive and seemingly unpredictable evolutionary variations in the phenotypic characteristics of organisms in the taxon. Thus, while it is possible to disjunctively specify (using polytypic definitions) the known variations in the diagnostic characteristics, the subsequent variations resulting from evolution cannot be fully anticipated. The most that can be explicitly given in a definition of a taxon is a polytypic definition which covers all the known varieties of the taxon, but does not cover the unknown later members of the species which may have different phenogypic characteristics because of evolutionary changes. The result of this is that if standard polytypic definitions are used, the class of organisms defined will be a proper sub-class of the taxon in question if the taxon is still evolving; and this proper sub-class generally will not meet the criterion for being a taxon of the taxonomic category. (B–7) thus is violated. Hence, polytypic definitions generally cannot be employed for still-evolving taxa in a taxonomy natural in virtue of meeting (B–7)

As was mentioned in Section I, David Hull has suggested that the problem can be overcome by adopting indefinite polytypic definitions for still-evolving taxa. E.g., one would define species T by saying

$\underline{x}$ is of species $\underline{T}$ if and only if $\underline{x}$ is $\underline{C}_1 \vee \ldots \vee \underline{x}$ is $\underline{C}_n \vee \ldots$,

where $\underline{C}_1, \ldots, \underline{C}_n$ are clusters of diagnostic characteristics which constitute sufficient conditions for being of each of the *known* varieties of organisms of taxon T, where it is necessary that specimens of any of the known varieties of T must meet one of the $\underline{C}_1, \ldots, \underline{C}_n$. In addition to $\underline{C}_1, \ldots, \underline{C}_n$, an indefinite number of additional *unspecified* clusters are included in the definition to cover unknown varieties evolution will subsequently bring into existence. Without quibbling over whether a partially specified definition is a definition, it should be noted that if such indefinite polytypic definitions are used to specify taxa in a phylogenetic taxonomic system, the system cannot meet condition (B–7). That it cannot is easily shown. For (B–7) to be met the definition of the taxa must be factually true. But as stated, an indefinite polytypic definition is not factually true or false: for if indefinite polytypic definitions are required, then it will be false that every individual of domain D which has the intrinsic property *being* a T (all the members of taxon T, since (A–1) is presumed to hold) will have the intrinsic characteristics in one of the specified clusters $\underline{C}_1, \ldots, \underline{C}_n$ – some varieties to emerge later will not. Whether the definition is empirically true will depend upon the unspecified clusters in the definition. Hence, the most that can be said of indefinite polytypic definitions is that *when made fully explicit they will be factually true or false.* But, an indefinite polytypic definition made fully explicit is nothing other than a polytypic definition. So if indefinite polytypic definitions actually are required in a taxonomic system, (B–7) fails for that system. Insofar as such systems are supposed to be natural in virtue of (B–7), the suggestion that indefinite polytypic definitions be adopted is unsatisfactory.

Since polytypic and indefinite polytypic definitions alike fail to yield phylogenetic taxonomies in biology which are natural in virtue of meeting (B–7), is it to be concluded that natural phylogenetic taxonomies are impossible? Since (B–7) seems the obvious sense in which phylogenetic taxonomies are supposed to be natural, this may seem the only reasonable conclusion. Nonetheless, I think the conclusions is unwarranted--that there is a perfectly good sense in which taxa can be defined in phylogenetic taxonomies so as to meet (B–7). This sense does involve a form of definition whose acceptance will require rather drastic alteration in the way taxonomies currently are conceived.

If a taxonomic system rationally denies the fixity of taxa, it must assume there is some underlying mechanism which brings about the alteration in taxa characteristics; for without the regularities of some such mechanism bringing order to the variation in taxa characteristics, it will be impossible to specify the varying taxa. And if taxa with varying characteristics cannot be specified for a given domain, the taxonomic system must be replaced by one with fixed taxa.[57]) Thus, if a phylogenetic taxonomy is feasible for a given domain, there must be some underlying mechanism which determines the variation in the possible diagnostic characteristics for taxa. This underlying mechanism, in principle at least, can be characterized by a scientific theory. Suppose we do have a theory which describes the variation in at least some of the possible diagnostic characteristics for taxa; this theory obviously must be relative to a set of definitions of taxonomic categories and a specified set of possible diagnostic characteristics. It is clear that such a theory can be used to define taxa in terms of the specified diagnostic characteristics--namely, by the definition of taxonomic category we have a criterion for being a taxon of category C. For each taxon T̲ of a category C̲ the theory describes the possible variations in the specified diagnostic characteristics. Thus, the theory specifies for each taxon all of the various possible combinations of the selected diagnostic characteristics which may be manifested by members of the taxon. Accordingly, whenever phylogenetic taxonomies are feasible for a domain D̲, it is possible to have a phylogenetic taxonomy consisting of a theory describing the variation in a set of diagnostic characteristics together with definitions of the taxonomic categories.[58])

An example will help clarify this. For biological speciation a phylogenetic taxonomy apparently is required (although some taxonomists would deny it). The reason for this is that taxonomy typically works in terms of morphological diagnostic characteristics, and if evolutionary theory is correct, these morphological characteristics will vary over time for a species. But these morphological characteristics are phenotypic reflections of genotypic variations in the species: The variation in morphological characteristics is a result of alteration in the genotypes of members of the species. Moreover, evolutionary theory (together with natural selection theory) characterizes the mechanism which brings about that genotypic variations. Thus, following the discussion of the previous paragraph, from the coupling of evolutionary theory together with a definition of the taxonomic category *species* a taxonomic *system* results which defines species genotypically. I have emphasized the systematic character of the resulting taxonomy to underscore the point that the resulting taxonomy is systematic in a way the taxonomies heretofore employed by biologists are not. For the resulting taxonomic system does not by itself explicitly specify the various species. Rather, it is a system of taxonomic laws applying to biological organisms.

To see clearly what this means, a comparison with physical theory is illuminating. Classical particle mechanics is a system of laws characterizing the behavior of systems of interacting particles. When it is applied to a physical system through use of auxiliary hypotheses, it does yield a description of the system in question. Specifically, when various auxiliary hypotheses specifying the initial state of the physical system (its configuration at some time t), various boundary conditions, etc.,59) are specified, then the laws of the theory (the "equations of motion") can be solved to obtain a special-case solution of the theory which is a law describing the behavior of the system in question. The special-case law *does* describe a particular system of interacting particles. An analogous situation prevails for evolutionary theory: the laws of the theory describe the sorts of genotypic variation which are possible for populations. Relative to the genotypic profile of a population at a given time, the theory describes the possible genotypic variations in that population over time.60) This class of possible genotypic variations in that population is not itself the species. Rather the species is that portion of this class which additionally meets the definition of being a *species* taxon – e.g., if the definition for sexually reproducing taxa were that a species is a population whose members consistently interbreed producing a reasonably large proportion of reasonably fertile offspring, then the species would consist of that temporal sequence of populations of genotypes such that the genotypes of the population do consistently interbreed producing a reasonably large proportion of reasonably fertile offspring. But this is just a sequence of populations of genotypes which have certain reproduction rates,61) and so the species will be the class of genotypes within the class of possible genotypic variations which have appropriate reproduction rates for a given time.62) Thus evolutionary theory together with the definition of taxonomic categories provides us with a systematic definition of taxa, wherein definitions of a particular taxon can be obtained by use of auxiliary hypotheses specifying the genotypic distribution of the taxon at some time together with other boundary conditions.63)

I imagine that many taxonomists may be willing to grant that such an evolutionary genotypic taxonomy is, in principle, possible, but would deny that it has any relevance to phylogenetic taxonomy. The most important reason given for this position probably would be the following: From the phylogenetic perspective taxonomy is an ordering of phenotypes, not genotypes, and since the slip between genotype and phenotype is loose, the sort of genotypic taxonomic system proposed will not translate into a phenotypic classification. To quote Simpson, "genetic evolution and somatic evolution are not identical or precisely parallel and. . . . it is somatic evolution that is directly pertinent in taxonomy."[64] While what Simpson says is true, the objection is misguided and turns on a failure to understand the nature of scientific systems. An illustration from physics will help indicate what the confusion inherent in the objection is. In classical particle mechanics the theory describes and provides a taxonomy of mechanical systems in terms of an underlying mechanism; these systems are described by the theory in terms of abstract, idealized parameters such as position and momentum under idealized conditions. For many purposes, these descriptions are inadequate. E.g., in engineering applications, it is desirable to describe and classify these systems as being, e.g., inclined planes of such and such dimension, angle of inclination, and friction coefficients, etc. For such systems, there are local laws which describe connections between these applied descriptions and the theoretical descriptions of the systems. These laws in effect say that if a mechanical system has such and such characteristics, then it corresponds to such and such system of particles whose current positions and momenta are. . . . These local laws are not operational definitions of the theoretical parameters; they are not definitions at all.[65] They are laws describing sufficient conditions for being such and such a sort of system. They are procedures for applying the theory to the phenomena, and have no definitional status whatsoever.

The situation is similar in phylogenetic taxonomy. Taxa are organisms; each of these organisms has a specific genotype and a specific phenotype. The phenotypic variations due to evolution in practice preclude giving an exhaustive phenotypic definition of a taxon; but it is possible to give an exhaustive genotypic characterization of taxa by specifying the taxonomic categories and the evolutionary laws at work. The portion of a taxon extent at any given time can be investigated with respect to phenotypic characteristics, and general phenotypic laws can be determined of the form

If $\underline{x}$ has phenotypic characteristics $\underline{C}$, then $\underline{x}$ is of taxon $\underline{T}$,

where $\underline{C}$ is a cluster of phenotypic characteristics. These laws in effect will be laws correlating phenotypic characteristics with genotypes; and they will be factually true or false. On the basis of these one is able to classify organisms on the basis of phenotypic characteristics; and doing so in no way requires knowing what the evolutionary future of the taxon is. At a subsequent point in the taxon's evolutionary history these laws will still be true, even though there may be no individuals which manifest the specified phenotypic characteristics. For new characteristic phenotypic combinations introduced by evolution, new laws may be discovered which are empirically true; and these may be used to classify organisms by taxon. Thus all the practical advantages of phenotypic classification are obtained, even though taxa are genotypically defined; and doing so in no way requires knowing the general mechanism whereby phenotypic characteristics are manifestations of genotypic characteristics.

It is worth comparing this proposal with Hull's indefinite polytypic definitional form of phylogenetic taxonomy. Hull's account defines a taxon in terms of an indefinite disjunction of clusters of phenotype characteristics; if fully specified, this will specify a class of organisms. Since these individuals have genotypes, it also indirectly specifies a class of genotypes which meets the taxonomic category definition and is consistent with evolutionary law. But the evolutionary law is not definitive of the taxon--the indefinite disjunction is. And we have seen that this makes the taxonomy non-natural in virtue of violating (B–7). In the alternative I've suggested, the evolutionary laws together with the taxonomic category definition specify the taxon by establishing laws which describe the variations in genotype possible for organisms of a given taxon; since (A–1) holds, if correct, these laws will be factually true. Hence, the taxon definition is. The various *specified* clusters which occur in Hull's indefinite disjunctive definition now function as the clusters $\underline{C}$ of phenotypic characteristics which occur in the above mentioned laws specifying *sufficient* phenotypic conditions for membership of an organism in the taxon. These laws, if correct, are factually true; they are not definitional. A moral is to be learned from the comparison: Whereas taxonomy has proceeded on the assumption that the definitions of taxa had to be in terms of the diagnostic characteristics used to classify the organism, this is not necessary. Rather, the criteria for definition need not be the same as those for its employment so long as laws are possible which guarantee that the criteria for its employment necessitate that the definitional requirements are met.

One last possible objection to this alternative requires consideration. I can imagine some taxonomists objecting to my alternative account of phenotypic taxonomies as follows: "We are willing to grant that for some areas, taxonomies of the sort you suggest are feasible, but they don't work in the biological realm. For, in the absence of full knowledge of the subsequent reproductive isolations which a species will encounter, it is impossible for even the special case taxonomic laws to give a characterization of the genotypes actually occuring in the species population at different times; the most that these laws can yield is a characterization of the potential genotypes which, given appropriately favorable environmental conditions, could be manifested in species organisms. But this is not identical with the class of genotypes which actually will be manifested by organisms of the species in subsequent evolutionary history of the species, and so the laws do not define the species, but rather only define a class of organisms (by genotypes) which contains the species as a proper sub-class. Thus, in the biological realm, your proposal does not define species." This criticism is not without substance. It is true that the special-case taxonomic laws, by themselves, do not directly specify the class of genotypes actually manifested by organisms present in the species; rather, they specify which genotypes will be present as a function of other parameters such as reproduction rate and satisfaction of the species category criterion. For any given time this gives the actual makeup when the appropriate values for these other parameters are supplied. And this is sufficient; for all that is required to define a particular species is that necessary and sufficient conditions for membership in the species be given; and the special-case taxonomic laws do this, stipulating, e.g., that an organism is of the species T if and only if its genotype is in set G, that the genotype has reproduction rate r, and organisms of that genotype consistently interbreed with other genotypes producing offspring whose genotypes have an average reproduction rate greater than r'. Thus the species itself *is* defined; only the definition is such that it cannot yield a test condition for species membership at time t unless augmented by some additional empirical data about the population at t. That is, the objection misses the point because it confuses the definition of a species with a test condition for membership in the species.[66]) On my proposal, the two are relatively distinct, with the empirical generalizations stating that organisms with phenotypic characteristics C belong to taxon T serving as the test conditions for membership (but not for non-membership) in the taxon. Again, the moral to be drawn here is similar to that drawn immediately above: Although taxonomists traditionally have proceeded on the assumption that taxa definitions must be capable of functioning as test conditions for taxon membership, they need not do so. And if the taxonomy is a reflection or manifestation of the workings of deep mechsnisms operating over the taxonomy's domain (as taxonomy is for the New Systematics), it probably is a mistake to insist that the taxa definitions also function, by themselves, as practical test conditions for taxon membership.

It is clear from the above discussion that if the underlying theory is factually true, the combination of it with an appropriate definition for taxonomic categorie yields a factually true set of empirical generalizations about the range of possible variations in organisms of a given taxonomic category. Thus, such taxonomic systems meet the following version of (B):

(B–8) A taxonomy is natural if (A–1) is met and it yields (relative to appropriate boundary conditions, etc.) a set of laws characterizing the range of variations in combinations of a specified set of characteristics which members of a given taxon may manifest; if correct, these laws will be factually true. The actual classification of individuals may be done on the basis of characteristics not occuring in these laws so long as there are factually true laws established to the effect that any individual having these characteristics also has a combination of characteristics within the variation allowed for the taxon in question.

With a little reflection it easily is seen that whenever a taxonomic system satisfies any of (B–1) – (B–7), it also satisfies (B–8). (B–1) – (B–4) are cases where the laws state that there is no (or virtually no) variation in the specified set of diagnostic characteristics; (B–5) – (B–7) allow variation, but require in effect that the taxa definitions be laws describing the allowable variation. Thus (B–8) constitutes the substance of (B): Any taxonomy which fails to satisfy (B–8) is conventional and, in an important sense, artifical.

Finally, the results of this and the previous sections can be applied to the answers to question 2) in Section I: Any taxonomy committed to answers a)-c), and some taxonomies committed answer d) to question 2, thereby are committed to (B–8), and hence to (A–1) and (A–2).

VI. Are category definitions empirically true or false?

Interpretation (C) requires that definitions of taxonomic categories be factually true. Clearly, this requirès that *being of taxonomic category C* be an intrinsic characteristic entering into the definition. As indicated in Section II, (C) admits of two plausible interpretations.

Interpretation (C') states stat being of taxonomic category C is a characteristic of the intrinsic characteristic *being* a T which determines the various taxa T in the taxonomic category. (C') thus presupposes (A–1); in practice it also presupposes (B–8). Such a taxonomy will assign a taxon T to taxonomic category C only if *being of taxonomic category C* is an intrinsic characteristic of the intrinsic characteristic *being* a T. In such a case, the appropriate form of definition for taxonomic categories will be an empirical issue, being the question what regularities govern taxa characteristics having the intrinsic characteristic *being of taxonomic category C.* If one holds this position in answer to question 4), the issue is not wholly empirical since it requires commitment to the metaphysical claim that characteristics can have characteristics; this metaphysical commitment would be unacceptable to a nominalist.

Interpretation (C") states that the definitions of the taxonomic categories, if true, specify intrinsic characteristics common to all of the taxa it comprehends. If taxa are themselves complex particulars, then taxa (as opposed to their members or the intrinsic characteristics *being* a T) will have intrinsic characteristics. (C") then requires that *being of taxonomic category C* be one of these intrinsic characteristics. (A–1) and (B) clearly must hold, so the definition will be that

x is a taxon of taxonomic category C if and only if
x has intrinsic characteristics. . .;

thus the definition will state an empirically true or false correlation between the occurrence of *being of taxonomic category C* and other intrinsic properties of taxa. The exact form of definition to be used (Aristotelian, polytypic, etc.),. . . will be an empirical matter. This interpretation is able to accomodate the biological, evolutionary, etc., definitions of species where these definitions are interpreted as stating the intrinsic relationship in virtue of which the taxon is the compound particular it is. Its viability depends primarly on whether taxa are particulars.[67)] If taxa are not particulars, it appears (C") is impossible since *being of taxomomic categoriy C* cannot be an intrinsic characteristic of the individuals in the taxon (the taxa are what are of taxonomic category C, not the individuals in the taxa– and it must be an intrinsic characteristic of some particular. If taxa are particulars, the taxonomic category definitions will be empirically true or false, and the exact definitional form is an empirical issue.

Thus, if (C) is viable, it must be either as (C') or (C"). I would hazard a guess that (C") is the more likely version, since if (C) holds it most likely will be in terms of the taxa being constituted by their members standing in certain relationship to each other, and if these relationships are intrinsic, the taxa will be complex particulars. Whether (C') or (C") are viable depends on the way the world is. If (C) does hold, there may be more than one taxonomic system for a domain in D which meets it. If neither is viable for a taxonomic system, (C) fails, and so the definitions of the taxonomic categories will be conventional, being nothing other than a decision to reserve certain terms to discuss collectively certain taxa. Even though (A–1) and (B) hold, (C) may fail for a taxonomic system. If the system is a standard taxonomy, this will diminish the usefulness and empirical significance of the hierarchal ordering. But so long as (B) holds, the taxa will be empirically significant, and thus not wholly arbitrary; the taxonomic system will be artificial only with respect to its hierarchal ordering.

VII. Artificial taxonomies are not factually true or false

So far we have considered three main different ways taxonomies can be natural in the sense of being factually true: (A), (B), and (C). It follows from our discussion that they are sucessively stronger requirements in the sense that (B) presupposes (A), and (C) presupposes (A) and (B); (B) is compatible with the denial of (C). Clearly any taxonomy which denies (A), (B), and (C) must, in a very basic sense, be construed as an artificial taxonomy; for such a taxonomy does nothing more than establish conventional names for referring to classes of individuals which share certain feature in common. This is not to say that such artificial taxonomies are wholly without use; for they are based on a kind of affinity, and for some purposes may be adequate. For example, I might introduce a taxonomy with a taxon comprised of all humans born since 1946 in Italy who immigrated to the U. S. at age 16 or older, play the concertina, are 6'4" or more tall, and have virtually no body hair. Such a taxonomy presumably will be artificial in virtue of (A) not holding; still, if I were a scholarship director at a large university, it conceivably might be a useful taxon for classifying applicants as to scholarship eligibility. A less bizarre possibility would be classification by statistical profile or statistical covariance–e.g., the sorts of taxonomies numerical taxonomists produce. Such taxonomies may or may not satisfy (A); and it is possible that useful taxonomies which do not will result. Useful as they are, in a certain basic sense such taxonomies are artificial.

Taxonomies which do not satisfy (B) in virtue of satisfying (B–8) or one of its stronger versions (B–1) – (B–7) also will fail to satisfy (A–1). The plausible taxonomies which fail to satisfy (B) invariably will be of the affinity type, and have the same sort of plausibility mentioned above in the discussion of taxonomies which fail to satisfy (A).[68]) Again, these are artificial in the sense of being mere conventions for usage of classificatory terms.

It seems to me quite clear from our discussion in Section I that, as used throughout most of the history of taxonomy, the notion of a natural taxonomy tacitly has required that (A) and (B) be met. In most cases, (C) probably has been incorporated in the notion. For phylogenetic taxonomies, it is absolutely crucial for the species category, and probably necessary for higher taxa. In any case, a very plausible notion of naturalness for taxonomies is that (A), (B), and (C) must be met. However, if a taxonomy fails to meet (C) but meets (A) and (B), it does not seem reasonable to defame it as being artificial; for it is artificial in a much weaker sense than those taxonomies which are purely conventional2). I don't think much more can be said in the abstract than the following in deciding which of these two senses of 'natural' is to be preferred in taxonomy: If a taxonomy's intended applications are such that which taxonomic category a taxon is in is to be empirically significant and not just a matter of convenience, then (C) must be met; otherwise it is sufficient for just (A) and (B) to be met. Any taxonomy which fails to meet (A) and (B) clearly is artificial.

At this point is may prove helpful if the discussion thus far is summarized briefly. Through a consideration of the history of taxonomic theory, it was found that throughout most of taxonomic history natural taxonomies have been understood as having an objective basis in nature. As such natural taxonomies are neither stipulative nor conventional systems for classifying organisms. To say that natural taxonomies have an objective basis in nature is excessively vague, and I have urged that this most plausibly can be construed as stating that natural taxonomies are factually true or false of the organisms or individuals in their domains – i.e., that the "definitions" of taxa are factually true or false assertions about the characteristics distinctive of members of the various taxa. Having suggested this interpretation, I then turned to an investigation of a number of kind of taxonomic system which historically have been advanced, and tried to determine what conditions must be met by each of these sorts of taxonomies if they are to be natural in this sense. It was determined that conditions (A–1), (A–2), (B–8), and possibly some of (B–1) - (B–7) or (C) must be met if the taxonomy is to be natural. In general, it was seen that whether a given kind of natural taxonomy was viable for a given domain was an empirical matter depending on the sorts of correlations holding between the various characteristics of the individuals. In particular, the question of the appropriate taxa definitional form for given taxa in a domain is a question of empirical fast--just as the question of the appropriate law-form for a given domain is a question of empirical fact. Indeed, we saw that under this interpretation of what it is for a taxonomy to be natural, taxonomies turn out to be a kind of scientific theory which must be confirmed much like any other theory.

That natural taxonomies are empirically true or false opens the possibility that none of the sorts of taxonomic theory proposed ever could be empirically true; however, if the interpretation of phylogenetic taxonomies given – (B–8) – is accepted this is highly unlikely. 69) Nonetheless there is a definite possibility that this is so. Such a possibility could be realized in two ways. First, it could be that for the taxa chosen (e.g., those used in the usual biological taxonomies), (A) fails – i.e., that *being* a T is not an intrinsic property of organisms. Second, it may be that any taxonomic system using standard taxa is such that no allowed definitional form results in an empirically true taxonomic system. Should either of these possibilities occur, the only taxonomic systems of the desired sort will be artificial or conventional in the sense discussed above. As many taxonomists today are convinced that only conventional taxonomies are possible, somewhat more systematic consideration should be given to conventional taxonomies than has been given above.

Suppose a taxonomic system employing taxa $\underline{T}_1, \ldots, \underline{T}_n$ is desired and (A–1) is denied for the taxa. Such a taxonomy clearly will be conventional in the following sense. If a membership in a taxon $\underline{T}$ is defined, say, as

For all organisms $\underline{x}$ ($\underline{x}$ belongs to $\underline{T}$ if and only if $\underline{x}$ is $\underline{C}_1$ or . . . or $\underline{x}$ is $\underline{C}_n$),

where $\underline{C}_1, \ldots, \underline{C}_n$ are combinations of diagnostic characteristics, the definition cannot be empirically true or false; for if it were, (B) would hold, which is impossible since (B) presupposes (A–1) and the latter has been denied. Rather, the definition can be nothing other than a decision to define any individual having $\underline{C}_1$ or . . . or $\underline{C}_n$ as being in taxon $\underline{T}$ and vice versa. Thus to say $\underline{a}$ is a $\underline{T}$ is to say nothing more than that a is $\underline{C}_1$ or a is $\underline{C}_2$ or . . . . or $\underline{a}$ is $\underline{C}_n$. The fact that the definition of $\underline{T}$ is just a convention for naming does not mean that the taxonomy will not have interesting theoretical properties. For suppose that the $\underline{C}_1, \ldots, \underline{C}_n$ have been chosen in such a way that there are a number of significant correlations between things which are $\underline{C}_1$ or . . . or $\underline{C}_n$ and other organisms, genetic characteristics, etc. Then a number of significant laws may be established about the members of taxon $\underline{T}$. E.g., it might be possible to establish various genetic novelties or traits which are characteristic of, or distinctive to, members of taxon $\underline{T}$. In such a case, the taxonomy will be non-trivial and scientifically valuable. This indicates that being an artificial or conventional taxonomy need not be demeaning. Indeed, it may be that only such conventional taxonomies are viable for a given domain.

VIII. Natural taxonomies as universal systems of taxonomy

So far I have interpreted the thesis that natural taxonomies are those which have an objective basis in nature as the requirement that natural taxonomies be factually true or false. Our historical survey suggests a second interpretation which needs to be considered. Since Linnaeus many taxonomists have maintained that natural taxonomies are those which take consideration *all* characteristics; presumably the idea is that when one looks at organisms in all their characteristics, one finds that they naturally fall into similarity groups in such a way that all the organisms in a given group are much more similar to each other than they are from organisms in other groups, and that the similarity groups are noticeably distinct from each other.

Such taxonomies which take into consideration all characteristics and classify on the basis of greatest affinity are claimed to be natural. Their naturalness is attributed to the alleged fact that, by taking into consideration all characteristics, the resulting taxonomy is a universal one useful for all branches of sciences; on the other hand, it is alleged that those which take into consideration only selected characteristics are useful only for special purposes and hence are artificial. This interpretation of what it is for a taxonomy to be natural seriously challenges the interpretation I have given in preceding sections, and so deserves serious consideration. As this interpretation is being advanced today by numerical taxonomists, I now turn to a consideration of their development of the thesis.

According to numerical taxonomists, "a 'natural' taxonomy is a general arrangement intended for general use by all scientists".70) As such, natural taxonomies convey more information than arbitrary or special classifications, in the sense that "a system of classification is the more natural the more propositions there are that can be made regarding its constituent classes."71) That is, "a classification serving a large number of purposes will be more natural than one which is more specialized and that the most useful and generally applicable classification will be the most natural one. . . . [A] system is the more natural the more propositions there are that can be made regarding its constituent classes."72) Thus in a natural taxonomy, "many generalizations can be made regarding a natural family of plants (e.g., with reference to distribution, chemical properties, wood structure, etc.,), whereas regarding an artificial group very few of such generalizations are possible."73) Thus, "a natural classification is that grouping which endeavors to utilize *all* the attributes of the individuals under consideration, and is hence useful for a very wide range of purposes."74)

As stated, this characterization of what it is for a taxonomy to be natural is compatible with condition (B), hence with (A–1). But it is clear that many numerical taxonomists take this to involve a denial of (A–1), hence a denial of (B): taxa are just conventional classes.75) So far as this definition of naturalness entails a denial of (B) and (A–1), it is just nonsense. First, it is not at all clear that the more similarities there are between individuals in a class, the more generalizations can be made. Suppose my class consists of two extremely similar individuals – e.g., identical twins. Does it follow that more generalizations can be made? At best, it is true in a very trivial sense: since virtually all their characteristics are similar, it is possible to generalize "all members of the class S have character C" for a large number of characters; and I certainly can make many more generalizations about specific characteristics for this class than I could for the class of all human beings. Does this make classification on the basis of being identical siblings more natural than to classify on the basis of being human beings? By the criterion of the more generalizations, the more natural it is, it apparently is. The point here is that if one measures on the basis of the number of different characteristics about which generalizations can be made for a class, then the smaller the class, the more similarities possible, and the more generalizations possible; the natural taxonomy, then, would be one which puts each individual into a separate taxon. This clearly is not a reasonable criterion of naturalness in classification. At the very least, if this is to be the criterion of naturalness, then it cannot be reasonable to demand that *all* characteristics be taken into account in determining affinity; rather the criterion must be that those organisms which share the greatest affinity in terms of a selected set of characteristics are to be classified together. But this is, in effect, to give up the whole program. For once selection of characteristics is allowed, there is no way to prevent different characteristics from being selected for different purposes, and naturalness will be relative to the intended application; but the greatest affinity definition is supposed to be one that is neutral with respect to intended application--which was the reason for requiring that *all* characteristics be taken into account.

It may be held that I am trivializing their claim. This may be. Numerical taxonomists nowhere give any indication how generalizations are to be counted or under what measure the number of generalizations is to be measured. There is no standard measure of the number of generalizations forthcoming from a system; indeed, there is a fairly extensive literature on the subject, with a notable lack of success in developing measures of the number of generalizations or systematic import of scientific systems.[76] As such, it seems to me that the burden of proof is on the numerical taxonomists to specify a reasonable measure of the number of generalizations which supports their claim. For other than the above sense of 'number of generalizations' there are no obvious measures which give any reason to believe that taxonomies which take into consideration all characteristics yield more generalizations. Even if a plausible measure could be found which did yield the result that such taxonomies yield more generalizations, it would provide no good reason for taking the classification as natural--even if we grant that natural taxonomies are those most useful to the scientific community as a whole. For what counts in science is not the number of generalizations, but rather the sorts of generalitations which are afforded. To use a terminology which is less popular than it once was, what counts is not the number of generalizations but the systematic import of the generalizations yielded.[77] Aristotle's dynamics employed a classification of dynamical systems which took into account a large number of affinities in characteristics; as such he was able to generate a large number of low-level generalizations about various kinds of systems. Aristotle's classification of dynamics thus was natural in the sense under consideration. On the other hand, Newton's dynamics classified dynamical systems merely on the number of bodies involved, their relative positions, and their respective momenta. The taxonomy itself afforded significantly fewer generalizations than Aristotle's; but the generalizations it did afford had great systematic import, and were ultimately much more important, much more useful to a wider range of scientific disciplines than Aristotle's larger number ever were. To turn this definition of naturalness back on its proponents, what historically has proven the most useful in science for a wide range of purposes have been generalizations having high systematic import; thus, if to be natural is to be useful for a wide range of scientific purposes, then the natural taxonomy is the one which yields generalitations with highest systematic import. If any one lesson is to be learned from the history of science, it is that the classifications which yield such generalizations are ones which use relatively few characteristics to classify. The issue, then, is not how many characteristics should be used to classify, but *which* few. The difference between scientific classification and common-garden classification is that the latter takes everything into account, whereas scientific classification works on the assumption that in nature different aspects of phenomena are orthogonal and hence can be treated independently from the rest of reality. Taxonomies which are natural in the numerical taxonomists' sense are so only in virtue of being pristine; to claim they are natural in any stronger or scientifically relevant sense is self-indulgent.[78)] I conclude that this interpretation of 'natural taxonomy' offers us no viable alternative to the account developed in preceding sections.

IX. Taxonomic systems and scientific theories

The thrust of Sections III–VII has been that whether a taxonomic system is natural or not is a function of whether it is factually true or not. In Section VIII I considered a current account of naturalness which tries to construe naturalness as being its scientific usefulness. While this attempt fails, it does make (albeit confusedly) a rather important point: The adequacy of a taxonomic system must be judged not only on its naturalness and truth, but also on its scientific usefulness. This is particularly important when one recalls that, depending on the world's make up, we have shown that there may be a number of factually true different and inequivalent taxonomies which are natural in virtue of meeting conditions (A–1) and (B) – and possibly (C). In deciding between such competing taxonomies, some assessment of their relative merits must be made on a basis other than their naturalness. In the last section, and obliquely toward the end of Section V, I have suggested that the criterion should be their usefulness – the scientific fecundity of the theories they can be incorporated into. On this criterion I find the phylogenetic approach headed in a more promising direction than numerical taxonomy. At least the former isn't trying to convert taxonomy into plant and animal sociology.

It should be obvious from this that I see a very close connection between taxonomy and scientific theories; in this section I want to sketch briefly and systematically what the connection is. First I must sketch briefly some basic features of scientific theories.[79)] Scientific theories have the task of describing, predicting, and (possibly) explaining a class of phenomena. To do so it selects and abstracts certain idealized parameters from the phenomena, and characterizes the phenomena in terms of just these parameters. The abstract replicas of phenomena so obtained are called *physical systems*, and characterize what the phenomena would have been if their behavior were solely a function of these idealized parameters. For example, the genetical theory of natural selection is concerned with evolutionary phenomena. It does so by considering populations of organisms as if they were populations of genotypes with various reproduction rates, crossover frequencies, etc., and ignores all other characteristics (including phenotypic ones) of the organisms. These idealized populations are the physical systems of the theory.

The configuration of a physical system at a given time is its *state*; the job of the theory is to give a precise characterization of the changes in state the physical system undergoes over time. The laws of the theory (the so-called "equations of motion") specify the general properties of state transition in such a way that if the configuration of a physical system is known for anytime, and appropriate boundary conditions are specified, a special case solution can be obtained which describes the state-transition history of that particular physical system. In terms of that special case solution it is possible to predict the subsequent states the physical system will undergo. For example, in the genetical theory of natural selection, the laws of the theory define genotypic distributions for the population over time as a function of reproduction-rate functions for the genotypes, cross-over-frequency functions, etc. The laws are stated as differential equations, and when the reproduction-rate functions, etc., are specified for a population a special case solution can be obtained by solving the differential equations. This solution, together with the specification of the populations genotypic distribution, etc., at time $\underline{t}$ (the physical system' state at $\underline{t}$),can be used to predict what state (genotypic distribution, etc.) the population will assume at any subsequent time $\underline{t}$.
As such, the various special-case solutions describe a class of possible physical systems; if this class is identical with the set of physical systems corresponding to phenomena (biological populations) which in principle could occur in the world, then the theory is empirically true; otherwise it is empirically false.

The abstracted parameters in terms of which physical systems are specified rarely are amenable to direct observation; thus, for example, in classical particle mechanics they are momenta, accelerations, forces, etc.; and in the genetical theory of natural selection they are genotypes, cross-over frequencies, reproduction rates, etc. Accordingly, to apply the theory to actual phenomena, it is necessary to adopt experimental procedures, which often are complicated and involve sophisticated apparatus, to determine the values of the physical system's parameters. These experimental procedures in essence are methods for converting observations and measurements on the observed phenomena into specifications of the values for the parameters which are used to specify physical systems. For example, in classical particle mechanics, the experimental procedures are employed to convert time, weight, and distance measurements on physical bodies into data about the various positions, forces, accelerations, and momenta of the interacting bodies in the physical system. In the genetical theory of natural selection, births and deaths of various genotypes, morphological characteristics, etc., are measured and then converted into characterizations of the genotypic makeup of the population, their reproduction rates, etc.,[80]) thereby determining the physical system's current state and boundary conditions. The experimental procedures for determining the values of the parameters specifying the physical system generally apply only to relatively homogeneous classes of phenomena; they are laws of rather limited generality applying only to restricted classes of phenomena. E.g., in the natural selection case, they might only specify that classes of organisms which have morphological characteristics C will have such and such genotypic distribution. The theory, then, has as its subject-matter the characterization of a class of physical systems of which the phenomena experienced and observed are manifestations, and the application of the theory to phenomena is by means of local laws describing the correlations between physical systems and their phenomenal manifestations.

Given this picture of theories, what implications does it have for taxonomy? First, it should be noted that the theory itself (if augmented by definitions of taxonomic categories) embodies a classification of physical systems. (The special-case solutions of the theoretical laws in effect define classes of physical systems; the definitions of taxonomic categories specify which classes are to count as taxa of what category). This taxonomy typically is highly abstract, classification being carried out in terms of a few abstract characteristics (namely, the parameters of the physical systems and boundary conditions). Since the physical systems manifest themselves as phenomena, this taxonomy thereby imposes a taconomy on the phenomenal manifestations of these physical systems-- a taxonomy which is in terms of the underlying physical systems, not their phenomenal manifestations. Thus, the genetical theory of natural selection together with boundary conditions and a definition of taxonomic categories imposes a taxonomy on organisms in terms of their phylogenetic properties, not their morphological characteristics. This taxonomy is wholly natural. The application of the theory to phenomena is in terms of local laws or experimental procedures which correlate various phenomenal characteristics with the parameters of physical systems; implicit in these laws is a classification scheme, since the laws are to the effect that such and such kinds of phenomena correspond to and are manifestations of such and such physical systems. Thus a taxonomy of phenomena is also needed. This taxonomy, however, is likely to be fragmentary, non-systematic, with overlapping taxa, etc. – due to technological limitations, inaccessibility of phenomena, and ignorance. Thus the phenomenal taxonomy in practice is likely to have relatively few of the nice properties taxonomists have demanded.

What implications are to be drawn from this for biological taxonomy? Insofar as it is the function of the scientific enterprise to generate comprehensive theories, the taxonomists, by their insistence that taxonomy must be morphological, have offered little of value to biological science.[81]) The problems they are working on and the solutions they find acceptable are of little use to theoretical biology. I would like to suggest that this is not so much the case because taxonomists do not wish their work to be of value to theoretical biology,[82]) but rather that taxonomists have had a very unclear and distorted idea about the nature of scientific theorizing and taxonomy's role and function in it.[83]) The above, hopefully, sheds some light on that role.

This is not to say, however, that the work of taxonomists has been without scientific merit. For the generation of comprehensive theories certainly is not the only legitimate function of science, and so there is no reason why taxonomy should concern itself exclusively with the production of taxonomies which can be incorporated into, or defined in terms of, comprehensive theories. There are several reasons why morphological taxonomies (as well as other sorts, e.g., cytological taxonomies) are needed. First, for many practical applications, even if a highly abstract theory-embedded taxonomy existed, it would be too cumbersome to employ, and simpler, less abstract taxonomies will be needed (much as classical physics continues to be used in most practical applications just because quantum theory is too complicated). An analogy is useful here. Chemistry initially was a kind of taxonomy of reactions (qualitative analysis). As physical and atomic chemistry developed, new taxonomies of a more abstract and theoretical nature were required (e.g., the periodic table); but this did not obviate the need for qualitative chemical analysis, and it still continues to be a respectable branch of chemistry. Indeed, in practical applications it is perhaps the most used. Similarly, while more abstract taxonomies are needed for incorporation into evolutionary theory, for many applications morphological taxonomies are also needed. An important point must not be overlooked here: the practical usefulness of qualitative chemical analysis is not sufficent to account for its continued survival. Essential to its continued use is the fact that it is compatible with, and in the main yields the same classifications as, the more abstract taxonomies incorporated into, e.g., physical and atomic chemistry. Analogously, to the extent that biology has comprehensive theories (such as evolution and natural selection theories) which embody non-morphological taxonomies, the morphological taxonomies which will be of general biological use will be those which are compatible with, and in substantial agreement with, the taxonomies incorporated into the accepted more comprehensive theories. Thus the instincts of phylogenetic taxonomy are sound; those of the numerical taxonomists are not--except to the extent that they are concerned to produce taxonomies which meet these requirements of compatibility.

A second reason why morphological taxonomies are needed and useful is that in a number of areas of biology there are no accepted comprehensive generalized theories. Any attempt to develop highly abstract taxonomies in the absence of such comprehensive theories would be premature and a mistake. This point can be made clearer in the following way: Although it involves a fair amount for oversimplification, a scientific theory consists in generalized laws holding between the taxa of a taxonomic system which uses the theory's parameters as diagnostic characteristics for defining taxa. Where the only existing theories or laws known are in terms of morphological features, morphological taxonomies are needed. But where more generalized cytological, evolutionary, genetic, etc., laws or theories exist, taxanomies are required which are formulated in terms of the theories' parameters. For the practical reasons mentioned above, morphological taxonomies are required in addition. If the further assumption is made that these various theories in different branches of science, if true, are mutually consistent and form a coherent whole, then the taxonomies they embody also must be consistent with each other, and not merely self-consistent; accordingly, there should be a morphological taxonomy which is consistent with all of these more abstract taxonomies. And it is such a morphological taxonomy which should be the goal of morphological taxonomy. For such a taxonomy will provide a biological lingua franca which can be used to relate the theoretical results of these various branches of biology on the organism level. This means, of course, that new theoretical developments in theoretical biology may require modifications in the morphological taxonomy.

In passing, I note that if the above claims are sound, a taxonomic analogue to the problem of theory reduction holds for taxonomy, and is of paramount importance.

X. Summary and conclusions

This has been a long, complicated, and involved paper, and so a short summary and conclusion is in order. I began by historically tracing a number of themes in the history of biological taxonomic theory. From this I extracted four basic questions, differing answers to which yield different positions on the nature of biological taxonomy (Section I). These questions and the answers given to them make it clear that the pivotal issue is the nature of the distinction between natural and artificial taxonomies, so I turned my attention to the natural-artificial distinction. Although there is general agreement that natural taxonomies have an objective basis in nature which artificial taxonomies lack, two basic different opinions prevail about what that objective basis is. One opinion is that natural taxonomies are factually true or false; the other is that natural taxonomies are those based on the largest number of affinites in all characteristics between organisms, and hence the most useful for science. In Sections II–VII I considered in detail the thesis that natural taxonomies were factually true or false. In Section II I presented an analysis of facts and empirical truth, and then used it to isolate three different ways in which taxonomies could be factually true. These were conditions (A), (B), and (C). Then in Sections III–VI I considered these conditions in detail, looking at a fairly exhaustive selection of the various ways in which these conditions could be met. In the process of doing so, careful consideration was given to the various proposed definitional forms for taxa and for taxonomic categories. From this investigation, some quite general results were obtained: (i) Condition (A–2) must be met by every taxonomy, whether natural or not. (ii) A more specific version of (A), namely (A–1), was obtained, and it was determined that this must be met by any natural taxonomy in this factual sense. (iii) Condition (B) in its various versions was shown to presuppose a basic condition (B–8), and to be natural taxonomies must meet (B–8). (iv) It was shown that questions of the appropriate sort of natural taxonomic system for a domain D, and issues about the definitional form to be adopted, are primarily empirical, though in certain cases metaphysical committments are required. In Section

VII, it was argued that taxonomies which fail to meet conditions (A) and (B) clearly are artificial in the very basic sense that they are nothing more than conventions to reserve certain terminology for referring to collections of individuals sharing a certain affinity. The theoretical employment of such conventional taxonomies was briefly considered.

Section VIII considered the second opinion on what it was for a taxonomy to be natural – that it was a taxonomy which classified on the basis of greatest affinity in all characteristics. It was argued that, as developed by numerical taxonomists, this criterion failed to guarantee that taxonomies meeting it are natural in any scientifically relevant sense. It was granted, however, that general usefulness of a taxonomy in a wide range of scientific enterprises is a legitimate and important factor in assessing the adequacy of a taxonomy. Consideration was given to the way taxonomies can have a wide range of scientific usefulness, and it was concluded that this was largely a matter of compatibility with generalized theories having a high degree of systematic import. Since such theories generally employ a very few highly abstract characteristics for classifying organisms, the thrust of the numerical taxonomists' notion of naturalness and usefulness is at odds with what is required for a wide range of scientific utilization. Phylogenetic taxonomy's orientation is sounder. Section IX considered the role of taxonomy in scientific theorizing and made certain generalizations about the contributions taxonomy can make to scientific theorizing, and what changes in biological taxonomy are required if these contributions are made.

Although the discussion here has been of taxonomic problems arising *in* biological taxonomy, the discussion has proceeded in terms of general epistemological and metaphysical considerations, and the vast majority of my conclusions are independent of any specifically biological considerations. As such my conclusions are general philosophical conclusions about the nature and role of taxonomy and classification in science in general, and not just in biological science.

Footnotes

1) I would like to thank Professor George Estabrook of the University of Michigan Botany and Zoology Departments and my colleague Professor Thomas Nickles for helpful criticisms of the draft version of this paper. I have also profited from various discussions with my colleague Professor Hugh Chandler and have been influenced by his paper "Three Kinds of Classes". Less immediately, I would like to thank Professor J. L. Heller of the University of Illinois for encouragement several years ago when I first became interested in taxonomy as an area of philosophical importance, and also for acquainting me with literature I otherwise would never have discovered.

2) Cf., e.g., Beckner, *The Biological Way of Thought;* Buck and Hull, "The Logical Structure of the Linnean Hierarchy"; Gilmour, "Taxonomy and Philosophy", and the various papers by David Hull cited below. I do not mean to suggest that because these authors limit themselves to peculiarly biological problems that their work is not of philosophical interest--for they are philosophically interesting and my own discussion is heavily indebted to them.

3) Although phrased differently, this definition of standard taxonomy is equivalent to that given for taxonomies elsewhere in the literature--e.g., by Beckner, *op. cit.,* p. 55, and by Hull and Buck, *op. cit.* I call such taxonomies standard taxonomies to distinguish them from other taxonomies or classification systems which do not share this hierarchal character.

4) Such definition is often referred to (somewhat incorrectly) as the definition of taxa names.

5) This list, of course, is not exhaustive of the theoretical issues which have been raised with respect to what it is for a taxonomy to be natural. E.g., the question whether different weightings should be given to diagnostic characteristics in taxa definitions often is raised in conjunction with "cluster" analysis approaches to taxonomy (e.g., the approaches of Adanson and the numerical taxonomists). Nothing I do here turns on such questions, and so they will be ignored here.

6) Strictly speaking, this is not correct as Aristotle sketches features of earlier systems (e.g., Empedocles' and Democratis' in *De Partibus Animalium*, I), but the accounts extant of these are very fragmentary, and so will be ignored here.

7) To my knowledge, Aristotle never explicitly uses terminology corresponding to 'natural taxonomy' and 'artificial taxonomy'; but he clearly has the distinction, since he warns against adopting artificial or non-natural taxonomies, e.g., in *De Partibus Animalium,* 643 a–b, where he discusses how classifications ought not to be made.

8) Most of Aristotle's taxonomic theory can be found in *De Partibus Animalium; Historica Animalium* contains some theoretical discussion (Bk. I) minus the metaphysics, together with an extensive classification of animals. The metaphysical principles employed are to be found scattered through his writings--e.g., *Posterior Analytics,* Bk. I, Chapter 4; Bk. II; *Metaphysica,* Bk. IV (Γ), and *Topica* 101b 39 ff. Good historical treatments of Aristotle's theory of taxonomy can be found in A. Hopwood, "Animal Classification from the Greeks to Linnaeus," and David Hull's "The Effect of Essentialism on Taxonomy: Two Thousand Years of Stasis" and "The Metaphysics of Evolution". Aristotle's plant taxonomies have been lost, but Theophrastus' work seems to embody much of Aristotle's own observations and to be based on the same theoretical principles. Theophrastus' works can be found in his *Enquiry into Plants* and his *Causes of Plants.* A good discussion of his taxonomy is to be found in T. A. Sprague, "The Evolution of Botanical Taxonomy from Theophrastus to Linnaeus." These articles by Hull, Sprague, and Hopwood provide good historical discussions of the various developments subsequent to Aristotle,

mentioned below, as do A. J. Wilmott, "Systematic Botany from Linnaeus to Darwin"; A. Hopwood, "Animal Classification from Linnaeus to Darwin," H. K. A. Shaw, "Post-Darwinian Development of Taxonomy (Botany)," and J. Smart, "Post-Darwinian Development of Taxonomy (Zoology)". A somewhat biased discussion can be found in R. Sokal and P. H. A. Sneath, *Principles of Numerical Taxonomy,* pp. 11–30; this volume has an extremely good index. Another excellent source is W. Whewell, *History of the Inductive Sciences.*

9) There is some dispute over whether this is so; e.g., Wilmott *op. cit.,* and Hopwood, "Animal Classification from Linnaeus to Darwin" disagree on whether this is so.

10) E. Mayr, *Systematics and the Origin of the Species,* p. 109.

11) Cf. C. Hempel, "Typological Methods in the Natural and the Social Sciences" for further discussion of it and examples from the social sciences.

12) Modern adherents to this position would include those who adhere to the *morphological definition* of species ("A species is a group of individuals or populations with the same or similar morphological characters").

13) That this is so is indicated by the fact that there seem to be few adherents to either position today; however, Mayr, *op. cit.,* devotes several pages (pp. 103ff) to their discussion, which indicates that these positions still had many adherents in the 1930's and 40's.

14) For detailed discussion of the shortcomings with each of these definitions, see David Hull, "The Logic of Phylogenetic Taxonomy," pp. 37–70. In his "The Effect of Essentialism on Taxonomy," Hull attempts to get around these difficulties with a disjunctive definition.

15) The notion of polytypic definition apparently was first introduced by Beckner, *op. cit.,* pp. 21 ff.

16) "The Logic of Phylogenetic Taxonomy," pp. 71–83.

17) *Ibid.,* pp. 115. It should be noted that this is not a cluster concept in the sense of Gasking, "Clusters"; cf. Hull "The Logic of Phylogenetic Taxonomy", pp. 83–105, for Hull's discussion why they are not.

18) Hull, "The Logic of Phylogenetic Taxonomy," p. 116.

19) Cf. Sokal and Sneath, *op. cit.,* p. 21.

20) Perhaps the best philosophical defense of this analysis of what it is for a taxonomy to be natural is by J. S. Mill in his *A System of Logic,* Bk, I, Ch. vii, and esp. Bk. IV, ch. vii.

21) Sokal and Sneath, *op. cit.*, p. 12.

22) *Ibid.*, Ch. 2; a discussion of some of these problems in their arguments can be found in Hull, "The Effect of Essentialism," Section 5A.

23) The answers to this question are somewhat vague; answer a) encompasses the range of answers given by those who give the various answers listed for question 2. Another form of the question is "What definitional form is appropriate for taxonomic categories"? In the recent literature, this question has been dealt with only on the lower taxonomic levels--especially the species and variety or sub-species levels. Modern taxonomic theory has given relatively scant attention to the higher taxonomic categories.

24) Cf. my "Facts and Empirical Truth". For brevity, the arguments used in developing and supporting this model will not be presented. Indeed, the entire model will not be presented, as a number of its features are not relevant to the issues under consideration. It should be mentioned that this model is heavily indebted to a number of suggestions made by Wilfrid Sellars in a stimulating series of conversations on facts.

25) Cf. his "Truth."

26) If one is committed to intentional particulars, then the account must be revised; cf. "Facts and Empirical Truth," Section IV. Since biological taxonomy for the most part is not concerned with intentional particulars, this complication is ignored here. Some terminology needs to be clarified. By the *extension* of a proposition I mean that portion of its meaning or propositional content which can be specified by giving its logical form and specifying the references of its subject and predicate terms; by the *intension* of a proposition I mean its sense (in Frege's sense). A language is *extensional* if the meanings of its propositions consists in their extensions; a language is *intensional* if the intensions of at least some of its propositions are parts of their meanings.

27) *Ibid.*, Section III; a sortal characteristic is an intrinsic characteristic whose extension is identical with that of some sortal concept or predicate of some language; a sortal concept or predicate, roughly, is one which specifies a *kind* of entity.

28) A word needs to be said about my use of the locutions 'is a T' and *'being a T'.* In presenting my analysis it is necessary to distinguish sharply between predicates used to attribute intrinsic characteristics and the characteristics attributed by them. The predicate expression 'is a chair' is used to attribute a characteristic to an object; the characteristic (property) attributed is that of *being a chair* or *chairness.* Whenever I am talking about the predicate expression (or predicate) I enclose the predicate expression in single quotation marks. Whenever I am talking about a characteristic I use such expressions as

being a chair

is a chair.

And when speaking generally or abstractly of predicates and characteristics I will use

and ' is a T'

being a T

analogously to refer respectively to the predicate expression and the characteristic it attributes. Thus my use of single quotation marks is in accordance with Quine's notion of quasi-quotation, where I am using single quotation marks in lieu of his "corners". These conventions for marking the distinction between predicate expressions and the intrinsic characteristics they attribute are used throughout the paper, and its central argument can be understood only in light of them

It is also important to realize that there is no necessary or intrinsic connection between the use of the word 'chair' and the intrinsic characteristic *being a chair,* and that with change in meaning 'is a chair' could be used to attribute a characteristic other than that of *being a chair.* It is also possible that the predicate 'is a T' in a language may attribute a complex characteristic (i.e., the joint occurence of C_1 and C_2, where C_1 & C_2 is an intrinsic characteristic). These observations enable us to forestall a possible source of confusion. Requirement (A) requires that to belong to a taxon T, an individual must possess the intrinsic characteristic *being* a T. All this requires is that there be a single intrinsic characteristic (simple or complex) possessed by all and only those individuals belonging to the taxon T. From the fact that I employ *'being* a T' to designate the characteristic it does not follow that the characteristic can only be specified using taxon name, T. It is perfectly possible that the characteristic *being* a T can be specified using some other (possibly more commongarden) predicate expression--so long as that expression, whatever it is, designates a single characteristic common to all and only those individuals in the taxon.

29) Cf. Bambrough, "Universals and Family Resemblances". All page references are to the reprinting in G. Picher, *Wittgenstein.*

30) *Philosophical Investigations,* Section 66.

31) *Op. cit.*, p 199.

32) *Ibid.*, pp 200–204.

33) I am grateful to Professor George Estabrook for pointing this out to me.

34) It is the case, however, that taxonomists who disagree on taxa definitions often can agree on what taxon a specimen belongs to; and taxonomists come to be able to recognize that a specimen is of, e.g., species *rosa multiflora* without checking any list of diagnostic characteristics. These phenomena suggest that *being* a T sometimes is and can be easily and directly recognized by competent taxonomists.

35) Cf. the author's "The Search for Philosophic Understanding of Scientific Theories", esp. Sections II-A, B and IV-B-2, for a comprehensive discussion of the results leading to this conclusion; for a particularly good critique of operationalism with specific reference to biology, cf. Hull, "The Operational Imperative".

36) Cf. "Facts and Empirical Truth", Section III.

37) "Facts and Empirical Truth", Section III.

38) To see that it is weaker, suppose, e.g., that one is committed to Aristotelian definition for taxa. Then x is a T if and only if x is $c_1, c_2, \ldots, c_n$ for some of intrinsic characteristics; this does not guarantee satisfaction of (A–1) although (A–2) is satisfied. To see this, suppose that the taxonomy is artificial in the sense that (A–1) fails. Then is a T is not an intrinsic property. One might object that this is impossible, viz.: if x is a T, then x is $c_1, \ldots, c_n$, and so x has the intrinsic property is c_1 & . . . & is c_n, so (A–1) is satisfied. But this is not generally true since there is no guarantee that the conjunction of intrinsic properties is itself an intrinsic property. Wheter it is depends on the constitution of the world. Cf. "Facts and Empirical Truth", Section I.

39) The exact number individuals must or can possess depends on the nature of the taxonomic system-whether it is standard, how many levels there are in the hierarchy, whether taxa may be overlapping in a given taxonomic category, etc.

40) Aristotle's own taxonomy was not, however, being the position that the intrinsic characteristics which were taxa were a combination of other intrinsic characteristics, which were the essential characteristics. Any individual who possessed a taxon intrinsic characteristic thus had the essential characteristics; it also had various other accidental intrinsic characteristics. The metaphysics of this version of essentialism seem unacceptable to taxonomists today, and so this view of essentialism will not be considered further.

41) For a discussion of the notion of causal necessity in the sense being used here, see A. W. Burks, *Cause, Chance, and Reason,* Ch. 5. Note that being an essential property is a modal characteristic in a nomic essentialism.

42) Chapters 1, 2. Achinstein's work has been anticipated in some respects by Chondler "Three kinds of Classes", and works by Dugold Stewart, T. S. Mill, William L. Davidson, and others in the 19th century

43) *Ibid.,* p. 25.

44) Cf., *ibid.,* Chapter 1, *passim;* for discussion that D_2 cannot be assimilated to polytypic definition, cf. pp. 27–28; other forms it cannot be assimilated to are discussed on pp. 28–31.

45) Whether this is possible is a difficult question, being part and parcel of the problem of specifying truth conditions for vague generalizations. The problem is akin to that of specifying the logic of semi-universal generalizations – i.e., the logic of the quantifier "for most x". It seems clear, however, that assertions involving semi-universal quantifiers are empirically true or false – even though their logic is not understood.

46) Aristotelian definitions can be construed as degenerate polytypic definitions; the discussion thus applies to Aristotelian definitions as well.

47) In fact, a major practical problem facing taxonomists is determining whether an organism is an anomalous specimen of an established taxon or is a new variety, species, or whatever. That this is a practical problem constitutes confirmation of the claims just made.

48) However, it should be noted that in those cases where polytypic definitions are required (even with fixed taxa), Achinstein's form of definition cannot replace disjunctive definitions, since his form of definition is incapable of separating the varieties in the required way. (Cf. *ibid.*, pp. 27–28 for related discussion.) Where varieties or sub-species occur, disjunctive definitions still are required – but with the disjuncts each being of Achinstein's definitional form.

49) Although it is beyond the scope of this paper to argue it fully here, I am of the opinion that this last alternative is essentially correct. Indeed, it is a natural corollary to the analysis presented below in Section IX.

50) If, however, the proposal in Section IX below is accepted, recourse to this quasi-universal form of definition will not be required.

51) Some of the possible cases where (B) doesn't hold already have been considered.

52) Intentional particulars are particulars which exist in virtue of being thought about, believed, or conceptualized; many philosophers deny the existence of intentional particulars.

53) This is not to say that there are not areas where it might be plausible to adopt such taxonomies; e.g., a case can be made that kinship systems and other social taxonomies are of this sort.

54) Whatever the merits of operationslism are (none in my opinion), it is clear that no operationalist taxonomy is compatible with the claim that taxa should have a philogenetic basis which is reflected in taxa definitions. As such, the philogenetic approach entails a rejection of operationalism and an acceptance of (B–7).

55) Cf. Hull, "The Logic of Phylogenetic Taxonomy," pp. 22–70, for a detailed consideration of their failures.

56) Cf. Hull, "The Effect of Essentialism on Taxonomy," Section 5–E; Section 5 also contains a discussion of the failures of the various definitions mentioned in Section I of this paper.

57) Indeed, some of the arguments in support of numerical taxonomy are exactly of this sort: it is impossible (in practice) to specify phylogenetic taxa, so a non-phylogenetic taxonomy must be adopted. Cf. Sokal and Sneath, *op. cit.*, Section 2.3.

58) This strongly suggests that *every* scientific theory either presupposes or contains (implicitly or explicitly) a taxonomic system as a sub-theory.

59) For discussion of auxiliary hypotheses, cf. H. Putnam, "Scientific Explanation," and "The 'Corroboration' of Theories," and the author's "What's Wrong With the Received View on Theories? "

60) In its present form, evolutionary theory probably cannot actually do this since it is not known, e.g., how to predict the range of possible non-lethal mutations which can occur in a given population. This does not count against the feasibility of the proposal--it only indicates that current evolutionary theory itself may be inadequate to do the job.

61) For a discussion of reproduction rates, cf., R. A. Fisher, *The Genetical Theory of Natural Selection,* Chapter II.

62) This suggestion may prove inadequate; whether it does is of no concern to the point I want to make here, which is to illustrate how evolutionary theory plausibly can be used to define phylogenetic taxa.

63) These boundary conditions might include, for example, reproductive barriers at subsequent times.

64) *Principles of Animal Taxonomy,* p. 91.

65) For a discussion of the literature showing why they are not definitions of any sort, cf. the authors "The Search for Philosophic Understanding of Scientific Theories"; cf. also Hull, "The Operational Imperative: Sense and Nonsense in Operationalism." For a discussion of what they are, cf. the author's *ibid.*, Section IV–E, and also his "What's Wrong With the Received View on Scientific Theories? " where they are what enables one to translate between phenomena and physical systems.

66) Such confusion has been characteristic of philosophy of science the first half of this century; cf. my "The Search for Philosophic Understanding of Scientific Theories", esp. Sections II–A, B; IV–B–1, C, E..

67) Recall here from Section II, above, that particulars need not be objects.

68) It is important to note that the various versions of (B) do *not* require that the taxa definitions be factually true, but rather only that they be capable of being factually true, and advanced as being putatively true. Thus the falsity of a taxa definition does not render it unnatural.

69) Cf. Section IX, where the analysis given there reinforces this judgment.

70) Sokal and Sneath, *op. cit.,* p. 12.

71) *Ibid.,* p. 19

72) *Ibid.*

73) Gilmour, "Taxonomy and Philosophy," pp. 466–467; it is fashionable among numerical taxonomists to attribute their account of natural taxonomy to Gilmour; cf., e.g., Sokal and Sneath, *op. cit.*

74) Gilmour, *op. cit.,* p. 472.

75) E.g., Gilmour, *ibid.,* is often cited approvingly as the source of this doctrine, and he makes it quite explicit that classes are conventional.

76) For a sampling of the problems, cf. Kemeny and Oppenheim, "On Systematic Power".

77) Cf. Kemeny and Oppenheim, *ibid.*

78) This is not to say that there will not be any scientifically reputable applications where a special purpose taxonomy working in terms of greatest affinity will not be useful.

79) Doing so is complicated by the fact that philosophers of science recently have rejected the previously standard analysis of theories ("theories are uninterpreted axiomatic calculi augmented by correspondence rule which give it a partial observational interpretation"), and no alternative analysis has won general acceptance. For a comprehensive account of the reasons the previously standard analysis (the so-called Received View) was rejected, cf. the author's "The Search for Philosophic Understanding of Scientific Theories"; a representative picture of current thinking about theories by philosophers of science can be found towards the end of that essay together with the papers and discussions in the author's *The Structure of Scientific Theories.* The analysis of theories which follows is one I've suggested in "What's Wrong with the Received View on Theories? "; for the exact sciences (which include the genetical theory of natural selection), I have worked the analysis out in careful detail in "The Meaning and Use of Models in Mathematics and the Exact Sciences", Chapter 2. For a short but comprehensive discussion of the approach see Section V–C of my "The Search for Philosophic Understanding of Scientific Theories".

80) More specifically, to determine the genotypic makeup, a random sample of organisms of different morphological types might be examined as to genetic makeup; the distribution of genotypes in the sample then is projected onto the population as a whole to determine its genotypic distribution.

81) This is underscored by the relatively low repute taxonomy is held in by biologists working in other areas of biology.

82) Indeed, there is ample evidence that they are concerned with its relevance. Certainly the whole program of the New Systematics and phylogenetic taxonomy is aimed at making taxonomy useful in theoretical biology. And it is very clear in the writings of the numerical taxonomists that they take its presumed usefulness to biologists working in other areas to be one of its chief merits.

83) Witness all the concern among taxonomists that taxonomy must be operational. No scientific theories of any substance exist (or ever did) which meet the requirements of operationalism. The adherence to operationalism is symptomatic of the gross misunderstanding of science and scientific knowledge which impeded progress in psychology for several decades earlier this century. Cf. Hull, "The Operational Imperative: Sense and Nonsense in Operationalism".

References

ACHINSTEIN, P., *Concepts of Science.* Baltimore: Johns Hopkins University Press, 1968.

ARISTOTLE, *Collected Works.* 12 vols. (ed. W. Ross). Oxford: Oxford University Press, 1928–52.

BAMBROUGH, R. "Universals and Family Resemblances", pp. 109–127 in M. Loux, *Universals and Particulars.* Garden City: Anchor Books, 1970; also on pp. 186–204 in G. Pitcher, *Wittgenstein,* Garden City: Anchor Books, 1966.

BECKNER, M., *The Biological Way of Thought.* New York: Columbia University Press, 1959. (Reprinted by University of California Press, 1968).

BUCK, R., and D. HULL, "The Logical Structure of the Linnaean Hierarchy". *Systematic Zoology,* 15 (1966): 97–111

BURKS, A. W., *Cause, Chance, and Reason.* Ann Arbor: University of Michigan Press, forthcoming.

CHANDLER, Hugh, "Three Kinds of Classes". *American Philosophical Quarterly,* 3 (1966): 1–5.

FISHER, R. A., *The Genetical Theory of Natural Selection.* New York: Dover, 1958.

GASKING, D., "Clusters", *Australasian Review of Psychology,* 38 (1960): 1–36.

GILMOUR, J. S. L., "Taxonomy and Philosophy", pp. 461–474 in J. S. Huxley, *The New Systematics* (Oxford: Oxford University Press, 1940).

HEMPEL, C. G., "Typological Methods in the Natural and the Social Sciences"; pp. 155–71 in Hempel, *Aspects of Scientific Explanation* (New York): The Free Press, 1965).

HOPWOOD, A., "Animal Classification from the Greeks to Linnaeus", pp. 24–32 in *Lectures on the Development of Taxonomy* (London: Linnean Society, 1950).

"Animal Classification from Linnaeus to Darwin", pp. 46–59 in *Lectures on the Development of Taxonomy* (London: Linnean Society, 1950).

HULL, D., "Certainty and Circularity in Evolutional Taxonomy", *Evolution,* 21 (1967): 174–189.

"The Effect of Essentialism on Taxonomy: Two Thousand Years of Stasis". *British Journal of the Philosophy of Science,* 15–16 (1955).

"The Logic of Philogenetic Taxonomy". Ph. D. Dissertation, Indiana University, 1964.

"The Metaphysics of Evolution", *British Journal for the History of Science,* 3 (1967): 309–337.

"The Operational Imperative: Sense and Nonsense in Operationalism". *Systematic Zoology,* 17 (1968): 438–457.

KEMENY, J., and P. OPPENHEIM, "On Systematic Power". *Philosophy of Science,* 22 (1955); 27–33.

LINNEAN SOCIETY OF LONDON, *Lectures on the Development of Taxonomy* (London: Linnean Society, 1950)

MAYR, E., *Systematics and the Origin of the Species.* New York: Dover, 1964.

MILL, J. S., *A System of Logic,* 8th ed., London: Longman's, 1965 reprinting.

PUTNAM, H., "Scientific Explanation" (Summary-Abstract) in F. Suppe, *The Structure of Scientific Theories* (Urbana: University of Illinois Press, 1972).

"The 'Corroboration' of Theories", in P. Schilpp, *The Philosophy of Karl Popper* (LaSalle: Open Court, forthcoming).

SHAW, H. K. A., "Post-Darwinian Development of Taxonomy (Botany)", pp. 60–79 in *Lectures on the Development of Taxonomy* (London: Linnean Society, 1950).

SIMPSON, G. G., *Principles of Animal Taxonomy.* New York: Columbia University Press, 1961.

SMART, J., "Post-Darwinian Development of Taxonomy (Zoology)", pp. 80–83 in *Lectures on the Development of Taxonomy* (London: Linnean Society, 1950).

SOKAL, R., and P. H. A. SNEATH, *Principles of Numerical Taxonomy.* San Francisco: Freeman, 1963.

SPRAGUE, T. A., "The Evolution of Botanical Taxonomy from Theophrastus to Linnaeus", pp. 1–23 in *Lectures on the Development of Taxonomy* (London: Linnean Society, 1950).

STRAWSON, P. F., "Truth", pp. 32–53 in G. Pitcher, *Truth* (Englewood Cliffs: Prentice Hall, 1964).

SUPPE, F., "Facts, and Empirical Truth", *forthcoming.*

"Meaning and Use of Models in Mathematics and the Exact Sciences". Ph.D. Dissertation, University of Michigan, 1967.

"The Search for Philosophic Understanding of Scientific Theories" in F. Suppe, *The Structure of Scientific Theories* (Urbana: University of Illinois Press, 1972).

The Structure of Scientific Theories. Urbana: University of Illinois Press, 1972.

"What's Wrong with the Received View on Theories? ", *forthcoming in Philosophy of Science* (March 1972)

THEOPHRASTUS, *Enquiry into Plants,* tr. A. Hart (Loeb Classical Library) Cambridge: Harvard University Press, 1916.

Causes of Plants, tr. R. Dingler, Philadelphia, 1927.

WHEWELL, W., *History of the Inductive Sciences,* 3rd ed., 2 vols., New York: David Appleton & Co., 1869.

WILMOTT, A. J., "Systematic Botany from Linnaeus to Darwin", pp. 33–45 in *Lectures on the Development of Taxonomy* (London: Linnean Society, 1950).

WITTGENSTEIN, L., *Philosophical Investigation.* tr. E. Anscombe. New York: Macmillan, 1953.

Rejane Bernier (Montréal)

Comments upon Prof. Suppe's paper on "Some philosophical problems in biological taxonomy and speciation"

While giving his exposé the title "Some philosophical problems in biological taxonomy and speciation", Prof. Suppe points out that he intends to stress the philosophical aspects of taxonomy. Many a biologist, working in their own field, have raised problems which, although deprived of a strict philosophical formulation, are nevertheless of a philosophical order by virtue of their content. The first two questions asked by Prof. Suppe (1. "What makes a taxonomic system natural"? and 2. "What definitional forms are appropriate for defining taxa and taxonomic categories"?) have called the attention of biologists since ages. The second of those problems has been debated since Aristotle whereas the first one was not clearly seen until the XVIIth century, following the overuse of a descriptive nomenclature in which the largest possible number of distinctive characters came to be included in the definition of any taxon.

Realizing that many naturalists have already discussed those problems of classification, Prof. Suppe attempts to discover what their main conceptions in the matter have been. He thus has recourse to the history of biology in order to give his investigation the proper background. In itself, this is excellent. What, however, appears to be more questionable is the interpretation that Prof. Suppe ascribes to the works of few of the naturalists he mentions.

Prof. Suppe does not show always a first-hand knowledge of the authors mentioned by him. I have never seen, in any text in the history of biology, that Harvey and De Graaf are listed among the XVIth century taxonomists. As a simple matter of fact, both lived in the XVIIth century, Harvey from 1578 to 1657 and De Graaf from 1641 to 1673 (assuming that the latter named really is Reinier De Graaf). As far as I am concerned and although I am fairly familiar with the work of both those authors, I am not aware that they have worked in taxonomy and I would be happy to know their works in that field if they exist. On the other hand, the works of Gesner, Bauhin and Fuchs, who all studied taxonomy and really lived in the XVIth century are not mentioned. Why?

According to Prof. Suppe, John Ray "maintained that all parts of the plant should be used in classification, and not just selected ones. For Ray, species were natural and immutable, but not determined by essences". To be sure, for Ray, the problems of classification did not have the same philosophical implications as for Aristotle. However, from just a few chapters of the *Historia Plantarum,* we can conclude that, far from considering all parts of the plants as necessary to classification, John Ray makes a first distinction between individual and specific characters. The latter are inherited while the former are not. Thus, for Ray, all variations in the number and colours of the flowers, in the size, in the leaves, in the colour of the root, in the colour, savour and appearance of the fruit, are not valid criteria for the species.[1]) Similarly, the place of origin, the time of blooming, the properties and the uses must also be rejected from the classification because they separe related species from one another. In Ray's idea, the possibility of distinguishing genera rests upon comparing flowers (chalices), seeds and origins of these[2]). In the same work, John Ray recognizes transmutation in certain species in a limited but indisputable measure (e.g. in the cabbage).[3])

I do not believe either that, for Linnaeus, the species are defined by all physiological and morphological characters, which is what Prof. Suppe asserts. According to him, considering the total number of characters leads to multiplying the species. In fact, Linnaeus has precisely been blamed for having reduced excessively, not increased, the number of species.[4]) In his *Fundamenta botanica* as well as in his *Philosophia botanica,* Linnaeus points out that same parts of the plants may vary in size, colour, odour, savour.[5]) He thinks that a classification should not be based on such parts but rather on those that are the most stable i.e. the sexual differences (number, relations, arrangement of the etamines and pistil).[6]) All parts of the reproductive system must be taken into account because they form the natural character of the plant. Thus, according to Linnaeus, systematics make possible the order of nature. In his first works, Linnaeus considers the species and genera only as products of nature but, later, after observing a number of specific changes, he comes to express the hypothesis that the orders might be the only categories produced by creation while the genera and species would have been formed successively in nature without direct intervention of the Creator.[7])

Also, the degree of familiarity with Lamarck shown by Prof. Suppe appears to me somewhat questionable. According to him, Lamarck did not develop his theory of evolution before 1812. How can one reconcile that idea with the *Discours d'ouverture* given by Lamarck at the Museum d'Histoire Naturelle de Paris in 1800 and printed as a preface to the *Système des animaux sans vertèbres* in 1801? It is strange to hear that Lamarck followed de Candolle (ca. 1819 *sic)* while his last work, *L'histoire naturelle des animaux sans vertèbres* was published starting in 1815 and the first volume of that work is precisely the one that contains all the lamarckian principles of classification.

All those chronological details are fairly secondary in comparison with the mistake made by Prof. Suppe when he claims that he can summarize Lamarck's thought by saying "All taxonomy is artificial". I have to make some remarks on this point. In the first place, the word "taxonomy" which, according to the Oxford English Dictionary dates back from 1790, was not used by Lamarck himself. This author wrote either classification or nomenclature, and it is true that he called both things artificial. But Lamarck also claimed that evolution took place in a natural order and that the "méthode naturelle" in biology is precisely intended to trace back the order sequence followed by nature in producing the living beings: "cet ordre est le seul qui soit stable, indépendant de tout arbitraire, et digne de l'attention du naturaliste"[8]). "Aussi l'ordre même de la nature, dans le règne animal, est maintenant esquissé, dans ses masses principales, d'une manière stable et satisfaisante. Les limites seules des classes, de leurs ordres, des familles et des genres sont encore exposées à l'arbitraire"9) Lamarck thus bases his natural method on the detection of natural relations which the living beings hold between each other.10)

I would like now to resume some of the themes which Prof. Suppe has developped at length. In the first place, let us examine the idee of the natural character of taxonomy. Prof. Suppe defines such a taxonomic system as the one that has an objective basis in nature. He admits, however, that the tenants of numerical taxonomy conceive the natural character differently. For them, the taxonomy is all the more natural as it is the more universal and can be utilized in a greater number of scientific disciplines. In his chapter VIII, Prof. Suppe criticizes very aptly the difficulties raised by the numerical taxonomy; in particular, he stresses the fact that, by considering *all* characters, one is not necessarily led to wider generalizations. He also notes that the important thing in science is not so much the number of generalizations but rather the sorts of generalizations which are afforded, their systematic import. Now, in science, the classifications upon which the generalizations with the highest systematic import are based are precisely those that rest a relatively few characters. The solution of the taxonomic puzzle does not depend on the number but on the choise of the characters that are being used; and these are few. This criticism, however, does not help us understand why Prof. Suppe refuses to recognize the natural character of the numerical taxonomy. In fact, this is rather surprising since the numerical taxonomy, like the standard taxonomies, possesses an objective basis in nature and takes into account, at least theoretically, not just a few but all available characters. Prof. Suppe position is probably due to the fact that such numerical taxonomies do not meet with the conditions which he has set as necessary for establishing a natural taxonomy i.e. those taxonomies deny A–1 which states that belonging to taxon A is a matter of empirical fact in the sense that being of taxon T is an intrinsic property the individual does or does not have. Furthermore, condition B cannot either be fulfilled (The definitions given for taxa, if correct, will be empirically true). In the case of a numerical taxonomy, the definition of taxa "does nothing more than to establish the convention that taxa name T will be used to refer to the largest class of individuals in domain D which stand in such and such an affinity to each other." That definition "fails to establish an empirically true or false correlation between diagnostics characteristics and being of taxon T." If neither A–1 nor B are fulfilled, the taxonomy is artificial and conventional (ch.IV).

For Prof. Suppe, several taxonomic systems are still possible although non-equivalent. Their distinction rests on the characters chosen for the establishment of the taxon. One may here wonder what Prof. Suppe is after. Does he simply want to show that a plurality of natural taxonomies are possible? If I understand him correctly, that seems to be his goal; but I fail to grasp why it is so important to insist so much upon that. We all know that we can take various points of comparison in nature and build up distinct classifications. It seems to me that what matters most on a philosophical level is not so much to know what the biologist can do theoretically or what he did in the past (invariance essentialism, nomic essentialism, ideal-type taxonomy) but rather what he is doing today. Although, in the past, and more especially in a fixist biology, it was possible to establish several distinct taxonomies because these were based on distinct characters, we must admit that contemporary taxonomists do not enjoy so much freedom. What mainly makes today's taxonomy consistent and gives it its unity (leaving aside a few points of individual disagreement) is the fact that, excepting perhaps some museum technicians, taxonomists ambition to build up not simply a taxonomy but a systematics, i.e. they attempt to describe the phylogenesis of the populations which they classify or of the specimens which they identify. So, for most biologists, the only serious natural taxonomy is the phylogenetic one.

Now, from a phylogenetic viewpoint, there cannot be several and different natural taxonomies. Evolution itself marks the road actually followed. A phylogenetic taxonomy must show in some way in its arrangements the role of the various structures during the constitution of lineage. Both the similarities and dissimilarities that can be demonstrated between groups are based on real, objective characters which have played a part in the course of evolution.

Prof. Suppe deals with phylogenetic taxonomy in Chapter V of his paper. He tackles the problem of defining the taxa and choosing criteria for establishing the taxonomic categories, taking into account the possibility for taxa criteria to vary with time. Having excluded the polytypic and indefinite polytypic definitions, Prof. Suppe tries and builds up a phylogenetic taxonomy, attempting to discover which mechanism determines the variations of the diagnostic characters. This mechanism can be considered a scientific theory. Prof. Suppe takes recourse, in that case, to the theory of evolution which could explain the future genetypic variations (through selection, reproduction rate, nature of the genotype and specific interfecundity) and would permit establishing a taxonomic system. The latter would, however, according to Prof. Suppe, be different from the one used by biologists. I must say that I remained perplexed at his attempt. It looks so theoretical that I wonder if a biologist has ever dealt with that question. The organism, which mutates randomly, evolves in so narrow a dependence with the environment that it seems to me to be very risky to try and forecast genotypic and phenotypic variations of species in the future. A discussion about how to predict the taxonomy of the future biosphere while we are still very ignorant of many existing species appears to me somewhat pointless. There is, in my opinion, another aspect of the problem of taxonomy which is related to evolution, that of the shifting of a character from one to another taxonomic level as evolution progresses. If, for example, one considers the taxonomic value of the emergence of the "tetrapod" character in a Crossopterygian, one must admit that this character at first appears as an individual property, acquiring progressively a subspecific value then becoming specific, generic, etc.. Why does that character take that importance and why does it move from one level up to another? Because, in a long period of drought, while most fishes were eliminated, those that had the ability to travel overland and conquest a terrestrial niche were at an advantage and reproduced more intensively, replacing the fish deprived of the "tetrapod" character. Thanks to the possession of that selective advantage, the tetrapod *Eusthenopteron* and its descendants have multiplied, undergoing further mutations in the course of time. As more important features have been added, the character "tetrapody" has been raised in the taxonomic hierarchy and thus has become the character of a super-class. It seems to me that, if one wishes to analize the problems of taxonomy, such cases are more important and have more concrete bases than the question of future specific variations.

Because it is a mirror of evolution, taxonomy cannot be independent of phylogenesis. Just as living organisms change, our knowledge too must change, i.e. it must reflect the objective world. There is no doubt that taxonomists are forced to revise their positions not only as the present-day livings evolve but also as our knowledge of fossils becomes more complete. A given *character* may be located either on the specific or generic or family level according to the kind of variations that the offspring has undergone. And our discovery of those variations may be more or less complete. We must remember that, as late as 1937, Johansson discovered fossil remains of animals which appeared to represent not merely a new species, genus, family, order, but truly a new class (Pogonophores).

Considering the changes of level that taxonomic characters may undergo according to the extent of our knowledge and the process itself of evolution, we can, in passing, raise another problem that Prof. Suppe has hardly touched. That is the problem of the difference which exist between the taxa of the various taxonomic categories as to their "naturality". The "natural" value of the species has already been questioned. Some biologists think that the species has no objective basis while many others recognize it as a real biological entity. As to the higher categories, excepting sometimes the genus, everyone admits their abstract, conventional, character. The same bony structure may be considered by some students as a family character, by others as on a sub-family and by still others as on a super-family level. That does not mean that those characters used to characterize higher categories are chosen artificially. They all correspond to an objective reality. But there is one aspect in them which is more or less arbitrary and that is the degree of importance that is given to each of them. This point too would be worth a more detailed analysis.

This remark on taxa and categories leads to another thesis in Prof. Suppe's paper, a thesis which, to me, seems to be very questionable.

In a first step, Prof. Suppe states that the world is composed of *particulars* which have intrinsic characteristics i.e. "properties they possess and relationships which they enter into with other particulars independantly of how anybody characterizes, conceptualizes, or conceives them". I accept that notion of properties and intrinsic relations but I do not accept that, later below, Prof. Suppe draws from that the following implication: "Whether an individual belongs to a particular taxon is a matter of empirical fact; hence, belonging to a taxon is an intrinsic characteristic".

Prof. Suppe grants no difference between the fact that an individual belongs to a species or any other taxon and the fact that a subject possesses a character or property. He thus puts on equal footing the facts of a flower being a rosa multiflora and of another object being a desk, being red, being a dog, being an electron. For him, as soon as an individual *a* is correctly classified as belonging to taxon T in a natural taxonomy, the characteristic *is a T* (or *belongs to T* or *is a member of T*) is an intrinsic character of *a.* I confess that I am unable to follow Prof. Suppe that far, unless I know what taxon T is made of. Even going back to the definition of the taxon as given by him i.e. a collection of living organisms, I cannot understand how the fact of belonging to T can be an intrinsic character of *a.* It seems to me that, in order to make that acceptable, T ought to be so to speak independent of *a,* whereas if T is a taxon which, admittedly, is drawn by the mind from a reality composed itself of numerous *a's,* I do not see how *a* can at the same time be the foundation of T and draw from it an intrinsic character. Taxonomy is a science that dictates and states the laws of classification and the criteria of the taxa. Excepting the species, the taxa are not real entities but groups made up by the mind from reality. Objectively there exists no such thing as a taxonomic system. The only thing that exists or has existed is a series of living beings which have succeeded one another in the course of time. One should not confuse the order of knowledge and the order of nature and I am afraid Prof. Suppe makes such a confusion. If T is worked by the mind, through abstracting common typical characters used as diagnosis on that level, than *a* cannot drawn an intrinsic character from its belonging to T. If, on the contrary, T is an objective reality, then we are no longer in an order of scientific knowledge or biological taxonomy.

I am willing to admit that an individual belongs to a taxon because it possesses a character which is common to all the members of that taxon but I cannot admit that this common character be precisely that of belonging to the taxon concerned. I see there a shift of the foundation of knowledge that appears to me inacceptable. In fact, where does the taxon come from? If it is composed from the characters of its members, then the community in which all the members share is precisely that character upon which the taxon is based, and it is not the fact of belonging to the taxon. If the taxon is not based upon a character, it can only be based upon the fact itself of belonging to the taxon: now this is impossible as long as the taxon is not established. How could it be established by taking as foundation that very fact that it belongs to the taxon?

I confess that such problems are puzzling to me and I cannot consider them as really serious and important in spite of the great vogue that they nowadays enjoy.

Prof. Suppe did not demonstrate anywhere in his exposé the *necessity* to admit that the fact of belonging to a taxon is an intrinsic character. He thinks that it is *plausible* that it be so. That word keeps coming back in his text. However, it seems to me very dangerous to found and elaborate a whole argumentation on that plausibility only.

Indeed, it is on that idea that belonging to a taxon constitutes an intrinsic property, that Prof. Suppe bases his definition of the taxa as scientific laws. We know that, for him, the mere fact belonging to a taxon is not a diagnostic character of *a*. There will be, according to him, a correlation between the fact of belonging to the taxon and the occurrence of characters $C_1 \ldots C_n$ (considered in another taxonomy as diagnostic characters). The definition of the taxon appears then as a generalization of that correlation; it must be empirically true or false.

Since taxon T is different from the diagnostic characters, what is it? If taxon *Rosa multiflora* does not include the diagnosis of *Rosa multiflora,* what does it include? How, then, can a *Rosa multiflora* be classified correctly in that particular taxon and not in another one? If T comprises the characters of *Rosa multiflora,* then, there is no necessity to establish a correlation between belonging to the taxon and possessing the proper characters of the diagnosis. But if, on the contrary, T does not include the characters of *Rosa multiflora,* there is no reason to say that *Rosa multiflora is a T.* In this case, to say that *a is a T* does not mean anything. And if *a is a T* means nothing and, therefore, the diagnosis establishes a correlation with that nothing, how do we obtain a scientific law? I feel quite lost! So being unable to accept that definition of the taxa as laws, I am not inclined to agree that biological taxonomy can make a scientific theory and the last ideas contained in Prof. Suppe's paper did not assist me much on this point.

I cannot help smiling when I read that taxonomists have a wrong idea of the theoretization and of the role and function which taxonomy can have in the same. The work produced by the taxonomists may not satisfy Prof. Suppe. But will Prof. Suppe's work satisfy the taxonomists better? I have the feeling that I witness a discussion between deaf people. Quite obviously, Prof. Suppe has no intuition of what the world of a biologist is, whether the latter is a systematicist or a morphologist. For example, he believes that, because taxonomy involves a considerable amount of morphology, its scope as a scientific theory is thereby restricted. Now, if there is at all a biological discipline in which morphology is a must, that surely is taxonomy, in which the practical problem is to classify material organisms, not abstract forms or quanta. It is, in my opinion, impossible to reach a true taxonomy of the living organisms without taking recourse to morphology at all levels (macroscopical, microscopical, cellular or even genic). In a recent article, Prof. Pirlot has shown that the fronteers of morphology cannot be restricted to the field of morphology.[11)]

It seems that the search in pure logics may lead to bypassing the reality of the true problems that a biologist has to face. The biologist does not allow himself to become entangled in such subtle details as those of logical analyses and, perhaps, that is why he can make science progress. . . Have logical analyses of those problems in taxonomy, whether they are written by Woodger or Gregg or Prof. Suppe, really helped the human mind progress in knowledge? I leave it to everyone to answer this question according to his own views.

Footnotes

1) "Plantas quae ex eodem semine ortum dicunt, & speciem suam satione iterum propagant, specie convenire superius docuimus; adeoque varietates illas in florum colore, germinatione & multiplicitate, in foliorum variegatione, radicis colore, fructus semini sive sapore aut etiam colore, specificae quam vocant diversitatis notas aut indicia non esse, conclusimus." (*Historia Plantarum,* Smith & Walford, London, 1963, tome 1, Ch. XXI, p. 42–43).

2) "Methodos autem illas quae plantas secundum locum natales, aut florendi tempus, vires & usus disponunt, rejicimus, quoniam omnes illae cognatas species separant, alienas consociant; eam autem quae a similitudine & convenientia partium praecipuarum, floris scilicet, & calycis, seminis ejusque conceptuali notas characteristicas generum sumit, amplectimur & ususpamus." (*Ibid.,* Ch. XXVI, p. 50–51.).

3) "Deinde cum dicimus in Plantis specie distinctis hanc ab illius semine non oriri, de eo quod plerumque & naturaliter contingit loquimur, non de eo raro & inusitate extra ordinem naturae. Nam ut inferius ostendemus, Plantarum nonnullarum semina degenerem & diversas speciei sobolem interdum producunt; unde colligimus dari in plantis veram speciem transmutationem", *Ibid.,* Ch. XX, p. 40–42.

4) "Linné eut, en outre, le mérite de réviser les genres et d'en réduire le nombre. Alors que Tournefort considérait comme des genres différents le prunier, le cerisier, le laurier-cerise, l'abricotier, Linné en fit autant d'espèces qu'il réunit dans le seul genre prunier. De même furent rangés dans l'unique genre poirier, le poirier, le pommier, le cognassier, le sorbier, l'alizier, qui devinrent des espèces voisines. Par ailleurs, Linné fit rentrer dans une même espèce les nombreuses variétés que ses prédécesseurs avaient considérées comme des types spécifiques. Il y eut cependant, et nous l'avons noté, spécialement en ce qui concerne les animaux, de manifestes exagérations dans la réduction du nombre des genres qui devinrent beaucoup trop compréhensifs", GUYENOT Emile, *Les sciences de la vie aux XVIIe et XVIIIe siècles. L'idée d'évolution,* Albin Michel, Paris, 1941, p. 103.

5) "259. Nomen specificum a partibus *non variantibus,* desumi debet; 260. Magnitudo species non distinguit; 309. Varietates monstrosas constituunt flores *multiplicati, pleni, proliferi,* plantae *luxuriantes, fasciatae, mutilatae:* in numero, figura, proportione & situ partium omnium, nec non saepe *Color, Odor, Sapor, tempus*" LINNE Charles, *Fundamenta botanica,* in *Systema naturae,* 4e ed., Paris, 1744; "260. Magnitudo mutatur a *Loco, Solo, Climate;* mutatur a copia alimenti in plantis, non minus quam in animalibus" *Philosophia botanica,* Stockholm, 1751.

6) "192. Character Naturalis omnes fructificationis notas, per singulas species convenientes recenseat, dissentientes vero sileat." *Fund. bot.;* "192. Omnes partes fructificationis examinandae, etiam quae visum eludunt, quamvis Microscopium adhiberatur (quod tamen rarissime adhibere necesse est), quum, ignota fructificatione, nulla de genere certitudo . . .", *Phil. bot.*

7) "PROBLEMA: Supponas D.O.O. in primordio a simplici progressum ad composita, a paucis ad plura! adeoque a primo vegetabili principio tot tantum creasse plantas diversas, quot *ordines naturales.* Has ordinum plantas IPSUM dein ita inter se a generando miscuisse, ut totidem exorirentur plantae, quot hodie existunt *genera:* Naturam dein genericas has plantas per generationes ambigenas, (quae structuram floris non mutant) inter se miscuisse, et multiplicasse in *species* existentes, quotquot possibiles, exclusis tamen e numero specierum ab ejusmodi generatione productis plantis *hybridis,* utpote sterilibus", *Systema naturae,* ed. cura J.F. Gmelin, 13e ed. Lipsiae, 1791, t. II, p. VIII–IX.

8) LAMARCK J.B.P.A., *Philosophie zoologique,* Chez Dentu et l'Auteur, Paris, 1809, p. 24.

9) *Ibid.,* p. 25.

10) Cf. *Ibid.,* ch. II, p. 39 sq.

11) PIRLOT Paul, Anatomie comparée ou morphologie évolutive dans les recherches contemporaines sur l'évolution, *Revue des Questions Scientifiques,* janvier 1971, 142 (1), p. 35 sq.

Guy Lafrance (Ottawa)

Le problème épistémologique de la classification dans les sciences humaines sociales

Le développement et la prolifération des sciences humaines sociales depuis la seconde moitié du XIXe siècle, bien qu'ils aient largement contribué à approfondir les connaissances sur le comportement humain, ont toujours soulevé des problèmes relatifs au statut de ces diverses et multiples sciences humaines. Parmi les différents problèmes posés, mentionnons celui de la distinction entre chacun de ces savoirs, celui de leur autonomie ou de leur interdépendance, enfin celui de leur classification.

Nous réalisons au point de départ toute l'ampleur de cette question, à laquelle nous avons moins la prétention de pouvoir fournir de réponses que le souci d'essayer d'y apporter quelque clarification. A vrai dire, la tâche que nous nous sommes donnée est d'abord celle d'une présentation historique des faits dont nous voudrions par la suite ou concuremment interpréter l'évolution. Il va sans dire, qu'aux yeux du philosophe, l'évolution et la prolifération des savoirs, dans quelque domaine qu'ils se produisent, soulèvent d'emblée des problèmes d'épistémologie qui, à son point de vue, conditionnent tout le problème de la classification.

Dans la première partie de notre exposé, nous présenterons les faits dans leur évolution historique; dans la seconde partie, nous présenterons les implications épistémologiques et philosophiques qui se dégagent de l'évolution des rapports entre les diverses sciences humaines sociales dans l'intention de pouvoir peut-être mieux cerner le problème de leur classification.

I

Si nous écartons du domaine proprement dit des sciences sociales les spéculations des philosophes depuis l'antiquité grecque sur les diverses formes de l'organisation politique, nous devons reconnaître que les sciences sociales actuelles ont eu des précurseurs immédiats parmi les penseurs et les philosophes du XVIIIe et du XIXe siècles. Condorcet et Montesquieu ont perçu la nécessité d'établir des études scientifiques sur le comportement social de l'homme vivant dans des sociétés diverses. *L'Esprit des lois* de Montesquieu offre un exemple incomparable à l'époque d'étude méthodique des moeurs, des lois et des coutumes des différents peuples. On trouve ramassé dans cette oeuvre gigantesque des études de philosophie politique, de philosophie du droit, de droit comparé, de géographie humaine, d'ethnographie et d'ethnologie comparée. On y trouve même, selon l'expression de Raymond Aron, "l'intention sociologique",[1]) puisqu'il aurait fait de la véritable morphologie sociale.

L'exemple de Montesquieu montrait déjà que la connaissance scientifique des différents éléments de la structure sociale humaine nécessitait le concours de plusieurs sciences. Une constatation analogue peut être faite au sujet des oeuvres de Saint-Simon, de Stuart Mill, d'Auguste Comte et de Herbert Spencer. Alors que la science de l'homme chez Saint-Simon devient une véritable histoire des classes sociales depuis le moyen-âge et qui se couronne par une vision quasi prophétique de l'avènement inéluctable de la classe des industriels et des producteurs; la sociologie d'Auguste Comte de son côté demeure profondément enracinée dans les cadres de la philosophie de l'histoire. Même le *Système de Politique Positive* n'est au fond qu'un chapître ajouté au *Cours de philosophie positive.*

Chez Herbert Spencer, la science sociale, tout en restant intégrée à sa philosophie synthétique et mécaniste, devait néanmoins, pour être véritablement science, maintenir le lien le plus étroit avec la biologie et répondre aux exigences de la science qui sont la généralisation sur laquelle on peut baser une interprétation qui ait au moins valeur approximative de vérité, et la prévision des résultats.[2]) En somme la science sociale de Spencer ne se distingue guère de sa philosophie organiciste inspirée des thèses évolutionnistes. Il n'en demeure pas moins que la sociologie spencérienne, bien qu'elle n'ait été qu'une formulation nouvelle de la pensée de Comte,[3]) eut une influence déterminante sur la sociologie française.

En effet, Spencer régnait en maître sur la jeunesse intellectuelle française de la seconde moitié du XIXe siècle. Il faisait figure de novateur et de prophète par sa manière d'introduire la thèse évolutionniste dans tous les secteurs de l'activité philosophique. Son plus ardent disciple en France fut sans contredit Alfred Espinas qui poursuivit avec acharnement l'idéal de fonder la sociologie biologique.[4]) Hélas! le projet de Spencer et d'Espinas, sans être vain fut sans lendemain. Il devait cependant donner lieu, au tout début du siècle, à des débats passionnés sur la constitution de la sociologie comme science autonome ou comme savoir essentiellement dépendant de la biologie ou de la psychologie.

Cette controverse devait ouvrir la voie à Durkheim et à son école. On nous permettra ici d'insister sur l'apport considérable de cette école dans l'évolution des sciences sociales. Nous retiendrons surtout le travail accompli pour délimiter le champ d'action de la science sociale ainsi que les premières tentatives de classification des déterminations ou branches de la science sociale conçue comme science architectonique des sociétés et des éléments de leur constitution.

Durkheim, après avoir étudié l'état de la science sociale en Allemagne avec le psychologue Wundt, s'attacha résolument au problème de l'autonomie de la science sociale. L'oeuvre première de Durkheim a consisté dans cette tâche êpistémologique de structuration et de véritable fondation de la science sociale qu'il voulait radicalement distincte de ce que l'on appelait alors la philosophie sociale. Durkheim, le premier, a considéré de front la question épistémologique de l'objet propre et de la méthode de la sociologie.[5]) C'est à partir de cette délimitation par Durkheim de l'objet propre de la sociologie à savoir: les faits sociaux considérés comme des choses, que la sociologie a commencé à se déterminer en des branches multiples ou en diverses sciences sociales en fonction de la distinction et de la classification des faits sociaux eux-mêmes. Durkheim a même entrepris une étude de la classification des faits sociaux et des représentations collectives.[6]) Mais du point de vue des sciences sociales elles-mêmes, la classification qu'il en a faites se réduit à trois catégories principales: la morphologie sociale, la physiologie sociale et la sociologie générale.[7])

La distinction entre morphologie et physiologie sociale peut être comparée à la distinction établie par Comte entre statique et dynamique sociale. La morphologie sociale, selon Durkheim, devait comprendre l'étude anatomique et matérielle de la société; c'est-à-dire l'étude de la base géographique des peuples dans ses rapports avec l'organisation sociale, ainsi que l'étude de la population, de son volume, de sa densité, de sa disposition sur le sol, etc. . . La physiologie sociale devait comprendre l'étude des fonctions et des manifestations vitales des sociétés et par conséquent se diviser en une pluralité de sciences particulières comprenant la sociologie religieuse, la sociologie morale, juridique, économique, linguistique, esthétique, etc. Enfin, la sociologie générale, selon Durkheim, serait une science synthétique, une anthropologie générale qui rassemblerait les conclusions qui se dégagent de toutes ces sociologies particulières. Cette science synthétique, Durkheim la considère comme "la partie philosophique de la science" dont la valeur dépend de la valeur des analyses dont elle résulte.[8])

Pour comprendre cette classification des sciences humaines sociales établie par Durkheim, il faut la rattacher à son postulat initial qui est celui de l'unité de la sociologie qui devait constituer comme le *corpus* ou le "système des sciences sociales".[9])

L'idéal et le rêve de la sociologie durkheimienne sera poursuivi par son disciple et collaborateur Marcel Mauss qui consacrera quelques études au problème de la classification des sciences sociales et de leurs rapports réciproques. La perspective épistémologique de Mauss face au problème de la classification dans les sciences sociales ne diffère guère de celle de Durkheim, sauf qu'elle élargit la classification en tenant compte du développement de nouvelles sciences sociales. "Ainsi, écrit-il, tout en se ralliant étroitement aux sciences qui l'ont précédée, tout en s'appropriant leurs résultats, la sociologie transforme leurs classifications. Il est à remarquer d'ailleurs que les diverses sciences sociales ont toutes tendu, dans les dernières années, à se rapprocher progressivement de la sociologie; de plus en plus elles deviennent des parties spéciales d'une science unique".[10])

A la différence de Durkheim toutefois, Mauss a dû tenir compte dans sa classification de la technologie, de la démographie, de l'anthropogéographie et de la géographie elle-même.[11]) Il a écarté de la classification des sciences sociales proprement dites, la statistique et l'ethnographie les considérant comme des sciences à part.[12]) Il a dû faire l'étude des rapports entre la sociologie et l'histoire,[13]) ainsi que l'ont fait Hauser et Simiand[14]) de même que Max Weber en Allemagne.[15])

Pour Mauss, la sociologie a tout à espérer des informations incontestables de l'histoire; mais ces deux disciplines ne cesseront pas pour autant d'être profondément distinctes parce que d'une part l'histoire "ne cessera jamais de décrire, de nuancer, de circonstancier", alors que la sociologie d'autre part tend à donner un système rationnel des faits dont elle n'a pas à faire l'histoire pourvu qu'elle connaisse les antécédents les concomitants, les conséquents ou si l'on veut tout le cadre social qui entoure les faits.[16])

Le principe de la distinction entre l'histoire et la sociologie d'après Hauser et Simiand, tient essentiellement, selon l'expression de Claude Lévi-Strauss, "au caractère comparatif de la méthode sociologique, monographique et fonctionnel de la méthode historique".[17])

De son côté Max Weber a insisté sur la solidarité et la corrélativité de l'histoire et de la sociologie en tant que sciences de la réalité humaine. Il n'est plus question pour lui, comme le pensait Durkheim, de faire de l'histoire une partie de la sociologie. Pour Weber, la causalité historique et la causalité sociologique sont des réalités complémentaires qui permettent d'expliquer et d'interpréter de manière compréhensive le comportement social et les institutions.[18])

Dans les études de classification de Mauss nous retrouvons encore une plus grande souplesse par rapport à Durkheim vis-à-vis la psychologie sociale; Mauss voit davantage une collaboration étroite entre ces deux disciplines distinctes plutôt qu'une intégration de l'une à l'autre.[19]) En outre, Mauss, sans renoncer au principe de l'unité de la sociologie, n'entretient plus le rêve durkheimien de cette science quasi englobante de tous les savoirs sur l'homme; il parle d'une anthropologie dont la sociologie ne serait qu'une partie.[20])

Nous retiendrons enfin de la classification de Mauss la vision presque étonnante du rôle que la linguistique serait un jour appelée à jouer dans les sciences humaines sociales. Par cette vision Mauss est peut être un des pères du structuralisme contemporain!

En 1927, Mauss disait de la sociologie linguistique, qu'elle "constitue, sinon, comme la morphologie sociale, une division fondamentale et se suffisant par elle-même, au moins une division des faits sociaux plus générale que les cinq autres divisions. Car la religion, la morale, l'économie, l'esthétique et la technologie viennent s'y cristalliser: elles se transmettent plus ou moins entièrement dans le langage et celui-ci a pourtant vis-à-vis d'elles une certaine autonomie".[21]) Le langage pour Mauss "porte à un très haut degré, souvent à un plus haut degré que toute autre pratique ou representation collective, le caractère singulier spécifique de singulariser chaque civilisation, chaque société: il est la chose d'une communauté".[22])
En 1934, reprenant le plan de la sociologie générale, Mauss verra dans la linguistique une science indépendante et tellement bien structurée qu'on aurait tort de l'insérer dans une sociologie générale. C'est par les travaux de Ferdinand de Saussure, de Marcel Cohen et d'Antoine Meillet, membre de l'école durkheimienne, que Mauss a pu voir dans la linguistique "l'une des sciences les mieux faites de tout le groupe des sciences humaines".[23])

De la linguistique il dira encore qu'"elle a fait des langues et du langage, et même, au-delà de ceux-ci, de toute une partie de la séméiologie, c'est-à-dire de la symbolistique, son domaine privé. Aucune intransigeance logique ne peut nous faire oublier la qualité spéciale des connaissances linguistiques actuelles. Parmi les sciences de l'homme, la linguistique est peut-être la plus sûre. Grâce à elle, et en lui-même, le langage est le seul phénomène humain dont nous sommes sûr; nous pensons le connaître, et nous le connaissons relativement dans toutes ses parties à la fois: d'abord par expérience interne et personnelle et par communication externe entre personnes se comprenant et devenant ainsi relativement homogènes; puis nous saisissons en un instant, par cette expérience – et à plus forte raison par cette science – les trois aspects du langage: physiologique, psychologique et sociologique. Et quand ces trois aspects ont été analysés, à part et solidairement, par les diverses parties de la linguistique, phonétique, morphologie, syntaxe et sémantique, ils l'ont été de façon si satisfaisante qu'il n'y a pas lieu – provisoirement – de séparer la partie sociologique des autres parties du langage. La langue est d'ailleurs évidemment plus soudée aux autres phénomènes humains que ne le sont les autres phénomènes sociaux, même totaux".[24)]

L'estime de Mauss pour la linguistique l'amena même à parler de la place de la sociologie à l'intérieur de la linguistique ainsi que du rôle de la linguistique dans la sociologie; ce qui est assez éloigné du concept premier d'une sociologie linguistique à l'intérieur du système des sciences sociales.[25)]

L'oeuvre de Marcel Mauss aura donc fait évoluer considérablement l'étude de la classification des sciences sociales depuis les premières tentatives de Durkheim en ce sens.

Si depuis Mauss l'étude du problème de la classification dans les sciences sociales ne semble pas avoir tellement retenu l'attention des chercheurs, les sciences sociales pour leur part ont enregistré des progrès considérables surtout du côté de l'ethnologie et de la linguistique. Outre les travaux de Lévy-Bruhl et de Claude Lévi-Strauss en France; c'est en Angleterre et aux Etats-Unis que l'ethnologie a connu ses plus grands développements avec les travaux de Boas, de Lowie, de Kroeber, de Mead, de Rivers, de Malinowski, de Radcliffe-Brown et de combien d'autres. Du côté de la linguistique ce sont les travaux de Benveniste, Martinet, Jakobson, Troubetzkoy, Chomsky et Bloomfield qui retiennent l'attention.

Toutes ces recherches ont eu pour résultat de relancer le débat autour du caractère scientifique et positif des sciences humaines sociales dans leurs structures et leur mode opérationnel; débat provoqué par le rapprochement de la linguistique et de l'ethnologie. Les résultats obtenus par l'analyse structurale des langues devaient faire naitre l'espoir de l'utilisation d'une telle méthode pour l'ensemble des sciences sociales, en supposant évidemment que les phénomènes linguistiques soient identiques aux faits sociaux.

L'ethnologie du même coup introduit, à la manière des linguistes, l'idée de la distinction entre une approche diachronique et une approche synchronique des phénomènes sociaux ce qui ne va pas sans soulever tout le problème des rapports entre l'ethnologie et l'histoire; à moins que l'on accepte l'idée d'une ethnologie qui soit à la fois synchronique et diachronique limitant son champ de recherches aux conditions inconscientes de la vie sociale et laissant à l'histoire l'étude des expressions conscientes.[26])

La linguistique – et surtout la linguistique structurale – a également apporté de grands espoirs à l'anthropologie sociale ou culturelle en lui laissant entrevoir l'accès possible à une méthode et à des résultats comparables à ceux des sciences exactes.

L'anthropologie, se voyant déjà au carrefour des sciences naturelles, des sciences humaines et des sciences sociales, suggère en même temps une conception nouvelle de la classification dans les sciences humaines sociales, dont l'anthropologie, cela va de soi, constitue le noyau.

Ainsi, selon la classification de Claude Lévi-Strauss (classification qui semble correspondre à une situation de fait dans les pays anglosaxons surtout), l'anthropologie serait comme la synthèse de l'ethnographie et de l'ethnologie. Alors que l'ethnographie correspondrait aux premiers stades de la recherche, c'est-à-dire le travail d'observation et de description; l'ethnologie représenterait un premier pas vers la synthèse, cette synthèse pouvant s'opérer dans trois directions, soit géographique, soit historique, soit systématique.[27]) L'anthropologie pour sa part constituerait la dernière étape de la synthèse; elle viserait "à une connaissance globale de l'homme, embrassant son sujet dans toute son extension historique et géographique; aspirant à une connaissance applicable à l'ensemble du développement humain depuis, disons, les hominidés jusqu'aux races modernes".[28])

L'anthropologie, en outre, entretiendrait des relations étroites avec la géographie, la technologie, la préhistoire, l'archéologie, l'histoire, la psychologie, la sociologie et surtout la linguistique; disciplines qui lui serviraient de modèles ou d'auxiliaires.

De cette classification il ressort que l'anthropologie se situerait nettement au carrefour des savoirs sur l'homme. Perspective qui n'est pas sans soulever de conflit avec la sociologie dont l'ambition va aussi dans le sens d'une totalité des savoirs sur l'homme ou selon l'expression de Mauss, dans le sens d'une connaissance des faits sociaux totaux. Un fait cependant demeure certain, c'est que l'anthropologie dite structurale ne saurait couvrir dans sa totalité l'objet même de la sociologie.[29])

Mentionnons enfin que dans ce vaste problème épistémologique de définition et de classification des sciences humaines sociales la tâche reste entière vis-à-vis des nouvelles sciences de la communication.

Devant ce tableau un peu rapide et fort incomplet de l'évolution des rapports entre les principales sciences humaines sociales, certaines lignes de faits se dégagent et plusieurs interrogations se posent.

Parmi les lignes de faits mentionnons en premier lieu le besoin d'unité et de collaboration éprouvé entre les diverses sciences humaines sociales à mesure que s'accroît le sentiment de leur interdépendance.

En deuxième lieu, le rapprochement constant des sciences humaines sociales avec les sciences naturelles grâce surtout aux techniques développées par ces dernières.

En troisième lieu, la crainte réciproquement ressentie par les spécialistes des différents sciences humaines de voir leur discipline subjuguée par une autre.

En quatrième lieu, l'absence ou presque d'une philosophie des sciences humaines sociales comparable à la philosophie des sciences et qui assurerait la synthèse des recherches particulières.

Ces lignes de faits, et il y en a certes d'autres, conditionnent, à notre avis, les formules plutôt arbitraires et diversifiées de classification des sciences humaines sociales. C'est pourquoi nous nous interrogeons sur les critères qui puissent assurer un fondement épistémologique valable à une classification rationnelle des sciences humaines sociales. C'est précisément ce travail d'épistémologie dans le sens de la délimitation de l'objet, de la méthode, du point de vue, du niveau et de l'intensité de la recherche dans chacune de ces sciences qui nous semble faire défaut. N'est-ce pas là une fonction qui revient à la philosophie et aux philosophes?

C'est pourquoi nous nous interrogeons encore, au moins en ce qui regarde le problème de la classification, sur la nécessité d'une philosophie des sciences humaines sociales.

Une fois dissipés les mauvais souvenirs laissés par l'ancienne philosophie sociale, il se pourrait fort que la philosophie dans sa fonction d'interprétation et de synthèse, soit appelée à jouer un rôle de premier plan dans ces sciences humaines. Nous ferions alors crédit à Claude Lévi-Strauss qui, il y a quelques années, affirmait que "l'origine philosophique de la sociologie française lui a joué, dans le passé, quelques tours; il se pourrait qu'elle fût pour l'avenir son meilleur atout".30)

Footnotes

1) ARON, Raymond, Les étapes de la pensée sociologique; N.R.F., Gallimard, 1967, p. 27.
2) Voir Herbert SPENCER, Introduction a la science sociale; 6e édition, Paris, Baillière, 1882, p. 48.
3) Spencer lui-même reconnaissait à Auguste Comte le mérite d'avoir été l'instigateur de la sociologie biologique: "C'est a M. Comte, écrivait-il, . . .que revient l'honneur d'avoir mis en lumière avec une précision relative la connexion entre la science de la vie et la science de la société". (Introduction a la science sociale; op. cit., p. 352).
4) De Espinas voir en particulier son ouvrage intitulé "Les sociétés animales" (Paris, Baillière,, 1877), et surtout un article paru en 1901 dans la Revue Philosophique sous le titre "Etre ou ne pas être" (loc. cit., p. 449 à 480) dirigé contre Gabriel Tarde et qui posait à toute sociologie scientifique le dilemme suivant: être biologique ou n'être pas.
5) cf. DURKHEIM, Emile, Les règles de la méthode sociologique, Paris, Alcan, 1895.
6) DURKHEIM, Emile, De quelques formes primitives de classification, dans Journal Sociologique, P.U.F., 1969.
7) Voir à ce sujet: DURKHEIM, Emile, Sociologie et sciences sociales dans La Science sociale et l'action, P.U.F., 1970, p. 137 à 159.
8) Ibidem, p. 152.
9) Ibidem, p. 146 et 151.
10) MAUSS, Marcel, Oeuvres, Editions de Minuit, Paris, 1969, T. III, p. 175.
11) Voir Ibidem, p. 180 et 191.
12) La statistique, selon Mauss, ne serait qu'une méthode pour observer les phénomènes variés de la vie sociale, alors que l'ethnographie aurait pour seule raison d'être "de se consacrer à l'étude des phénomènes qui se passent dans les nations dites sauvages" (Ibidem, p. 174-175).
13) Ibidem, p. 168 à 174.
14) HAUSER, H.,L'Enseignement des sciences sociales, Paris, 1903. SIMIAND, F., Méthode historique et science sociale, Revue de Synthèse, 1903.
15) Voir les Essais sur la théorie de la science, Paris, Plon, 1965, (trad. de J. Freund).
16) MAUSS, Marcel, Oeuvres, op. cit., T. III, p. 169-170.
17) LEVI-STRAUSS, Claude, Anthropologie structurale, Paris, Plon, 1958, p. 3
18) WEBER, Max, Essais sur la théorie de la science, op. cit.
19) Voir Oeuvres, T. III, op. cit., p. 269 à 302.
20) Ibidem, p. 246.
21) Ibidem, p. 193.
22) Ibidem, p. 192.
23) Ibidem, p. 444.
24) Ibidem, p. 310-311.
25) Ibidem, p. 312.
26) Voir à ce sujet Claude LEVI-STRAUSS, Anthropologie structurale, Paris, Plon, 1958, p. 3 à 33.
27) Ibidem, p. 387.
28) Ibidem, p. 388.
29) Voir à ce sujet l'article de André G. HAUDRICOURT et Georges GRANAI: Linguistique et sociologie, dans les Cahiers internationaux de sociologie, Vol. IXI, 1955, p. 114 à 129.
30) LEVI-STRAUSS, Claude: La Sociologie française, dans La Sociologie au XXe siècle, Ch. XVIIe, p. 545, publié sous la direction de Georges GURVITCH, P.U.F., 1947.

Patrick A. Heelan (Stony Brook, N.Y.)

The logic of changing classificatory frameworks

The context of assumptions in which I am working comprises those counter-positions to classical logical empiricism, established by such authors as N. R. Hanson, P. K. Feyerabend and T. S. Kuhn, such as the absence of any hard distinction between observational and theoretical language, the validity of multiple explanatory viewpoints, the existence of both continuous trajectories of theory development and discontinous trajectories representing revolutionary episodes in the history of science or culture.

I shall be concerned with descriptive linguistic framework, the subject matter of this study. A linguistic framework is a projection at a given epoch in language of a functioning intentionality-structure.[1] A framework is *descriptive* whenever it purports to make realistic classifications or discriminations among its objects. For the purposes of this inquiry, a descriptive framework will be designated by a symbol of the form L_A, where A is the *context* of the language, and L_A is the open set of sentences that can be used to make factual discriminations or classifications in the horizon of the context A. The context A is the set of subjective and objective conditions necessary and sufficient for the appropriate use of L_A (i.e., for statements in L_A to be true or false). Any given L_A is an entity in history, since it is the linguistic projection at a given epoch of the intentionality-structure ingredient in A. Thus, over a period of time, there will emerge a sequence of descriptive languages L_{A_1}, L_{A_2}, L_{A_3}, and so on, each one a projection of the same intentionality-structure, the later in the sequence being more adequate projections than the earlier. It is supposed that an invariant functioning intentionality-structure such as is supposed to underlie the sequence L_{A_1}, L_{A_2}, L_{A_3}, and so on, operates in the process of development to correct inadequacies in the linguistic projection. I call the sequence L_{A_1}, L_{A_2}, L_{A_3}, and so on a "tradition."

For example, the sustained search for an adequate mathematization of motion beginning in the late middle ages with the Merton and Paris schools, and developing through Galileo, Kepler and Newton to Einstein in the twentieth century constituted a tradition. The contexts of this search – roughly the "internal steering field" of Radnitzky -- which are judged in retrospect from the latest stage of development of the research tradition, is a paradigm of normal science of the kind described by T. S. Kuhn[2] – the complex of institutionalized public criteria in which a tradition is entrenched and which carries it forward from a less adequate to a more adequate stage of expression.

Among the units I have called "descriptive linguistic frameworks," some pairs of these will be related to others in the way successive frameworks within a common tradition are related. I shall call this relationship one of "implication" between frameworks. Other pairs will not be related directly to one another but each will be related to a third framework in which the two traditions merge. Yet other pairs will not be related at all, either directly or indirectly. The relation of implication is a linear or partial ordering of frameworks which I shall symbolize by the functor "$\rightarrow$ " and which bears the following interpretation: $L_X \rightarrow L_Y$ whenever what can be said appropriately and truly in L_X can be said appropriately and truly in L_Y, but not necessarily vice versa. The relationship of implication between linguistic frameworks is a logical and not a temporal relation.

There are many problems connected with the analysis of the relationship of implication between frameworks, but these I shall reserve for fuller treatment elsewhere. However, a brief commentary on the phrase "whatever can be said appropriately and truly in L_X" is in order. The phrase refers to the use of sentences formulated in L_X to make statements about possible discriminations in the context X. The sentences are appropriately used, whenever they are used in their proper context--that of X; and when so used to make statements, the statements are either true or false. Sentences in L_X connote the concepts of L_X which are all functions of (the hermeneutical circle of) L_X. Sentences in L_Y connote the concepts of L_Y which are all functions of (the hermeneutical circle of) L_Y. When L_X and L_Y are different frames, the concepts of L_X and the concepts of L_Y are – on first scrutiny at least – incomparable. Whenever L_X implies L_Y, however, certain sentences used appropriately in X can also be used appropriately in Y. The concepts invoked by these two uses are not univocal, since L_X is not L_Y; nor are they equivocal, since L_Y is a development of L_X. They are "analogous concepts," and the sentences as used in L_X and L_Y are "analogous predications" of the same sentence. The structure of this kind of analogy will have to be explored further, but in another paper.

Though the relationship of implication between frameworks is a logical and not a temporal relation, I shall be interested both in synchronous and diachronous frameworks. In the former case, we are concerned with the degree of unification through implication that can be established among the diverse descriptive frameworks in use at a given epoch; for instance, between Aristotelian type frameworks and scientific frameworks, neurophysical frameworks and psychological ones, microbiological frameworks and macrobiological ones, and so on. The relation of implications leads to two types of structures: a linear ordering when the frameworks are more or less adequate projections of the same intentionality-structure, and a partial ordering constitutive of a non-distributive lattice when the frameworks are complementary. A third possibility exists – that no ordering of the kind an implication exists between the frameworks; then the frameworks are simply incompatible. In the latter case – of diachronous frameworks – the most usual case is that where the temporal ordering parallels that of the logical ordering. We can then speak of a formal logic of developmental framework transpositions. The relation of implication leads to the construction of two models – a linear ordering characteristic of a single tradition developing in history, and a partial ordering constitutive of a non-distributive lattice for two or more complementary traditions developing in a dialectical way. There is a third possibility that no ordering of the kind of an implication exists: then the incompatible traditions develop separately without a dialectical relation to one another.

I shall consider in greater detail three cases: (i) the development of a single tradition, (ii) an example of incompatible traditions, and (iii) the lattice structure of complementary or dialectically developing traditions.

(i) Development of a tradition. In the development of a single tradition, $L_{A_1} \rightarrow L_{A_2}$. This relation implies that the two frameworks L_{A_1} and L_{A_2} make a common set of discriminations within the horizon of L_{A_2}. To understand in what sense this is meant, consider the example already mentioned above of Kepler's Laws L_{A_1} and Newton's Laws L_{A_2}.

A sufficiently complex heuristic structure can start a tradition of inquiry that manifests itself in and through a historical family of conceptual frameworks. The movement to give a mathematical account of motion that started in the middle ages with Bradwardine and others, continued through Galileo, Kepler, and Newton and found its most developed expression in Einstein's general theory of relativity, was such a tradition. The theories of motion it generated cannot be opposed and contrasted with some alleged basic descriptive account that remained historically invariant but must be opposed and contrasted with one another, for although they use the same vocabulary and sentences and make some common discriminations in their domain of inquiry, they also come in conflict. Newton's Laws (L_{A_2}), for example, give the same results as Kepler's (L_{A_1}) for a single planetary system, but they yield different results for a multi-planetary system like the Solar system since Newton's Laws explain observed interplanetary gravitational perturbations but Kepler's Laws do not. Since two theories cannot be simultaneously held if they yield inconsistent results in a domain where the difference is significant, Newton's Laws came to replace Kepler's in the historical development of mathematical theories of motion. The framework transposition exemplified is one where both frameworks are concerned with the same type of problem and involve a common, though analogical, set of descriptive predicates. Everything said truly of natural objects in L_{A_1} (Kepler's frame)-- but in retrospect from L_{A_2} (Newton's frame)-- can be said truly in L_{A_2}, but not vice versa. The sequence of historical frameworks L_{A_1}, L_{A_2}, and so on, that constitute the development of a tradition, then are a linearly ordered set under the implication relation " $\rightarrow$."

(ii) Incompatible frameworks. A different problem arises where conceptual frameworks using the same vocabulary and sentences employ, nevertheless, incompatible categorizations or classifications. This is the realm of conventionality in theory and observation. There can be no common domain of discriminations made by such frameworks.

An example might be the description of geometrical relationships between things that appear as everyday objects. In ordinary language those relationships are based upon a rule of congruence (a rule for deciding when two lengths are equal) that involves measurement by transported rigid rulers. This rule of congruence gives a Euclidean space to everyday objects – the space to which we ordinarily belong and which carpenters, architects, and others use in the exercise of their profession. We have in fact become so accustomed to the Euclidean character of our everyday world that we are ready to believe that God made the world to be Euclidean. But the everyday world is Euclidean only in so far as man uses a ruler or some equivalent form of length measure. Other rules of congruence exist that yield different geometrical descriptions of the world. For example, it has been shown by Rudolf Luneburg that a binocular visual comparison of lengths under standardized conditions gives a kind of visual space that turns out to be Lobachevskian (a constant curvature space of negative curvature) 3). Objectivity (in the sense of intersubjective validity) for the new spatial experience is gained by involving a sufficiently large sample of normal people in structuring their spatial affirmations in this way and teaching them to speak the language appropriate to the description of objects in visual space. That such ordering of space is in fact more than a laboratory experiment is suggested by post-impressionist painting. In a study of van Gogh's painting, I have postulated a close relation between the space represented and the artist's awareness of the anomalous (non--Euclidean) structure of binocular visual space, and I have proposed an explanation for the widely-recognized "distortions" of Van Gogh's space on the mathematical principles appropriate to the new space.4)

A mathematical analysis of the visual forms which ordinary rectilinear physical objects, like tables, chairs, floors, walls, ceilings, etc., would have in the perception of an observer habituated to constituting his space in this way can be computed from the geometrical formulas. Such an observer would notice for instance, that the floor plane tilts upwards towards the horizon, that distant objects are brought forward, that nearby objects protrude towards the observer and appear in inverse perspective and that straight lines are bent in certain predictable ways. To people habituated to a Euclidean interpretation of their perceptual experience these anomalies appear as *perceptual illusions.* Some of these illusions have been noticed in the course of experiments on human perception, for example, with points of light in a dark room; others are in common experience such as car headlights at night or the heavenly bodies in the night sky. The peculiarities I have mentioned, are found represented in the pictorial space of Vincent van Gogh's paintings and to a certain extent also in those of Paul Cezanne. These painters, who were not geometers, learned, it seems, to speak a new spatial language, a language based upon a binocular experience of length congruence. This new linguistic proposal was not an arbitrary one, if my interpretation is correct. It was rooted in a geometrization of space by visual means, resulting in a new kind of space which seems to have replaced classical Euclidean space as a form of the real world in the linguistic usage of the two artists mentioned. For them, classical Euclidean space – the space in which we cope with practical survival problems – became presumably a domain of mere appearances; reality for them was non-Euclidean.

Meaning, language and conceptual framework, then though conventional, are not arbitrary, since they grow out of human experience. There may, however, be no way of inferring all the untapped possibilities of human experience, past, present and future, and actual experience may be the only profile of human potentialities we can concretely get to know.

One might argue that a conceptual framework must conform to some set of basic facts established as such, say, by a Supreme Intelligence, all antecedent to human experience. But unless the Supreme Intelligence is conceived of as a supremely intelligent man (contacting the world therefore through a sensitive body) its description of facts cannot in general be normative for human experience.

The example I have used of two incompatible geometrical conceptual frameworks, Euclidean and Lobachevskian, illustrates two possible forms of the human experience of space. These two frameworks cannot, however, be combined into one description without incoherence. They are two separate incompatible spatial descriptions, and not different aspects of one thing. The root of their incompatibility lies in the fact that they are embodied in two mutually exclusive sets of heuristic behaviors.

Let me introduce some new terms and specify more exactly the meaning of some terms already used. *Heuristic structure*[5]): A heuristic structure is the anticipation of a known unknown; that is, it is the structure of systematic inquiry guided by a set of canons and embodied in a pattern of human exploratory behavior. It is a form of intentionality-structure. A heuristic structure gives meaning to the descriptive language in which the inquiry is formulated. When the form of the unknown has been explicitly sketched out, we find it lies on the circumference of a closed set of semantically linked descriptive predicates called a hermeneutical cicle. *Horizon*[6]): A horizon is the set of possible objective (in the sense of intersubjective) facts attainable by and through a heuristic structure equipped with a descriptive language. *(Descriptive) Language:* A language is the set of descriptive statements which in the formal mode map the material facts or discriminations belonging to a horizon. The mapping is a 'picturing' of facts or discriminations in language, a 'projection' of a horizon of factual human experience into a domain of public signs (sentences). I take a descriptive language, then, to be the open set of descriptive factual statements that can be constructed using the limited linguistic resources of a descriptive language, namely, names (or other referring expressions), descriptive or explanatory conjugates that give the basis for classifications and the logic of sentences, predicates, and quantification. *Linguistic context:* The necessary and sufficient conditions subjective and·objective, for the use of a descriptive language.

The two cases we have considered are the replacement of one horizon by another more adequate than the first within the same tradition of inquiry, and the co-existence of multiple though mutually exclusive horizons standing for a variety of valid projections or classifications of human experience, constituting, therefore, different and incompatible traditions.

(iii) Complementary frameworks and dialectical development.
A third kind of framework relation is what I shall call 'the complementarity of horizons.' The term 'complementarity' is taken from quantum mechanics and expresses that peculiar relationship found in quantum mechanics, for example, between the two descriptive languages of (exact) position and (exact) momentum. The term was applied by Niels Bohr to the contextual character of description in quantum mechanics: given certain conditions, position language was the appropriate one; given other conditions, momentum language was the appropriate one. The conditions that constitute the linguistic context are related to measurement. Thus the theoretical inability in quantum mechanics to measure exact position and exact momentum simultaneously makes it impossible to have one linguistic contect in which both (exact) position and (exact) momentum language can be used. However, if we are willing to admit the use of inexact position and momentum predicates (the inexactitudes to be governed by Heisenberg's Indeterminacy Principle), then the complementary languages of exact position and exact momentum appear as limiting cases of a more general quantum mechanical descriptive language of inexact concepts. Complementary languages, then, are context-dependent descriptive languages which, although conditioned by mutually exclusive linguistic contexts, do not constitute incompatible traditions, and serve, in some sense, as expressions of lower restrictive or limiting viewpoints within a higher more comprehensive viewpoint. The broader and more inclusive descriptive language of the higher viewpoint stands to the lower complementary languages in a way that will be stated more precisely and formally below.

The paradigm case of complementarity is exact position language L_a and exact momentum language L_b in quantum mechanics. Both L_a and L_b have their necessary and sufficient conditions: L_a, that the quantum system be in contact with a position-measuring environment (instrument plus suitable boundary conditions); L_b, that the quantum system be in contact with a momentum measuring environment. Precise position language L_a is the linguistic projection of precise position measuring contexts $\underline{a}$; precise momentum language L_b is the linguistic projection of precise momentum measuring context $\underline{b}$. Let it be noted that L_a and L_b are subsets of a broader, more inclusive quantum mechanical kinematic language $L_{a \oplus b}$ in which the space-time description of a quantum system can be formulated, even if the environment should exclude the possibilities of a precise position measurement or a precise momentum measurement.[7]) Precise position language L_a, precise momentum language L_b and the language of the higher viewpoint, $L_{a \oplus b}$, constitute, with suitably defined complements, what is called a non-uniquely complemented lattice of language. Let us then define the condition under which two languages are complementary: L_a and L_b are *complementary* in $L_{a \oplus b}$, if and only if, L_a, L_b, and $L_{a \oplus b}$ constitute part of a non-distributive lattice of languages under a partial ordering to be defined more accurately below. What I have done here formalizes, in my view, the significant element of what is meant by 'complementarity' in physics.

In contrast to the account of complementarity I have just given in term of an enriched and transformed synthesis of the complementary languages and a variety of contexts in which this language is appropriately applied, is the traditional account of Niels Bohr in which the complementary languages are not internelly transformed and synthesized but are used separately subject to a controlling judgement made from a metacontextual point of view. In this latter account, $L_{a \oplus b}$ is assumed not to exist: only L_a and L_b exist, each being a different subset of classical physical language. The account I have given as my own agrees better with Heisenberg's original expressed in his first paper on quantum mechanics, than with Bohr's view which he later adopted as his own.[8]

Let us suppose that there exists a language L'_a, which is a development of L_b. and a language L'_b which is a development of L_a (the common lower bound of the lattice) constitute the kind of non-uniquely complemented lattice sketched in Figure 1 below.[9]

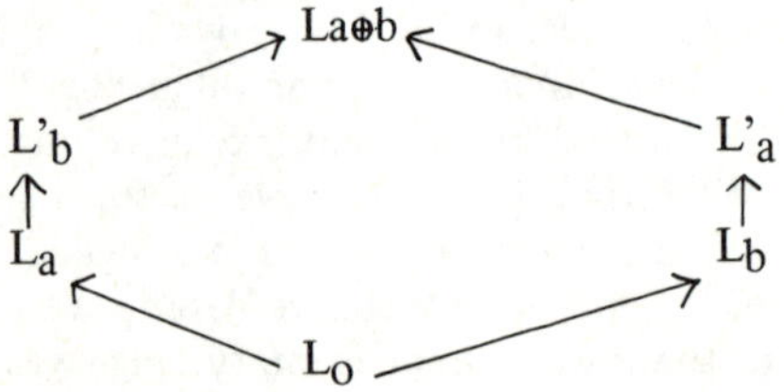

Figure 1: A non-Boolean lattice for two complementary languages L_a and L_b.

The relationship represented by the arrow is a transitive reflexive relation and is called a 'partial ordering' or 'implication.' Whenever any two elements of the lattice, say L_x and L_y, are in the partial ordering relation, symbolized "$L_x \rightarrow L_y$", then we say that L_x *implies* L_y. Moreover, a least upper bound (l.u.b. or sum) $L_x \oplus L_y$, and a greatest lower bound (g.l.b. or product) $L_x \otimes L_y$, of L_x and L_y are defined in the following way: $L_x \oplus L_y$ is the l.u.b. of L_x and L_y if and only if both L_x and L_y imply $L_x \oplus L_y$ and $L_x \oplus L_y$ is the lowest such element in the diagram with that property. $L_x \otimes L_y$ is the g.l.b. of L_x and L_y if and only if $L_x \otimes L_y$ implies both L_x and L_y and $L_x \otimes L_y$ is the highest such element in the diagram with such a property. From these definitions it follows that $L_x \rightarrow L_y$ if and only if $L_x = L_x \otimes L_y$ and $L_y = L_x \oplus L_y$. Moreover (in the case being considered) every element of the lattice has a complement represented by a prime with the properties such that $L_x \oplus L'_x = L_a \oplus_b$, $L_x \oplus L'_x = L_o$, and $L''_x = L_x$. The lattice in the diagram is non-distributive, since

$$L_a \oplus (L'_b \otimes L_b) = L_a$$
$$(L_a \oplus L'_b) \otimes (L_a \oplus L_b) = L'_b$$

but

$$L'_b \neq L_a$$

This kind of lattice differs from a Boolean lattice where the operations of sum and product are distributive.

The elements of the lattice are not descriptive kinematic sentences of quantum mechanics, but a set of six distinct descriptive quantum mechanical languages, each treated as an indivisible element of the lattice. The partial ordering (symbolized ' $\rightarrow$ ') between two languages L_x and L_y relates higher or lower viewpoints by a new form of implication, one that relates languages and not sentences: $L_x \rightarrow L_y$ if and only if whatever "can be said" in L_x "can be said" in L_y. The l.u.b., then, of two languages L_x and L_y is the lowest language L_z such that whatever "can be said" in L_x or L_y "can also be said" in L_z. This corresponds to the lowest higher viewpoint which comprises both lower viewpoints. The g.l.b. of any two languages L_x and L_y is the highest language L_w such that whatever "can be said" in it "can also be said" in L_x and L_y.

A logic that relates languages rather than sentences I shall call 'a context logic,' for each language has its unique context, the set of necessary and sufficient conditions for its valid use; thus a logic of languages is also a logic of contexts, a context logic.

The lattice of six distinct languages taken in the material mode can be re-interpreted in the formal mode as a logic for propositions in a meta-language I call 'the meta context language.' The meta-context language describes linguistic contexts and names the languages appropriate to the context. In this meta-context language, the logical sum ('p or q') of the two sentences 'p' and 'q' is mapped on the l.u.b. of the two sentences, that is, on '$p \oplus q$'; the logical product ('p and q') is represented by the g.l.b., that is, by '$p \otimes q$'; negation ('not-p') is complementation. The partial ordering of 'p' and 'q' ('$p \rightarrow q$') is implication. The correct locus of the non-classical logic of quantum mechanics for example is, contrary to the received view, not in a quantum mechanical object language which is most probably subject to classical logic, but in that language I have called 'the meta-context language of quantum mechanics.'

The extension of context logic from descriptive or classificatory languages, say, to philosophical languages is also possible. Bypassing the difficult problem of defining what constitutes a philosophical language, let me simply postulate that to every coherent philosophical *corpus* C, there is a unique philosophical language

L_c. I shall suppose that there are criteria for deciding when a set of sentences constitutes a coherent philosophical *corpus* and will consider the Aristotelian and the Platonic-Augustinian tradition as differing and subsequently synthesized in the philosophy of Aquinas. The Aristotelian would be represented by a historical sequence of philosophical languages L_{a_i} where i indicates the members of the sequence. Similarly, the Platonic-Agustinian tradition would be represented by a historical sequence L_{b_i} of philosophical languages.

Now it is the contention of Gilson, Geiger, Fabro, and others that Aquinas synthesized both these traditions in an original way. Let me try to formulate this thesis using the notions of context logic sketched out above.10)

Taking a scissors to the *corpus* of Aquinas' writings, let us cut out all those passages that have an Aristotelian flavor. To these, we shall add a sufficient number of passages also of the *corpus* but of a neutral character to constitute a coherent body of philosophical writings. Then this subset of the Thomistic *corpus* taken in itself apart from the rest of Thomas' writings would constitute, objectively speaking and independently of the subjective intention of the author, an Aristotelian *sub-corpus* which could be artificially carved out of the same Thomistis *corpus.* Let L_a be the Aristotelian sub-language and L_b the Platonic-Augustinian sub-language of these *sub-corpora.* Then L_a and L_b are sub-languages of Thomas' unabridged philosophical language $L_{a\oplus b}$. The Gilsonian thesis would then claim that $L_a \rightarrow L_{a\oplus b}$; and $L_b \rightarrow L_{a\oplus b}$, where L_a and L_b are the lower viewpoints relative to $L_{a\oplus b}$, the higher viewpoint. The Gilsonian thesis, however, would make the additional claim that Thomas created an original synthesis of the two traditions transcending both and each enriching the other with new insights. In other words, the synthesis fulfills the condition that $L_{a\oplus b} \neq L_a \cup L_b$ (that is, $L_{a\oplus b}$ is not the set theoretic union of the sentences of the original languages), since more "can be said" in $L_{a\oplus b}$ than "can be said" in L_a or L_b. L_a, L_b, and $L_{a\oplus b}$ obey the relations of the context logic discussed above where L_a and L_b are the lower viewpoints and $L_{a\oplus b}$ is the higher viewpoint. $L_{a\oplus b}$ is then a philosophical language in equilibrium, more or less, with the critical philosophical experience of Aristotelians, Augustinians, and Thomists.

Other examples of language pairs that may well fulfill the partial ordering relation of complementarity are physics and biology in relation to biophysics, mental and physical languages relative to ordinary language, esthetic and functional languages relative to the philosophy of value, and so on.

The relationships exemplified by complementarity are logical relationships and as such do not involve a historical or genetic ordering of the elements. Some of the examples, however, suggest that a genetic and historical relationship may in certain cases parallel the logical one; that is, the logical arrow ' $\rightarrow$ ' may sometimes coincide with the direction of historical evolutionary development. Indeed the fulfillment of the logical relation may even turn out to be the most reasonable criterion to judge "true" (authentic) evolutionary development from the "merely apparent" (or inauthentic).

The passage from two disjoint and in some respects mutually antithetical traditions to a synthesis can be decribed in the following way, which I should like to propose as a logical analysis of a progressive dialectical development. At time t_o, two research communities, one representing the Aristotelian and the other the Augustinian traditions, enter into dialogue. Let L_{a_o} be that member of the sequence of Aristotelian languages spoken initially by one research community, and L_{b_o} be that member of the sequence of Platonic-Augustinian languages spoken initially by the other research community. During the course of the subsequent dialogue L_{a_o} comes to be replaced by L_a, and L_{b_o} by L_b (both replacements in the same tradition) in such a way that a new language $L_{a \oplus b}$ is formed satisfying the complementary relationships: $L_a \rightarrow L_{a \oplus b}$, $L_b \rightarrow L_{a \oplus b}$ but $L_{a \oplus b} \neq L_a \cup L_b$. $L_{a \oplus b}$ is spoken by everyone. It is a language in which Aristotelians, Augustinians, and Thomists can all share, each understanding the other while maintaining -- if they wish -- the separateness of their respective traditions. It is the non-Boolean character of the complementarity relation that guarentees that the synthesis $L_{a \oplus b}$ is richer than the antitheses L_a and L_b taken together but in mutual contextual isolation. $L_{a \oplus b}$ is the outcome of a progressive dialectical development involving two complementary traditions represented by the sequences L_{a_i} and L_{b_i}.

An illustration of progressive dialectical development in the history of science is given by the history of the notion of *species* in biology. Linnaeus followed the essentialist tradition, traceable back to Aristotle, according to which each organism possessed a historically unchanging specific form, embodied in a type specimen, and shared only by individuals of the species. Without such a fixed specific form, it was thought impossible to classify organisms. Darwin, on the other hand, postulated a continuously varying specific form, varying with individuals and over the course of time, with natural selection operating to sustain some variations and to eliminate others. For Darwin, consequently, specific forms were subject to a gradual and continuous alteration in the course of time, making distinctions between species a matter of drawing arbitrary lines across a continuum of forms. The New Systematics in biology, by defining species with respect to a biologically interbreeding historical population, unites what is significant and relevant in both of these concepts. The phenotypic (or Aristotelian) description was enlarged to include a range of type specimens representing the population and related to one another through frequency distributions. As long as the environment remained unaltered, the specific description remained fixed, as for instance, in the case of some archaic biological forms, trilobytes, for instance. If the environment, however, changed, then selection pressures worked with other natural causes to bring about gradual modifications of phenotypic characters. Thus, a species that is fragmented into geographically isolated subspecies subject to differential selection pressures, eventually becomes so many different species. Thus, if the environment is stabilized the specific form can be treated as Aristotelian; if the environment is altered, the specific form changes according to the Darwinian notion. What was needed to unite those two extremes and make them conceptually coherent, was to make the term "species" refer to a historically continuous interbreeding population at equilibrium – or tending to equilibrium – with its environment. The new notion, moreover, is richer than the two former notions, for it allows both wide and narrow variations in specific characteristics but in a controlled fashion, it allows for fast or slow rates of changes in specific forms, and it gives precise criteria for interspecific discriminations whenever these have evolved from the same species. Thus, the three notions, S_A (Aristotelian), S_D (Darwinian) and S_{NS} (New Systematics) satisfy the lattice relation of Figure 1: the lattice in non-Boolean, since S_{NS} is richer than S_A and S_D (The complements S'_A and S'_D are, in general, merely theoretical entities that serve to guarantee that S_{NS}, the l.u.b., is richer than the Boolean union of S_A and S_D.)[11]

Another illustration of progressive dialectical development in the history of science is given by the history of theories of optics. A summary of the history of optics especially useful for my purpose is given by Mario Bunge in his volume *Scientific Research II,* on page 47.[12] Let us name the theories listed by Bunge in chronological order of appearance: L_a, L_b, L_c, L_d, L_e, L_f. L_g, L_h, L_i, L_j, L_k. Then this set can be partially ordered with the aid of Bunge's table (Table I, Stages of Optics) by an implication relation of the following kind: $L_x \rightarrow L_y$ whenever all the optical problems that can be solved in L_x can also be solved in L_y, but not necessarily vice versa. Under this ordering relation, the set of optical theories have the logical structure schematized in figure 2 of a non-distributive lattice ($L_a, \ldots, L_f$) on top of which is an incomplete non-distributive lattice ($L_f, \ldots, L_k, \ldots$) that would be completed if a l.u.b. for L_h and L_k were added. There is no doubt that the absence of such a l.u.b. is felt keenly by physicists as an imperfection in physics, and that research is being directed towards a unification of those theories. It will be remembered that in order to get the set of optical theories to fulfill the stipulation for dialectical development, I re-interpreted the implication relation between frameworks to mean "whatever problem can be solved in L_x can be solved in L_y but not necessarily vice versa." To what extent this form of the relation coincides with the form used in the rest of this paper will be considered critically and at length in another paper.

Figure 2:
A combination of partial and linear orderings of optical theories in their historical sequence, taken from Table 1.

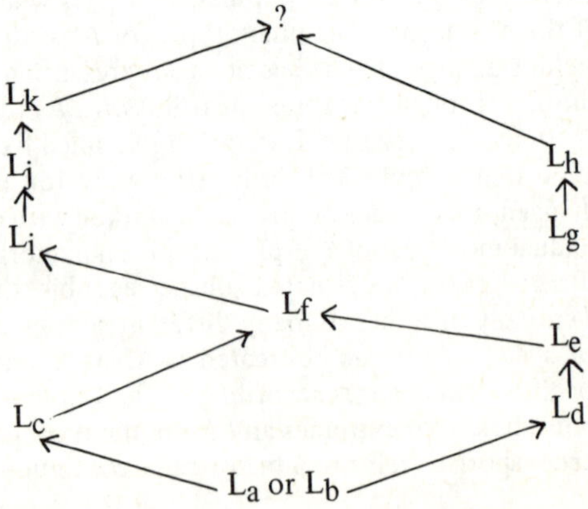

The possibility of extending the notion of complementary traditions from two to three or more complementary or mutually antithetical traditions is suggested by the observation that the property of forming part of a non-uniquely complemented lattice is not restricted to just two complementary traditions but could involve three or more. The structure of such a lattice with eight elements is represented in Figure 3 below.[13)]

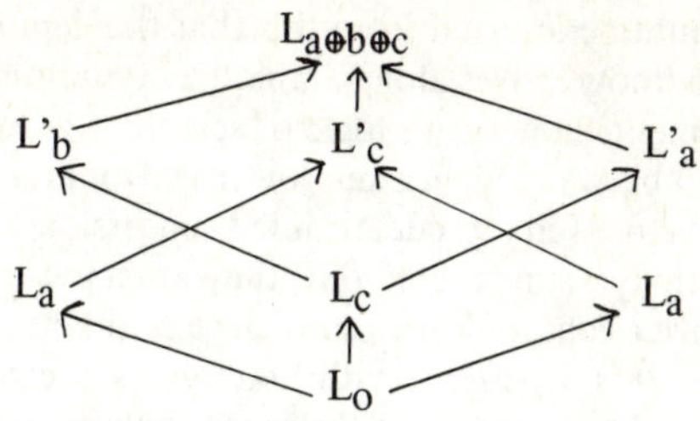

Figure 3: A non-Boolean lattice for three complementary languages L_a, L_b and L_c.

We are then led to the proposal of defining a progressive dialectical development as one that involves various historical traditions so ordered as to constitute a non-distributive lattice by a partial ordering of the analogical kind that can exist between traditions where the partial ordering also follows the direction of the passage of time.

I shall end this discussion by asking: Is it plausible that scientific description is moving towards the ideal of a unique, perfect and definitive description of the physical world?

On the one hand, the doctrine of complementary frameworks which shows how it is possible for rival frameworks to come together in a synthesis which contains each and is larger than both, renders plausible the view that the outcome of a historical dialectic between complementary scientific theories might well be cumulative, leading to successive syntheses in the sense proposed above, rather than to a discontinuous non-cumulative sequence of historical "paradigms." On the other hand, the earlier discussions strongly suggest two things: (a) the contingency of human patterns of life, and (consequently), (b) since a human pattern of life is used to explore and so to constitute a World, the conceptual framework we use to give meaning to a pattern of life is to some degree contingent and conventional. These considerations render implausible the claim that science, cumulative and progressive as it may be in the sense of the preceding paragraph, is moving towards a unique and definitive goal independently of the path the history of science takes to reach it.

Is scientific progress indefinite? Among the answers attempted is that of Lonergan who does not give a categorical answer to the question.[14)] On the one hand, he admits there is in science always a further decimal point to be conquered, in the conquering of which a framework transposition may take place. But with respect to radical theoretical revision, he makes the following statement: "Theories can be revised if there is a reviser. But to talk about revising the revisors is to enter the field of empty speculation in which the name, revision, loses its determinate meaning." There is, then, in his view, an upper limit to revision built into the heuristic structure of human inquiry: the basic structure of human inquiry cannot itself be revised. I cannot be inferred from this that the deployment in time of the disinterested desire to know moves along a uniquely determined trajectory of evolutionary development. Lonergan's logic of scientific growth, which defines growth by the relation between higher and lower viewpoints, does not restrict the higher viewpoint to the next in a predetermined succession of viewpoints. Since real novelty and creativity are not ruled out, they are a part of what a philosopher and historian of science might look for in his close acquaintance with the history of science and culture. It is consistent with Lonergan's doctrine in *Insight* that the history of science should turn out to be the creative development by man of evolutionary sequences of complementary frameworks, a view put forward and held by T. S. Kuhn and P. Feyerabend in their most recent publications.[15]) Science, then, like music, would be in a certain sense a human composition; music exploiting the potentialities of a man-made sound, science exploiting the potentialities of man-made physical interactions with nature.

Perfect science, like perfect music, may not be the least upper bound of all lattices, but merely the l.u.b. of all surviving historical lattices. One might well find among the "blind alleys" of science, or in certain esoteric theories, incompatible traditions dropped because they were not able to be synthesized in the dominant historically developing lattice. Or one might find a variety of lattices developing historically side by side, though incompatible in principle with one another.

From the foregoing, it is clear that there are a variety of logical models at hand to understand inter-framework relationships and especially developmental transpositions between frameworks in history. The task of using these models practically in problems of classification, has yet scarcely been begun.

Typical facts and laws	Ray optics (HERO)	Hydrodynamic aether theory (DESCARTES)	Corpuscular theory (NEWTON)	Longitudinal wave theory (HUYGENS)	Transversal wave theory (FRESNEL, CAUCHY, GREEN)	Electromagnetic theory (MAXWELL)	Electromagnetic theory without aether	Electromagnetic theory with curved space	Electromagnetic theory with electron theory	Quantum electrodynamics and nonrelativistic Q. M.	Quantum electrodynamics and relativistic Q. M.
1. Rectilinear propagation	x	x	x	x	x	x	x	x	x	x	x
2. Reflection	x	x	x	x	x	x	x	x	x	x	x
3. Refraction	x	x	x	x	x	x	x	x	x	x	x
4. Extremal travel time	x	x	x	x	x	x	x	x	x	x	x
5. Dispersion	x	x	x	x	x	x	x	x	x	x	x
6. Superposition				x	x	x	x	x	x	x	x
7. Double refraction				x	x	x	x	x	x	x	x
8. Decrease of speed in transparent media				x	x	x	x	x	x	x	x
9. Diffraction				x	x	x	x	x	x	x	x
10. Interference				x	x	x	x	x	x	x	x
11. Doppler effect				x	x	x	x	x	x	x	x
12. Polarization					x	x	x	x	x	x	x
13. Radiation pressure			x			x	x	x	x	x	x
14. Anomalous dispersion									x	x	x
15. Invariant speed light							x	x			x
16. Change of frequency in gravitational field								x			
17. Light scattering									x	x	x
18. Blackbody spectrum										x	x
19. Photoelectric effect										x	x
20. Compton effect										x	x
Alphabetic designation	L_a	L_b	L_c	L_d	L_e	L_f	L_g	L_h	L_i	L_j	L_k

Table 1: Stages of Optics
(From M. Bunge, Scientific Research II, p 47)

Footnotes

1) The terms "language", "language framework", "intentionality-structure", "context", "projection" and "horizon" are to be understood for the purposes of this paper in the sense in which they are used by the author in his papers, "Horizon, Objectivity and Reality in the Physical Sciences", *Internat. Philos. Qrtly,* 7 (1967), pp. 375–412, "Complementarity, Context-dependence and Quantum Logic", *Foundations of Physics* 1 (1970), pp. 95–100, "Classical Logic and Quantum Logic: Their Respective Roles", *Synthese* 21 (1970), pp. 2–33, and "The Logic of Framework Transpositions", *Internat. Philos. Qrtly,* 11 (1971), pp. 314–334. A brief discussion of these terms will be found later on in this paper.

2) Gerard Radnitzky, "Science Policy and Research Planning: Some Theoretical Models", Institute for the Theory of Science, University of Gothenburg, Sweden, 1969. Thomas S. Kuhn, *The Structure of Scientific Revolutions.* Chicago, University of Chicago Press, rev. ed. 1970. I am overlooking the controversy about the exact sense of the term "paradigm" in Kuhn.

3) Rudolf Luneburg, *Mathematical Analysis of Binocular Vision.* Princeton, 1947.

4) Patrick A. Heelan, "Toward a New Analysis of the Pictorial Space of Vincent van Gogh"

5) Heelan, "Horizon, Objectivity and Reality in the Physical Sciences", *op. cit.* note 2.

6) *Ibid.* The term is also used by Lonergan who defines it as "the maximum field of vision from a determinate standpoint", cf his "Metaphysics as Horizon", *Cross Currents,* 16 (1966), p. 489.

7) This has been shown by the author in "Complementarity, Context-dependence and Quantum Logic" *op. cit.*, and "Classical Logic and Quantum Logic: Their Respective Roles", *op. cit.*

8) The title of Heisenberg's first paper on quantum mechanics was, "On the quantum theoretical re-interpretation of mechanical and kinematical relationships", *Zeit. f. Physik* 33 (1925) 879. Bohr did not develop his theory of complementarity until at least one year later, and Bohr's first public account of it was in the course of a lecture given at Como in 1927 on the occasion of the centenary of Volta's death. The text is reprinted in *Atomic Theory and the Description of Nature* (Cambridge Univ. Press, 1961). For a justification of the author's views, see the two papers cited in n.7 above, and *The Observable: Observation, Description and Ontology in Quantum Mechanics* (forthcoming).

9) For the properties of lattices, see G. Birkhoff, *Lattice Theory,* 2nd ed. Providence, Rhode Is. American Math. Soc. 1948.

10) See, for example, E. Gilson, *The Christian Philosophy of St. Thomas Aquinas.* London, 1957: L.B. Geiger, *La participation dans la philosophie de S. Thomas d'Aquin.* Paris, 1942: C. Fabro, *Participazione e causalita.* Turin, 1960.

11) The details of the above analysis of the notion of *species* were worked out by Conor F. Duggan in an unpublished doctoral dissertation. "Karl Jordan and a Conceptual Revolution in Taxonomy: 1895–1910" (National University of Ireland, University College, Cork, 1971).

12) Mario Bunge, *Scientific Research II.* New York, Springer-Verlag. 1968.

13) Antonio de Nicolas, The State University of New York at Stony Brook, has found an illustration of this lattice in the implicit philosophy of the Rigveda. Of his *Four-dimensional Man.* Bangalore, 1971.

14) Bernard F. Lonergan, *Insight.* London, Longmans, 1957. Pp. 303–4.

15) For instance, in their contributions to *Criticism and the Growth of Knowledge,* ed. by I. Lakatos and A. Musgrove, Cambridge Univ. Press, 1970.

E. J. Ashworth (Waterloo)

Classification schemes and the history of logic

Logic is one of the most important means of classification we have, for it enables us to appraise our reasoning by drawing the distinction between valid and invalid inferences. Its aim is a simple one, and easily stated, but when we get down to the task of specifying under precisely what conditions a true premiss set will entail a true conclusion, it seems that a whole range of different types of classification is necessary.

Logicians commonly start by drawing the distinction between informal or natural languages and formal or artificial languages. Even at this point, divergent interpretations are possible. One can argue with the early Wittgenstein that natural language has a hidden ideal structure, which it is the task of the logician to uncover; or one can argue with the later Wittgenstein that natural language involves a series of games with different structures, any one of which the logician can choose to present as a formal language. Given both the complexities of natural languages and the variety of formal languages which have been developed, the latter interpretation is by far the most pausible. Once the notion of a formal language has been isolated, one can go on to draw the distinction between syntax, or the study of the relations of signs among themselves, and semantics, or the study of signs as interpreted, as having meaning and as being true or false. In turn we can obtain the notion of different types of logical calculi. For instance, a propositional calculus has one set of signs with certain limited transformations of these signs, and it is interpreted by the assignment of truth values to its constituent parts; whereas a quantificational calculus has a more elaborate set of signs with transformations to match, and it is normally interpreted by means of the assignment of members or sets of members of domains to its constituent parts.

The teacher of logic is often tempted to claim that these types of classification are integral to the study of logic. This is true when logic is seen as the foundation of mathematics, but to say that only through these distinctions can one sensibly talk about valid and invalid inferences is a much larger claim, and a more dubious one. I intend to look at selected aspects of the history of logic in order to throw some light on the problem of just what kinds of classification are necessary to the isolation of valid inferences, which I take to be the true task of logic. In particular, I shall look at the definitions of valid inference offered by the Scholastic logicians of the late fifteenth and early sixteenth centuries, since this is the historical period with which I am best acquainted.[1])

Before I look in detail at some of the differences between scholastic logic and modern symbolic logic, two red herrings must be dealt with. The first red herring arises from the attempt to distinguish scholastic logic from modern logic in terms of its objects. It has been argued that whereas modern logic is concerned exclusively with formal languages, scholastic logic was a so-called rational science, whose objects were second intentions or second-order concepts like 'genus', 'species', and 'syllogism'.[2]) This claim overlooks two facts. First, there was a significant group of scholastic logicians who argued that logic was an *ars sermocinalis* or linguistic art, just like grammar and rhetoric.[3]) Second, people's views about the nature of logic (and there were more than the two I have mentioned) did not affect their analysis of the relations between propositions and the variety of inference forms. Whatever they said about the true object of logic, they were unable to avoid approaching logic through language.

Some people might wish to object at this point that scholastic logicians were psychologists, that they approached logic through the study of mental operations. It is indeed true that a number of scholastics (and non-scholastics) of the period in question equated the stages of logical investigation with the three operations of the mind.[4]) The logic of terms was said to correspond to the simple apprehension of concepts, the logic of propositions (i.e. their structure and relationships) to the making of judgments, and the logic of argumentation to reasoning. However, these claims never added up to a genuine psychologism. That there were three operations of the mind was taken to be a fact, and a fact of some interest; but no attempt was made to argue that validity could ce ascertained through the investigation of mental operations. More generally, scholastic logicians made it clear that they viewed logic not as an empirical but as a normative discipline. They did this by distinguishing between natural logic, or the multitude of ways in which people actually do reason, and artifical logic, or the body of (relatively) isolated inference forms which was the proper subject of investigation. They would have scorned the notion that they were committed either to the study of the thought forms of the population at large or to the study of all of the thought forms of any sub-set of the total population. They were concerned with deductive arguments, not with inductive arguments, or with hunches, intuitions and all the other means which human beings sometimes use to arrive at a conclusion.

The second red herring is the question sometimes raised as to whether scholastic and symbolic logic are two logics or one.[5]) At first blush it may seem that any answer to this question will play a crucial part in settling the wider question of what kinds of classification a satisfactory logic should embrace, but unfortunately the question is a misconceived one, so that no answers to it will be of any help. The two-logic theorists seem to fasten on two points on particular.[6]) First, they consider the relationship of the propositional calculus to the quantificational calculus, and then argue that propositional logic is not basic to syllogistic in the way that the propositional calculus is basic to the quantificational calculus. It is certainly true that scholastic logicians did not isolate these two calculi, for they did not isolate any formal system. However, they clearly used what we now call propositional rules in syllogistic, for instance when reducing syllogisms of the second and third figure to the first. Nor is there any reason to suppose that had they been presented with the two systems in question they would have denied that the propositional calculus was included in the first-order quantificational calculus without the reverse being true (which is all I mean when I say that the propositional calculus is basic).

Second, the two-logic theorists consider the interpretation of universal and particular propositions with respect to existential import.[7] It is true that scholastic logicians did not accept the interpretation basic to the standard first order quantificational calculus. Instead of saying that universal propositions are true when the subject class is empty, they said that they were false. This gave them a perfectly respectable means of validating those inferences which they wished to validate, such as the inference from an A proposition to an I proposition, even if it did involve them in some difficulties. For instance, they could not accept contraposition, or the inference from "All S is P" to "All non-P is non-S" without insisting that the terms have reference and that they not refer to the universal class. However, one can only go on from this to argue that there are two logics by overlooking the equally significant differences of interpretation between various quantificational systems. The standard first order quantificational calculus is based on the assumption that the theorems are interpreted only in non-empty domains, but one can set up another more restricted calculus whose theorems are valid even in the empty domain. The standard first order quantificational calculus with constants is based on the assumption that all constants denote, but one can set up another more restricted system in which non-denoting constants are perfectly admissable. Indeed, these alternative systems have a big advantage over the standard first order quantificational calculus in so far as those inferences whose validity rests on existential presuppositions can be syntactically isolated from those whose validity is not so dependent. But the main point I wish to make is that since the investigation of different existential presuppositions leads one to the conclusion that there are not two but many logics, we have here no useful way of distinguishing scholastic from modern logic.

The principal feature of scholastic logic is that it was concerned solely with a natural language, Latin, and not with an artificial language which could be viewed either as interpreted or as uninterpreted.[8] Latin sentences already carried an interpretation which could not be overlooked or ignored. They either had a clear and accepted meaning, or they could be given such a meaning through careful analysis. In general they either had an obvious truth-value, or it could be specified under what conditions they would have a given truth-value. For instance, "Socrates was a man" was always taken as true, whereas "John is sitting" was taken as true when uttered under certain circumstances. There were problem cases, of course. Sentences with non-referring subjects like "Santa Claus wears red" and "Santa Claus weighs 320 pounds" did not have obvious truth-values, but this could be dealt with for the purposes of logic by making the arbitrary decision that all such sentences be assessed as true when affirmative and false when negative. Other sentences like "This is false" when said of itself could not so quickly be labelled as true or false because of the paradoxical results, but methods were developed for dealing with these too. Thus we have a situation in which semantic interpretation is prior to the isolation of inferences, with resulting constraints upon the type of inferences which could be viewed as valid. There was no possibility of varying the interpretation in order to accomodate a different set of rules, or of varying the rules in order to accomodate a new interpretation.

That scholastic logicians started with an ordinary language affects the types of logical classification they developed in three ways. In the first place, they included within logic a set of distinctions which would normally be viewed, in part at least, as preliminary to logic. I have in mind here the distinctions made under the general heading of the logic of terms. These have four functions. Firstly, they isolate the group of admissable symbols by distinguishing meaningful from meaningless utterances, or at least, they attempt to isolate such a group. The more acute logicians realized that there were difficulties because of the relationship between what is uttered and what is intended by what is uttered. 'Baf' is normally just a nonsense word, but a man could use it to express derision, in which case it would have meaning of a kind. Secondly, the distinctions pick out the ways in which meaningful symbols can be classified. The most important of these was the distinction between categorematic and syncategorematic terms, where the latter correspond to logical connectives and operators. Thirdly, the distinctions made under the heading of supposition theory pick out the ways in which meaningful symbols can be combined and some of the transformations which are possible. For instance, one can say "Man is species" but not "Socrates is species"; one can transform "Man is an animal" into "This man is an animal and that man is an animal and the other man is an animal" but one cannot so transform "Man is species". Fourthly, the notion of material supposition is used to distinguish various levels of language, and use from mention. In "Man is a word" and "Man is an animal is true" both 'man' and "Man is animal" are said to have material supposition.9)

In the second place, they did not draw any clear distinction between syntactical and semantical matters. The nearest they same to making such a distinction explicit was when they subdivided valid inferences into those which were formally valid and those which were materially valid. By a preliminary definition, two inferences were said to have the same form when they had the same copula or connective, the same quantity, quality and relation of terms, and the same acceptance or supposition of terms. In the light of this definition, an inference was said to be formally valid when all inferences with the same form were valid, and materially valid when at least one other inference with the same form was invalid. Thus materially valid inferences held only by virtue of the meaning of the terms, whereas the validity of formally valid inferences was quite independent of the meaning of the constituent terms.

In the third place, the use of a natural rather than an artificial language meant that no clear distinction was drawn between various types of inference. It was recognized that "P and Q, therefore P" was different from a categorical syllogism, and that a modal syllogism was different from either of the first two inferences, but there was no attempt to assign each inference to a different system, or to say that one system was basic to another. This is an inevitable concomitant of the absence of a formalized syntax and of a formalized semantics. Normally types of inference are distinguished from one another in terms of the syntactical structure of the system to which they belong and in terms of the type of semantic interpretation deemed appropriate to that system. Or, if one is dealing with inferences couched in ordinary language, they are distinguished from one another in terms of the systems in which they can be shown to be formally valid. For instance, a relational inference is assigned to the polyadic predicate calculus because even if it is valid, it cannot be shown to be valid in the monadic predicate calculus or in the propositional calculus.

We must now see what happens in scholastic logic when the attempt is made to specify what constitutes a valid inference.[10]) Does the absence of all the classifications which are today considered an essential part of logic radically affect the logician's task of assessing arguments and distinguishing what is legitimate from what is illegitimate?

I must begin by noting that scholastic logicians used *propositio* in a special sense. A *propositio* was not some special kind of entity which was common to sentences in different languages or to different type-sentences in the same language, all of which referred to the same state of affairs. Rather it was a token-sentence, a particular group of sounds or marks or thoughts which existed at a particular time and place. That this was so, complicated matters in two ways. In the first place, it was possible to conceive of an inference which failed because of the non-existence of one of its parts. In the second place, some propositions like "No proposition is negative" were such that their existence produced a paradoxical situation. How these complications affected the definition of a valid inference, we will now see.

The starting point for the discussion of validity was an examination of the most obvious definition, "An inference is valid if and only if it is impossible for the antecedent to be true and the consequent false."[11]) Some people accepted this as adequate, but others felt that it expressed neither the necessary nor the sufficient conditions for validity. There were two reasons for rejecting it as a necessary condition. In the first place, it was claimed that where self-reference was involved, one could sometimes have a valid inference with a true antecedent and a consequent which was not true. For instance, "Every proposition is affirmative, therefore no proposition is negative" was accepted as valid, even though the consequent could not be true. To be true it must exist, and the moment it comes into existence, it falsifies itself.[12]) In the second place, such a definition overlooks the requirement of existence. An otherwise valid inference like "Socrates runs, therefore a man runs" could have a true antecedent and a consequent which was not true because it did not exist.

One might object at this point that it makes no sense to speak of an inference which has an antecedent but no consequent, but the claim may easily be explicated in terms of the type-token distinction. If an inference is defined as a set of type sentences such that whenever the tokens of the antecedent sentences are true, then the token of the consequent sentence is true, one can indeed argue that there may be situations in which some but not all of the tokens exist, and hence that the definition is inadequate. In any case, whether sense can be made of the matter or not, it is not important, because the final definitions of valid inferences were set up in such a way that the existence of the constituent propositions was not required. This had to be the case if the difficulties produced by such propositions as "No proposition is negative" were to be avoided.

Just as the definition in terms of truth was thought to fail in the task of giving a necessary condition for validity, so also it was thought to fail in the task of giving a sufficient condition for validity. It seemed to allow a number of obviously unacceptable inferences like "No proposition is negative, therefore man is an ass". One can argue, it was said, that it is impossible to have a true antecedent and a false consequent here because if the antecedent is assumed to be true then it is, as an existent proposition, false, and if the consequent is false, then the falsity of the antecedent follows by an acceptable rule. Yet the inference is clearly invalid, because it violates the rule that an impossible proposition cannot follow from a possible proposition.

The solution to the problem was sought in an examination of the relationship between the truth and the possibility of propositions. A good account was given by Jacques Almain who argued that it does not follow that because a proposition cannot be true it is impossible, or that because it cannot be false, it is necessary. He went on to say that one can have two propositions which are compossible, like "No proposition is negative" and "God exists", even though they cannot be true together; and one can have two propositions which are synonymous, though only one is true, as where "There is no spoken proposition" is both spoken and written. Finally, he said, a proposition like "Some proposition is negative" can be false even though its contradictory, "No proposition is negative" can be false even though its contradictory, "No proposition is negative", cannot be true. The import of these claims is that, a proposition is possible when the state of affairs it refers to is possible, but it is possibly true only when its existence as a token sentence does not conflict with the state of affairs described.

In accordance with these distinctions, a number of authors turned toward a definition of validity in which the notion of possibility was clearly applied to states of affairs rather than to the truth of propositions. One version was given by the Spaniard, Celaya. He still held that an inference was valid if and only if it was impossible for the antecedent to be true and the consequent false, but he drew a distinction between two senses of 'true' and the two corresponding senses of 'false'. The antecedent could be true in the sense that it signified things to be as they in fact were, or it could be true in the sense that it signified things to be as they were by means of a true proposition. Only the first sense is applicable here. However, the more usual version of the new definition of validity dropped all reference to truth. It was now claimed that an inference was valid if and only if it was impossible for whatever the antecedent signified to be the case without whatever the consequent signified also being the case.

This new definition was not without problems of its own. In the first place, existence was still seen as a source of difficulty. One does not wish to accept the inference "God exists, therefore some proposition is indefinite" as valid; yet if one admits that things are as signified by the antecedent, one is comitted to the claim that the antecedent proposition exists, which in turn renders the conclusion true, so that it is impossible for things to be as signified by the antecedent without their being as signified by the consequent. The escape route was the addition of an extra clause, "or as can be signified". One can agree that things are as can be signified by the type sentence "God exists", without committing oneself to the existence of any particular token. Other awkward cases like "No proposition is negative, therefore some proposition is negative" can be ruled out by the same device, whereas legitimate cases like "Every proposition is affirmative, therefore no proposition is negative" are not affected.

In the second place, semantical self-reference had to be guarded against. Logicians cited the following two examples: "This consequent does not signify things to be as they are, therefore this consequent does not signify things to be as they are", and "This inference is valid, therefore man is an ass". The first case seemed to be an instance of the valid rule leading from synonym to synonym, yet when the consequent referred to is its own, the antecedent signifies (or can signify) things to be as they are, but the consequent does not. The second case seemed to be invalid, for the impossible was derived from the possible, yet it was not possible for the antecedent to signify things to be as they are without the consequent also so signifying, for it was not possible for the antecedent to signify things to be as they are at all. These problems were dealt with in a very simple way. Two clauses were added to the definition of a valid inference: that it should not invalidate itself, and that the consequent should not have "does not signify things to be as they are" appended to it.

In the third place, since propositions were regarded as token sentences, another type of problem arose from the arbitrariness of the connection between a set of words and a state of affairs. That is, a sentence like "An Ass runs" could be reinterpreted to mean "God exists", or it could be taken to refer to just one aspect of the state of affairs that it signifies, namely that an animal is running. These possibilities could wreak havoc with an otherwise valid inference, or at least with an inference whose validity was dependent upon the semantical rather than the syntactical properties of its constituents. Hence, various clauses were added to the definition to ensure that propositions retained their original and total meaning. This final addition completed the definition of a valid inference.

With this complex and subtle definition of valid inference in hand, scholastic logicians set about the task of sorting out different kinds of valid inference, and they did so with enormous success. Not only did they list a great number of rules which would now be assigned to the propositional calculus, but they gave a great many rules dealing with modal and temporal operators, and with inferences from quantified to non-quantified propositions whose variety and subtlety is impressive even today. Moreover, their definition was couched in such a way that they were also able to discuss those non-formal inferences which are often of more interest to philosophers than the formal ones. Had they used modern classification schemes, they would have been unable to do this in the context of logic.

Are we now to conclude that elaborate classification schemes are irrelevant to the pursuit of logic, so long as we have an adequate definition of a valid inference? The answer to this question will depend in part on how much one wants out of logic. If one wishes to study the metalogical properties of formal systems, to obtain a complete set of rules, or to relate logic to mathematics, scholastic logic is necessarily inadequate. However, if one wishes to classify those inferences which are used in ordinary language, then one can argue that an elaborate classificatory apparatus combined with the development of formal systems will be a hindrance rather than a help. Even the simplest sentence contains subtleties which will be lost in symbolization. Moreover, there is the grave problem of which system to choose when one is symbolizing and assessing an inference. This problem has two facets. First, one may pick a system which is inadequate to one's purposes. If one attempts to show that a relational inference is valid in terms of the standard monadic predicate calculus, one will fail. Yet one has not proved that the inference in question is not valid. Second, one may pick a system whose standard interpretation is alien to one's purposes. A logician who wishes to show that '––P', therefore 'P' holds would be ill advised to choose the intuitionist propositional calculus. Similarly a logician who wishes to show that " 'Fa' therefore '(Ex)Fx' " should not choose a version of the quantificational calculus which admits non-denoting constants. The more systematic one's approach to formal logic, the more arbitrary the choise of system seems to be, and hence the less relevant to the normal day to day task of assessing arguments. Scholastic logic, on the other hand, seems perfectly adapted to normal requirements. It is both unpretentious and powerful; it does not violate normal intuitions; and it is non-arbitrary. Or so one might think.

However, let us look a little more closely. What are we to make of the following claims? "An impossible proposition implies any other proposition." "A necessary proposition follows from any other proposition." "If you come to me I will turn you into an ass" is true provided that you do not come to me." "All chimeras are chimeras" is false because there are no chimeras, but "No chimeras are chimeras" is true for the same reason." The first two examples, the paradoxes of strict implication, follow straight from the definition of a valid inference. The third example is a consequence of the truth-functional interpretation given to promissory conditionals. The last examples are a consequence of the arbitrary decision to save the square of opposition by counting all affirmative propositions with non-referring subjects as false. Yet none of the examples corresponded to the normal intuitions of the sixteenth century. They all gave rise to acrimonious debate, and were accepted only because of the exigencies of the desired system of rules and the desired interpretation of that system. Thus even the scholastics, operating within the framework of ordinary language, were forced to make some of the arbitrary decisions which people tend to blame modern logic for. One may still prefer scholastic logic to modern logic for various reasons, but that it enshrines a true and completely non-arbitrary system of picking out valid inferences cannot be one of them.

In the last resort, the presence or absence of modern classification schemes in logic does not make so much difference as one would like to think.

Footnotes

1) I intend to use the term 'scholastic logician' more narrowly than is proper, to refer just to those men whom I am concerned with.

2) This view of logic derives from St. Thomas Aquinas, and it is emphasized today by authors like Veatch and Wade. For references, see note 5.

3) This view was the standard one in the early middle ages, and it was retained in the sixteenth century by both Nominalists and Humanists.

4) This claim was made popular by St. Thomas Aquinas.

5) See H.B. Veatch, *Intentional Logic: a Logic based on Philosophical Realism* (New Haven, 1952); H.B. Veatch, *Two Logics: The Conflict between Classical and Neo-Analytic Philosophy* (Evanston, 1969); F. C. Wade, Introduction to John of St. Thomas: *Outlines of Formal Logic* (Milwaukee, Wisconsin, 1955).

6) Veatch makes various other claims. For instance, he says that logic is concerned solely with intentions; and he says that scholastic logic is a 'what-logic' whereas modern logic is a 'relating-logic'.

7) Wade, p. 22, goes so far as to claim that scholastic logicians denied the existential import of particular propositions, since they were concerned not with things but with concepts (intentions). This is false.

8) When I say that Latin is a natural language, I am using the word 'natural' as the opposite of 'artificial'. Scholastic logicians themselves drew a distinction between a natural language which was common to all men and a conventional language whose symbols were arbitrarily chosen. Any spoken language, including Latin, was conventional. A natural language in the scholastic sense was a purely mental language, composed of concepts whose relation to things was neither arbitrary nor conventional

9) The device of quotation marks was unknown. However, words or phrases having material supposition were often prefixed by 'ly'. For further details about the theory of supposition, see my paper, "The Doctrine of Supposition in the Sixteenth and Seventeenth Centuries," *Archiv für Geschichte der Philosophie* 51 (1969) 260–285.

10) For detailed references, see my paper, "The Theory of Consequence in the Late Fifteenth and Early Sixteenth Centuries," *Notre Dame Journal of Formal Logic* (forthcoming).

11) The antecedent could be either a single premiss or a set of premisses.

12) Since propositions were viewed as occurrent entities, a situation in which God had destroyed all negative propositions was easy to imagine. That is, there was no difficulty attached to the assumption that "All propositions are affirmative" was true.

C. A. Hooker (London, Canada)

The impact of quantum theory on the conceptual bases for the classification of knowledge

§ 1. Introduction

There are many aspects to the classification of human knowledge – I shall concern myself only with a rather idiosyncratic subclass of the already rather specific class of classifications that occur in the physical sciences and in quantum theory in particular. I am referring, for example, to the classification of a phenomenon as involving individuals rather than being a continuous process, as being macro rather than micro, as being objective rather than subjective. But how this apparently scattered set of classifications are related will appear in the sequel.

By the conceptual bases for the classification of knowledge I understand those fundamental features of our conceptual scheme which make possible the classifications (hence distinctions) which we do make (in the physical sciences in particular).

The question to which I am going to address myself, therefore, is: "What impact has quantum theory had on the fundamental features of our conceptual scheme pertaining to those distinctions crucial for physical science (and for modern physics in particular)? "

To answer this question we must first get a clear grasp upon the central notions of "classical" (i.e. pre-quantum and pre-relativistic) science and trace back from thence to the fundamental features of the classical conceptual scheme in which those notions are embedded; we shall then be in a position to evaluate the impact of the quantum revolution on them. After that, specific questions can be treated in detail without the risk of losing perspective upon the whole.

§ 2. The classical conception of science

Classical science proceeds on a tacit assumption common to virtually all western philosophy, namely that the form of the world is independent of man's knowledge of it.[1]) That is, that ontology is independent of epistemology (in a general sense, cf. footnote 1). One strong tradition (the Realist tradition) has it that the world exists, in all its fullness, independently of man *qua* knowing mind (i.e. the physical role of the body is allowed for), thus explaining why ontology is independent of epistemology, namely because it is logically prior to epistemology. This tradition encapsulates the mainstream of scientific thought. But even the maximally contrasting tradition (the Phenomenalist tradition) – at least in its British-style empiricist phase – holds that those contents presented to mind which together constitute experience of the world have their form and order independently of that perceiving mind in the sense that there is no conscious control over their form and order (at least as pertaining to empirical matters, else how are they to be empirical?).

We can restate this common tradition in another way: There is no epistemological content intrinsic to the use of a conceptual scheme. That is, a conceptual scheme is neutral, as it were, with respect to the world – it does not prejudge the world, it does not prejudge what exists nor how existing thing relate. One can formulate sentences in any given conceptual scheme and these, if they are contingent (i.e. if they are descriptive and not logically true), may be true or may be false – which they are is determined by the way the world is. Any other conceptual scheme would also have descriptive sentences formulable in it and which of these is true would again depend on how the world – that *same* world against which the sentences of the prior conceptual scheme were measured – is. There is to be no intrinsic connection between the conceptual scheme employed and the nature of the world in the sense that different conceptual schemes might bring with them different worlds, or that the one, existing world forces a unique conceptual scheme upon us. True, some concepts and some conceptual schemes will be more suited to the world than others, but that is a different story; it is part of that laborious process by which man learns to fashion useful descriptive machinery in his head, where it has no intrinsic relation to reality, and through painstaking experiment learns to formulate sentences (theories) within it which do have a relation to reality (namely they are contingently true of it).

Kant's philosophy represents an important instance of the contrasting alternative to this position. Kant held that perception and conception were intimately related, that to perceive the world in a certain way was to conceive of it in a certain way. The world 'initself' was unknowable, but the world-knowable was a function of man's conceptualizing. Hence conceptual structures come to possess intrinsic epistemological content. Kant attempted to capture this content explicitly in his synthetic *a priori* truths, but the actual situation is at least more complex than Kant allowed.[2]) The tradition against which Kant was reacting was the Empiricist tradition associated with Locke-Berkeley-Hume. We have already noted that this tradition subcribes to the no-intrinsic-epistemological-content view.[3])

It is against this background – and on the non-Kantian side – that the classical conception of science functions. I have elsewhere summarized its main features in nine theses and these may in turn be summarized as follows:

C_1 ($C1_{1\to 3}$): Physical reality is divisible into conceptually distinguishable elements, all complex objects consist in definite structures of the fundamental elements which are their constituents. Moreover, each element of physical reality is precisely and exhaustively describable within the classical conceptual scheme of physical attributes and objects of attribution at definite spatio-temporal locations. [4])

This formulation skids over what is already an important problem for classical physics, namely the existence of components of a quantity obeying vectorial addition laws (e.g. forces, fields, waves and velocities). The problem may be stated thus: Given any resultant vectorial quantity, there is an infinite number of ways in which that quantity may be mathematically resolved into components. Furthermore, there actually exist classical physical systems which seem to behave as if every one of these possible resolutions was physically applicable (though of course only one can be tested at a time). In what sense then can these infinitely various mathematically possible resolutions be said to exist? We shall have a little more to say about this later on, for the moment I note only that this problem is restricted to vectorial quantities and does not afflict those entities characterized by non-vectorial properties, for example, the atomic reductions of macroscopic objects.

C_2 ($C1_{4 \to 7}$): Physical theories correspond to, or "mirror", the world when they are adequate, i.e. every relevant element and attribute of the reality has a corresponding counterpart in the theory. Statistical theories represent less than complete knowledge of the state and behaviour of an ensemble of systems.

Statistical theories are regarded as applying to collections of perfectly definite elements in those cases where a full knowledge of the states of each member of the collection in some one or more respects is either impractical or not useful to obtain.

C_3 ($C1_{8 \to 9}$): Knowledge is gained of the states of physical systems by the making of measurements on the systems. Physical systems exist and evolve independently of the presence of observers, *qua* observers.

A measurement is a straightforward physical process of interaction between a measuring instrument and a measured system, which is to be given a theoretical treatment exactly on a par with that given any other physical process, the outcome of which is directly related to the feature of the system under investigation in a known way. Moreover, it is assumed that, their physical bodies aside, the existence and character of physical processes in general is not directly dependent upon their being observed, *qua* act of intellectual cognition, by observers.

The general conception of the physical world conveyed in the preceding statements will no doubt be familiar to the reader. *It is a measure of the revolution brought about by the advent of the quantum theory that every one of these claims has been challenged.* What is involved here is no small alteration of the detail of our general conception of reality but a revolution in the very foundations of that conception, touching every aspect of our understanding of the world, ourselves and the relations between the two. The story of how the central characteristics of the quantum theory lead away from the classical conception of reality has been told many times before, and most adequately by Niels Bohr, the great founder-philosopher of the quantum theory. [5]) I shall retell it only very briefly here.

Classically, a particle, that is a material object whose dimensions are small compared to the other relevant spatial distances, is characterized by a sharp spatial definition and a definite trajectory through space in time. The particle has momentum and energy and these quantities are also therefore sharply localized in space and time. By contrast, a wave-field cannot be sharply localized. The energy and momentum of the wave are distributed throughout its spatial extension and are therefore also not sharply localized. It follows that the wave has no definite trajectory through space in time, but rather propagates through a finite spatial volume in time. In addition, waves obey a characteristic wave equation, linear combinations of the solution of which are also waves and solutions of the wave equation (superposition principle). In consequence we obtain the characteristic wave phenomena of interference, diffraction and polarization, none of which are associated with particle motions. In short, there is the greatest possible contrast between the two types of phenomena.

The striking thing about quantum theory is its marriage of these two notions. The fundamental equation of state is a wave-like equation, the Schrödinger equation, solutions ψ to which formally characterize the state of micro systems. ψ obeys a linear superposition principle and exhibits patterns corresponding to the characteristic wave phenomena of interference, diffraction and, with spin included, polarization. Despite this, energy and momentum do not have continuous ranges of values and whenever a suitable position determination is carried out, one always finds the energy and momentum (in the form of a 'particle') sharply localized in space. In the light of this, ψ is given a statistical interpretation as follows: Let ψ be expanded in terms of a complete orthonormal set of eigenfunctions $\phi\lambda_i$ (x), corresponding to the operator of some relevant physical quantity, with eigenvalues λ_i;

$$\psi = \sum_i \delta_i \, \phi\lambda_i (x) \qquad \int_{-\infty}^{+\infty} \phi^*\lambda_i (x) \; dx = \delta_{ij}. \tag{1}$$

Then the probability that the quantity in question has the magnitude λ_i is $|\delta_i|^2$. This is essentially the Born Rule for interpreting ψ. Thus while the *distributions* of individual micro processes in a large ensemble of similar processes show all the characteristics of a classical *wave* phenomenon, the *individual* processes show all the characteristics of classical *particle* phenomena. (Actually this is too crude and sweeping an assertion, but we shall leave it stand for convenience of exposition of the central ideas.) Moreover, the individual micro processes can be carried out so that no more than one system at a time is involved. During that time, therefore, and before the actual culmination and macro registration of the process, the normal description of the micro system has the formal features of a classical wave, displaying self-interference etc. Yet upon a measurement being made, say of position, one obtains a highly localized particle-like result. There is, moreover, a deep-going connection between the particle-like energy and momentum and the wave-length and frequency of the corresponding wave, namely:

$$p = {}^{h}/_{\lambda}, \; E = h\nu \tag{2}$$

In consequence of this connection and the localizability characteristics of waves, we have the Heisenberg Uncertainty Relations:

$$\Delta p \cdot \Delta q \gtrsim h, \quad \Delta E \cdot \Delta t \gtrsim h \tag{3}$$

which sets mutual limits to the accuracy with which the conjugate variables p,q; E, t can simultaneously be determined.

In consequence of the striking differences between classical and quantum mechanics, physicists have proposed that the classical view of reality outlined above be abandoned in one or more respects. Let us briefly try to gain some feeling for the extent of the proposed changes.

One school of thought, closely associated with the name "Copenhagen Interpretation" and under the leadership of Heisenberg has argued that it is experimentally meaningless to assign, simultaneously, precise values of conjugate quantities to the same micro systems. Moreover, that the values of physical magnitudes which a system displays are not attributes which it possesses prior to a measurement of that quantity, rather they are made to appear at the same time as, and because of, measurement.[6)] Moreover, because of the finite quantum of action, any measurement disturbs the system in an utterly unpredictable way. But this disturbance is such that the Heisenberg Uncertainty Relations are always satisfied. [7)] One who views quantum mechanics from within this general framework can hardly avoid rejecting, or at least seriously modifying, all the principles C_1, C_2, C_3.

There has also been a definite tendency to supply a subjectivist interpretation of quantum theory. This may arise in a thorough-going way as the assertion that quantum mechanics describes only our knowledge of physical systems and not the systems themselves. But since this is a view now largely rejected and one which makes no commitment to an account of physical reality, I shall set it aside. A subjectivist content to the theory has, however, been encouraged by the theory of measurement, especially as formalized by von Neumann (28, Chapter VI – see especially London and Bauer 27, Schlegel 33 and Wigner 36). It has seemed plausible to say that a measurement is not completed, and the actual values for the outcome not realized, until the act of conscious observation on the part of the observer. Under these conditions C_2 and C_3 will be rejected (and if the remainder of the interpretation is taken over from the former position, the remaining principles will also be rejected).

Finally there is Bohr's position (cf. 2, 3). Bohr holds to the view that the descriptive language of physics is, and will remain, fundamentally classical, but claims that there is now a limitation upon the simultaneous applicability of certain classical concepts to a single situation. This is brought about because the existence of the indivisible quantum of action radically alters the usual distinction between subject and object, the experimental situation being one indivisible whole. Under these circumstances the physical conditions required for the applicability of the classical concepts having to do with conjugate quantities cannot be jointly realized. *In the sense intended in the classical context,* Bohr rejects C_1 and C_2 (his is not a subjectivist position), though there is a clear sense in which Bohr retains his own version of C_1 and C_2. [8)]

This very brief sketch of positions has revealed the radical departure from the classical conception of reality which have suggested themselves to physicists and philosophers in consequence of the advent of the quantum theory.

In contrast to this movement, Einstein (2) and with his co-workers Podolsky and Rosen (8), together with some other physicists (for example Blokhintsev, Bopp, Popper) constitute a small group of physicists and philosophers who are determined to treat quantum theory as a species of statistical mechanics, many of them hoping ultimately to reinstate the classical conception of reality. This latter goal clearly means regarding the quantum theory as a statistical theory of the classical sort, or discarding it altogether as false (or both). Quantum theory must then offer either an incomplete (and/or an inadequate) description of physical reality.

I shall summarize briefly part of the main argument of the essay from which the contents of this section have been drawn by saying that it would appear at the present time that there are powerful arguments in favour of the view that quantum theory is both an adequate physical theory (in the sense that it is at least capable of predicting the statistical results actually found in experiments) and that there is no known way to satisfactorily interpret it as a statistical theory over a realistic ontology in the classical spirit. This conclusion, which underlines the genuineness of the quantum revolution, thrusts us back onto the approaches already discussed, all of which have deep-going implications for our organization of empirical knowledge. In the following I shall try to discuss some of those implications – at least as far as they are now understood, for I am convinced that we have hardly begun to understand the conceptual consequences of the quantum revolution.

§ 3. The metaphysics of science

Lying behind the conceptions of physical reality and physical theory – the two are inseparable – outlined above lie general principles which are fundamental to our entire conceptual scheme, principles which have remained virtually unaltered until the twentieth century and which I believe have governed the development of physical theory until the present day. In fact it is only the depth of the revolutions wrought by twentieth century physics that have begun to teach us how to perceive and critically examine that conceptual structure. Let us first outline it – and again I shall borrow briefly from my 21. I shall commence by listing the major elements in our "classical" scientific conceptual scheme in the form of terse descriptions in order of decreasing generality:

(M_0) The subject/object (ego/world) distinction.

(M_1) Substance(s) in space and time, possessing properties.

(M_2) Numerical identity of individuals, qualitative indistinguishability of universals.

(M_3) Determinateness of properties, spatio-temporal and others.

(M_4) Laws of Logic.

(M_5) Two specific realizations of M_1: (i) atoms and the void, (ii) plenum (aether, field) and no void.

(M_6) The Causality/Harmony distinction.

This specification is terse to the point of frustration, but one could fill a book merely expanding on it.

Of course not all of these principles can be regarded as independent of one another, our conceptual scheme has a very high degree of coherence. Thus, for example, it is senseless to attempt to discuss the distinction M_0 independently of M_1 and M_2, since what makes possible our regarding ourselves as distinct from the remainder of the universe is precisely our ability to regard ourselves as an object, i.e., a substance possessing properties, located in space and time and distinguished both numerically and qualitatively from the remainder of the universe regarded as objects located in space and time. But insofar as these various features of our single conceptual scheme are themselves the distinguishable points of emphasis in the discussion of the fundamental conceptual impact of quantum theory it is worthwhile to single them out in the above fashion.

In the other essay (21) I emphasized (briefly) the ancient and perennial nature of these elements. Here we shall be more interested in the two ontology-types offered by M_5 and in their relations with the other elements.

The first thing to which I wish to draw attention is the manner in which these elements undergird the classical conception of science sketched in § 2.

The general relationship can be stated very easily: The classical conception of science is a *specific illustration* of the general metaphysical principles. The interest lies in the details of this specific illustration.

The distinction M_0 is the basis for the classical presupposition of the prior independence of the world *vis-à-vis* our knowledge of it which was pointed out earlier and which is explicitly embodied in the epistemology of C_3. Of course it is equally possible to understand this distinction in a Kantian sense and the need for this particular understanding, we shall see, is what Bohr claims quantum theory emphasizes.

The group $M_{1,2,3,4}$ spells out the fundamental structure of the world as we conceive it, M_4 specializes to the actual development of western science. The concept of a substance is at least this: The concept of a subject of predication. Moreover, material substances have, fundamentally, spatio-temporal location. Normal material individuals (i.e. physical objects) are spatially bounded substances of which at least some (other) predicates are truly predicable. In addition, there are those special individuals which fill all space and which are therefore unbounded, if space is – I refer to the physical fields (the gravitational, electro-magnetic and nuclear fields). Our concept of substance and property is such that we distinguish between qualitative and numerical identity. We admit the possibility that there exist material objects for which all predicates (except some spatial predicates) are truly predicable of any one if and only of they are also truly predicable of any other, and yet these objects, being two or more in number, are distinct. Predicates which are not coextensional in respect of the objects are those spatial predicates referring to the locations of the objects, for it is precisely the possibility of having otherwise indistinguishable objects at distinct spatial locations that makes possible the distinction between qualitative and numerial identity, or, as I should prefer to say, between qualitative indistinguishability and numerical identity. This latter distinction will in general only be possible where predicates are precisely delineated, hence M_4. Moreover, M_4 is part of the drive for mathematical precision in our descriptions of things.

As M_5 reminds us, there have been two great realizations of the foregoing schema: the field and the atom. These two realizations differ profoundly in their conceptual organization and we shall shortly examine some of those differences. For the moment I want only to draw attention to this remarkable fact: *It was the genius of classical science (especially post-Newtonian science) to have managed to keep these two components carefully separted from one another.* In this way the fundamental differences between the two obscured and the inherent conceptual tensions in classical physics covered over.

In the twentieth century these two great metaphysical schemes and their correlative conceptual and logico-mathematical structures met twice, in optics and in atomic theory.

The *first* interaction occurred when light was found not to behave as a wave should (unique velocity relative to its *plenum of propagation* – the aether) nor yet to behave as a particle (the *ballistic* hypothesis was refuted). Out of this situation relativity was born. And we discover that the change from the Galilei space-time group to the Lorentz space-time group necessitates a change from a theory which is permitted to be either local or non-local to a necessarily local theory where all mechanical action is continuous and the natural ontology is that of the field. Indeed, the incorporation of the particulate or atomistic aspect of matter is still a chief stumbling block to Einstein's unified field programme.

The *second* interaction occurred in the atomic domain and centres around the 'duality' of matter and light, i.e. around the fact that matter and light behave in ways which can be partially, but only partially, incorporated into the atomistic framework and partially, but only partially, incorporated into the plenum framework. Out of this circumstance quantum mechanics was born. And we note that this change was one from a theory which was permitted to be either local or non-local to a necessarily discontinuous theory which has an intimate relationship to non-locality. In the quantum theory, what with its wave (plenum) equation and statistical (atomistic) interpretation, we find the latest confrontation of atomistic and plenum concepts.

A necessary prologema to the understanding of the situation and difficulties of modern physics must surely be the study and understanding of the development and character of these two great metaphysical schemes, their correlative conceptual (descriptive) structures and their relations (or lack of them) with one another.

The principles M_1 through M_6 give the fundamental structure of the physical ontology. That fundamental structure is reflected in the fundamental syntactic (or deep gramatical) structures of our (natural and) scientific languages (in the so-called western world).[9]) Or, for those who are more linguistically sensitive, we may say that M_1 through M_6 express the 'descriptive metaphysics' culled from the fundamental structure of our (natural and) scientific languages – cf. Strawson 34. Classical scientific theories reflect this fundamental structure, as expressed for example through C_1.

M_4 gives the structure of the truth-preserving moves in the language whose fundamental structure has already been decided by its conforming to the structure laid down by M_0 through M_3. From this point of view M_4 might be considered a relatively superficial feature of scientific language – i.e. superficial relative to M_0 through M_3 – for it does not determine the fundamental conception of the world (the fundamental structures of language) but only an additional structure built upon these as foundation. Though this would again be too simple-minded a view (the deeper and shallower structures are intimately related), there is some point to the distinction, as the proposals for a non-standard quantum logic show.

The distinction M_6 has an important history in western science, but I shall have little to say of it here.

§ 4. The metaphysics of the atomistic and plenum schemes

To gain some idea of the depth of the origins of the contemporary interpretational problems of quantum theory it is necessary to briefly review some relevant features of the deeply contrasting metaphysics of these two schemes (if the squeamish reader prefers, of the fundamental conceptual structures of these schemes). Here I shall be summarizing portions of my more extended examination in 22.

In the atomic schema the fundamental individuals are the atoms (here I use the generic sense of the term), a plurality of substances each localized in space. All other individuals are spatio-temporal structures of these fundamental individuals (I shall refer to them as derivative individuals). Likewise, there are fundamental and derivative properties. The intrinsic properties to the fundamental individuals are the fundamental properties and these are atomic or quantized (i.e., there is only a finite number of values for their magnitudes) just in case there are only a finite number of kinds of fundamental individuals. All other properties (i.e. all properties of derivative individuals) are reducible to spatio-temporal structures of these fundamental properties. All change is reducible to spatio-temporal rearrangements of fundamental individuals. All laws (in particular, the laws of change) are reducible to fundamental laws concerning spatio-temporal relations among fundamental individuals and the fundamental properties only.

The situation with the plenum schema is in maximal contrast to this. There is only one fundamental individual, the plenum, and it is present everywhere. All individuals consist of spatio-temporal distributions of properties. There is no fundamental/derivative contrast inherent in the schema, though scientific experience may lead us to introduce a distinction between some complexes of properties as being more 'fundamental' than others (just in the sense determined by their respective roles in theories). Similarly, there is no intrinsic distinction between fundamental and reducible properties, all properties are, prior to actual scientific experience, on an equal ontological footing with one another. All change is basically qualitative change, i.e. change in properties. (Thus even motion is to be viewed as the systematic disappearance and re-instantiation of clusters of properties in such a fashion that the cluster sequence forms a spatio-temporal trajectory.) The laws of nature will concern fundamentally the relations amongst properties and property-complexes and again, prior to any roles which they might play in actual scientific theories, all such relations are on an equal footing.

There is, then, the deepest possible contrast between the two kinds of schemes, between the two kinds of metaphysics. One could go on developing this contrast at considerable length, but I shall develop only three more features of it which are pertinent to the discussion of quantum theory.

The first of these concerns the notion of continuity and discontinuity. We already have the following results: there is a fundamental element of discontinuity in the atomic schema, namely the substantial discontinuity introduced because the fundamental individuals are spatially bounded substances and the fundamental property discontinuity introduced because each kind of fundamental individual has only one value for each of its intrinsic properties. In fact, were space and time themselves quantized there would be no element of continuity whatever in the atomic schema (for then all change would likewise be discontinous and so would every reducible property) and in fact the assumed continuity of space and time in classical science provides the only entrance of continuity into the atomic schema (and then only at the level of derivative individuals and changes in derivative individuals). We might say that the 'natural', certainly the homogeneous, development of the atomic schema is one in which space and time are quantized and discontinuity reigns supreme.

In contrast, the 'natural', certainly the homogeneous, situation with the plenum schema is one of total continuity. A detained examination leads to arguments with the following conclusion: if space and time are continua, then physical states are well defined if and only if every change (whether in properties or in spatio-temporal location) is also continuous. (Actually this argument holds good for the atomic plhema as well.) On the other hand we must notice the following important fact: if we do introduce a quantized or discontinuous space and time to the plenum schema (combinations of a continuous space with a discontinuous time or *vice versa* turn out not to be very attractive) then we do introduce a fundamental element of discontinuity into that schema. If space is quantized then the spatial distribution of properties is also quantized (though not of course the values of those properties). Similarly, a quantized space and time leads to discrete or quantized changes. In this case, *and only in this case,* the characteristics of the plenum schema approach somewhat those of the atomic schema. Since the 'marriage' of these two schemas lies at the heart of quantum theory I believe that the preceding considerations suggest most strongly that only under the conditions of a quantized space and time will we be able to erect a conceptually coherent quantum structure in which all states are precisely defined and all processes well understood.

The second feature of the two schemas which I wish to discuss here concerns the algebraic structures which both find natural to employ. Lying at the heart of the atomic schema is a set – theoretic structure. There is an important reason for this. Fundamental individuals can be neither created nor destroyed in the true atomic schema. One counts them, and counts groups of them, using the ordinary laws of arithmetic. The algebraic structure of groups of fundamental individuals is therefore that of a power set algebra. The same applies to derivative individuals, essentially because derivative individuals are reducible to spatio-temporal complexes of fundamental individuals. (At least we can always apply such counting procedures to the fundamental components of derivative individuals, regarded-as-collections-of-their-components, and we can apply the same processes to derivative individuals themselves regarded-as-structures so long as the structures are stable and resist interpenetration.) This set-theoretic structure provides the natural foundation for the phase-space representation of the state of a collection of fundamental individuals in classical physics. States in the phase-space themselves have the algebraic structure of a Boolean algebra which is in turn isomorphic to a power set algebra.

In the plenum schema, however, there are no such natural basic individuals to form the basic for counting. The most natural way to represent the state of the field has turned out to be the use of vectors in a vector state space. But vectors obey different laws of combination than do countable individuals. In particular, the natural algebraic structure of a vector state space is that of the non-boolean lattice of subspace of the space. Once again we face the consequences of this contrast in quantum theory, for there we attempt to deal with individuals but to assign them states represented by vectors in a vector state space (in Hilbert space).

Finally, there is the question of the criteria of identity in the two schemas. We can begin to see the fundamental contrasts between the atomic and plenum schemas over this issue by considering the question of the possible relation among individuals in the two schemas. In the atomic schema the possible relations among fundamental individuals is clear cut – it is one of total exclusion. In fact fundamental individuals obey the following exclusion principle: E.P.: no two fundamental individuals can occupy the same spatio-temporal location. Only of derivative individuals do we allow interpenetration (and only then through interpenetration of their respective spatio-temporal structures, not of their substantial components). In actual cases even this interpenetration is avoided because it usually turns out that the forces among fundamental individuals which are responsible for the dynamical stability of the derivative individuals are also sufficient to prevent the interpenetration of derivative individuals (at least by like individuals).

The situation is quite different in the plenum case. There is nothing intrinsic to the idea of a quality complex which prevents the mutual interpenetration, or superposition, of such quality complexes. To have this happen would be to have no more than the following happen: systematic changes in the pattern of quality instantiations occur which such that two quality complexes approach one another from differing spatial locations and merge into one another. (Perhaps later re-emerging as two, perhaps not.) Such merging is a commonplace among such entities as, e.g., waves in a fluid, and fields (gravitational, electromagnetic, nuclear) associated with several different sources – but these are plenum phenomena. Again, of course, we shall discover that some quality complexes do not superpose (namely, just those referring to the usual physical individuals) but there is nothing inherent in the notion of such complexes which prevents their superposition, it lies only in the physical relations which experience teaches us hold among them. Now the concept of superposition, for that matter the concept of being wave-like, is central to the peculiarities of the quantum mechanical treatment of matter.

Thus the differences between the atomic and plenum schemas in this respect are crucial for understanding quantum phenomena, for examble, for understanding the peculiarities associated with the so-calles 'identical particles'. Thus we may begin to sense that the real origins of our perplexities over the quantum treatment of the so-called 'idential particles' stems from the fact that quantum denizens share some of the properties characteristic of the individuals of the atomic schema (i.e., of atoms or particles) but also have some characteristics which are totally foreign to this schema and which can find a natural home only within the plenum schema (namely their ability to 'merge').

It behoves us, therefore, to look at the fundamental problems of identity within the two schemas. Within the atomic schema, the fundamental identities are those of its fundamental individuals and these are determined just be the spatio-temporal trajectories of those individuals. (Qualitative likeness and difference is irrelevant here since all fundamental individuals of the same type have exactly the same qualities.) Fundamental individuals obey the following identity principle:

I_1: A continuous spatio-temporal trajectory constituting the successive locations of a substance also constitutes the identity of the individual whose trajectory it is.

In addition, there is the question of the identities of the derivative individuals. Of these we may make two distinct judgments: those concerning substantial identity and those concerning morphic identity (i.e., identity of an individual *qua* structure of a certain sort). Clearly questions concerning substantial identity are reducible to those concerning the identity of the fundamental individuals constituting the derivative individual in question. But on occasion we judge an individual to be the same (derivative) individual just in case it continues to possess the same structure (e.g., cathedrals). In this case it is the continuous spatio-temporal trajectory of a spatio-temporal structure which constitutes the identity of the derivative individual in question. In all cases we can see that questions of identity are effectively reducible entirely to spatio-temporal considerations.

By contrast, the fundamental identity in the plenum schema is that of qualitative sameness or difference. All other identities are reducible to these or spatio-temporal terms. Qualitative identity is in maximal contrast since the same quality may be instantiated at many different spatial locations at a single time whereas this is not possible for the fundamental individuals of the atomic schema. The identities of all individuals, i.e., of all spatio-temporal quality complexes, in the plenum schema has two aspects, that pertaining to the sameness or difference of a qualities which are instantiated in the quality complexes and that pertaining to the spatio-temporal distributions of the qualities in the complexes. All quality complexes obey the following identity schema:

$I_2(1)$: Two quality complexes are identical (are instances of the same quality complex) if and only if they are component qualities are, pair-wise, instances of the same qualities and their spatio-temporal structures are identical.

Now to construct individuals within this schema, we need to distinguish between quality complexes which are structurally and qualitatively indistinguishable but having distinct locations. Thus we have the following identity principle for individuals:

$I_2(2)$: A continuous spatio-temporal trajectory for a quality complex which is everwhere the same along the trajectory constitutes the identity of the individual whose trajectory it is.

If to the identity principle I_1 in the atomic schema we were to add a doctrine of natural kinds or of essences, so that to remain the same individual (as opposed to the same substance) certain qualities had also to remain the same then we would achieve an identity principle in that schema which was quite close to that for individuals in the plenum schema. (And the way in which in ordinary language these two disparate metaphysics are intertwined has been, I suppose, one of the strongest reasons for wishing to introduce such doctrines.) Of course we may also wish to introduce such doctrines in an appropriate form to the plenum schema for in general we employ a criterion of identity which is more relaxed than $I_2(2)$ for we allow some properties to change without destroying the identity of the individual. But these shades of judgment, in either schema, are not my interest here. I wish only to set out the fundamentals of the two schemas.

Now that we have some appreciation of the depth of the disparity between these two fundamental conceptual-methaphysical schemes which underlie physical science we can turn with profit to a description of the general problems of quantum theory.

§ 5. The mysteries of quantum theory

The essential mystery of quantum theory is embodied in many now 'classical' experiments. Of these, the two-slit experiment will suffice for our purposes. As indicated in the diagram below, a source S of quantum individuals (one usually says 'particles') is available which impinge upon a screen A having two slits a_1, a_2 in it and those individuals passing beyond the screen A finally impinge upon an absorbing screen B where their arrival is detected in some fashion.

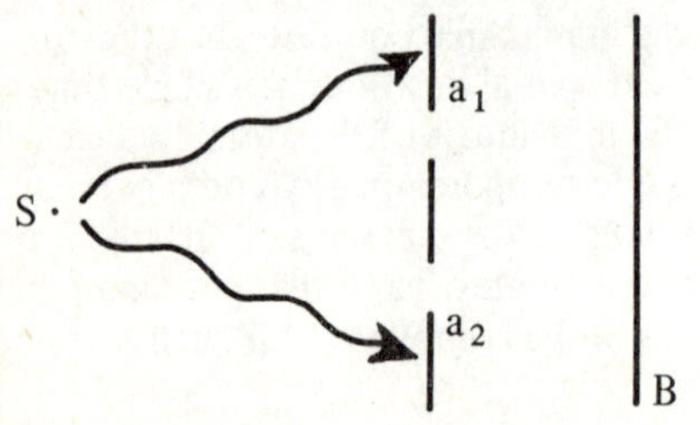

We suppose the experiment conducted such that the combination of the known energies of the quantum individuals and their times of arrival at the screen B force the conclusion that no more than one quantum individual in is the apparatus, (i.e., between S and B) at a time. Under these circumstances numerically single, spatially localized, quantum absorption events are detected at the screen B. The pattern of these events with both slits open, however, has all of the mathematical properties of an interfering pair of wave trains issuing simultaneously from the two slits. Therin is illustrated the mystery of quantum theory.[10)] This mystery can effectively be summarized as follows:

(P_1) We have a clear conception of only two basic kinds of physical ontologies, that provided by the atomic schema and that provided by the plenum schema. These two basic schemas are not compatible, i.e., they cannot be applied simultaneously to one and the same piece of physical reality.

(P_2) There are some features of quantum experiments which strongly urge the formulation of our quantum descriptions and quantum theory within the atomic schema (e.g., the numerically single, spatio-temporally locallized absorption events on screen B in the two-slit experiment).

(P_3) There are some features of quantum experiments which strongly urge the formulation of our quantum descriptions and quantum theory within the plenum schema (e.g., the distribution pattern of absorption events on the screen B in the two-slit experiment).

(P_4) A consistent explanation of entire quantum experiments is not possible within the atomic schema. (e.g. one cannot explain the distribution pattern on the screen B within the atomic schema, for within that schema it is impossible for numerically single individuals either to pass through more than one slit at a time or to interfere, either with any other such individual or with themselves, in such a fashion as to produce the desired distribution pattern of events at the screen B.)

(P_5) A consistent explanation of entire quantum experiments is not possible within the plenum schema. (e.g., it is not possible to account in a convincing physical fashion for the appearance of spatio-temporally localized absorption events at the screen B using only inherently non-localizable wave trains.)

We are thus at a complete loss as to how to obtain a clear conception of the ontology of the quantum world and this, at bottom, is the heart of the mystery of quantum theory.

From the very beginnings of quantum theory there had, of course, been thinkers who had attempted to formulate a consistent and physically plausible quantum theory entirely within one or other of the atomic or plenum schemas. Obviously if this could be achieved we should at once dispel all of the puzzlement surrounding quantum phenomena and know, at least in principle, precisely how quantum phenomena were to be precisely described. Schrödinger has, from his earliest involvement with the subject, persistently tried to reformulate the quantum theory from entirely within the plenum schema. His initial position was that there was nothing but real matter waves moving in a continuous physical field. These denizens are, of course, plenum individuals par excellence. Although his views have since been modified in various ways, he still clings essentially to the hope and persuasion that a formulation of quantum theory along these lines will some day be successfully concluded. On the reverse side of the coin lie such men as Landé and Popper who are attempting to formulate quantum theory entirely within the atomic schema. Popper, e.g., holds that there is nothing but atomic particles (generic sense) within the quantum domain and that all physical phenomena are essentially to be accounted for in terms of interactions amongst these.

For either attempt there are the peculiar problems raised by those features of quantum theory which seem to belong to the other fundamental scheme. Thus for the protagonists of the matter wave and the like there is the mysterious 'collapse of the wave packets' on interaction between such matter waves so as to produce spatio-temporally localized events.[11]) This feature of quantum theory, which fundamentally originates with the attempt to understand quantum phenomena from within the plenum schema, has since achieved a notorious life of its own within the so-called 'measurement problem' as such. The protagonist of the atomic schema, on the other hand, must cope with the peculiar distributions of localized events which occur. He has been apt to respond only very rarely by attempting to introduce unusual forces or other such physical mechanisms to the situation; his more usual response being to suggest that the theory offers a less than complete description of the quantum circumstances. In fact, both sides have been wont to write off various aspects of the theory as pertaining merely to our epistemological situation, thus saving them the embarrassment of having to take them physically literally. (For the plenum schematist the collapse of the wave packet becomes a purely informational affair, while this also is true of the atomic schematist for he suggests that quantum theory offers only the distributions of the probabilities of physical outcomes for a given situation based upon our existing knowledge of the circumstances.)

I shall have no more to say of these approaches to quantum theory in the present essay for they throw no more light on the conceptual structure of the situation nor on the task we have set ourselves. But it would be a serious error to thence conclude that these attempts have no philosophical or physical merit and are not to be taken with the utmost seriousness. Already we can begin to sense the following lesson here: If one has a situation where disparate conceptual schemes simultaneously are in use, as one does in quantum theory, then the incoherence thus introduced can be eliminated in favour of only one of the schemes just in case one is prepared to adjust the epistemology of the theory accordingly so as to remove the elements of the other schema from functioning at their original level. Thus the atomic schematist give the ψ wave a purely probability status, its behaviour no longer representing a literal physical behaviour, while the plenum theorist is ultimately forced to do the same, within the context of a new theory of the real matter wave (such as de Broglie's, see 4, 5, 6). This already shows us a glimpse of the powerful relations there are between conceptual schemes (metaphysics) and epistemology in science.

In sharp contrast to the foregoing approaches to quantum theory there are the approaches of Niels Bohr and of the so-called quantum logicians. These two approaches contrast with those described above because neither of them attempt to reformulate quantum theory within one of the *existing* fundamental schemas, rather both attempt to alter the logico-conceptual structure of physics in order to resolve the quantum mystery. Let us turn now and examine these two approaches.

§ 6. Quantum logic.

The quantum logician proposes to make a change in the basic logical structure governing the relations among assertoric sentences of the physical language in order to remove the quantum mystery. In short, he proposes a major modification of M_4 of the fundamental metaphysics. The *intended* net result of this modification is to be the formulation of quantum theory entirely within the atomic schema (suitably modified with respect to its logical structure, but otherwise unaltered). I shall now recapitulate briefly the development of the doctrine.

Let us consider for a moment the classical phase space presentation of the classical theory, for example, for a system S. The phase space is so constructed that the specification of a location in it exhaustively describes the physical properties of S (at the time of its location there). In other words, an exhaustive collection of true propositions concerning the system is determined by specifying its location in phase space. All other mechanical propositions pertaining to the system are known to be false. Of course the propositions asserting the system to have various physical properties will stand in logical relations to one another. And in fact the entire set of propositions, true and false, which may be asserted of the system will form a complete logical structure which can be specified by extracting its Lindenbaum-Tarski algebra. (This latter is the algebra of classes of logically equivalent propositions within the logical system.) This logical structure is a Boolean structure (i.e. the structure of ordinary, or classical logic) i.e., the algebraic structure is that of a distributive, ortho-complemented lattice.

Now the descriptions of which we have been speaking so far are maximally precise descriptions, i.e., no more precise assertion concerning the physical properties of the system in question can be made than by asserting one of the propositions of the form "The location of S in phase space is R". There are, in addition, less precise propositions. (Thus compare "The velocity of the system, S, is between v_1 and v_2" with "The velocity of S is v_3".) A collection of such vaguer propositions would specify the location of the system to within some finite volume of phase space only. This finite volume of phase space is represented exactly by the *set* of points (locations) it contains and these in turn represent the physical possibilities for the actual state of the system. The structure of a phase space itself is appropriately captured by the power set algebra of the sets of its points (locations). This is the algebra pertaining to the relations among finite volumes of the phase space.) The logical structure of the set of all propositions pertaining to the system, both maximally precise and less than maximally precise, is still the classical Boolean structure.

And this is as it should be, for then we know the mathematical structure of phase space and the logical structure of the propositions which correspond to it are exactly suited to one another. (The Stone representation theorem shows that the two are isomorphic to one another.) Thus far the extraction of the logical, or logico-algebraic, structure of classical physics.[12])

It is not unnatural that we should want to try the same approach when we turn to quantum theory. In quantum theory, the space corresponding to classical phase space is the Hilbert space of a physical system. Specifications of the physical state of a quantum system, maximally precise or not, correspond to specifications of geometrical regions of the Hilbert space, just as they did in classical physics. At this point the analogy breaks down, however, and quantum and classical theory part company. The maximally precise specification of a system in quantum theory corresponds, not to a point location in the Hilbert space, but to a one-dimensional subspace of the Hilbert space. And less than maximally precise specifications of a quantum system correspond, not to sets of locations in the phase space, but to higher dimensional subspace spanned by all of the vectors representing the physically possible maximal specifications of the actual state of the quantum system. Moreover, no specification in quantum systems turns out to be maximally precise in the classical sense. The best that can be achieved, the sets of assertions corresponding to a one-dimensional subspace of the Hilbert space, are assertions which are maximally precise in respect of a subset of the set of physical characteristics of quantum systems, these subsets containing only what are known as compatible properties (their corresponding operators commute). The upshot of these divergences from classical theory is that the algebra of the logic of quantum propositions, corresponding to the Lindenbaum-Tarski algebra of a classical system, is that of a non-distributive, orthocomplemented lattice. This algebra is in fact isomorphic to the algebra of linear subspaces of a Hilbert space (or of any other linear vector space). Because of this structure the quantum logic cannot be embedded into a Boolean logic.[13])

To see how the return to a formulation entirely within the spirit of the atomic schema is intended to be effected through the adoption of the new quantum logical structure, let us consider a ficticious physical system having only two properties, say position and momentum, and each of these having only three allowable distinct values. We may then conveniently represent the physical state of our system, call it S, in terms of a three-dimensional Hilbert space which can be conveniently 'coordinatized' using two different sets of three mutually orthogonal vectors, those corresponding to the three possible position values for S and those corresponding to the three possible momentum values for S. These are, respectively, q_1, q_2, q_3, and p_1, p_2 and p_3 in the diagram below.

Physical states of the quantum system are represented, with maximal precision, by a suitable vector (more precisely, by the one dimensional subspace which the vector spans) in the above Hilbert space. To obtain our set of propositions we shall make the following associations: The null space (the origin) of the Hilbert space is to be

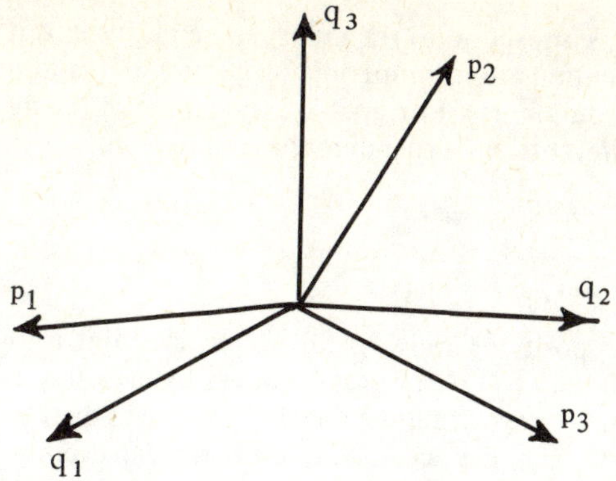

associated with the always-false (contradictory) proposition, the entire Hilbert space is to be associated with the always-true (tautologous) proposition, the one-dimensional subspace spanned by the vector q_1 is to be associated with a proposition Q_1 asserting that the system S has the location q_1 and so on for the other one-dimensional subspaces. The proposition Q_1 or Q_2 is to be associated with the smallest subspace spanned by the vectors q_1 and q_2 and so on for other such combinations, while the proposition $Q_1 . Q_2$ is to be associated with the intersection of the subspaces spanned by the vectors q_1 and q_2. We have then the following immediate results. The propositions

Q_1 or Q_2 or Q_3, P_1 or P_2 or P_3

are always true. This is the fundamental result which supports the claim that we are always able to assert of a particular quantum system that it is in one or other definite position and has one or other definite momentum at any time. It is therefore the foundation of the view that quantum systems are fundamentally atomic in nature and characterized by exactly the same physical properties as characterize atomic systems within classical physics. On the other hand we know that maximally precise quantum specifications are not classically exhaustive, i.e., they do not offer precise specifications of all classical properties. In our present case, for example, to assert Q_1 is to offer no information whatever concerning the truth values of the P assertions even though the assertion Q_1, corresponding to a one-dimensional subspace of the Hilbert space, is maximally precise. According to proponents of this view, this is to be explained by pointing out that the assertion Q_1 represents our *knowledge* of the system at a given time *as a result of a measurement* and that at that time we are simply unable to say what the precise value of the momentum of the system is. This is nicely borne out in the logic, for the proposition

$Q_1 . (P_1$ or P_2 or $P_3)$

is a perfectly well-formed proposition in the logic and is in fact equivalent to the proposition Q_1. The proposition supports the view that we are able consistently to assert that even while the system has some definite location it also has some definite momentum (though unknown) at the same time. On the otherhand we do know that all propositions of the form

$Q_i . P_j$ $(i, j = 1,2,3)$

are always false, for they correspond to the null space, and this is as it should be since position and momentum are incompatible properties in quantum theory (their operators do not communte). Fortunately, the failure of the distributive law for quantum logic prevents our expanding the proposition

$$Q_1 . (P_1 \text{ or } P_2 \text{ or } P_3)$$

to obtain

$$(Q_1 . P_1) \text{ or } (Q_1 . P_2) \text{ or } (Q_1 . P_3)$$

for the latter of these propositions is always false, corresponding to the null space, while the former is equivalent to Q_1. The same applies to any other combination of P and Q propositions. By preventing the distributive move, therefore, it would seem that the quantum logician has accomodated the incompatibilities among physical properties of the quantum system consistently with the treatment of quantum individuals as classical atomic individuals.

Actually, the situation is not quite so simple and elegant as the preceding would have us believe. According to the foregoing story, every proposition pertaining to a particular quantum system is, just as in classical physics, either true or false. The quantum atomic systems are, just as in classical physics, always in perfectly definite physical states. There are just two differences between quantum and classical physics; on the one hand we are forced to remain ignorant of some aspects of the physical state of a quantum system while we are gaining information concerning other aspects of this physical state, and on the other hand we are prevented from making moves in the logic which would be incompatible with this quantum mechanical situation by the failure of the distributive law. Unfortunately, the semantics of quantum logic proves to be more complex than this. For it follows immediately from the work of Kochen and Specker (17) that it is impossible to consistently simultaneously assign truth values to all of the orthogonal bases of a three-dimensional Hilbert space such that in any one basis exactly one one-dimensional vector in the basis is assigned the value True and the remainder are assigned the value False. But to recapture the idea that each quantum individual possessed a definite value for the magnitude of each relevant physical quantity as determined in the atomic schema we should have wanted precisely the foregoing impossible result. The semantics of quantum logic is still under intensive investigation and relatively little is yet known about it. (There is, for example, the related, but somewhat different, approach by van Fraassen (11, 12) and quite recently there has been suggested the development of a new conception of state analogous to that in classical physics but fully specified just by the maximal set of true consistently permissible within quantum theory – see, for example, Jauch and Piron 25.) In any case, we should be prepared for the result, already suggested by Friedman and Glymour (13), namely that in the bivalent (True, False) semantics appropriate to quantum logic it may not be possible to maintain that for every quantum system every proposition has a precise truth value. If this is the case then the aim of reformulating quantum theory so as to entirely recapture the spirit of the classical use of the atomic schema would have failed.

Whatever the outcome of this complex issue the significance of the quantum logical approach to quantum theory for our interests in this paper will not have been lessened. This approach proposes a profound alteration in the logical structure of the conceptual scheme of physical description in order to remove the mystery of quantum theory. If the original goal of this approach is achieved, and we are able to obtain the normal bivalent semantics in which every proposition is either true or false, then it will have been shown how a deep-going indeterminacy at the quantum level can be entirely resolved into a merely epistemological limitation consistent with a theoretically perfectly determinant quantum reality through the introduction of a modification in the logical structure of physical description. If this goal cannot be achieved, and such results as that of Kochen and Specker suggest that it cannot, then we shall be forced to concede that the mystery of quantum theory originates from a much deeper level of the theory still – deeper than anything so superficial as epistemological limitations produced by the 'disturbance of the system' in observation and deeper then can be eradicated by changing even the logical form of discourse. It will suggest rather that the origin of our difficulty lies, not with M4 of the fundamental metaphysics, but with M5, that it lies in the attempt to marry the disparate atomic and plenum schemas.

If fact, the very Hilbert space representation of quantum systems represents this marriage quite clearly. On the one hand, the maximally specific states correspond to sub-spaces of Hilbert space and are related to each other vectorially and this corresponds, as we have seen, to the natural representation in the plenum schema, a representation which is alien to the set theoretic representation natural to the atomic schema. On the other hand, with one-dimensional sub-spaces, or vectors, are associated definite physical properties such as particular positions, momenta and so on and these give rise to physical state descriptions which are characteristic of the atomic schema (this is especially obvious in the case of those descriptions asserting a localized position for the system) and which are alien to the plenum schema. I pointed out in § 4 these two schemas are related in deep-going ways to continuity and discontinuity and that the attempt to combine continuous space or time with discontinuous changes in other physical characteristics (or vice versa) requires, for consistency, the introduction of a deep-going indeterminacy. In the light of this it would not be at all surprising to find that the indeterminacies of quantum theory are ineradicable within the quantum logical approach which does not attempt to deal directly with these two conceptual structures but only, as it were, to 'ease their marriage problems' by changing some of the rules of the game. Such a conclusion, which would seem to be suggested by the investigations of Friedman and Glymour, would only reinforce the view that we have here to do with a clash between the two fundamental conceptual structures which we employ as the basis for the classification of scientific knowledge.

But we have yet to consider the philosophy of Niels Bohr, so I shall turn to a discussion of his position forthwith.

§ 7. The quantum philosophy of Niels Bohr

Bohr is unique among quantum theorists and philosphers, for he develops a synthesis of the two disparate conceptual structures (i.e. those of the atomic and plenum schemas) which transcends both of them. Bohr's philosophy has some remarkable affinities with that of Kant for Bohr essentially rejects what I earlier described (Section 2 above) as the traditional epistomology of Western science. I shall first briefly outline and explain Bohr's specific quantum philosophy and then draw out the more general consequences.[14)]

The essential content of Bohr's philosophy of the quantum domain may be summarized in the following four doctrines:

B_1: The results of all experiments are to be described using only the resources of the conceptual framework of classical physics.

B_2: The applicability of a physical concept to a particular physical situation (à fortiori the applicability of classical concept to physical situations) is dependent upon the relevant physical conditions obtaining in that situation.

B_3: Action is quantized for all micro-processes and in consequence any micro system under investigation is indivisibly connected to the macro measuring instruments through which we investigate it, and connected in definite ways characteristic of the quantum theory, e.g. $E = h\nu$, $p= h\lambda$, etc.

B_4: Because of B_1, B_2, and B_3 there is an inherent limitation on the simultaneous applicability of classical physical concepts to the same physical system under the same physical conditions. Which concepts are applicable to a given system depends upon the entire physical situation in which the system is located, including, in particular, the measuring apparatus involved.

The foregoing doctrine entails that there is a corresponding limitation upon the classical physical descriptions of experimental results which are applicable to a given experimental situation. Let us consider briefly the concepts which are most commonly used by way of illustrating his position. According to Bohr, if it is to be relevant to, or applicable to, a physical situation, the concept of position requires a rigid body which defines the spatial reference frame and some manner in which the object whose position is to be determined can be brought into relation with this rigid reference frame. Thus, for example, one might determine the location of an object by passing it through a narrow slit in a rigid diaphragm rigidly attached to the spatial reference frame. By passing the object through the slit one has placed it in a determinate spatial relation to the rigid reference frame. The concept of momentum, on the other hand, logically requires that the object in question be brought into some dynamic relationship with the spatio-temporal reference frame. Thus, we may measure the momentum change in a particle by allowing it to interact dynamically with an object elastically attached to the rigid spatial reference frame in such a fashion that the subsequent motion or position of the object (depending upon the measurement technique) can both be determined and is sufficient to determine the object's momentum (change). But, as Bohr remarks, one cannot, *logically* cannot, simultanteously have the same instrument both rigidly and elastically (i.e. non-rigidly) attached to the same spatial reference frame. Thus it proves impossible to simultaneously realize the condition relevant for the applicability of the concepts of momentum and position to the same object. It is thus impossible to apply those concepts simultaneously to the same object. It is thus impossible to provide simultaneous physical descriptions of that object employing these concepts in a categorical fashion.[15]) This example offers a fairly clear illustration of what Bohr intended by the B_2 and B_4 doctrines. (The doctrine B_3 simply assures that we cannot reduce the incompatibility to negligible significance, i.e. cannot offer arbitrarily precise simultaneous descriptions of the sort forbidden by Bohr's analysis, simply by making the interactions between the object and the measuring instrument – slit or whatever – as small as we please.) It remains to have something to say about Bohr's doctrine B_1 and this will lead us in to the heart of Bohr's Kantianism and to the more general significance of his philosophy.

The Kantian-like elements of Bohr's philosophy begin with Bohr's doctrine B_2. This doctrine teaches that the application of concepts to the world is tied in a very intimate fashion to the conditions obtaining in the world. The very circumstances which make a measurement possible also then serve to define the concept to be applied in the statement of the measurement outcome. We can know no more of the world than the combination of our empirical concepts and the realizable macroscopic situations permit. This intimate connection between our conceptual scheme and the realizable states of the world is a thoroughly Kantian idea. In fact, Bohr specially extends the notion of perception to include the general measurement circumstances which we can realize and then states very clearly, what is a thoroughly Kantian position, that the ways in which *perception* of the world occurs are intimately related to the ways in which we are able to *conceptualize* the world, i.e. to the kinds of descriptions which can, in principle, be offered (or are well-defined in the circumstances). And indeed for Bohr the use of the particular conceptual structure comes to have the appropriate epistomological content. Thus the use of the classical descriptive structure will, for Bohr, tacitly assume that $h = o$, with all that implies. And for us to discover that $h \neq o$ is nothing less than for us to discover that a certain kind of conceptual structure is inappropriate, conceptually and logically appropriate, to the description of the world.

The intimate connection between conceptual schemes and the conditions under which they operate culminates in Bohr's doctrine of the role of the subject/object distinction. According to Bohr the very existence of a language of unambiguous communication (= objective communication, for Bohr) presupposes that there is a clearly defined matter about which we communicate. But we can obtain a clearly defined object of communication only if it is separated in a well-defined way from the communicating subjects. Bohr summarizes his doctrine with these words:

"... a close connection exists between the failure of our forms of perception, which is founded on the impossibility of a strict separation of phenomena and means of observation, and the general limits of man's capacity to create concepts, which have their roots in our differentiation between subject and object." 3, p.96.

This doctrine leads naturally on into the question under what circumstances one is able to effect an unambiguous distinction between subject and object. Bohr answers this question itself unambiguously: one can effect this distinction in our world only at the macroscopic level. Under these circumstances the appropriate descriptive concepts to employ are those of classical science. Hence the doctrine B_1. All of our descriptions of experimental outcomes *must* be couched in terms of the concepts of classical science (i.e. in terms of our common sense concepts, suitably refined) because it is only under those macroscopic circumstances where classical concepts are well-defined that we are able to effect the clear separation between object and communicating subject that is necessary for unambiguous physical description.

Bohr therefore attempts to transcend the specific conceptual frameworks characteristic of the atomic and plenum (field) schemas, for him there are not really just particles, or just waves, there are only concepts appropriate to the description of phenomena-in-particular-kinds-of-macroscopic-circumstances. In Bohr's philosophy there is a shift of emphasis away from questions concerning the nature of the fundamental ontology of the physical world towards the epistemology of physical science and the conceptual structures which embody our knowledge of the physical world. It remains true that the descriptive concepts which we do find appropriate to those macroscopic situations which are realizable fall more or less into two groups – namely, those, as we should say in other contexts, which pertain to the atomic and plenum schemas – but these two groups of concepts are used in a *complementary* fashion according as the macroscopic circumstances dictate that first the one and then the other group of concepts are the appropriately applicable group. The world remains what it is, and always was, a mystery to the mind of man if regarded as a "thing-in-itself". Science remains what it is and always was, an attempt at systematic description of the pattern of phenomena as apprehended by conceptualizing, perceiving man. There is in that sense, for Bohr, no peculiar quantum world, no peculiar task for quantum physics, no major disruption of the fundamental continuity of science in the passage from classical to quantum physics.

This is not yet a very sharp formulation of Bohr's doctrine from the specific point of view of the structure of quantum theory. (It is too much in the tradition of Bohr's own semi-mystical way of writing!) We can, however, develop the position as follows. What Bohr is proposing is a *meta*-linguistic ordering structure – based upon his sematical analysis of the nature of physical concepts (cf. B_2) – which regulates the use of object language sentences in physical contexts. We have then at the meta-linguistic level a *language of contexts* such that to each distinct physical context is assigned a unique set of sentences of the object language. These latter sentences are just those physical-descriptive sentences well-formed in that context. Thus consider the set $b = \{C_i\}$ of physical contexts and the set $L = \{p_i\}$ of physical-descriptive propositions. We define a function $F: b - P_L$, P_L the power set of L, such that p_j is well-formed in $C_i = p_j \in F(C_i)$. Within $F(C_i)$ propositions form a Boolean logic. Moreover, b itself has a structure, some contexts being more precise, some more restrictive, than others and some included within others as specialized cases. I shall postulate that b has the structure of a non-distributive lattice L_b. It is also plausible to suppose that the $F(C_i)$ form themselves also into a non-distributive lattice, in fact into a lattice isomorphic to L_b.[16]) Now we can specify exactly the structure of the language of context by specifying the structure L_b, L, F, L_{P_L} . Specifying this structure is what is required to carry Bohr's approach to completion – it has not yet been carried out (though heelan's work goes a considerable distance in this direction, see section 8 below). In particular, it is clear that quantum theory itself will be needed to specify both L_b and L_{P_L} and there remains the question of the relations, if any, between L_b, L_{P_L} and L_Q the non-distributive, orthocomplemented lattice of sub-spaces of a Hilbert space. In any case, we have now achieved our goal of obtaining a precise formulation of Bohr's position on the conceptual structure of quantum theory.

Bohr then has been able to resolve the mysteries of quantum theory, not ideed by reconciling the two disparate conceptual structures pertaining to the atomic and plenum schemas, nor by attempting to eliminate one in favour of the other, but by transcending both, unifying them in a meta-linguistic theory of conceptual structures and their role in science. Thus by shaking himself free from the tacit resuppositions of the major current of Western scientific thought, and only through this revolutionary expedient of introducing a major shift in our epistemology of science, has Bohr been able to produce an harmonious conception of quantum theory.

§ 8. On the relations between the quantum logical and Bohrian philosophies of quantum theory

Despite their apparent desparateness, there is a relatively intimate connection possible between the quantum logical and Bohrian approaches to quantum theory. The most obvious feature which they share in common is that they both exclude certain sentences from the quantum theory as ill-formed. The quantum logician achieves this by altering the inference rules of his logic, the Bohrian achieves the same end by ruling certain concepts as undefined in certain circumstances. But though the quantum logician aims only at recapturing the spirit of the atomic schema for quantum theory, the Bohrian aims at encompassing all of the traditionally used descriptions in the quantum domain, those pertaining to the plenum schema as well as those pertaining to the atomic schema, in a comprehensive, rational pattern. In fact, one of the questions very rarely treated by those who take the quantum logical approach is what they would do with those descriptive remarks whose terms are drawn from the plenum conceptual scheme, such as "the electron has a wave length of   microns".

Now the principle on which the Bohrian distinguishes those sentences which are ill-formed is really fairly clear and straight forward. Those sentences which are compatible, i.e. *logically* compatible (with one another), are just those which are associated with the same, or similar, experimental (macroscopic) arrangements. For, remembering that it is the macroscopic circumstances which determines which concepts will be defined, for any given *type* of macroscopic circumstance a certain cluster of concepts will be well defined. All descriptive assertions embodying these concepts will therefore be well formed and allowable in the Bohrian schema. All other descriptive sentences will de disallowed.

The principle of rejection of sentences in the quantum logical system is less clear on the surface of it. But quantum logicians also define a notion of compatibility which, whatever its other peculiar features may be, turns out to distinguish as those propositions compatible with a given proposition just those propositions which either refer to other values of the same physical magnitude as that which is involved in the given proposition, or to values of algebraic functions of the magnitude referred to in the given proposition, or to logically weaker combinations of the foregoing propositions. In essence then it selects out the same group of propositions as the Bohrian selects out by referring to experimental arrangements. The quantum logician then simply forbids certain ways of combining propositions drawn from sets of propositions which are not compatible. This amounts to a similar, if slightly less strict, restriction to that which the Bohrian applies.[17)]

This relation between the two approaches has been strikingly brought out in a slightly different development of a quantum logical approach by Heelan.[18)] To (briefly) outline Heelan's approach we begin by noting that the non-distributive lattice of the quantum logical propositional system can be built up by 'pasting together' boolean sub-lattices of this lattice. Each boolean sub-lattice contains only mutually compatible propositions. Heelan then makes a nice, and important, distinction between experimental propositions and theoretical propositions, i.e. between descriptions which are couched entirely in terms referring to the observable features of the macroscopic apparatus and circumstances and descriptions containing theoretical terms drawn from quantum theory. He then points out that the set of compatible propositions forming a complete boolean sub-lattice can be regarded as the complete set of macroscopic descriptions of possible experimental outcomes pertaining to a given experimental situation. These descriptions then correspond to the experimental descriptions, using the terms of classical physics, which the Bohrian countenances for that situation. Now such sets of descriptions can be related to one another only through the use of theory. Heelan regards the relating, theoretical sentences as belonging to a *meta* language for which the macroscopic observational descriptive sentences form the object language. In the meta language, then, the sets of comparable experimental propositions are related together in the fashion determined by quantum theory. This pattern turns out to be the quantum logician's non-distributive, non-boolean lattice. This relation is the counterpart to Bohr's use of the quantum theory as a mathematical algorithm with which to relate the classical experimental descriptions. And quantum theory is here given a similar role, namely that of a meta linguistic ordering pattern, to that which the Bohrian assigned it, namely that of a transcendent conceptual schema within which to combine in a coherent way the conceptual structures, and the descriptions couched in terms of these structures, pertaining to the atomic and plenum schemas.

Now Heelan asserts, I think without proof, that the metalinguistic structure of context languages is isomorphic to the lattice of sub-spaces of a Hilbert space. I am inclined to think, therefore, that his will not be the last word on this approach to understanding Bohr. The reason for this is rather simple: I doubt that the physical contexts, which are the central physical entities for Bohr, form a lattice – if they form a lattice at all – which is isomorphic to the lattice of sub-spaces of a Hilbert space. I doubt this just because I doubt that L_b has an orthocomplement at all.[19]) If this is true it would lead one to suppose that more must be said to accomodate Bohr's approach, which hinges on what is *physically* possible in differing *physical* contexts, with the quantum lattice approach. Conversely, since it would be surprising if the sets of sentences which were well-formed in various distinct physical contexts formed a certain kind of lattice whereas the context themselves did not, it suggests that Heelan's assumption of isomorphism should also be looked at more closely. But for the moment, since the answers to these questions remain uncertain in my mind, I shall continue on as if Heelan's approach provided the entire answer (as indeed might yet prove to be so).

A few more brief comments on Heelan's approach must suffice. Like the Bohrian approach, Heelan would seem to have little or no difficulty in incorporating descriptive sentences employing concepts drawn from the plenum schema into his approach. Although he does not explicitly discuss such examples, limiting himself to descriptions drawn entirely from within the atomic schema as do most other quantum logicians, I suppose that a sentence such as "the electron stream has a wave length λ" would be regarded by Heelan as one of his experimental propositions to be part of the boolean sub-lattice of propositions pertaining to the appropriate situations (e.g. diffraction experiments). Since Heelan, like Bohr, can deal with any and all macroscopic situations which turn up – they simply represent so many more contexts within which boolean sub-languages operate – there seems to be no reason why he cannot straightforwardly incorporate these kind of descriptions in quantum theory. By the same token, Heelan seems not to be in difficulty over the consistency of the semantics of his quantum meta language in quite the same way in which the other quantum logical approaches find themselves in difficulty. (But then the Bohr-Heelan approach does not attempt to reformulate quantum theory within a single ontology – atomical plenum – conceived in the classical spirit.) For Heelan, the quantum meta language is no more than a language of contexts, a language which specifies sub-languages of physical descriptions which are to be used in various contexts. Although it is not self-evidently consistent (at least not to me) there is certainly no way in which the constructions which lead to trouble for the other quantum logical approach, e.g. that by Kochen and Specker, could be applied to Heelan's system, for the

Kochen and Specker construction involves the simultaneous consideration of distinct bases for Hilbert space but for Heelan these are simply associated with different macro contexts and hence the corresponding groups of assertions would not be simultaneously applicable to the same physical system.

Finally I return again to the distinctive role which Heelan assigns to quantum theory. The quantum theory plays a role of a *meta* linguistic device for ordering our physical descriptive (sub) languages. (As such, Heelan's approach is to be sharply distinguished from that of other quantum logicians – whatever the other similarities might be.) For Heelan then quantum theory – and for that matter any other physical theory – becomes, or at any rate provides (and in the case of classical theory, provides only tacitly), a conceptual structure for the ordering of the coherent descriptions we give of varying physical contexts. As such, each physical theory plays a fundamental role in the determination of the conceptual bases for the classification of physical knowledge.

§ 9. Reflections and conclusions

I began this paper with a question: "What impact has quantum theory had on the fundamental features of our conceptual scheme pertaining to those distinctions crucial for physical science? " Let us now see what has been provided by way of an answer to that question.

From my point of view the most important contribution of quantum theory to our understanding of the conceptual basis of our physical theorizing has been the way in which the very difficulties of that theory have forced us to return to the fundamental conceptual systems which undergird our physical theorizing, in particular, the way in which it has forced us to recognize the fundamental differences between the conceptual schemas pertaining to the atomic and plenum ontologies. It was the genius of classical physics to employ both of these schemas but to keep them carefully segregated from one another. In this fashion, most of the difficulties which could arise from employing both simultaneously were either avoided or thoroughly obscured. (There were indeed some instances of this potential conflict which did not quite escape obscurity, eg. the problem of the self-energy of the electron in electro-magnetic field theory, however these were not seen to be merely the tips of vast conceptual iceburg problems but were rather regarded as problems lying, as it were, only in the surface mathematics of the subject.)

This ingenious segregation of conceptual structures practiced by classical science has had, however, several unfortunate consequences. In the first place, it has encouraged a certain superficiality of approach to the foundation of physical theorizing on most of our parts. Specifically, there has been a wide spread belief – still wide spread in many scientific circles – that the physical content of a theory lies solely, or almost solely, within its specific dynamical equations (or other mathematical laws governing the physical states concerned). Thus physicists were lead to concentrate on Newton's equations of motion, or on Hamilton's equations, and encouraged to ignore the significance of the fact that classical mechanics was couched in a phase space language at all. It is responsible for the fact that only very recently has it been recognized that the canonical formalism long ago introduced into classical mechanics is in fact an abstract structure of much more general significance than its particular use in that theory. Correspondingly, it has encouraged physicists to view the essence of quantum theory as being captured in the content of the dynamical equation (Schroedinger's equation). Again only very recently, and largely under the stimulus of the quantum logical approach, has it become recognized that virtually all of the general features of quantum theory of significance in the physical discussion of its meaning stem from those features pertaining to its formulation within a Hilbert state space at all.

The difficulties of understanding quantum theory have finally forced us, however, not only to recognize the overriding significance of the abstract methematical scheme within which physical theory is formulated, but driven us even further to recognize the profound ways in which conceptual structures enter into physical theorizing at very deep levels. The forcing of this awareness is the greatest contribution which quantum theory has made to date to our understanding of the role of conceptual structures in the classification of scientific knowledge.

But before continuing with this theme, let me briefly look at those approaches to quantum theory which have been mentioned in the preceding remarks from this point of view. We have said that, on a deep level, the significant difference between quantum and classical physics is that, whereas the latter largely maintained a careful segregation between atomic and plenum conceptual schemas and their related mathematics, quantum theory attempts a marriage of the two. We have said enough, I hope, to make it seem plausible that such a marriage will be a very difficult, and improbable, affair-certainly an undertaking almost unique in the history of physics. From this perspective those approaches to quantum theory which seek to reduce its interpretation wholely within either the atomic, or the plenum, schemas cannot seem very promising avenues for research. At best, we may suppose, they will succeed in truncating the quantum theory so that what remains as physically significant no longer attempts any conceptually significant marriage of the two different schemas. (Such, at bottom, are the positivist views of a Heisenberg in which meaningfulness and/or epistemic value are suitably restricted.) Though I do not preclude the idea that such an approach might succeed, it seems to me highly improbable that it will do so in the bounds of the existing quantum theory and it seems an approach of simply insufficient depth to cope with the subtleties posed by that theory. (Moreover, it would even be a shame, from a purely philosophical point of view, if the approach did succeed. For it would divert our eyes from the quantum theory itself in its 'old form' which has so much of philosophic important to teach us.)

Likewise, it seems to me, the quantum logical approach is also too superficial – at least in its most direct and naive form. I cannot escape the impression that, without due sensitivity to the deep conceptual origins of the peculiarities of quantum theory, the quantum logician seeks simply to manipulate rules of inference so that embarrassing sentences cannot arise. This smacks of being a thoroughly ad hoc procedure when stated in this bold fashion. Not only so, there is no good reason to believe that it will succeed, for it seems reasonable to think that if the conceptual conflicts run deep enough then no simple juggling of the rules of inference will be sufficient to avoid the difficulties to which that conflict leads. This view is in fact born out by the preliminary results of Bub-Glymour-Friedman. I have the impression that an adequate and convincing quantum logic will have to be based on a much deeper understanding of the conceptual structure of quantum theory than quantum logicians have hitherto seen as necessary. (Though I must also add that, in the hands of mathematicians and not philosophers, the exploration of the quantum logical, or algebraic approach to quantum theory has lead already to many powerful insights into the mathematical structure of that theory.)

But quantum theory does have higher unexplored reaches. Thus we might ask, not for a naive bivalent logic with 'fiddled' inference rules discussed in an ordinary, boolean metalanguage, but for a thoroughly quantum logical approach in which the metalanguage itself is also non-boolean. Such an approach, which is only one among several unexplored alternatives, doubtless holds considerable promise and it would be premature to pass judgement on these at this stage.

There is one approach to quantum logic, however, which does, on the face of it, seem very promising and on which I shall venture to pass some judgement here. That approach is Heelan's and its promise lies precisely in its being a formal counterpart to Bohr's philosophy of quantum Theory. This latter feature of Heelan's approach is also what allows me to pass some judgement on it, for we now have a perspective in which to see the significance of Bohr's approach to quantum theory.

The genius of classical physics, as I said, was to segregate uses of the atomic and plenum conceptual schemes. The means by which this was done, however, were only tacit at the classical stage. I believe we can understand Bohr as having attempted to make explicit for quantum theory the means of effecting that segregation. In this way one indeed achieves a consistent quantum theory. Moreover it is, in an important sense, the same kind of way of achieving consistency as classical physics employed (albeit tacitly), and it is this fact which allows Bohr correctly to claim that, from his point of view, quantum theory is a rational generalization of classical physics. The Bohr-Heelan revolution then is the one obvious way to avoid difficulty in quantum theory, namely to specify at the metalinguistic level rules which will insure the segregation of sentences pertaining to the two distinct conceptual schemes.

Undoubtedly, the enterprise is successful. Moreover, viewed from this perspective, it is a very agreeable surprise that the solution should throw so much light on the significance of the distinction between a Kantian and a non-Kantian approach to epistemology on the one side and on the significance of the formal structure of the metalanguage of a physical theory on the other. Even if we could see no alternative to this approach we should still be in a rewarding position, for Bohr's approach throws into relief the depth of the penetration of our conceptual schemes into the informative structure of physical theory by concentrating on the epistemological content of those structures as they feature in our physical theories.

And yet I am again left with an uneasy feeling. Bohr's resolution of the difficulties is an obvious one and an obviously successful one – and yet it remains somewhat unsatisfying. Do we have no other choice, one wonders, than to be stuck with our two disparate conceptual schemas, destined only to go on producing increasingly complex metalanguistic devices for insuring that they do not conflict with one another within the mathematical structures which we employ? This approach has the consequence that mathematics becomes very rapidly divorced from physically significant conceptual content. Whilst our descriptions of physical ontology, our classifications of physical phenomena, remain where they have been for the last two thoussand years our abstract mathematics goes soaring up into the heights of complex Hilbert space and beyond. In classical physics there was an intimate connection between the mathematical structures employed (that pertaining to classical phase space and to real linear vector spaces) and the conceptual structures (atomic, plenum structures) which gave physical content to the mathematical assertions. In quantum theory this close tie was already broken apart under the Bohrian approach, the quantum mathematical formalism being regarded as no more than an algorithm for computing those results which can't, at the appropriately low level, be captured with one or other of the old conceptual schemes. And it has now become customary on the part of physicists and philosophers alike to contemptuously dismiss 'thinking in pictures' and to press on with formulation of even more abstract mathematical structures asserting, in this Bohrian tradition, that the physical contents lie only in their ground-level contact with observation. But it is not simply a matter of 'thinking in pictures' which makes some of us uneasy about this approach. Rather is it the fact that we are impressed by the intimate relations which have existed in the physics of the past between the mathematical structures employed and the conceptual structures which lie behind them and we are uneasy lest these insights be lost, indeed lest the goal of achieving a rationally comprehensible physics be overlooked altogether in the proliferation of abstract mathematical structures. The alternative which recommends itself is a careful study of the relations between conceptual structures empolyed in quantum theory and the mathematical structures of that theory in the hope that a unified conceptual-mathematical scheme will some day emerge.

On reflection, it would seem then that quantum theory has taught us nothing final concerning the conceptual basis of the classification of knowledge. What it has taught us, I think, is the depth to which conceptual structures do penetrate into our physical theorizing and the profound way in which those conceptual structures do provide a basis for the classification of physical knowledge.

Footnotes

1) It is interesting that this apparently does not hold true in all philosophies of the seemingly most obvious characteristic of the world, namely its existence. For Phenomenalistic philosophy holds that the existence of the world is certainly not independent of men's minds. But, wishing to do justice to our obvious epistemological circumstance and the development of science, Phenomenalistic philosophy nevertheless retains the tenet that the form of the world is independent of men's minds. I shall not attempt here to explore the sense of strain which exists between these two principles.
2) As shown, for example, by Reichenbach's investigation 31, which is one of the few that really count against a Kantian position in a way that the Kantian himself ought to recognize – most attacks on the clarity, or intelligibility etc., of the notion of a synthetic *à priori* are carried out from a much more general background (usually, in English philosophy, from within the Empiricist tradition) which is so fundamentally in conflict with the Kantian that they can at most be taken as proofs that within the framework of (often tacit) assumptions made no category of synthetic *à priori* truths is possible.
3) or almost so, since if the world imprints simple concepts on the *tabula rasa* of the infant human mind, as was usually held, then at least the possession of these simple concepts evidently bring with them the knowledge that they are instantiated in the world. But beyond that, the human mind created to itself more complex concepts from these crude beginnings at will and what was created certainly had no (more) intrinsic relation to reality.
4) The $C1_{1\rightarrow 3}$ in parentheses indicate the corresponding assertions in my 21.
5) The story can be found, for example, in the opening sections of $_3$, Essays 1 and 2 and 2, Essay 1 and especially Essay 4. See also the essays of Feyerabend 9, 10 , Heisenberg 19, Putnam 30 and those in Rozental 32. See also Jammer 24.
6) See Heisenberg 17–20 and his essay in Rosental 32; Bohm in his earlier book, 1, also took a similar position on the latter issue.
7) The reader will note that the spirit of this last sentence clashes with the three before it. Actually, what one finds when reading Heisenberg leaves one uncertain as to whether what is meant is a *semantic* assertion (it is semantically meaningless to ascribe a precise position to a system with a precise momentum etc.), a *physical* assertion (a system with a precise momentum has no precise position etc. – the position measurement creates a position, destroying momentum precision in the process), or an *epistemological* assertion (a momentum measurement destroys all knowledge of position). It is not my purpose, or obligation, to clarify this particular interpretation of quantum theory. Actually we shall see towards the end of the paper that, once quantum theory is understood from a certain point of view, this situation in which Heisenberg's approach finds itself is none too surprising.
8) Only a brief summary exposition of Bohr's thought will be given in what follows, but a more complete exposition can be found in my 21.
9) For a part of the reason for maintaining a careful cultural relativism – at least until we know more about ourselves, see my 23.
10) This is in fact the only portion of the two-slit experiment which need be related. It is usually added that in the case where one of the two slits is closed the distribution of individual quantum absorption events at screen B changes to resemble that produced by a stream of ordinary, classical particles impinging on the screen A. But in the first place this is only approximately true, at least according to quantum theory, since in the case where the slit diameter is comparable to the de Broglie wave length associated with the quantum individuals in question we should also expect to see a distribution pattern at the screen B analogous to the

single slit diffraction pattern of classical optics; and in the second place the wave mathematics used to provide an explanation of the double slit distribution pattern is equally adequate to provide an explanation of the single slit distribution pattern.

11) I do not mention here the argument about the ψ function being complex and 3N dimensional (for N particles) as I want to take the *general idea* of a real matter wave seriously as a potential explanation of the phenomena and prescind from the specific mathematics of contemporary quantum theory.

12) For pointing out to me the possibility of this rather elegant way of approaching the logico-algebraic structure of classical physics, and in particular for drawing my attention to the profoundity of the Stone representation theorem in establishing the correspondence between the logical structure and the structure of classical phase space, I am largely indebted to my colleague Professor J. Bub.

13) The literature pertaining to this approach to quantum theory is increasing very rapidly at present, a fairly complete list can be found in my 21. A philosophically informative, though not rigorous, treatment of the problem can be found in Putnam 30, while a very thorough treatment of the mathematics of the approach can be found in Varadarajan 35.

14) For a more extensive development and examination of Bohr's philosophy see my 21.

15) I enter this last qualification to avoid all of the irrelevant logical niceties of sentence construction which might be put forward as technical objections to the unqualified assertion.

16) I leave the reader here to consider the details, and the merits, of these two assumptions. I note, however, that I have nowhere claimed that these lattices are the same as the lattice of sub-spaces of a Hilbert space – indeed, there is strong reason to believe that they will not be the same. For that matter, I do not even guarantee that such lattices will prove to be orthocomplemented! I am at present working on these ideas and hope to be able to publish the results shortly. But see also my comments on Heelan's position in what follows.

17) It is slightly less strict because, as the illustrations of two sections back clearly show, there are some combinations of propositions drawn from sets of propositions which are not compatible (e.g. some combinations of momentum descriptive and position descriptive propositions) which the quantum logician does allow as forming well-formed sentences in his system. The Bohrian analysis on the other hand allows no such combinations at all as well-formed.

18) See Heelan 14, 15, 16. I wrote this paper before I was aware that Professor Heelan would attend this conference. Doubtless he will explain his own approach in more detail than I show here.

19) Again this is not the place to argument the matter, I leave the reader to consider how the complement of a physical context or situation might be defined and whether it could possibly be unique.

Bibliography

1. Bohm, D. *Quantum Theory.* Prentice-Hall, N.J., 1951.

2. Bohr, N. *Atomic Physics and Human Knowledge* Wiley, N.Y., 1958.

3. Bohr, N. *Atomic Theory and the Description of Nature.* Cambridge U.P., Cambridge, 1961.

4. Broglie, L. de. *Non-Linear Wave Mechanics.* Amsterdam: Elsevier, 1960.

5. Broglie, L. de. "Thermodynamique Relativiste et Mecanique Ondulatoire." *Annals Institute Henri Poincaré, 9* (1968), pp. 89–108.

6. Broglie, L. de. "The Reinterpretation of Wave Mechanics." *Foundation of Physics, 1* (1970), pp. 5–15.

7. Einstein, A. *Ideas and Opinions.* Crown Publishers Inc. N.Y., 1954.

8. Einstein, A., Podolsky, B., Rosen, N. "Can Quantum-Mechanical Description of Physical Reality be considered Complete? " *Physical Review, 47* (1935), pp. 777–780.

9. Feyerabend, P.K. "Problems of Microphysics". In R. G. Colodny, Edited, *Beyond the Edge of Certainty,* University of Pittsburgh Press, Pittsburgh, 1965.

10. Feyerabend, P.K. "On a Recent Critique of Complementarity". Philosophy of Science: Part I, *35* (1968), pp. 309–331; Part II, *36* (1969), pp. 82–105

11. Frassen, B. van "A Formal Approach to the Philosophy of Science" (Part II) *Pittsburgh Studies in Philosophy of Science,* forthcoming in Vol V 1971.

12. Frassen, B. van "Semantic Analysis of Quantum Logic" in *C. A. Hooker (ed) Contemporary Research in the Foundations of Quantum Theory* Dordrecht-Reidel, 1972.

13. Friedman, M., Glymour, C. "If Quanta had logic", Mimeographed, Princeton University, 1971.

14. Heelan, P. "Quantum and Classical Logic: Their Respective Roles". *Synthese,* 21 (1970), pp. 2–33.

15. Heelan, P. "Complementarity, Context D Dependence, and Quantum Logic", *Foundations of Physics, 1*(1970), pp. 95–110

16. Heelan, P. "The Logic of Framework Transpositions" *International Philosophical Quarterly,* Vol. 21 (1971), pp. 314–334.

17. Heisenberg, W. *The Physical Principles of the Quantum Theory.* Dover, N.Y., 1930.

18. Heisenberg, W. *Philosophical Problems of Nuclear Science.* Faber and Faber, London, 1952.

19. Heisenberg, W. "The Development of the Interpretation of Quantum Theory". *In Pauli, W.: Niels Bohr and the Development of Physics.* Pergamon Press, London, 1955.

20. Heisenberg, W. "The Representation of Nature in Con-Temporary Physics". *Daedalus, 87* (1958), pp. 95–108.

21. Hooker, C. A. "The Nature of Quantum Mechanical Reality: Einstein vs. Bohr", to appear in *Pittsburgh Studies and the Philosophy of Science,* Vol. 5, 1972.

22. Hooker, C. A. "Physics and Metaphysics: A Prologema for the Riddles of Quantum Theory", to appear in *C. A. Hooker (ed) Contemporary Research in the Foundations of Quantum Theory* Dordrecht, Reidel, 1972.

23. Hooker, C. A. *Empiricism and Conceptual Change.* Forthcoming.

24. Jammer, M. *The Conceptual Development of Quantum Mechanics.* McGraw-Hill, N.Y., 1966.

25. Jauch, J.; Piron, C. "On the Structure of Quantal Proposition Systems", *Helvetia Physica Acta,* Vol 42, (1969) pp. 842–848.

26. Kochen, S.; Specker, E. "The Problem of Hidden Variables in Quantum Mechanics", *Journal of Mathematics and Mechanics, 17* (1967), pp. 59–87.

27. London, F.; Bauer, E. *La Theorie De L'Observation en Mecanique Quantique.* Hermann et Cie, Paris, 1939.

28. Neumann, J. von *Mathematical Foundations of Quantum Mechanics,* translated by H. P. Robertson, Princeton Univ. Press, 1951.

29. Putnam, H. "A Philosopher Looks at Quantum Mechanics". In Colodny, R. G. Ed. *Beyond the Edge of Vertainty,* University of Pittsburh Press, Pittsburgh, 1965.

30. Putnam, H. "Is Logic Empirical? " In Cohen, R. S. Ed. *Proceedings for the Boston Colloquium in the Philosophy of Science 1966–1968* Humanities Press, N. Y., 1969.

31. Reichenbach, H. "The Theory of Relativity and A Priori Knowledge" University of California Press, Los Angeles, 1965.

32. Rozental, S. *Niels Bohr,* Interscience Press, N. Y., 1967.

33. Schlegel, R. *Statistical Explanation in Physics: The Copenhagen Interpretation.* *Synthese, 21* (1970), pp. 65–82.

34. Strawson, P. F. *Individuals* Methuen, London 1959.

35. Varadarajen, V. S. *The Geometry of Quantum Mechanics.* Van Nostrand, Princeton, 1968.

36. Wigner, E. P. "The Problem of Measurement". *American Journal of Physics, 31* (1963).

Suman Datta* and Jason E.L. Farradane (London)

A psychological basis for general classification

Knowledge is a phenomenon of the mind; the mind is our consciousness of events. The mind creates its own imagery or construct of the external world on the basis of sense data created by physical stimuli; such stimuli may be external to the body, or arise from the body (including the brain), or possibly from the mind itself if it is postulated that the mind is something different from the brain. This latter distinction is not however of importance for present considerations, since it is the characteristics of knowledge and our manipulation of it, and not its origin, with which we are concerned. The sense data (percepts), or combinations of them which we may call sense data patterns, are retained in the mind by what we call memory; despite the popular belief, often expressed, that nothing is ever completely forgotten, there is much evidence that most experience is rapidly forgotten unless reinforced by repetition or association mechanisms. The commonest form of association is that provided by language, whereby a given sound (or its written equivalent) is associated with a particular sense data pattern of different origin. This also aids our capacity for recognition of, or distinction between, different sense data patterns, which can then be named 'concepts'. The word 'concept' may be applied to a single sense data pattern, such as a physical entity or an abstract entity, or to various levels of abstraction which we call 'classes'; there is also a more complex, heterogeneous type. In many cases, however, a single sense datum or a connected group of sense data is recognized as a part only of many possible concepts, and is not treated as itself a concept unless a complex level of abstraction is added; such data I will call 'pre-concepts' and are exemplified by properties, for which we linguistically use adjectives, whereas the abstract complex becomes a concept denoted by a noun; consider, for instance, the words 'red' and 'redness'.

Reality cannot be apprehended directly, but is the construct of the mind from the sense data. Wherever such words as 'external' or 'physical' are used, they must equally be understood as being mental constructs. Cases of direct holistic sensing of 'reality' have been claimed, for instance by so-called extra-sensory perception, or by visions or revelation, but these cases are neither regularly reproducible nor proven; they can possibly be explained in other ways, and are anyway rare enough to be disregarded at present. Because we do not yet know enough about the means

---

*) Part of this work was carried out by Mrs. Datta for her Ph.D. thesis in Information Science.

by which the mind handles the higher levels of abstraction, we cannot provide such explanations, or even deal with belief and faith to an adequate extent. Such difficulties do not however imply that abstractions, even of a philosophic order, are different from other types of concept in principle; they are merely less and less derived from the influences of stimuli external to the brain.

The external world remains, however, a construct of the mind on which, by the verification of predictions, we can place increasing reliance. Such a construct does not, moreover, verge on solipsism, since the stimuli of apparent external origin are independent of our will. As Dingle wrote in 'The Scientific Adventure' (p.261): "When I look into the sky, I do not see what I choose; I see what I see." Conditions of self-deception or hallucination often occur, but in such cases the experience of other observers will contradict, or throw doubt upon, the first experiences, which are unverifiable and limited to the one individual. Other people form part of our construct of reality, and the information we receive from them must agree with our own experience for either to be acceptable. Ultimately, truth is definable only as a consensus of experience among a totality of minds. To organize knowledge we need such a definition of truth in order that the organization, or classification, shall be generally acceptable. We also need an understanding of the way in which more abstract forms of knowledge are produced, and here induction plays a large part. What is defined as 'logically true' is usually taken to mean a statement derived by accepted deductive rules; inductive thinking, which is innate in man, reaches conclusions by means of other rules, which have been less studied and understood, and are often disputed, but the success of such thinking, demonstrated by the test of predictions from the generalizations, provides a consensus which becomes equally acceptable as truth. Indeed, a deduction which is not just a tautology can be made only from some initial generalization, which must have been derived inductively in the first place. Different people, with different backgrounds of experience, will have different accumulations of knowledge, but these differences are largely dispersed by communication between individuals or groups, or by education or other social pressures. Residual differences in content of knowledge do not however imply differences in basic methods of thinking. Likewise, there is no reason to suppose that the basis of thinking is different in different subject fields.

Any classification cannot therefore be an ordering of entities, activities, or other elements of an external reality; it must be an organization of our mental concepts or constructs of external entities and elements. We have no evidence, and can have no evidence, that any groups or relationships exist in 'Nature'; at any point in a search for such evidence, we can still find only our mental constructs. We create 'classes' by the selection of certain characteristics or criteria which seem useful for providing some form of order, for its own sake or for desired applications, at a given time. The desire to *find* order is one of the basic human drives. Unless we can create some order we cannot overcome our fears of the unknown, we cannot free ourselves from the burden of having to re-interpret reproducible experiences afresh each time they are encountered, and we cannot utilize the essential human capacity to build upon experience. The order so established is a mental ordering, and the characteristics of the classification obtained will be a reflection of the properties and activities of our minds.

Classification is however more than the grouping of concepts by selected characteristics. It is concerned also with the relationships between concepts or groups of concepts; the perception of relations is another fundamental activity of our minds. These relations are derived from our recognition of a previously encountered concept, or concepts, from our realization that two concepts have some sense data (percepts) in common, or from the more developed capacity, at a relatively late stage in a child's development, to associate two concepts which have no percepts in common. These three stages of 'clarity of perception' combine with the three stages of associative power (awareness, temporary association, fixed association) to provide all the relationships we use in thought.

There is some evidence that concepts are of four basic types only. Guilford[1]) suggested that these types could be 'figural', 'symbolic', 'semantic', and 'behavioural'; the last was only a postulate. This omits 'properties'; to some extent properties are pre-concepts, but measurable properties become full concepts, and may have been overlooked because they are the measures in psychological investigations and not the direct objects of attention. One may compare the use of 'time' as a measure in physics, whereas we know very little of the nature of 'time'. We have preferred an empirically derived set of basic concept types, which nevertheless are not too dissimilar to Guilford's (when his definitions are considered), namely: entities, activities, 'abstracts', and properties. These basic types have been found sufficient to encompass all simple concepts; 'simple' here means concepts which can be defined as being of one category alone. Each category will comprise concepts of varying degrees of complexity and it may be suitable, though again empirically, to arrange such concepts in order of complexity, if such order can be discerned. It may perhaps also be seen as the 'part-whole' relationship. Guilford also showed that heterogeneous concepts could be produced when two or more different basic concepts types were somehow interconnected within one, more complex concept. The principles of such complex interconnections are not yet known, and need investigation; such heterogeneous concepts are numerous and easily identifiable when their definitions are considered. For example, a weed is a plant (entity) with an abstract connotation of undesirability, and an author is a person (entity) with the acticity of writing. The major disciplines, used as 'main classes' in earlier classifications, are in fact heterogeneous and are usually combinations of all four basic concept types; for example, 'chemistry' comprises chemicals (entitities), reactions (activities), theories (abstracts) and properties. Such heterogeneity accounts for the difficulties of achieving anything like uniform subdivision in earlier classifications.

On the basis of the above considerations, it is suggested that, since knowledge comprises the concepts and their interrelations, any ultimately enduring scheme of classification must be based upon the manner in which the mind forms and manipulates its concepts. The psychological evidence, obtained experimentally by various workers has already been discussed by Farradane in various papers[2]), and need not be repeated here. It yields the system of relations and the concept organization already adumbrated here.It is in good agreement with other psychological work on the learning process, in which the associative mechanisms, which are evidenced very early, are gradually combined with the mechanisms for achieving increasing clarity of perception, until the most developed form is the application of the cause-and-effect relation. The formulation of class concepts is an inductive process, and induction also becomes an essential part of human mental development. Deduction is a faculty which is acquired only somewhat late in a child's progress, and rounds off the normal 'scientific method' of natural learning by experimentation, with trial and error; induction and, finally, prediction for test, are also later skills. There has been no evidence to suggest that the mental mechanisms so developed during childhood will change in adult thinking. The stock of concepts acquired in the experience of each individual varies enormously, but the mechanisms by which they are handled will be the same in all normal individuals.

The degree of intelligence of the individual, either in the form of the degree to which he perceives the stimuli which produce sense data, or in the form of his ability to apply his mental mechanisms, will of course also effect the knowledge and *apparent* thinking processes of that individual. In certain subject fields, such as some of the arts, lack of exact knowledge means that the more developed processes are not often applied, and knowledge remains largely a matter of associations. In other fields, and even in some graced by the title of a 'science', terminology and definitions are still so inexact that classification is also hindered. These difficulties should not influence us to change the principles of classification to be applied, although they may long delay the provision of adequate classification in such subjects.

In addition to the psychologically-derived principles of classification there are also principles which have been derived from the study of classification as a subject, and which concern the representation of a classification. Classifications suggested by the Greeks (such as the 'tree of Porphyry') and still largely used in libraries have been of the 'hierarchic' or family-tree type in which main classes of the recognized disciplines, put in some postulated order, are successively broken down into smaller 'classes' and eventually, perhaps, single concepts. The mixture of principles of subdivision, by different types of relationship, was not often realized. Scientific classifications, on the other hand, especially in biology, have been almost purely generic in type, constructed by 'synthesis' from the particular to the general. With the development of information retrieval problems in detail, the principles of 'synthetic' classification have been more fully developed; the original principles have been reversed, the synthetic method is fully applied, and the major disciplines, which are the most complex concepts, are reached last. The synthetic method also enables other problems to be solved, and it can now be seen that five requirements should be satisfied for clear structuring. There should be only place only for each concept (term); there must be complete freedom of interpolation and extrapolation with the advance of knowledge; compound or complex subjects should be represented by synthesis of simpler concepts; the relations beteen concepts in compoun or complex subjects should be expressed explicitly; and the different lines of relation should be kept separate. These conditions can be fully satisfied by the method of classification new to be described.

The skeleton of the classification can be shown in diagrammatic form with the different lines of relation represented in the different directions.

Fig. 1: Concept Organization

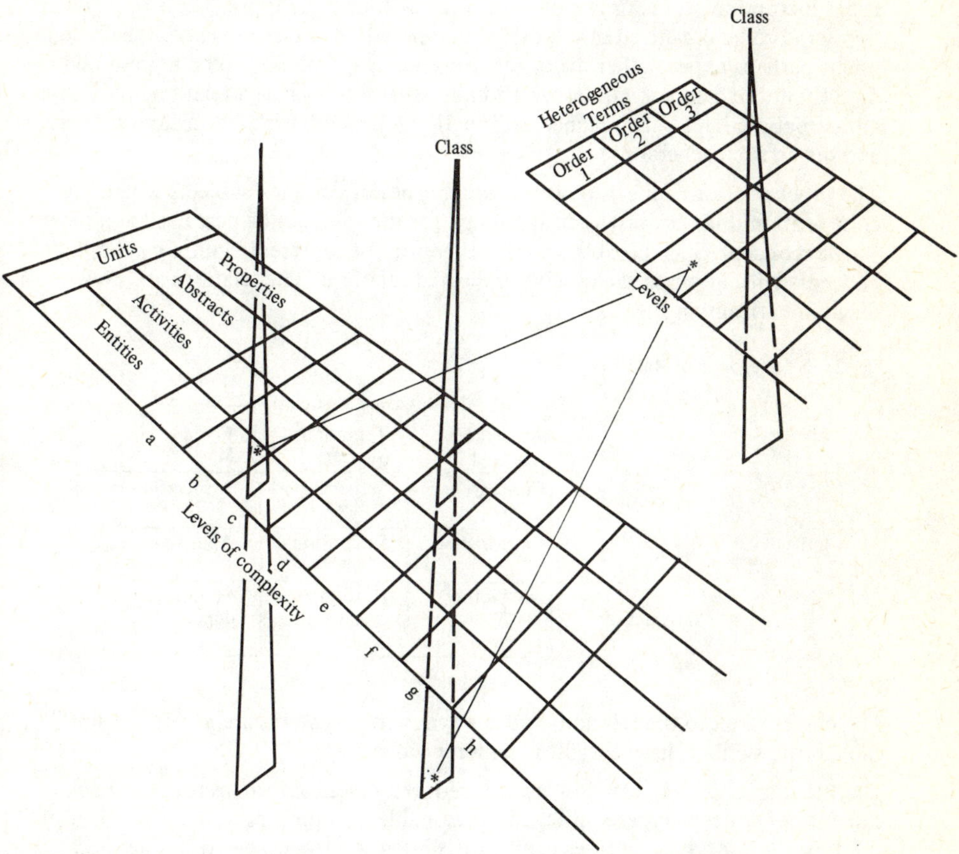

The first major division is into the four basic concept types. The mind makes use of the part-whole relation and the generic relation, both of which are derived by fixed associations of 'not-distinct' concepts (having sense data in common), to such a preponderating extent that it seems desirable to represent them clearly and fully in the diagram. The basic concepts can therefore be arranged in order of complexity, which is the part-whole relation, along one direction. At each level of complexity it will usually be possible to discern classes, representable in a third direction, each with its own levels of grouping. Finally, the heterogeneous concepts, formed from concepts of more than one type, must be represented in a second diagram, equivalent to further 'directions' or dimensions. Such heterogeneous concepts will exhibit their own initial groupings by the number of basic concept types incorporated, i.e. orders of two, three or four basic types. These orders will furthermore be organized into levels of complexity and classes. Theoretically, one might perhaps expect other major relations used by the mind to be represented directly in the diagram, but 'association' is usually too vague a relation, and cause-and-effect, perhaps better named as 'functional dependence', is too sophisticated and not in such general use.

The other relations appear, however, in all types of complex subjects where the combination does not proceed far enough for the result to be perceived as a new 'single' concept. Such complexes must therefore be represented only by the initial concepts with interposed explicitly indicated relations. The relations are shown diagrammatically in Fig. 2.

Fig. 2: Categories of Relation

Increasing Association →

		Awareness	Temporary Memory	Fixed Memory
Increasing Clarity of Perception ↓	Recognition	Concurrence /θ	Self-activity /*	Association /;
	Not-distinct	Equivalence /=	Dimensional /+	Appurtenance /(
	Distinct	Distinctness /)	Reaction /–	Functional dependence /:

The names given to the relations and the typewriter symbols are purely nominal, though the symbols have a slight mnemonic value.

The theoretical outline can now be reduced to a practical form as follows. Each category of concept types appears, on reasonable assumptions, to be divisible into sub-types, for instance, entities can be divisible into Physical entities, Chemical entities, Living entities, and Artifacts; Activities can similarly, but with less certainty, be divided in Physical activities, Living activities, Physical abstract activities, and Mental abstract activities. The Abstract concepts are conceived as divisible into Physical Abstract, Symbolic Abstract, Behavioural Abstract, and

Space and Time. Properties, on careful consideration, have been divided into Pre-concept Properties, and Measureable Properties, each further divisible into Physical, Behavioural and Abstract. The completely abstract forms of properties will of course be listed under Abstracts. These divisions provide different columns in the diagram. The levels of complexity have so far been difficult to identify: with entities the order looks easy to achieve; with activities and abstracts possible levels are still not clearly discernable, and further psychological investigations will be needed. Properties, curiously, exhibit four interesting possible levels of general apllication. By adding different types of coding symbols for each concept type, for each level, for each class, etc., an exact unique code can be attained for any given concept, and combinations of these will no doubt enable heterogeneous concepts to be uniquely coded. The following tentative sample schedules will indicate the detailed representations eventually to be produced.

Fig. 3: Entities

ENTITIES

A Physical	B Chemical	C Living	D Artefacts
a. Elementary particles. e.g. electron, meson	a. Elements. e.g. sodium	a. Self-reproducing compunds or complexes. e.g. virus	a. Separated natural materials. e.g. (cane) juice
b. Atomic. e.g. atom, isotope	b. Radicals, Ions. e.g. ammonium, sulphate	b. Cell components. e.g. mitochondria, chromosome	b. Derived single product. e.g. screw, paper, brick. sugar
c. Molecular. e.g. molecule	c. Chemical compounds. e.g. barium-hydroxide	c. Cells. e.g. cell	c. Assembly of deverived single products. e.g. book, machine, wall
d. Molecular assemblage. e.g. liquid. solid	d. Chemical complexes. e.g. resin, cellulose	d. Tissues, etc. e.g. skin, blood	d. Complexes of assembled, products. e.g. factory, railway, house
e. Ordered molecular assemblage. e.g. crystal, mineral	e. Chemical aggregates. e.g. bile, milk	e. Organs. e.g. lung, flower	
f. Complex physical aggregate. e.g. rock, mountain, river		f. Organ system. e.g. digestive system	
		g. Whole living entity. e.g. tree, cat, man	

Fig. 4: Activities

ACTIVITIES

E Physical	F Living	G Physical Abstract	H Mental Abstract
a. Simple. e.g. moving, vibrating, floating	a. Simple physiological. e.g. breathing	a. Simple. e.g. increasing	a. Simple. e.g. counting, numbering
b. Complex e.g. slipping, bolling cutting, grinding, purifying	b. Complex unit physiological. e.g. digesting, secreting	b. Complex. e.g. using, deteriorating	b. Combined. e.g. reasoning, calculating, evaluating
c. Series of complex activities. e.g. analyzing, manufacturing	c. Combined complex physiological activities. e.g. metabolizing, reproducing (biol.)	c. Combined complex. e.g. renting	c. Complex e.g. planning, governing
	d. Physiological with mental. e.g. walking, singing, remembering, thinking		

Fig. 5: Abstracts

ABSTRACTS

I Physical Abstract	J Symbolic Abstract	K Behavioural	L Space	M Time
e.g. rays energy, heat, light, sound, pressure	a. Unit level. e.g. amount, size, number, letter b. Combination of units, e.g. word, formula, graph, ratio c. Complex. e.g. sentence, structure, circle, cube d. Higher level. e.g. scheme, mechanism, procedure, reaction, composition	e.g. love, hate, pain, feelings, use	e.g. position (above, below, place, etc.) and geographical names	e.g. time before, after, and named time periods

Fig. 6: Properties

PROPERTIES

Levels	PRE–CONCEPTS PROPERTIES			MEASUREABLE PROPERTIES		
	N Physical	O Behavioural and Physiological	P Abstract	Q Physical	R Behavioural and Physiological	S Abstract
a. Simple	e.g. dense, solid viscous, elastic, conductive	e.g. blind, curious	e.g. red, circular	e.g. density turbidity viscosity, elasticity, conductivity	e.g. blindness	e.g. redness, circularity
b. Comparative	e.g. hard, dry, hot, homogeneous, fresh	e.g. fluent, intelligent, extrovert	e.g. heavy, large, old, beautiful, pure, intense	e.g. dryness, humidity, specific gravity, freshness	e.g. fluency, intelligence	e.g. purity, intensity, beauty, weight, brightness
c. Interactive	e.g. acid, alkaline, porous	e.g. viable, neutrotic	e.g. sweet, popular, compatible	e.g. acidity alkalinity, porosity	e.g. viability	e.g. sweetness
d. Interactive with change	e.g. toxic filterable, sterile, soluble	e.g. painful	e.g. stable, unstable	e.g. toxicity filterability, solubility, decolorizing power	e.g. painfulness	e.g. stability

The levels shown in the diagrams are at present only tentative. Further investigation is needed to establish an order more certainly in accord with human thinking. Furthermore, although it is easy to suggest levels for the entity concepts, it is more difficult for the activity concepts, and very difficult for the abstract concepts; some other principles may have to be sought.

In accordance with Fig. 1, one can conceive the existence of classes (with various groupings in each) at any level of any concept type. The representation implies the highest class term at the top of the triangle, and smaller groupings at stages down the triangle, e.g. one triangle might be Mammalia, broken down, as in zoological classification, to individual species or varieties at the bottom. (The drawing of the triangles to cut the plane of concept types has no significance and is only for convenience of drawing.) The mental mechanisms by which we produce heterogeneous terms have also not been elucidated, and their ordering is also as yet not possible. We can in fact only *suggest* heterogeneous terms of each 'order'. Thus terms formed from two basic concepts types may be: planet (earth-mass plus orbiting), dust (matter plus measurable physical property), author (man plus activity) etc. Terms formed from three basic types may be: diet, disease, fossil, etc. Terms formed from all four types will include most major disciplines, e.g. physics, chemistry, psychology. There will be different levels, e.g. star, galaxy, universe, and there will be classes such as classes of stars.

The relations shown in Fig. 2 are each a category of relational situations between concepts, there being perhaps theoretically four types of each, according as to whether the two concepts to be related are concrete and concrete, concrete and abstract, abstract and concrete, or abstract and abstract; one concept is always more the object of attention than the other, and hence the order matters. The part-whole and generic relations are both manifestations of the 'appurtenance' relation (the names of the relations are only for convenience). Thus writing A /(B means that 'B' is part of 'A', or is, say, a species of the genus 'A'. As noted, before, however, these two relations form so important a part of our thinking that they have been represented directly in the schedules. The 'fixed' relations are all very important, since they are those best remembered in our knowledge, but 'association' is too vague in most cases, and 'functional dependence' (or causation), being the result of our most developed thinking mechanisms, is relatively too rare to be given special representation in the schedules. The other categories of relation are common, but do not provide clear 'lines' of concepts suitable for the classification schedules. They seem, however, to be involved in the formation of the heterogeneous concepts.

The meanings of the relations have been elucidated partly theoretically, but are most easily demonstrated by practical examples:

Concurrence, A/θ B means A in the presence of B, and can be used for bibliographic form, e.g. Chemistry/θ Encyclopaedia.

Self-activity A/*B describes the intransitive verb situation, e.g. Bird/*Migrating, a process (as distinct from an operation) such as Sugar/*Crystallizing, or the dative case situation where there is an indirect object, e.g. feeding food to a rat is written as Rat/* Food/– Feeding. There are other possible meanings.

Association A/;B expresses mere association, e.g. Prison/; Disgrace, or abstract properties, e.g. Cathedral/; Beauty, or the agent of a process, e.g. Cutting/; Knife. It also can distinguish a past process from an actually occurring process, e.g. Clothes/; Washing means clothes which have been washed, in contradistinction to Clothes/– Washing, clothes that are being washed.

Equivalence A/= expresses some degree of equality, or the relation of the use of one thing as another, e.g. Molasses/= Fodder.

Dimensional A/+ B expresses position in space or time, e.g. Book/+ Drawer, or states, e.g. temperature, electric charge, or speed, e.g. Crystallizing/+ Rate. It also covers *variable* properties of amount or size.

Appurtenance /.C expresses the generic relation, the whole-part relation, intrinsic properties and intrinsic ingredients.

Distinctness A/)B is rarely required to express mere difference, but is also applicable to relations of substitutes and imitations, e.g. Diamond/) Glass.

Reaction A/– B covers the action of any thing or activity on another thing or activity, e.g. Metal/– Acid, Water/– Purifying, Drying/– Controlling. Such actions are operations, not processes.

Functional Dependence A/: B expresses A causing B, or B arising out of A, e.g. Earthquake/: Destruction, Author/: Book, Wheat/: Bread. An interesting use is to distinguish a non-separated part, e.g. Plant/(Seed, from the case when it has been separated, e.g. Plant/: Seed.

In all cases, careful semantic analysis of the meaning of a statement is necessary, since linguistic forms are very unreliable guides (compare the meanings of 'colour of a rose', 'sport of kings' and 'book of the film'). The above examples by no means cover all the possible applications of each relation, but experience with analysis in a wide range of subjects has shown that the relations are sufficient to cover all situations, and that with only short periods of practice good consistency of representation can be achieved. The differences in linguistic expression in different subject fields make it necessary to elucidate and standardize the representation of the relations involved in each case.

For more complex subjects, two-dimensional representation proves necessary, e.g. Water /(Density. Note that the common

```
          /–
          Heating
```

linguistic form: 'effect of heating on the density of water' is semantically misleading, as it is the water which is heated. There may also be 'circular' situations,

```
e.g.      Sugar/( Moisture /– Analzying
             /;                /;
          Angle of Slip /– Measuring
```

In the expression of more complex subjects, much larger and more complex diagrams are easily constructed to give exact expression of the meaning, or exact *classification.*

The last sentence emphasizes the fact that classification is not just an ordering of single concepts in groups; the representation of complex subjects (unless there is actually available *one* word meaning the whole complex, such as are the heterogeneous concepts) requires diagrammatic representation of interconnections by explicit relations. This form of general classification, it will be seen, can include *all* concepts, whereas many older classifications hardly contain abstract concepts or activity concepts or property concepts. These may perhaps be thought unnecessary, but it is at least desirable that a general classification scheme should be *able* to include all concepts. If less complexity is required, it should be possible to make appropriate selections from the theoretical whole.

It may be worth noting that in tests of the methods in the field of information retrieval, where complex subjects were indexed and the system tested for the retrieval of entries (records by recourse to the questioners, extremely good retrieval was achieved, with very little recovery of erroneous matter. Further work is in progress to discover, experimentally, several other aspects of human thinking, such as jumps (omission of intermediate concepts) made in the expression of ideas, and the way people interconnect statements, especially inductively. The results can be elucidated by means of the methods of semantic or relational analysis described above. The methods also offer scope for many other avenues of research bearing on the principles of general classification.

References

1. J.P. Guilford: Amer. Psychologist, 1959, **14**, 469–479; Ann. N.Y. Acad. Sci., 1960, **91**, 6–21.

2. J. Farradane: J. Docum., 1955, **11**, 187–201; Inf. Stor. Retr., 1963, **1**, 3–11; Inf. Stor. Retr., 1967, **3**, 297–314.

Phyllis A. Richmond (Mayfield Village, Ohio)

Review of.: Suman Datta and Jason E. L. Farradane. A psychological basis for general classification

The basic definitions in this paper are:

concept = a single sense data pattern, such as a physical entity, or an abstract entity, or to various levels of abstraction which we call classes.

pre-concept = a single sense datum or connected group of sense data recognized only as part of many possible concepts, and not treated as a concept in itself unless a complex level of abstraction is added. This pre-concept is exemplified in the idea of properties.

The difference between pre-concept and concept is the difference between the words "red" and "redness".

Reality is considered a construct of the mind, truth a consensus of experience among a totality of minds. "To organize knowledge we need such a definition of thruth in order that the organization, or classification, shall be generally acceptable. We also need an understanding of the way in which the more abstract forms of knowledge are produced and here induction plays a large part." Datta and Farradane emphasize that, though people have different cumulations of knowledge and different interpretations, these differences can be dispersed by communication. When you stop to consider how often in talking with friends and colleagues you say, "Do you mean this? " or "No, I meant that . . ." or "What you say applies in the special case of . . ." and so inter-communication clarifies and sharpens the semantic content of our thought. Classification set down in the brief, literally telegraphic form of tables has to be interpreted in terms of context or hierarchy and often misses some of the angles which turn up in the give and take of direct communication.

The Datta and Farradane hypothesis' is that classification is an organization of mental concepts or constructs applied to an external world. This becomes apparent when one reads the magnificent list of classes from " 'a certain Chinese encyclopedia' " given by Michel Foucault in the preface to his book, *The Order of Things:* " 'animals are divied into' (a) belonging to the Emperor (b) embalmed (c) tame (d) sucking pigs (e) sirens (f) fabulous (g) stray dogs (h) included in the present classification (i) frenzied (j) innumerable (k) drawn with a very fine camelhair brush (l) etcetera (m) having just broken the water pitcher (n) that from a long way off look like flies".''*

* Foucault, Michel. *The Order of Things; an archaeology of the Human Sciences* (New York, Pantheon Books, 1970). Translation of *Les Mots et les Choses.*

The human necessity of making order is shown when a hitherto blind person gains sight and at first sees only a blur of colors, or in the concentration it takes to hear tones of music in another culture, or to see the Eskimo's dozen kinds of snow. The idea of relations is stressed and interconnectedness, as well as finding a place in memory for new things by connecting them with old. Induction is important in the learning process for concept relating as well as for concept formation. The *mechanism* for handling concepts is fairly constant among people although the store of them varies greatly from person to person. The multidimensional character of concept organization is visualized in a chart of three dimensions. Parts of this have been reduced to tabular form in Figures 3, 4, 5 and 6. In Figure 3 "Entities" and Figure 4 "Activities" each of the four types shown are organized by what is essentially the integrative level idea. Each higher level contains and improves upon the one below it. In most cases, it could not be broken down again into exactly the same components. A cat killed and dissected is no longer a cat. Figures 5 and 6 cover ground only recently developed and the categorization of abstracts and compounds is interesting.

Difficulties of a linguistic nature are discussed in connection with relationship factors. Finally the authors come to the conclusion, which seems to be commonplace as one deals with classification structure, that two dimensions simply are not enough for clarity and diagrammatic representation of some sort is necessary. I have students who begin with arrowgraphs of words in thesauri to illustrate relationships, procede through faceted classifications, seeking a suitable formula for showing relationships, and end up with physical models made of wires, tags, plastic balls, candy drops and so forth, always trying to indicate visually what is going on in their minds in the formation of class concepts and their ties one to another.

If a general classification is to contain *all* concepts it will, as Datta and Farradane explain, have to include abstract, activity and property concepts, which are pretty much lacking now. Further research is planned on leaps, which, as Whewell put it, go "to the top of the stair" rather than step by step.** There are leaps in thinking as one talks, leaps in unintended directions while reading or talking or taking a bath (mental jumps unintended by the speaker or author, that is). This is a completely different approach than that of Fairthorne, who despairs of a landscape where all concepts are visible at once, to be caught and recorded in a grand classification scheme. It is possible that the emphasis on psychology of learning and thinking, the "how" of it, may contribute to the "aboutness" of discourse, but it also seems highly probable that this must occur in a temporal manner as described by Fairthorne. One must look both inside and outside the individual in *time* to begin to reproduce the classificatory process and how to map it.

** Whewell, William. *Novum Organon Renovatus, Being the Second Part of the Philosophy of the Inductive Sciences,* 3rd ed. (London, J. W. Parker, 1958) Bk. 2, p. 114.

Added comments on verbal exposition

1. Is the work on evaluation of his system (such as precision rates) available for public verification?
2. Is the experimental work in psychology acceptable to professional psychologists? Where is it published? The type of material covered suggests this should be in psychological journals, provided it meets or improves upon current standards for the design of an experiment. It would appear otherwise that it may add to the pseudo-science of "information science".

J. L. Jolley (Great Missenden, Bucks. U.K.)

The Holotheme
A Paper submitted to the First Ottawa Conference on the Conceptual Basis of the Classification of Knowledge

In offering this paper to the Conference I must record my gratitude to Robert Fairthorne, who has undertaken to present it in my absence. I have admired his work ever since I first encountered it, and have enjoyed each opportunity of discussing our common problems in the field of information handling ever since we first met. This, I recall, was about fifteen years ago, at an imposing, unwieldy and very, very boring conclave concerned with the standardisation of the eighty-column punched card. We kept our spirits up by passing ironic notes to each other. On that day he encouraged me greatly, as he has done from that time on. I have learnt much from him; and now I am indebted to him yet again.

Of course, the nurse who presents the baby to the apprehensive or eager relatives cannot possibly be held responsible for its constitution. She may even dislike the look on its face, the sound of its voice, the shape of its ears. I hope this is not true in the present case. But, however this may be, I am more than happy to leave my mental offspring in such professional hands.

J. L. Jolley

Summary

Human knowledge is here regarded as a set of mental models of an external reality, accompanied by other models known to be fiction and by yet others whose status in this respect is disputed.

A long-term experiment was carried out, in which ideas were put in constructive order: that is, an order in which the relation of part to whole dictates the position of an idea. During the first part of this experiment ideas corresponding to passive entities (objects or substances) were studied. When constructive chains of these were made, regularities appeared. These revealed that the elements of human knowledge fall into eight integrative levels, corresponding to eight repetitions of the regular pattern found in the chains. Each level was found to contain two ranks, each of eight formative grades, these being the steps by which entities of a lower level come together to form entities of a higher.

The experiment was continued with a view to finding what other categories of notion existed, and seven of these were found, making eight in all. All of these were found at every grade in every level. They were named 'semantic types'.

It is emphasised that this process can be repeated by anyone who wishes to do so, and it is expected that the same results will be achieved.

A binary code was developed in order to supply notation for the array of types, grades and levels, and the second stage of the inquiry began. This took the form of a search for a mathematical theory which would represent the findings of the first stage. Set and number theory were found to supply the desired pattern, and could themselves be arranged by means of a sequence of binary code numbers based on the properties of relations. These code numbers were isomorphic to those already developed to represent the grades and levels, and consequently were the numbers which would have been allotted to the notions of mathematics on the basis of the pattern of grades and levels alone.

This support continued consistently upwards from semigroups through groups, rings and fields to vector spaces and algebras. It is believed to justify the binary code as a suitable model für the classification of the elements of knowledge at the present time.

The whole pattern of knowledge, and this representation of it in particular, are here named 'the holotheme' (the whole set). Names are supplied for the semantic types, the integrative levels and many of the formative grades including all of these as they apply to objects and substances. A typical series of questions is given, directed to finding the place appropriate to any idea by determining its relation code sequence.

Many uses are suggested for the pattern, including applications to co-ordinate indexing, to the development of academic curricula, and to the classification of mathematics. It is also suggested that the pattern may be identical with the so-called 'deep structure of language' which is currently the subject of investigation by specialists in linguistic theory.

The Paper

Standpoint

I propose to uphold the view that the elements of human knowledge can be placed in an order which is infinitely hospitable, easily taught, independent of personal opinion, and based on a simple repetitive pattern which is already widely known. To do so, I must display such an arrangement and discuss its properties. If I am in error, no audience is better qualified than this to reveal my mistakes. If I am right, there is no audience I would more gladly persuade to explore the pattern further, or which could make a greater contribution to its development.

To start, I should explain the standpoint from which I consider my subject. I regard the ideas or notions which make up human knowledge as falling into two classes: those which are mental models of regular patterns we find in the world about us, and those which are mental recastings of these models. The rearrangements are fiction, the originals are fact, and on the whole we can tell the difference. To the models we give names, and from the names we develop languages. Our command over the real world increases as our models of this become more accurate; our languages grow in efficiency as we increase the truth with which they convey the models we have made. The most useful arrangement of human knowledge will be obtained by forming the most exact model of reality. This will not exclude fiction, but will give it special treatment, and (so far as agreement may be reached) it will also provide a distinguishing mark for notions whose truth to reality is still a matter of argument.

Such a most-exact model will bear directly on linguistics, since it will be the model which speech must represent and may well turn out to be the so-called deep structure of language under another name. It will also have continuing relevance to other disciplines, much as the periodic table underlies the study of chemistry. No doubt its gapa will always be more important than its areas of completeness, for they will correspond to those places where research is still to be done. It will have especial value for those whose work is to catalogue and to classify what we currently know. The documents they index and file will deal with topics which have a place in it, and the retrieval systems they create will be the more efficient for being more closely moulded to it.

I suspect that a most-exact model of this type will also display great economy and coherence of pattern, so that it may be of considerable use in the field of education. It is likely to set the major subjects of the school and college curriculum in an obvious, understandable, sensible relationship to each other, so that the bewildered scholar may the more readily see where his studies fit in the vast field of knowledge offered to him.

So much for my standpoint, except for this: exploration and adjustment of such a pattern may never be completed, but I think we may now know enough to get the outline right. It is to this matter that this paper is directed.

A history of the experiment

It seems clear that the development of a universal ordering of ideas must be based on a study of the notions we believe correspond to the real world about us. This must supply rules whereby the notions may be placed in positions which they cannot help but hold. So far as is possible, personal opinion must be ruled out as a reason for placement: the only help must be that which comes from a careful inquiry into the structure of the ideas themselves.

My concern with this problem started in 1959. It was a spare-time affair, much interrupted, and it comprised three stages. During the first, which lasted for several years, I examined ideas taken from large numbers of classifications and word lists which I encountered in my daily work as an indexing consultant. My aim was to find what major varieties of idea could be distinguished, starting with as few opinions on the matter as possible, and consequently avoiding all other writings on the subject. To embark on a course of research without making a study of existing information may seem unusual at least, and the more censorious may consider it wickedly spendthrift of precious time. In principle I agree. Yet in this case it was necessary, for a secondary goal of the study was to find out what would happen if exactly this course were taken. Most of the arrangements of ideas I met in daily use were no more relevant to my needs than the doctrine of signatures is to the classification of plants. I had to start as innocently as I could, and to compare results with those of other workers when results were to hand.

The outcome of these years' work was a general pattern which appeared to be complete, repetitive and internally consistent; but it was based entirely on observation and cried out for a formal theory, for mathematical underpinning. I looked for this, and the second stage of my inquiry then began, quite suddenly, in 1965, when I realised that set theory provided the pattern I sought. I fastened on this and worked upon it, referring back and forth between the textbooks and my observations, trying to develop a simple, understandable terminology with which to talk of what I was about. As my confidence grew, I began at last to read other people's works on the matter, and to relate them to my own experience.

By 1967 I was satisfied with the formalism, and the present stage began. This is concerned with consolidation, drawing conclusions, finding applications and developing rules of the sort which may one day grace a textbook on practical holothemics, by which is meant the study of the whole set of ideas we may form. Typical rules are concerned with finding the position occupied by an idea within the general pattern. Such rules can be reduced to a series of choices between alternatives which come in pairs and are mutually exclusive. Consequently they can generate series of binary digits, and these are well suited to embodiment in the memory banks of a computer. This property may have a special appeal for those who are concerned with handling information by means of electronics.

Rules of this sort place ideas according to four main groups of properties. These determine a notion's integrative level, formative grade, semantic type and speciality mark. They must now be explaines – first in general, and then (except for the speciality mark) in detail.

The general pattern of ideas

An integrative level is a consecutive sequence of sixteen degrees of complexity of structure. There are eight such levels, all being well known as the domains of major sciences. To place an idea in its level, three choices are needed, but in practice these appear in the disguise of a single choice between eight alternatives.

The sixteen degrees within each integrative level are the formative grades. These are the steps by which the ideas of one level build up into those of the next. Four choices are needed in order to decide on a notion's formative grade, and these are based, as I hope to show, on four familiar properties of relations. This is not surprising, for if ideas are to be given unique positions in some array by means which are not arbitrary, they must be placed according to their nature, and their nature arises from their internal structure. The relations between the parts of this structure are therefore on central importance.

The formative grades fall into two groups of eight, so that each integrative level contains two ranks, one composed of the simpler of the two groups and one of the more complicated. Each grade contains examples of eight semantic types. These are major varieties of idea, whose existence has called for the development of the different parts of speech – nouns, verbs, adjectives, and others. An idea's type, like its level, is often obvious, but in difficult cases the routine of asking three separate questions may be called for.

The remaining matter with which this type of placement-rule is concerned is the allocation of a speciality mark. This is the only choice which may not always, in theory at least, be decidable by examining a definition of the idea to be placed. It is to do with the distinction between plain ordinary certainties about the structur of our world, which are accepted quite generally, and ideas which are held by at least one group of sane people to correspond to no reality. These are special ideas, which may be known to be fantasy, or may be hypotheses awaiting proof, or may be unprovable though many people assert them to be true. This sort of speciality is nothing to do with abstraction. Numbers are abstract, but they are generally accepted none the less, and from this point of view they are counted as real. Religious tenets, by contrast, are special. Believers may be certain that they are real, and this indeed may be so; but there are also unbelievers. The proper way to accommodate both groups is to give special status to their ideas.

Integrative level, formative grade, semantic type and speciality mark: these are the framework round which a structure of human knowledge may be built. Having dealt with them in general, I must now turn to a more particular treatment.

Chains of notions

The essential process in the investigation of integrative levels is that of making chains of ideas such that each earlier idea is a constructive part of each later idea, just as bricks may be parts of houses. As this example may suggest, this is best-done by considering material bodies, objects, entities. These may here be called 'things'.

The following is a chain of things:

a proton
an atomic nucleus
a carbon atom
a methyl group
a phospholipid molecule
a lipoprotein membrane
a mitochondrion
an animal cell
a muscle fibre
a muscle
a heart
a cardiovascular system
a human being
a degreasing section
a painting department
a production line
a factory
a neighbourhood
a town
a county
a nation
an alliance
the United Nations

It is interesting to form other chains and to compare them, setting similar notions together. For example, an electron might occupy the same line as a proton, and a plant cell might appear on the same line as an animal cell. When this is done, some chains will be found to have gaps in them; others will seem to have a superabundance of ideas. Here is a comparison of chains:

a carbon atom	a magnesium atom
a methyl group	⋮
a phospholipid molecule	a chlorophyll molecule
a lipoprotein membrane	⋮
a mitochon drion	a chloroplast
⋮	a chloroplast layer
an animal cell	a plant cell

It is clear that each chain calls for an expansion of the other.

When enough chains of this sort have been made, regularities appear. At intervals, the sorts of idea which form the basis of important disciplines occur. Sub-atomic particles; atoms; molecules; organelles; cells; organs of the body; entire plants or animals; organised human groups; nations. This list is not complete, but it offers a starting point. At an early stage in my study, I named ideas of this sort 'units', and I wondered whether there was a general property by which they could always be recognised.

The answer, it seemed, was provided by the concept of homeostasis, as first applied in the field of biology. All the units possessed this property to a high degree: given favourable circumstances, they remained in balance with their surroundings. If not too severly damaged, they mended themselves. Even atoms, lacking electrons, trapped them if they could; and if a trade union lost its general secretary it appointed another. Units demanded completeness, and completeness appeared to be judged by degree of ability to survive.

However, from this point of view some units fared batter than others. Molecules were more independent than atoms: they did not rush into partnership as readily as did atoms, although they too could interact. Cells were more independent than the organelles they contained, which mostly relied on the cells for survival. Leaves and flowers might die but the plant could still continue. An industrial company might close down a department, a state might extinguish old divisions, but the company and the state could both live on.

I remember my surprise when I saw that the more and the less dependent units alternated along the entire length of any chain within which no units are omitted. Try it and see. This is the sort of regularity which encourages anyone looking for a pattern, because it appears without being forced, and justifies the procedure of simply gazing at one's problem inquiringly.

My own first list of units ran as follows:

Photons
PARTICLES
atoms
MOLECULES
organelles
CELLS
organs
PLANTS AND ANIMALS
departments
ORGANISATIONS
local governments
NATIONS

I noticed with amusement that it was notably organic. It did not deal with artefacts. Also, I observed a second hidden pattern in the chain. Every fourth unit seemed in some way to be eminent, to be a culmination of the ideas which went before. Physics with chemistry aspired to molecules; the life sciences rose to the study of living beings; sociology's highest unit was the nation, although of course the pattern went on to alliances and empires. As a special example, I thought of the range of interest of my family physician. The study of medicine covers the range from the organelle to the human being and a little beyond. Of course, there is the province of social medicine to be remembered, and the province of pharmaceuticals at the simpler end of the chain, but the concentration is upon the four types of unit in the middle of my list.

The pattern which was emerging looked binary, and it was hard to resist the thought that a list containing twelve units was incomplete. There ought to be sixteen. The positions held by the other four might, of course, come after that held by nations: but this view had little to recommend it compared with the alternative, that they came before photons and were, in fact, to do with mathematics. I remembered reading speculations that physicists might one day have to make matter out of space, and I guessed that two units concerned with geometries might precede the photons and particles of sub-atomic physics, while two units of even greater simplicity occupied the two positions at the very start of the chain. I therefore added the following to my list:

members of sets
FULL SETS
points
CLOSED GEOMETRIC FIGURES

This was pure hypothesis, but it gave me a very persuasive arrangement. Gingerly, I equated each pair of units with an integrative level, and noted that as a result I could say that space appeared after one level had been traversed, mass with time and energy appeared after two levels, and life appeared after four. The pattern was certainly binary: here were the powers of the number two.

Formative grades

In the early stages of the study I paid scant attention to mathematics, despite my conclusion that set theories and geometries occupied the two lowest levels. I was more concerned to discover what lay between one unit and the next, and this involved examining ideas from levels above the mathematical, which seemed a little too abstract for my purpose. I began by considering those ideas which came immediately after units in my chains. There was an immediate result: my collection of these contained a marked preponderance of what I called 'assemblies': groups of units in which no precedence could be found. The two electrons in the innermost orbit of an atom afford an example; so do the members of a leaderless group of people involved in a general discussion; so do a pair of eyes, the atoms in a hydrogen molecule, the sides of a hexagon, a pair of cuff-links, a swarm of midges, the voters in an election when each has one vote only.

The notions which followed assemblies also had something in common. Sequence, precedence, order appeared in them. Examples are afforded by a production line in a factory, the digestive system of an animal, the rainwater system of a house. I found the name 'system' was so frequently used in connection with them that for a time I adopted it; but in due course it began to seem better to emphasise the seriality of the collections of units involved, so I began to use the word 'series' for notions of this type.

When units, assemblies and series had been removed from my chains of ideas very little remained. That little, however, was of considerable interest. It consisted of reciprocal or interactive series (I thought of them as assemblies of series), and it introduced an effect of feedback. Examples are: the afferent and efferent nervous systems taken together (and generally, with the brain and the spinal cord, called the central nervous system); the transmission, ignition, braking and other systems, taken together, of a motor vehicle without its bodywork (I have often seen these driven past my window on their way from the engineer to the coachbuilder); and the two series of unit parts which between them make two-way conversation possible by telephone. At first, after some hesitation, I called these reciprocal series 'combines'; but when the word 'system' became free as a result of my adopting the title 'series' for the unidirectional type of notion I was able to use it for this more complex variety. It chimed in well with current practice: general systems theory is concerned with interactive series. Indeed, notions of this sort are central to the study of cybernetics. It is hard to overrate their importance.

When sufficient series are brought together in this way, the resulting complex system is sufficiently interactive to produce a new unit, ensuring its homeostasis by the co-operation of its trains of elements.

This method of examining ideas reveals four steps, including a lower unit, assembly, series and system, before a higher unit is reached, and since I had arranged my integrative levels so that each contained two units I concluded that each level contained eight steps. I found it helpful to think of these as formative stages, stages in the formation of higher things from lower. I also found it useful to distinguish between the upper and the lower set of four stages in each level by calling a unit of the lower rank a subunit, an assembly of the lower rank a subassembly, and so on. With this convention, an atom became a molecular subunit, and the alimentary canal of a human being became a biomorphic sub-series.

The study of the formation of higher ideas from lower can go one step further. To see how this comes about, consider the distinction between an atom of copper and copper as a substance. One is an object, the other is a large collection of such things, an idea without bounds, copper atoms ad infinitum. The same situation is found at other levels: electricity is a boundless collection of electrons, space is a boundless collection of points, sulphuric acid is a boundless collection of molecules of that substance, yeast a boundless collection of yeast plants. Further, not all substances are simple in the sense of being made of units of one sort only. Others are mixtures: air is an example. Yet others have an even more complex internal structure – consider plywood. In the case of plywood the substance is bounded in one dimension but no specific bounds are set in the other two.

There is a significant comparison to be drawn between units and assemblies, on the one side, and simples and mixtures, on the other. Units and assemblies are concerned with objects; simples and mixtures are concerned with substances.

There is no necessary order about the internal structure of simples or mixtures; but if laminar substances are considered then an order can be found: in the case of three-ply plywood this is clearly first ply – middle ply – last ply. Suppose laminates are equated to series, accordingly: can we then find something which may be equated to systems? Laminates are bounded in one dimension: perhaps materials which are bounded in two dimensions might fill this higher role. Examples are afforded by hosepiping, rope, cinematograph film. These are of indeterminate length, but breadth and thickness are assumed. After trying various names for them, I ended by calling them 'runs'.

To take this to a conclusion: the next step may be to consider materials bounded in three dimensions. These, however, appeal to us directly as objects. They are clearly units, as indeed they should be, since units follow systems.

This line of argument led me to divide each formative stage into two, one part serving to accomodate objects and the other to accomodate materials. These parts I named 'grades', and it is from this division that the sixteen grades of each integrative level are derived.

Artefacts

Consideration of materials – wire, rubber and the like – led me to become concerned with the placing of artefacts in my chain of integrative levels. Books, cups and saucers, chairs, pianos, roofing tiles, bottles of weedkiller, the Venus de Milo, transistors, jet engines, football boots, electric shavers and battleships all required a home. It was a land of pitfalls. I knew, for example, that an idea was not more advanced than another idea simply because it physically contained that other idea: tortoises are contained in shells, but they are more advanced than the shells which contain them. Again, I knew that physical size was nothing to go by: the sun, for all its immensity, is not as advanced as a boll weevil. I suspected that a hinge, incorporating the facility of movement, might be higher than the door and door frame which it served to connect, however complex these might be, so long as they were rigid in themselves.

In the end I set up the hypothesis that the two levels between that of molecules and that of communities were occupied, on the mechanic side, by two pairs of units (with their appropriate assemblies, series and systems) as follows:

single-piece parts
MOVING PARTS
engines and other organs of machines
MACHINES

At the level of communities the two branches of the chain came together: a farm, for instance, consists not only of people, other animals and plants, but also of buildings, tractors and harvesting machinery.

After experimenting with more chains of ideas I concluded that this was satisfactory, though it seemed that unitary completeness could not be judged in the light of homeostasis on the mechanic side. Instead, it was something to do with fitness for purpose. None the less, I believed I could recognise a unit artefact when I saw one. My lawnmower ceased to be one when its handle broke away.

Names and notation for the levels

After this course of inquiry it became possible to identify integrative levels with sufficient precision to enable them to be named. They could also be numbered, and for convenience I did this in both decimal and binary. The result, after I had tried out various alternative titles, was as follows:

level 0 :	set-theoretic: binary code sequence 000
level 1 :	spatial: binary code sequence 001
level 2 :	subatomic: binary code sequence 010
level 3 :	molecular: binary code sequence 011
level 4 :	cytomechanic: binary code sequence 100
level 5 :	biomorphic: binary code sequence 101
level 6 :	communal: binary code sequence 110
level 7 :	national: binary code sequence 111

The decimal and binary numbers offered a basis for a more developed notation. In the case of the binary numbers I found it convenient to place a point after the notation for the level, and to follow this with a zero (for the lower rank of the level) or a unit (for the upper). This was followed in its turn by two more binary numerals - 00 for a unit, 01 for an assembly, 10 for a series and 11 for a system. Finally, in the fourth position after the point, the numeral 0 indicated that an idea was an object whilst the numeral 1 showed that it was a substance.

Thus an assembly of molecules, such as the double helix of deoxyribonucleic acid, bore the code sequence 011.1010. Atoms appeared at position 011.0000, followed by the simple substances or pure elements they formed, at 011.0001.

This completes the detailed treatment of integrative levels and formative grades. Semantic types remain to be examined: to these I shall now turn.

Semantic types

The study of semantic types is not dissimilar from that of integrative levels: major varieties of notion are identified and set aside from a collection of notions, variety by variety, until none remain. To carry out this part of my inquiry I took words at random from a dictionary, laid aside all those I recognised as the names of things, and looked carefully at the remainder. Two sorts of idea were easy to find in this residue. There were words which stood for qualities (blue, triangular, cheap, northern) and words which stood for phenomena (rain, wind, the apocalypse, a conference, a surprise). I tried to relate these to the ideas I knew to be things, and concluded that things were passive whilst phenomena were active, and that things were entitive whilst qualities were attributive. I felt that I could fairly be accused of achieving dim glimpses of the glaringly obvious; and yet at the same time the two dichotomies, passive and active, entitive and attributive, put me well on the road to defining three varieties of idea which I had never seen singled out in this way before. Things were passive and entitive; qualities were passive and attributive; phenomena were active and entitive.

Put like this, the argument led me to think there must be a form of idea which was active and attributive. It seemed that, if passive attributive ideas were the properties of things, then active attributive ones might be the properties of phenomena – something to do with the manner in which the phenomena took place. This led me to remove many adverbs from my list of randomly chosen words (large numbers of nouns and adjectives had already been set aside), and it also caused me to look again at many adjectives I had thought were properties of things. For example, in the phrase 'that factory is productive', the adjective 'productive' is not strictly descriptive of the factory, but of the phenomenon of using the factory, of manufacturing. Thus I learnt that the part a word may play in speech is no sure guide to the type of notion it represents. It is only a general indication of this, though often a good one.

At this stage, almost all the words remaining in my list were verbs. It took me a long time to realise what I am sure others could have seen in a flash, namely that verbs represent a sort of idea which connects other ideas, that they are to do with relations. My readings in logic had made me familiar with the distinction between a term (adjectival or substantive) and a relation. Here I had a third dichotomy.

Thus at last I found myself concerned with the study of relations. Relations give structure to terms. Those which are passive and entitive inform things: examples are being-contained-in, being-bent-round, being-attached-to. Passive attributive relations give form to qualities, active entitive relations are embodied in phenomena. Active attributive relations are to be found in active attributive terms, the properties of phenomena, which for lack of a better name may be entitled 'modes'.

When I had set aside all the words in my collection which appeared to refer to one or other of the four types of relation or of the four types of term, the only remaining vocables were irrelevant to my task. They included two interjections, the indefinite article, and a word of uncertain meaning culled from Shakespeare, namely honorificabilitudinitatibus. It seemed, therefore, that my set of semantic types was complete. Those which were terms included things, qualities, phenomena and modes; names were still needed for those which were relations. I did not find much help in the literature on the subject. The most widely agreed word in English was 'operations', a name which was applied to any relation of the active sort. In many textbooks on mathematics, passive relations were known simply as 'relations', the passivity being implied. For these I adopted the title 'conditions'. I then after several trials with words whose other meanings often led me astray, concluded that entitive conditions could well be called 'structures' whilst those of the attributive sort could be named 'states'. For entitive operations I used the word 'actions' and for attributive operations the title 'activities'.

Since binary code sequences can be allotted to integrative levels and formative grades, it is perhaps natural to consider allotting them to semantic types also. The three dichotomies provide a basis for this, the only problem being that of deciding upon an order of precedence for them. On the ground that all ideas embody relations, which must therefore come first so as to be available, I symbolised relations by means of a zero in the right-hand position of the code sequence, whilst terms were represented by unity. Similar arguments gave attributes precedence over entities, and passive ideas precedence over those which are active.

Names and notation for the types

The resulting set of semantic types, with their names and code sequences, was as follows:

passive attributive relations (states): 000
passive attributive terms (qualities): 001
passive entitive relations (structures): 010
passive entitive terms (things): 011
active attributive relations (actions): 100
active attributive terms (modes): 101
active entitive relations (activities): 110
active entitive terms (phenomena): 111

In passing, I wondered whether these code sequences were amenable to addition. The rule 0 + 0 = 0, 1 + 0 = 0 + 1 = 1 + 1 = 1 seemed appropriate. For example, a phenomenon could be thought of as a thing which embodied an active entitive relation, and the rule duly gave 011 + 110 = 111.

Full code sequences

The code sequences for semantic types may be adjoined to those for integrative levels and formative grades. For example, the code sequence for a nation is 111.1000.011, which may be interpreted as a national (111) unitary (1000) thing (011). Finally, a speciality mark may then be added after yet another point: zero for an idea which is generally accepted to correspond to reality and unity for ideas of special status, fictional or questioned by at least one section of responsible opinion. Canada then achieves code sequence 111.1000.011.0; Ruritania gains sequence 111.1000.011.1.

These code sequences may be shortened if decimal notation is used. The sequence 111.1000.011.0 then becomes 7.08.3.0, which may be further reduced to 70830 by omitting the points. Such a sequence retains the significance of its elements: for example, 7 stands for the national level whilst 08 stands for a unit and 3 for a thing. If significance may be foregone, except for that which attaches to the code sequences taken as wholes, further shortening is possible. There are 2048 different sequences of eleven binary digits, and there are more than this number of ways of choosing three letters from an alphabet of twenty-six letters, so a simple trigraph code may be used. For my part, however, I think that this notation, if adopted, might prove to have gone too far along the road towards brevity at the expense of significance.

Relation of types to levels

The identification of eight different semantic types leads to the question, is the pattern of integrative levels and formative grades the same for all of them? If type codes are added to grade and level codes as I have just described, such equivalence is assumed to be the case. So far as I can see, this view is justified by the facts. It is possible to make chains of qualities, phenomena, modes, structures and the like, such that the simpler precede the more advanced, and the series of grades and levels duly makes an appearance. The distinction between objects and substances has its parallel in the realm of qualities in the form of a distinction between what may be called positionals and what may be calles extensives. Positionals cannot be divided; extensives can. There is sense in talking of half an inch, but not in talking of half the North pole. In a similar way, phenomena may be divided into events and processes, modes into datals and duratives. An event occurs at a given time; a process takes place over a given period, which may of course be unbounded in our thought. This corresponds, on the active side, to an object being found in a given place and a substance occupying a given space, which may be unbounded in our mental image. A practical instance is to be seen in the contrast between the process we call fire and the event we call the oxidation of an atom. This example occurs at the molecular level; to illustrate the matter at a higher place in the pattern the distinction between the communal phenomenon of trade and the communal event of making an exchange may be taken. Other instances are traffic and the passage of a single vehicle, war and the firing of a single gun.

Rather than treat each semantic type in detail, I propose to offer a few further illustrations and then to proceed to the mathematical underpinning. Each semantic type has its equivalents to unit, assembly, series and system. In the realm of modes, for instance, we are familiar with such measurables as miles per hour and centimetres per second. These are complex ideas made out of simpler notions. They correspond to series: precedence is important in them. To see this, note that miles per hour is not the same as hours per mile.

Specialists in work study examine trains of events leading to desirable ends and develop methods of reaching the same ends with less effort. Here we encounter typical series of phenomena in the upper biomorphic rank. Another example can be found when a chemical works goes on stream and substances move from vessel to vessel whilst undergoing a sequence of reactions. I should remark that this occurs at molecular level even though the factory in which it takes place is at the level of communities. If substances (for example, sulphuric acid) are subsequently recovered for recycling then the whole cycle of phenomena achieves the status of a system. The Krebs cycle in a living cell is another case of a molecular system within an object of a higher level.

Relations display patterns akin to those of the terms in which they are embodied. The command structure of an army, for example, is a pattern of relations, and so are large numbers of political theories. At the spatial level and below, complex arrangements of relations are found in mathematical formulae, with place-holders to show where the terms (usually numbers) may be inserted. These group the terms into assemblies, series and systems of many sorts. Simple examples are the formula for an arithmetic mean, the pattern made by the elements of a rectangular matrix, and the hypergeometric series.

Generics and collectives

While I studied semantic types I found many taxonyms, words naming notions which group other notions into classes. I distinguished two varieties of these, the generics and the collectives. To give instances immediately: in respect of a poodle, the notion of a dog is generic and that of a pet is collective. Iodine is generically a halogen but may be collected with other substances as an antiseptic or as a staining agent.

Although generics and collectives of all semantic types appear possible, most of those for which we have names are concerned with objects or substances: that is, with things. Generics are always closely concerned with the construction of the notions they bring together. Iodine is a halogen because of the number of electrons in the outer orbit of each of its atoms. Other elements with the same structure in this respect are also halogens. Iodine is a thing, and so is an electron, and indeed it seems that generic ideas always group ideas of the same semantic type as themselves. The construction of a thing, after all, makes use of other things, and that of a quality makes use of other qualities.

By contrast, a collective idea groups notions which are of a different semantic type from itself. An example is the concept of a weapon. Weapons are things grouped according to a phenomenon, namely their use. Antiques, antiseptics, ironmongery, roofing materials, stock-in-trade, paint, professors and public relations officers: these, too, are collectives. We live in a world of them.

Generics are easy to place in the pattern of integrative levels. They occupy the position held by the notions they bring together. Collectives, however, tell another story. A rock may be used as a weapon but the concept of a weapon is not fully developed at the molecular level where the rock is found. The proper home for a collective is the lowest grade at which its definition is complete. For example, the notion of a weapon may need an idea from the upper biomorphic rank as one of its constituents. If this is the most advanced idea required, then this is the concept which determines the place which the notion of a weapon must hold.

The holotheme: results of the study of words and meanings

This completes the brief account of semantic types. It may now be useful to summarise the view I had formed of my subject by the time I had finished this part of my inquiry. Human knowledge, I concluded, is concerned with notions which fall into one or other of eight clearly defined integrative levels each of sixteen formative grades each of eight semantic types each of which may be generic or collective to a greater or less degree. Further, some of these ideas correspond to reality, some may do so but are challenged in this respect by respectable bodies of opinion, and others are agreed to be imaginative only. The complete pattern of all ideas may be called the holotheme since other words which might be appropriate are pre-empted for use in representing other meanings by other disciplines. Binary code sequences can be formed in such a way that they act as titles for positions in the holotheme, and the pattern they reveal can be discovered by anyone who follows the course of action I have just completed. It is a constructive pattern, based on the way ideas are made out of simpler ideas. The more certainly those ideas which correspond to reality can be identified, the more consistent does that part of the pattern which imitates reality become. For example, the four elements of mediaeval cosmology are earth and water, air and fire. These do not fit into the pattern at the molecular level as neatly as do atoms, and anyone who tries to use them as subunits of this level will find that the result is to interrupt the major design of the holotheme here. This matter may be dealt with by treating the mediaeval elements as one of the sorts of idea which are on the imaginative side and allotting them a speciality mark accordingly. The holotheme is a pattern of ideas, not of statements; it is no more than the raw material of knowledge; it contains places for lions and for bravery and even for the-bravery-of-lions, but no place for the comment that lions are brave. Problems about statements, about how we know what we know, are not its concern.

It is time to turn to the mathematics.

The development of a theory

Students of set theory become familiar with various properties of relations. Since the holotheme seems to be most neatly arranged if account is taken of how notions are constructed, and since this is to do with the relations embodied in the notions, the properties of relations cannot help but interest those who examine it. The relations dealt with in set theory hold between subsets of a main set of terms which is called the full set. The full set is taken to be one of its own subsets, and the empty set (the set with no terms in it) is also taken to be a subset of the full set. Every term in a full set is said to be a member of that full set, and it is also a member of many of the subsets. Some of the relations between these subsets of terms are passive, being conditions; others are active, being operations. Examples of conditions are overlap, absence, presence, containing, membership, identicality. Union, intersection and complementation are examples of operations. These relations may be examined by means of a binary code sequence bused on the properties they possess.

It is convenient to deal with conditions first. Four properties are relevant, of which the first is reflexiveness. To place this property first is not an exercise of mere personal choice: reflexiveness can be defined by the use of one subset only, and since no relation can be defined by the use of no subset this makes reflexiveness the first definable property. A condition is reflexive if it relates a subset (or, more generally, any term whatever) to that same subset (or, more generally, that same term). Identicality is reflexive because any subset is identical with itself. Absence is not reflexive because no subset of which we are thinking is at that time absent from our thought. Whatever we may symbolise by the letter A, the absence of A is not A. If in fact the letter A stands for a subset, and if the letter R stands for a relation, we may say that reflexiveness is the condition in which A R A.

The second property is symmetry, which calls for two subsets if it is to be defined. An example is overlap: if A and B are two subsets which have a common member then they are said to overlap each other: A overlaps B and B overlaps A, a condition which we may represent by saying that A R B and B R A.

The third property is transitivity, which requires three subsets for its definition. The relation of containing is transitive: if A, B and C are three subsets and if A contains B whilst B contains C then A contains C. We may say that we have a transitive relation when A R B and B R C and A R C.

The fourth property differs from the others in that it does not hold between subsets of a full set but between entirely separate full sets – full sets of different varieties, without any common member. So far as I know, this between-set property has no standard name. I have found it convenient to call it transversiveness. The condition of membership is transversive, being between totally different full sets. For example, a man may be a member of the subset of men with blue eyes. Whatever this subset may be, it is certainly not a man. None of the subsets of the full set of men is a man. The full set of subsets has no member in common with the full set of men. This effect is familiar to those who know how an index may be represented by a full set of items (things indexed) crossed by a full set of features (characteristics which the items may possess). Such a network is called a data field, and in it the two full sets – the items and the features – are visibly transverse to each other.

Relation codes

Armed with these four properties, the student of the holotheme may form a relation code. Each relation may be represented by a four-digit binary sequence which has symbol 0 at its right-hand end if the relation is not reflexive, and 1 if the reverse is the case. To the left of this, lack of symmetry is show by 0 and symmetry by 1. Next, 0 stands for intransitive relations and 1 for those of the transitive variety. Finally, at the left-hand end, 0 signifies that the relation occurs between subsets and 1 that it is transversive, occurring only between full sets.

These properties are easy to check against the behaviour of any relation when deciding upon its code sequence. For example, once it is known that the relations of inclusion is allowed to cover the case of something including itself, it is simple to check that inclusion is not transversive, is transitive, is not symmetric and is reflexive. Its code sequence is therefore 0101. The condition of containing, sometimes known as proper inclusion, differs from this only in being irreflexive. Its code sequence is 0100. The following is a short list of some of the conditions dealt with in set theory, from absence (which is the simplest) up to membership, which is the first transversive relation:

relation	**relation code sequence**
absence	0000
presence	0001
disjunction	0010
overlap	0011
containing	0100
inclusion	0101
mutual exclusion	0110
identicality	0111
membership	1000

from the viewpoint of holothemics, the outstanding importance of this collection of relation code sequences arises because it is exactly the same as the first nine code sequences for formative grades in any integrative level. It can be extended as far as code sequence 1111 (suited to a relation which is transversive, transitive, symmetric and reflexive). This produces a collection of sixteen code sequences in all, and in the set-theoretic level provides a complete set of sequences covering the whole range of grades, from lowest to highest. Thus the binary pattern, fundamental to set theory, is thereby basic to the holotheme itself.

On the active side the same effect is found. Active relations, operations, may be distinguished as idempotent if they leave the entity on which they work unchanged. Idempotency corresponds to reflexiveness, and may be shown by the use of one subset only. The property corresponding to symmetry is commutation: two subsets are needed in order to show it. The operation of bringing together two subsets is commutative: the same result is achieved no matter which of the two we start with. If the two subsets are disjoint, having no common members, although they are both taken from the same full set, then this operation may be called adjunction. It is not idempotent because a subset cannot be adjoined to itself.

The active property corresponding to transitivity is non-association. Speaking without rigour, this means that order matters. In the realm of numbers, substraction is not associative: the result of (5–3) – 2 is not the same as the result of 5 – (3–2). As it happens, adjunction does not possess this property. It is associative.

If an apology is needed for treating of matters well known to my audience it is freely offered. The behaviour of relations is common property these days: it is taught in primary schools. I include the simple explanations of symmetry, transitivity and the rest for the sake of completeness, and to show the order in which they make their appearance in the pattern of notions I describe.

Here is a short list of operations of set theory, corresponding to the first seven formative grades:

relation	relation code sequence
complementation	0000
constancy	0001
adjunction	0010
union (and intersection)	0011
relative complementation	0100
restriction	0101
nand (and nor)	0110

This list, though not quite as long as the list of conditions, is sufficient to display the way the pattern grows. I should perhaps note that restriction is the operation of choosing, from a single subset, the members of that subset, or if a second subset is available then choosing those entities which are members of the first but not of the second. Example: choose the man – but if any have red hair, then leave those out. Nand is short for 'not and', and is complementary to intersection; nor is short for 'not or' and is complementary to union. This last operation is also known as rejection or as the Sheffer operation.

Sets are basic entities. Their qualities are numbers, which are basic attributes. Tables of relations between numbers can be constructed, corresponding to the lists of relations between subsets and sets. The following is an example. It is not complete, for it displays only the reasonably well-known conditions and operations of simple arithmetic; but it extends over the whole sixteen grades of the lowest integrative level, by way of compensation.

relation (passive: condition)	(active: operation)	relation code sequence
being negative	succession	0000
being positive		0001
	addition	0010
	cyclic addition	0011
being-greater-than	substraction	0100
being-greater-than-or-equal-to	cyclic subtraction	0101
		0110
equality		0111
being fractional	inversion about unity	1000
being whole		1001
	multiplication	1010
	finding-the-highest- -common-factor-of	1011
	division	1100
being-properly-divisible		1101
		1110
	averaging	1111

So much can be said about the pattern of relation code sequences in the set-theoretic level that it is hard to tell where to begin. The following seven paragraphs are no more then a brief indication of its richness.

The framework of mathematics

The study of the framework of relations which is found in mathematics has resulted in a series of increasingly complicated arrangements being described and named. Each consists of a set of terms together with a set of operations and conditions suited to them, and each higher pattern of terms and relations includes its predecessors of lower degree. Briefly, a semigroup is an arrangement in which addition is possible; a group permits both addition and subtraction; in a ring, addition, subtraction and multiplication can be carried out; in a field, addition, subtraction, multiplication and division are all available. This is a very bare statement of the situation, for there are groups based on multiplication and division as opposed to addition and subtraction, so that from one point of view a field can be thought of as a pair of suitably connected groups. Be that as it may, it is clear that the first four code sequences of the set-theoretic level provide for a semigroup (for two, in fact – one containing positive numbers and the other containing numbers which are negative). The next four provide the additional requirements for a group. The next four add the extra facilities needed for the formation of a ring, and the last four conclude the matter by catering for the needs of a field. In this case the field is that of rational numbers.

This however, is not the whole story. Rational numbers are the most advanced kind of number required in order to carry out addition, subtraction, multiplication and division with the certainty of always obtaining an answer; but they are by no means the most advanced kind of number which is it possible to have. Nor are fields the most advanced type of mathematical construction. There are, for example, vector spaces. Anyone who examines the matter will find that these have a home in the spatial level, as do the new relations they embody.

A vector space is a collection of elements which form a group for a commutative operation, and which can be handled by having a field applied to it: for example, the polynomial $5x^4 + 3x^3 + 9x^2$ can be multiplied by three-fifths, this being a rational number applied by multiplication to the polynomial after being taken from the field in the set-theoretic level. This is the simplest example of the apllication of the concepts of the set-theoretic level to more complicated matters. When an economist calculates the national income and relates it to the national product he is applying level 000 to level 111: the income and the product are national, but the arithmetic is fundamental.

Another aspect of the set-theoretic relation code sequences can be seen if a single grade tetrad is chosen and the relations of different types which possess it are studied.

For instance, sequence 0100 applies to the state of being-greater-than, the action of subtraction the structure of containing and the activity of relative complementation (forming the subset of things which possess one attribute but not another, when some of them have both). These relations are closely connected to each other: for example, it can be said that the conditions permit the operations: if one number is greater than another then subtraction of one from the other does not yield a zero answer. The tables of relations possessing the various code sequences can be filled out by chasing correspondences in this way.

A further point of interest is the order of definition. Later relations can be defined by the use of earlier ones. I did not expect that this would happen when I chose the order in which to place the relations' properties in the code sequences. The effect, however, seems to be found throughout. In the case of structures, we start with absence (0000). The absence of absence is presence (0001). The relation between an absent subset and a present subset is disjunction (0010). The absence of disjunction is overlap (0011) The relation of a subset found in an overlap to either of the overlapping subsets is being-contained-in (like containing, this relation carries sequence 0100).

Similar sequences of definition can be found on the attributive side. Averaging (or, more exactly, calculating the arithmetic mean) calls for addition and division, and occurs (as it should) later than the latest of these.

The Peano axioms for the formation of the natural numbers hold a highly suitable place in this lowest level: they come at the very start. They embody the principle of making numbers by succession, starting from zero. The code sequence of the quality zero is 0000, and so is that of the action of forming a successor. Inversion about zero also carries code sequence 0000, so it seems possible to form all negative natural numbers without leaving grade 0000, and then to invert them, reaching grade 0001 and the positive natural numbers by this means.

The ideas of units, assemblies, series, systems, simples, mixtures, laminates and runs appear in level 000 although they are more readily recognised in higher levels. The notation which divides each formative stage into a grade containing objects and a grade containing substances provides a zero in the fourth position of its tetrads when objects are considered, and a unit when substances are meant. The notation developed from the study of set theory uses the zero to stand for notions which are irreflexive, and a unit for those which are not. If the hint thus provided is taken, then the property of reflexiveness is suited to substances. A similar comparison of the two sources of the notation relates symmetry to assemblies and mixtures, in which the constituent ideas display no order of priority, whilst transitivity becomes manifest as the property of laminates and series, in which order is to be found. As a simple example, grade 0010 may be taken. It contains the relations of adjunction and addition, which are fitted to the assembling of objects. Its reflexive companion, grade 0011, holds overlap, the typical condition of a mixture, in which two or more substances are to be found mingled in one region.

The last sequence of the set-theoretic level, 000.1111, is followed by 001.0000. Here the triad of the spatial level appears for the first time, followed by the tetrad belonging to the lowest grade of any level. The new set of sixteen grades offers places for points, lines, planes and volumes and the figures they contain, matrices, convergent series, real and complex numbers, and all the conditions and operations which apply to such more advanced ideas. It appears to build up in the familiar way, with polynomials playing the part which is played by integers in level 000. Real numbers appear in its lower rank and complex numbers in its upper. Complex numbers can be represented in many ways: for example, they may be shown as ordered pairs of real numbers; they may be shown as a particular type of matrix; they may be shown in the form a + ib where a and b are real numbers and i is the square root of minus one. They may also be shown as a particular sort of ratio between polynominals, and here the way polynomials may imitate integers is again revealed: integers form rationals by this means, and rationals, like complex numbers, occupy the upper rank of a level.

The condition of being a being-a-power of a rational number (the most advanced type of number available when the spatial level is encountered) carries relation code sequence 001.0001, the reflexiveness arising because any number is a power (the first power) of itself. The powers, multiplied by suitable coefficients, become the terms of polynominals, and indeed can be considered to be the simplest form of polynomial in their own right. The plus symbol which may connect different terms in a polynomial represents a relation which is subtly different from that of addition in level 000. It occupies grade 001,0010, and is a sort of precursor of the relation which connects the elements of a vector in a space of two or more dimensions. Having achieved this condition, its repeated application can be used to form approximations to irrational numbers (the series of numerals used in, say, the decimal representation of a number is only a polynominal in disguise). The concepts of being a least upper bound and of being a greatest lower bound may here be called into play, and carry code sequence 001.0011 (compare with highest common factors and lowest common multiples one rank below).

So it goes on. Extracting the root is one of the operations of grade 001.0100, and it will be recalled that very many roots are irrational. Here is the heart of the matter so far as a large class of real numbers is concerned. The non-associative effect shown in the grade tetrad may be compared with the similar effect one rank below, where division carries code sequence 000.1100 and is concerned with the formation of rational numbers. It may also be compared with the situation two ranks below, where subtraction carries the sequence 000.0100 and is intimately bound up with the formation of integers. Returning to root extraction, it may be noted that this provides an essential operation in the forming of a geometric mean (at grade 001.0111 – compare with the position of the arithmetic mean one rank below).

In the lower spatial rank, it becomes possible to play with formulae and with their representations on graphs. The representations (parabolas, hyperbolas, catenaries, circles, cubes, pyramids and the like) are of course on the entitive side, and may be expected to occupy the same grade as that held by the formulae which describe them. When the higher rank of the level is reached, matrices arrive, and the mathematics of movement copies the shearings, rotations, translations, reflections, dilations and so on which can be applied to the shapes which are now available amongst the entities. In the end, high in the level of space and close to the subatomic level, the tensors and spinors and complex matrices and quaternions are found, amongst which the subatomic physicist hunts for descriptions of his fundamental waves and particles.

Relation codes and the holotheme

As I explored the pattern of relation code sequences, I became aware that I was not constructing a model to give a mathematical form to my arrangement of everyday ideas, but was merely describing the layout of the lowest integrative levels. The model I sought, and the network of ideas to be found in set and number theory, were one and the same thing. I looked at the symbols for relations between terms and found none that were not one or other of my four semantic relation types: I invite my audience to do likewise. Nor did I find any term which was not a passive or active entity or a passive or active attribute. As I followed the series of relation codes upwards through the level of space and into the subatomic level where matter, time and energy appeared, I began to feel I was doing no more than unroll, with great labour, a chart of an ocean which was already well-known to everyone but myself.

At the same time I knew that other workers at the trade of organising knowledge had reached opinions at variance with mine. Most of these provided a pattern which was far less regular than my set of relation code sequences, a pattern whose very complexity seemed more in keeping with the bewildering ramifications of knowledge than did my rigid binary scheme. Certainly a binary framework is suspiciously neat. I can understand the view that, perhaps unwittingly, I started with it and forced a large number of notions into the pigeonholes it provides. However, I do not think this is so. I knew nothing of set theory when I began, and the notation I used in order to handle entities and attributes in the early days of my inquiry was based on the alphabet. The pattern of noughts and ones came later, and I find that this helps to persuade me of its validity.

When I began to read the literature on the subject I found evidence of a widespread wish to base an integrative level on the mind and its creations. This was made manifest in many different ways. For example, the notions of mathematics might be given a home above the level of human beings on the ground that a mind is needed in order to form or to comprehend such impalpable matters. Alternatively, a psychological level might be postulated, above the level of biology and below that of society, with mind as its principal entity. I think this approach is mistaken. As a set of notions, the entire holotheme is a mentefact, if this name may he used to cover ideas which correspond to reality as well as to the special ideas which are admitted to be fiction or whose basis in reality is at any rate disputed. This being so, I prefer to place the notions where their internal make-up tells me I should. As for the mind, it seems most consistent to treat it as a phenomenon, the brain at work, amazingly skilful, flexible, creative and powerful, yet susceptible to interference as are the other functions of the body.

Rules for placing ideas

Be that as it may, I found that the placing of ideas in the holotheme was made easier when I described what I meant by the words representing them and followed this by analysing the definition in a methodical way. Here is a typical series of placement rules directed to this end:

"First, write down your definition of the meaning of the word you wish to place.
Next, consider the notion to which it refers or any representative of the notions it groups together.
In respect of this, ask:
. . . is it passive (0) or active (1)?
. . . is it attributive (0) or entitive (1)?
. . . is it a relation (0) or a term (1)?
From your answers, write down the code sequence for the semantic type.

Next, ask:
. . . is it general (0) or special (1)?
From your answer, allot the speciality mark.

Next ask:
. . . what is the highest unit which it contains or to which it refers?
. . . is this unit in the lower (0) or the upper (1) rank of a level?
From your answers, write down the code sequence for the integrative level and rank.

Next ask: is it, or does it embody, a relation which is:
. . . intransitive or associative (0) or otherwise (1)?
. . . non-symmetric or non-commutative (0) or otherwise (1)?
. . . irreflexive or not idempotent (0) or otherwise (1)?
From your answers, write down the code sequence for the notion's formative grade.
Now combine the code sequences to produce the full sequence for the idea's level, grade, type and mark. Convert this to decimal or to such other notation as has been decided upon.

Next ask:
. . . is this sufficient, or is further specificity required? If the latter is the case, is the idea collective or is it on the other hand generic or specific?
If further specificity is called for, apply these placement rules again according to the special instructions for treating collectives or generics."

This is a long series of choices, but experience in making them soon leads to speed: the specialist sees the integrative level and the formative grade at a glance, and the semantic type is equally readily obtained – especially since the definition thus analysed has already covered this ground. The choices become obvious, and the placement rules are only called upon in difficult cases.

Advantages of the pattern

It may now be time to ask what gains are made by using a pattern of this kind. I believe there are many. Here are some examples, starting with a fairly specialised field of work and then dealing with more general applications.

My own daily task is the development of co-ordinate indexes – that is, of means of putting items of all types away in store after allotting suitable features to them in a retrieval device which is later used to co-ordinate a subset of the features and to carry out a search for the items possessing these. A major problem in setting up a co-ordinate index is the development of a suitable list of features, and this is especially tricky when the items are documents and the features are their subject matter. Any pattern which identifies synonyms and separates the meanings of homonyms, as this does, is a welcome aid. Other advantages gained from using relation code sequences are: they supply an automatic key to part-whole relationships, which may often be combined with purposive relations (clay for bricks for walls for houses); they are independent of language – Julius Caesar fits at position 101.1000.011.0 in any tongue whatever –; they offer an assurance that no types of meaning are overlooked; and they provide an effective discipline for defining the words of the thesaurus or feature list, so that secular drift of meaning may be controlled. Finally, no matter how language may change, the use of such a pattern may ensure that, once an idea is well defined, it is defined for keeps.

The notions used in a co-ordinate index may be of any degree of complexity. They are normally fairly simple, but there is nothing in principle to prevent one such feature being 'the effect of gift stamp trading on the profit ratio of all-night automobile service stations' and another being 'the factors affecting labour retention under overtime conditions'. It is clearly possible for both of these features to appear in a co-ordinate index against a suitable item, and although it stretches the imagination a little it is possible to conceive each feature as possessing a single-word name in some specialised vocabulary. The holothemic pattern must provide a place for complicated notions of this kind. In principle this is possible without strain. The placement rules will depend on the methods used for assembling the components of the complex notions (for example, the order of the levels may be chosen: number before place, place before time, time before chemical composition, and so on). Developing such rules is meticulous work. Rather than dilate upon this, however, I should like to point out that large numbers of classifications place items in order in accordance with a single complex feature of just the kind I have described. The techniques for building this up and the rules for permitted permutations of its elements (if it is recorded in different forms in more than one place) can be reconciled with the methods used for deciding where it fits in the holotheme. Consequently the pattern of levels, grades and semantic types may turn out to be of considerable value to national and international subject indexes covering the whole field of knowledge. In offering what appears to be a complete set of places for ideas and showing the relations between them it provides a powerful means of developing consistent rules for forming complex notions out of simple ones.

A second example of the use of the relation code may be taken from mathematics. The code arises from the behaviour of sets and numbers, and seems to provide the subject with a natural pattern which displays it easily. Some years ago, when I was newly satisfied that the set-theoretic level provided me with an exact theory for the holotheme, I was asked to draft a textbook on the behaviour of sets and numbers for limited circulation to schools. The aim was to provide a series of exercises to show the properties of reflexiveness, symmetry, transitivity and the like with the aid of a collection of specially-punched feature cards. Punched feature cards have arrays of numbered positions on their faces, each of which represents an item which may have the feature represented by the card. If the item does in fact possess the feature, its position is punched. Consequently the card displays a subset of the full set of all the items indexed, and may be used for demonstrations of union, intersection, disjunction, overlap and other relations found in the lowest integrative level. The properties of numbers can also be shown: for example, the holes in a card may be counted, hidden at random, related to the total possible, and so on. The experience of producing the textbook made me sure that the code sequences of the set-theoretic and spatial levels could be used as a basis for the classification of mathematics and for a full-scale presentation of the subject to all who study it.

A further application of the code may be found if its application to mathematics is extended to the remainder of the school and university curriculum. The holothemic pattern is a map of knowledge. As interdisciplinary studies grow in importance in centres of advanced learning, and as schoolchildren find themselves more frequently taught by means of projects carried out with the aid of resource banks, the older divisions between subjects begin to break down. When this occurs a new set of guidelines is needed to replace the vanishing boundaries. The framework of levels and grades described in this paper may well serve such a purpose. If it proves valid, it will remain firm no matter how the temporary partitions within it are arranged and rearranged to suit academic needs.

I recall a simple example of this in the realm of curriculum development. It is possible to divide the subjects taught at the average school for teenagers into those which involve physical and mental skills – woodwork, languages, trigonometry, swimming – and those which are concerned with describing a pattern – the organisation of government, the sequence of history, the structure of the atom, the construction of the human body. Obviously the two types overlap: languages involve patterns of grammar, the study of biology involves handling a microscope. However, if the largely descriptive element of education is considered on its own, it may be compared subject by subject, with the levels of the holotheme. This is a revealing exercise. When I carried it out in respect of my own children I found that their schooling dealt with the structure of things at every level except the communal. Sets and shapes were dealt with by mathematics; particles, atoms and molecules by physics or chemistry; cells and living beings by biology; the structure of government both national and local was cared for by various combinations of history, civics and current affairs. By contrast, the structure of any large administrative body below that of local government was conspicuously absent. It was no teacher's duty to explain that an industrial company has a quality control section, a cash flow system, a maintenance department, a time office, besides the more obvious production lines and sales offices. Companies, clubs, professional associations, universities, trade unions, are of course the main units of the communal level, their departments being subunits. Incidentally, I suspect that a very useful analysis of management science could be based on the layout of the communal level. So far as the school curriculum is concerned, I concluded that in my children's case at any rate the organisations in which they are most likely to earn their livings are the ones least likely to have been explained to them, even by the harassed and probably press-ganged careers adviser.

These notes on applications of the relation code pattern are, I suppose, a very small sample of its possible uses, and (moreover) a sample biassed by my own special interests. I should like to conclude by making a few more comments arising from my own work or amusement. Before doing so, I should like to express the hope that this audience will agree that the pattern I have presented does in fact possess the properties I listed at the start. It is infinitely hospitable: every place in it can be subdivided as may be necessary. It is easily taught: few problems are met with in learning a list of eight levels or in listing eight semantic types or in displaying a set of formative grades which is common to every level. The properties of relations, on which it is based, may not be common ground to all adults today, however well educated; but the teachers in our primary schools are ensuring, through the medium of the so-called new mathematics, that a foundation is laid on which the series of levels I have described will seem a natural edifice. It is independent of personal opinion: given the rules, an Arabic-speaking Eskimo living in Peru will, from standard textbooks, place the concepts of photosynthesis, Tarzan of the Apes, the omega minus particle and the acceleration of derived demand in the same positions as will a Bantu-sepaking Russian in Bengal: not because he or she has referred to the schedules or the decisions of a central authority but because that is where the notions fit. It is based on a simple repetitive pattern: little is simpler than binary, and binary is certainly widely-known. I should add that the pattern came of itself during my inquiry. I started out with no sort of idea as to what I might find. If I had begun sixty years ago, when little was known of set-theory or the contents of the subatomic level or the things to be found inside a living cell I should have had a harder time of it, and might well have ended in boredom or despair turning to chess or politics or fishing. I was fortunate in my decade, which gave me the materials for the work, and in my family, who allowed me to play with symbols when I should have been moving the lawn, and in my local authority, whose many public libraries provided me with books by the dozen.

Comments and conclusion

When I look at this whole structure of thought I realise how much of it has been found or checked by the process of comparison. I remember the pleasure with which, having guessed that units, assemblies, series and sytems might be found between molecules and the organelles of cells, I encountered an article on the mitochondrion. This was a subject on which I was quite ignorant, and the diagrams displaying the unit lipid molecules assembled into membranes, with the input and output series making this body the powerhouse of the cell, together forming a system, gave me a special sense of discovery. Also I recall seeing a film on television, made by time-lapse photography, which had the strange effect of showing people in the guise of a substance – pumped methodically through traffic lights by an unseen force, dispersing on leaving a subway – as if the city breathed them as a lung breathes molecules. For amusement, I checked the integrative levels. The number of grades between a city and its people is the same as the number between a lung and its air. I thought: the city, breathing people in and out: what a tremendous poetic image. I was often one of those people in the city of London. I am certain there is an exciting literary study to be done on the similes of the levels. It is so like the Elizabethan doctrine of correspondances. I found myself reading Orchesographie, the world as a dance, and thinking of the links between phenomena. World-views based on the active side of the pattern concentrate on the interactions between these, and are unified by general system theory whose province (in the language of the holotheme) is every active notion, every concept whose code sequence is of the form xxx.xxxx.1xx.x. This is exactly half of all the available sorts of idea. The word 'system' is used here in the same sense as it is used in the rest of this paper, since the discipline of general systems theory concentrates or cybernetics, feedback, equifinality, the behaviour of assembled series, notions of the form xxx.x11x.1xx.x.

Such random skimming over the map of knowledge reminds me of a pond skater. How far can that lightweight insect see into the deeps of its pond? For my part I am still fascinated by the dimples and ripples of the water's face; I have not learnt how to consider it more profoundly. I speculate: should gravity be treated as if it were three-dimensional surface tension in the lower molecular grade? Is it an effect of space trying to become as spacious as it can after mass has made an appearance? What is the connection between mass, length, time and energy on the one hand, and things, qualities, modes and phenomena on the other? It is odd how these quartets pair off – mass in things above the spatial level, length somehow related to scales on which qualities may be measured. In each quartet, any three can define the fourth. Can a sort of slide rule be made, graduated with the operations of arithmetic, such that if its origin is placed anywhere in the holotheme, taking the occupants of that place as units, then the mathematics appropriate to the treatment of higher notions will appear on the rule opposite them? This would certainly put group theory correctly opposite to the omega minus particle whose existence it predicted. How would it affect the mathematics of road traffic treated as a liquid flow? Is money to be thought of as social energy, its tiniest coin being equated to a photon? An individual can control a nation, yet living beings are to be found two levels below the national: control is here exerted upwards: does this supply an answer to the question, where is the concept of deity to be placed? The very highest place in the holotheme seems unsuitable for such an idea, since the pattern of levels is based on a part-to-whole relation, and to use position 111.1111.111.1 would imply that godhead is constructed out of interlocking sovereign states of the earthly type. On the other hand, since upward control can be found at biomorphic level, the concept of an eternal being of infinite power may realily find a home there, supported by the view that relationship with deity is personal.

Is this the stuff of philosophy? If so, then for mundane reasons I have entered an unknown land. Sirs, I am a mere journeyman come to your court, a weaver of patterns out of such threads of knowledge as I can find. This pattern is called the holotheme. I hope it pleases you.

References

The idea that the universe of knowledge may be arranged in an order corresponding to the increasing complexity of its elements is a commonplace in philosophy. I suspect that the problem has never been to discover if it is possible, but has always been to see how to get it right. In an appendix, B. C. Vickery ('Classification and Indexing in Science', Butterworth, London) gives twenty-four arrangements of the major sciences, starting with that of Hugh of St. Victor in 1140 A. D.; virtually all of them show a strong awareness that a constructive sequence exists. I have not been able to find an instance of the name 'integrative level' for a degree of complexity in this sequence earlier than 1937, when Joseph Needham gave the Herbert Spencer lecture at Oxford under the title 'Integrative Levels: a Revaluation of the Idea of Progress'. This is printed in his book, 'Time, the Refreshing River' (George Allen and Unwin, London). A footnote to the lecture gives references to earlier uses of the word 'level' in this sense. For a history of the speculation, the practical inquiry and the slow growth of understanding in this affair of how the world is made up, I found Stephen Toulmin and June Goodfield's 'The Architecture of Matter' (Hutchinson, London) fascinating reading, and its brief epilogue furnished a statement of the mid-twentieth-century position which gave me considerable impetus. They write: 'The most far-reaching outcome we can hope for from twentieth-century matter theory. . . is a common system of fundamental concepts, embracing material systems at every level.' This, of course, is what a study of the holotheme may supply.

A work of immediate relevance to the development of such an arrangement is Oliver L. Reiser's 'The Integration of Human Knowledge' (Porter Sargent, Boston). This I found of especial value when I was looking at the connection between the higher ranks of mathematics and the lower ranks of physics. For many rules about how the higher levels depend upon the lower and yet control and organise them, James K. Feibleman's paper in the British Journal for the Philosophy of Science (May, 1954) should be consulted. Alex B. Novikoff's 'Concept of Integrative Levels and Biology' (Science, 2 March 1945) should also be noted.

In this paper I have concentrated on ideas of the passive sort, but the pattern itself shows that there are as many notions of the active as of the passive type. As an example of work which tackles the active side I should mention Ludwig von Bertalanffy's 'An Outline of General System Theory' (British Journal for the Philosophy of Science, 1950). This article presents the behaviour of living beings and their organs in terms of systems of equations which correspond exactly to such systems as they would appear if placed in their proper formative grade. Another book of special application to the active side is W. Ross Ashby's 'Design for a Brain' (Chapman and Hall, London). W. Sluckin's 'Minds and Machines' (Penguin, London) must also be noted, as must 'A Model of the Brain' by J. Z. Young (Oxford University Press).

I cannot leave this part of my subject without listing Gilbert Ryle's very impressive work 'The Concept of Mind' (Hutchinson, London). I found this especially relevant when I was looking at the arguments for and against the existence of an integrative level based on the mind as a special sort of entity.

So far as basic logic and mathematics are concerned, I must list G. Birkhoff and S. MacLane's 'A Survey of Modern Algebra' (MacMillan, London and New York), R. Stoll's 'Sets, Logic and Axiomatic Theories' (Freeman, San Francisco and London) and, for its special emphasis on the behaviour of various mathematical operations, M. Bruckheimer, N.W. Gower and R. E. Scraton's 'Mathematics for Technology' (Chatto and Windus, London, and Clark, Irwin, Toronto). 'Basic Mathematics' by R. G. D. Allen (MacMillan) is a work to which I have often had recourse, and Irving Adler's 'The New Mathematics' (Dennis Dobson, London) must be noted as the source of my earliest knowledge of numeric groups, rings and fields. It sets out to give a simple, clear description of a well-defined subject. In connection with it, I should note how very often I have found really good works of untalkative but direct explanation went to the heart of that part of the pattern about which I consulted them.

I am very conscious that in collecting information to help my inquiry I have read part or all of many hundreds of works, including virtually every copy of the 'Scientific American' and the 'New Scientist' for a decade. I have no notes as to the source of many of the helpful ideas I have found in these documents. There is, I suppose, nothing to be done here except issue a blanket expression of gratitude to cover them all. The thanks are no less warm for being directed to many different places, some of them unknown.

My own previous publications on the theory of integrative levels are few. There is a letter giving the outline here expanded in the 'Journal of Documentation', Volume 20, Number 4, December 1964. There is a contribution to the 'New Scientist' on 'Relation Codes: an Ordering of Ideas' (16 May 1968) and a paper printed in the 'Classification Society Bulletin' on 'the Pattern of Meaning' (1967). There is 'Punched Feature Cards in Set and Number Theory', circulated to members of the Midlands Mathematical Experiment in 1969. Finally, there is a chapter in my book 'Data Study' (World University Library – Weidenfeld and Nicolson, London; McGraw Hill, New York and Toronto; Hachette, Paris; Kindler, Munich; Meulenhoff/de Haan, Antwerp; Il Saggiatore, Milan; Guadarrama, Madrid; and Heibon Sha, Tokyo). In these the terminology differs in two or three particulars from that which I now use; but the pattern described is the same.

Appendix

Summary of Nomenclature for the Levels

There are:
Eight integrative levels, each containing a higher and a lower rank;
Eight formative grades within each rank, making sixteen within each level;
Eight semantic types within each grade.
Also, a distinction is made between special ideas and others.

Formative grades come in pairs, the pairs being formative stages; consequently each rank contains four stages, and each stage contains two grades.

The levels are: set-theoretic, spatial, sub-atomic, molecular, cytomechanic, biomorphic, communal, national.

The pairs of grades within each rank are: units and simples, assemblies and mixtures, series and laminates, systems and runs. These names apply to things; other names may be suitable for other semantic types.

The semantic types are: states, qualities, structures, things, actions, modes, activities, phenomena.

ADDENDUM to 'The Holotheme',
paper presented on behalf of J. L. Jolley to the First Ottawa Conference on the Conceptual Basis of the Classification of Knowledge, October 1971.
By Robert Fairthorne (North Farnborough, U.K.)

PRESENTATION

The paper describes a long and careful inquiry into the ideas for which people have found it useful to coin words. It concludes that the pattern of human knowledge, as displayed by these ideas, is binary.

This conclusion is not a facile compliance with the fashion of the day, following the computer revolution which has made the binary pattern popular. The author made strenuous efforts to resist reaching this opinion. At each step, however, the problems encountered were more tractable if a dyadic treatment was applied to them than if any other pattern know to the author was adopted.

The heart of the matter is the concept of a relation code. Such a code places relations in positions corresponding to their properties, and it places terms in positions corresponding to the relations they embody. It seems to be the natural basis for any arrangement which locates ideas according to their internal construction. It is eminently teachable but cannot be applied until clear definitions of the notions on which it is brought to bear are available.

Relation codes are not abstruse, being based on the simple and well-known properties known as reflexiveness, symmetry and transitivity or on these properties' active counterparts. These properties have an inherent order: reflexiveness calls for only a single term before it can be defined; symmetry calls for two; transitivity calls for three. The use of this order allows eight different types of relation to be placed in sequence.

The method can be shown on a blackboard or overhead projector by writing a column of code sequences, fairly well to one side of the board or screen, starting at the top with 000 and ending with 111, these being the numbers from nought to seven in binary notation. The properties of reflexiveness (or idempotency), symmetry (or commutation) and transitivity (or non-association) may be shown by units in these sequences, starting with the right-hand position for reflexiveness or idempotency. As described in the paper, simple conditions and operations of set and number theory may then be placed in columns against their respective code sequences, one column being used for number conditions, the next for set conditions, the third for number operations and the last for set operations. An intelligent audience has little difficulty in choosing the right position for any relation of this sort.

When enough relations have been placed, the fact that they form a series of increasing complexity becomes clear. So do the similarities between relations bearing the same code number. An example not mentioned in the paper is concerned with the way in which conditions permit operations. If two subsets are disjoint (010) we may adjoin (010) them, and find the cardinal number of the resulting subset by adding (010) the cardinal numbers of the two original subsets. Again, if one subset is contained in (100) another then its cardinal number is less than (100) that of the other and may be subtracted (100) from that of the other to produce a result other than zero. This operation corresponds to that of forming the relative complement (100) of the smaller with respect to the larger.

In the full pattern of notions described in the paper, conditions precede operations and are therefore available to permit the operations which correspond to them.

The sequence of relation code sequences provides an elegant basis for the treatment of sets and numbers. It was used in this way in the author's privately-circulated "Punched Feature Cards in Set and Number Theory".

The principle of defining more complex ideas by means of those which are simpler may be shown by the use of this relation table, and the opportunity may be taken to introduce the names 'formative grade' and 'formative stage', each stage containing two grades. For example, the first stage contains grades 000 and 001. These names may be justified by reference to the formation of higher ideas out of lower. Also, the table makes it possible to explain the meanings of the names 'unit', 'assembly', 'series' and 'system' as applied, respectively, to ideas which embody passive relations bearing code sequences 000, 010, 100 and 110. Such ideas are irreflexive are refer to objects. Their reflexive counterparts, 001, 011, 101 and 111 are appropriate to substances, although these are of a rudimentary nature at so fundamental a level as that of set theory.

It is possible to dwell for some time on set and number theory. The paper omits much: for example, at this level it concentrates on relations and not upon the terms which embody these. It may be helpful to draw a lattice for a full set of three members, showing all the subsets of the set related by the condition of containing. Such a lattice looks rather like a cube standing on one corner. The opposite corner, which is the cube's highest point, may be shown as occupied by ∅, the empty subset. Below this, the three subsets A, B and C may be shown. Next come the three subsets AB, AC and BC. Finally, the subset ABC – the full set taken as one of its own subsets – occupies the lowest point. Such a lattice is a complete representation of the internal structure of a full set of this variety. The empty subset ∅ is a typical unit. A subset such as AB is a typical assembly (of A and B, which are adjoined). The subsets A, AB and ABC form a series. Finally, the subsets A, AB, AC and ABC form a system. This collection of four is known as a filter: more generally, a filter consists of all the subsets of a given full set which contain a given member of that full set. It is of interest to note that the filter in question appears as if it were one face of the cube. This is typical of systems: unlike series, there is an element of completeness about them, but nevertheless they are not the whole figure, the whole new unit.

If the audience is asked to form the relation code sequence for division it will prove to be the same as that for subtraction; and a similar exercise will reveal the code for multiplication as the same as that for addition. To achieve a distinction between the more advanced operations and the simpler pair, a further property of relations must be introduced, so that the code sequences contain for digits instead of three. This new property is transversiveness, the property of holding between complete full sets as opposed to within them. Multiplication then gains the sequence 1010 and is distinguished from addition (0010). The new property, of course, produces eight further relation code sequences.

At this point it is possible to talk of an upper and a lower rank of an integrative level. The sixteen code sequences now available span the level of set theory; the lower rank consists of the first eight, and the upper rank of the remainder. The way in which the whole level contains the requisites for a mathematical field, as described in the paper, may be noted.

Thus the relation code grows into a code suitable for representing integrative levels, and in fact (by the use of three further positions to the left of that occupied by transversiveness) handles eight of these.

This is the second major concern of the paper. The eight integrative levels found in nature are described and named in it. In the author's view, this is definitive: there are eight levels, and these are they. If he is right, then the pattern of grades, stages, ranks and levels provides a remarkable unification of the sciences, showing how they are interrelated, how they depend upon and serve one another. It is possible, in less modest moments, to feel that here, once and for all, is the arrangement of knowledge which corresponds to reality. At times of greater humility a mood of dismay intervenes. There is so very much work to be done, even if we seek only to fill in a corner here and there, to hang a few panels on the framework, to try to adapt the gaunt outline to fill some practical human need.

It is characteristic of the whole pattern that mathematics underlies and supports it. Mathematics occupies two levels in all: that of set and number theory, which may be given the code sequence 000, and that of space, which may be given sequence 001. If a full stop is used, as in the paper, to separate the level code from the code sequences corresponding to ideas within the level, then mathematics may be said to reach from 000.0000 to 001.1111. Perhaps the enormous representational power of binary patterns arises because they are the first patterns to appear. They are found beneath all others.

Above the levels of mathematics there are two levels occupied by physics and chemistry. These, the subatomic and the molecular levels, are followed in their turn by two levels occupied by the life sciences and by those advanced mechanical aids from gumboots to aeroplanes which simulate and extend the powers of living beings. These are the cytomechanic and biomorphic levels. Finally, two levels concerned with the social sciences are reached – the communal and the national. These are filled by the vast range of notions to do with organised communities, omitting only those communities known as families, and similar simple groups, which occupy the upper rank of the biomorphic level.

This entire range of ideas, the whole sweep of human knowledge, is here called the holotheme. The name 'universe', which at first sight might appear apposite both because of its meaning and because of this widespread use, is far too often regarded as referring to an aggregate of entities, leaving attributes on one side, and forgetting about relations altogether. It is interesting to note that the most spectacular recent advances in knowledge have occured where the levels of the holotheme meet – in molecular biology, where the upper molecular rank encounters the lower rank of the cytomechanic level, and in subatomic physics, where the lower subatomic rank fits on to the higher rank of the level of space.

At the risk of making the subject boring by repetition, it may be mentioned that the arrangement described in the paper seems to be common to effective classifications of scientific notions. These place ideas in order by reference to their parts: that is, to notions which occur earlier in the holotheme than do the ideas being classified. For example, the periodic table places atoms (units of the lower molecular rank) in order according to the number (lower set-theoretic rank) of their protons (upper subatomic rank). Botany and zoology make use of taxonomies based on the same principle. This of course is the principle of placement according to structure.

The third matter to which the paper gives special attention is the study of semantic types. Eight of these are described, and these are held to be the eight main types of notion which can be distinguished in any sufficiently large random collection of ideas. The types exist as a result of three dichotomies which are found in the holotheme – that between relations and terms, that between attributes and entities, and that between the passive and the active. The author uses these to extend the binary code sequences so as to place the semantic types in order and to give each a place in every grade of the entire pattern. This is achieved by allotting the numeral zero to passivities, to attributes and to relations in three-digit code sequences whose positions record these properties in that order from left to right. Thus 101 stands for an active attributiv term, 110 for an active entitive relation and (by attaching these code triads to the right-hand end of the grade sequences) a concept such as addition achieves the code sequence 000.0010.100. The earliest of all notions transpires to be nullity, the passive attributive relation which is embodied in the numeric quality, zero.

The paper contains a general justification for the order chosen for the semantic types. A particular example may be given. Consider the set-theoretic operation of following, which in its most rudimentary form provides the empty subset with a following subset of one member. Where does the idea 'one' come from, for use in this definition? It arises from the numeric operation of succession (as given in the Peano axioms), which is available because numeric operations precede operations on subsets – because attributes precede entities.

The provision of the last placing device in the code sequences described in the paper seems to be needed for the sake of completeness. No sort of novelty can be claimed for the use of such a thing as the speciality mark. It would be pleasent to leave it out, since it is not the province of the arranger of ideas to decide which are fact and which are fiction. Nevertheless, if the basis of his arrangement is the ascertained construction of ideas which are almost universally believed to correspond to reality, then he will find notions which cannot be placed because their correspondence with reality is questioned or because they break an extensive pattern or because their internal construction is in doubt. He must then place them as nearly as he can, and note their special nature. Do quarks exist? Perhaps; but at any rate the particles of which they may prove to be constituents are real enough. So particles are mundane; but quarks are special. If the existence of quarks is proved, then nothing will be easier than to remove the speciality mark now placed against them.

To conclude, it may be fun to try placing two notions which have not been dealt with in the paper. The first of these is causality. The author remembers it well, because he was challenged to place it after a lecture on the holotheme which he gave during the early stages of his work.

Causality is a relation. If it is agreed that it is first found when mass, energy and time are available, then it is in the lower rank of the subatomic level. If we may also agree that it is immediate, then it is neither commutative nor idempotent, and associativity is irrelevant. Its position is therefore 010.0000.110.0, the last four digits indicating that it is an active entitive relation whose status is undoubted.

Of course, if our definition of causality allows for distand effects by way of intermediate results, then the code sequence is 010.0100.110.0, the tetrad (0100) showing the associativity of the relation. These sequences may with advantage be compared with those for immediate and mediate following, and for immediate and mediate succession, in the lower set-theoretic rank.

The other notion is the notion of a notion. First, a definition is required. Reaching into the deepest recesses of the author's ignorance, suppose that an idea is regarded as a pattern which is recorded and maintained in the brain. The brain is an organ – a subunit of the biomorphic level. To be within the brain, in any physical sense, is to be below this level, and therefore in the cytomechanis level at the highest. Suppose then that the pattern we call an idea is believed to be made of a subset of brain cells which between them take on a particular condition which represents the idea. The sensation we know as thinking then consists in reading these conditions and manipulating them. There are, of course, other views as to the physical accompaniment to the mental image we call an idea, so the code sequence will certainly require a speciality mark. However, if the above view is accepted, then an idea is an entitive condition in the upper cytomechanic rank, and its formative grade depends on the assumptions made about the arrangement of the cells in question.

If all this appears to be too physical, it should be recalled that there is nothing to prevent the use of the well-known words for mental states – feelings, emotions, sensations, opinions and the rest – in any written representation of the matter. There is no reason whatever for the user of an index based on this analysis of the holotheme to feel that he or she is on unfamiliar ground.

This may perhaps serve as a closing remark. The object of the work is to produce an arrangement which corresponds as well as possible to affairs as we know them, in the belief that the most serviceable pattern will be the most exact image of reality. If the effort succeeds, one of the signals of its success will be that it feels familiar, so that people say to each other, 'I can't see what's new in it. It all seems quite ordinary to me'.

Derek Austin (London)

A conceptual approach to the organisation of machine-held files for subject retrieval

It is, at first glance, hard to see an obvious point at which the work of the philosopher impinges on the design or use of a classification scheme as understood by the documentalist. Before beginning this paper I checked the definition of philosophy which appears in Webster's Dictionary, hoping for at least a glimpse of correspondence with the central themes of classification research. Philosophy, we are told, is "A search for the underlying causes and principles of reality"; it is also "A quest for truth through logical reasoning".

These preoccupations with truth and reality should, without doubt, serve as the mainspring of any professional activity. Unfortunately, they just will not do for the librarian or documentalist, who has to cultivate a sense of detachment from such matters. Only in this way can he apply his professional skills effectively to the indexing and coding of subjects which he personally regards as non-truths, as well as to those subjects which happen to measure up to own conception of reality. In fact, the librarian's concern with the contents of documents cannot, if he is to fulfil his necessary role as an intermediary between written records and their users, extend to the truth or otherwise of their contents. This matter of detachment or neutrality will be reconsidered later in other contexts, and will emerge as a major factor in the design of any classification system which covers the whole of knowledge.

Despite this fundamental difference, however, it is possible to detect two areas of interest which the philosopher and the classificationist have shared at one time or another. Like classical philosophy, library classifications have traditionally been concerned with the organisation of a universe of knowledge into systems or classes, and students of librarianship are still taught, as an historical background to their craft, the principles of Porphyry and the systems of Aristotle and Bacon. The principal library classifications in use today all incorporate main classes of knowledge in one form or another, and I suspect that this must continue to be the basis on which libraries are going to be organised, even in the age of the computer. There is, however, a shift of interest in recent approaches to documentary classification; a swing of attention away from an organised universe of knowledge and towards the systematic organisation of a universe of concepts. In the sense that this involves us in categorial systems and the combining together of specific concepts to express the sense content of documents, it might be said that our work bears some relationship to both analytical philosophy and linguistics. But this relationship is tenuous, to say the least. We are concerned in an entirely heuristic way with the development of working systems rather than the refinement of explanations, and we hang by a very thin thread on the skirts of these 'higher' disciplines.

Librarians have been concerned with words and their semantic content for as long as they have been organising libraries into classes. Words arranged alphabetically in the subject index have always been the principal route to information in documents, the catalogue serving as the necessary complement to the classified order of the shelves. Cutter's 'Rules for a dictionary catalog' of 1876 still serves in many respects as a working tool, and includes directions for recognising the most significant word in a compound phrase when this is used as a subject heading, 'significance' depending upon grammatical factors such as adjectival or prepositional construction. This concern with words is not, of course, old when measured against a philosophical time scale. In a recent broadcast interview Sir Karl Popper [1]), pointed out, rather disparagingly, that "worrying about words and their meanings is one of the oldest pastimes in philosophy. Plato repeatedly mentions that the Sophist Prodicus was interested in the different meanings of words". What is relatively new in library science, however, is a concern with the systematic organisation of terms into semantically related categories, principles of division being applied to a universe of concepts and not, as in traditional classifications, to a universe of knowledge. In many respects we have had to rediscover those principles which were known intuitively to Peter Roget over a hundred years ago. In the past twenty years or so a number of specialised thesauri have been produced, covering various fields such as engineering, medicine and the construction industries, in which terms are organised to a greater or lesser degree according to their meanings. Taken altogether this represents a considerable effort; one which is nevertheless justified by the need to organise and make available the sense content of a rising tide of information in documents, a phenomenon sometimes referred to as the 'information explosion'. It has become increasingly obvious that traditional library classifications are rather blunt instruments so far as access to knowledge in records is concerned, a fact which has been further highlighted by the availability of mechanised information files and the advent of the third generation of digital computers.

It is worth considering why traditional classification schemes have proved to be inadequate as document retrieval tools, especially in a mechanised environment. In this, as in other fields, performance is a matter of the relationship between function and structure. The principal function of a traditional library classification is the organisation of large collections of documents regarded as so many physical entities. The structural basis of this organisation is a scheme of main classes, such as biology, philosophy, physics and literature, which altogether cover the universe of knowledge as represented in published writings. This presupposes that knowledge possesses known boundaries, that the territory within these boundaries can be divided into separate autonomous states or main classes, and that every subject owes allegiance to one main class. The first approach in the design of these schemes is therefore directly to the universe of knowledge, which is broken down into mutually exclusive classes by the application of various principles of division, this process continuing through the creation of sub-classes and so on down to the point at which the individual concept is identified and set within the context of its class. At each step in this sequence a notational symbol representing the basic class, such as an ordinal string of letters or numbers, is altered or extended in some way, either to show the introduction of a new principle of division or else coordinate membership of the sub-class so formed. In this way a hierarchy of symbols is established which can be transcribed on to documents or written on catalogue cards, so that a logical order is imposed upon the entire collection.

There is no doubt that this is a very efficient way of organising the stock of a library, and it is hard to see how a division of knowledge into mutually exclusive classes could be bettered for this purpose. The system has obvious faults, but it also possesses the two great virtues of ruggedness and simplicity. A system of main classes cannot, however, always provide an equally satisfactory route between the terms of an enquiry and the subject content of documents. Perhaps the most obvious disadvantages of a traditional library classification from this point of view springs from the basic concept of main classes *per se.* A great many documents no longer fit readily into traditional disciplines, so that it is frequently difficult to decide, except in an entirely arbitrary way, to which section of a general scheme a subject should be assigned. The example which comes most readily to mind is the increasing volume of literature on space research, which covers a number of the traditional disciplines ranging from apparently unrelated fields such as medicine and psychology to the obvious fields of engineering and astronomy. Inevitably this means that documents which are of interest to a particular group of users are going to be scattered across a range of different disciplines, and there is no way in which the system can draw these distributed relatives together. A classification based on main classes also possesses a certain rigidity of structure which impairs its ability to cope with advances in knowledge. It is, for example, frequently difficult to fit new aspects of an expanding subject field into the printed schedules of an existing scheme of notation, nor is it possible for the classifier to anticipate how these advances will be dealt with when they are admitted into the scheme.

A further feature of a classification scheme based upon main classes springs directly out of its original approach to the universe of knowledge. Since the whole field of knowledge is fragmented down in a series of steps to the point at which a particular concept is named within the context of its class, it follows that the same idea is bound to be identified by various different class numbers which are scattered throughout the classification. A concept such as 'Water', for example, will be notated in various quite different ways depending on whether it has been considered as a chemical substance, an economic resource, a geographical feature and so on. This is restrictive in the sense that a document on 'Water' has to be assigned to one of these classes even though such a narrow viewpoint was least in the mind of the author. It also leads to bad programming economics when these systems are considered as means for coding the subject content of documents in machine-held files. For this purpose it would clearly be more efficient if we could identify a single conception such as 'Water' by a once-and-for-all symbol which serves as its retrieval 'label' anywhere in the subject file.

The inability of a main class system to deal with specific subject enquiries can be explained most simply in the following way. A discipline-based classification and its parallel scheme of notation sets out to identify the compound subject as concisely as possible and as a single entity. From the viewpoint of library organisation it is important that the symbol which represents a composite subject should indicate the shelf position of a book or the place of a card in the catalogue without becoming either so long or so cumbersome that it ceases to be acceptable for this purpose. But it must then follow that these systems can only occasionally identify as independently retrievable elements the separate parts of a composite subject, even though this is the way in which users of an information service invariably state their enquiries; that is, as questions made up of separate terms formed into phrases which take no account of whether or not the subject so constructed corresponds to some place in the schedules of a particular classification scheme. In order to match these enquiries in any satisfactory way we need a system which gives us the ability to file and search strings of terms which have been constructed with something like the same degree of freedom that the enquirer himself possessed, without concern for the length of the resultant string and despite the fact that such a system would be over-complex as a library organising tool. The classificationist therefore faces a dilemma of some magnitude. On the one hand the librarian requires a relatively simple, robust and subject-expressive system which is capable of organising his library as a collection of physical objects. The information officer, on the other hand, needs a concept-specific system which allows him to match the terms of a user's enquiry against a file of data representing the subjects of his document collections, and consisting of separate concepts organised into strings and held in manually-operated or machine-readable files.

Something of a compromise solution to this problem is found in faceted classifications such as the Universal Decimal scheme and the Colon Classification. These schemes are based on main classes which are fragmented down in the traditional way, but the concepts identified by notation at each step in the division of a class can then be used as components in the building of various composite subjects, including those which are not especially provided for in the printed schedules. Like so many compromise systems, however, these schemes are not entirely satisfactory either for library organisation or for subject retrieval. Although founded on a system of main classes, which is the basis of library organisation, the class numbers created through this synthetic process are frequently so long and cumbersome that they lack the simplicity which is needed for shelf ordering purposes, and one hears of librarians who truncate these numbers for shelving purposes, then use an entirely different system for document retrieval. Nor are these schemes fully effective as coding systems for use in machine-held files. Just as in the more traditional 'enumerative' classification, an individual concept may be identified by more than one symbol depending upon the main class from which it has been selected, and this, as noted earlier, leads to poor programming economics. This may not be a serious problem if we are dealing with a collection restricted to a special subject field which mainly falls within a single class, but it is of some consequence in a general collection which covers the whole of knowledge, and this is the field with which I am principally concerned in this paper. Secondly, these schemes prescribe that concepts should be cited in a predetermined order when class numbers are being constructed from their elements. This is not of itself a bad thing – some form of grammatical prescription in string writing is a necessary factor in an machine-based system, not only to ensure that enquiries put to the system can be made to match the strings already held on the file, but also to distinguish clearly between the different subjects which can be made out of the same component terms. In faceted classifications, however, the prescribed order of concepts is geared to the use of these schemes as tools for library organisation, which means that concepts have to be cited in a sequence which reflects their relative significance as shelving factors without taking any account of the effect which this might have on the meaning-fulness of a subject string.

Perhaps for these various reasons it has come about that library classification, by which I mean the systematic organising and coding of the sense content of documents, has acquired a bad reputation as a subject retrieval tool. People engaged on research into classification now tend to avoid this word altogether, and claim instead that they are studying an 'Information retrieval language' or a 'Concept coding system'. In those fields where rapid access to the content of recent literature is of economic importance – that is, in engineering and other branches of the hard sciences – systems have been developed which have aptly been called the abandonment of classification. Notation has been dropped altogether as a coding or ordering device, and use is made instead of words as they appear in documents. Terms are generally checked against a standard list of approved words, ensuring at least that synonyms are entered under a common name, but this is not mandatory, and in some cases terms are used exactly as they appear in the record to hand, no guidance being given as to the kinship which exists between different but semantically related words. Not only is semantic organisation abandoned, but no attempt is made to indicate the syntactic relationships which link these terms together into a subject statement which summarises the content of a document. Any term which is reckoned as a useful route or 'keyword' to the document is entered on a card, together with a short citation or a code which identifies the document in which the term occurred. Supposing, for example, that a document identified as AM747 dealt with the subject 'Fatigue of roller bearings in aircraft engines'. This document code would then be transcribed on to cards which are separately filed under the headings 'Fatigue', 'Roller bearings', 'Aircraft' and 'Engines'. It is possible that a compound term like 'Roller bearings' might itself be broken down into its component parts, each being entered on a separate card. In using the system an enquiry is broken down into component terms in a similar fashion, and these are matched word by word against the file on cards. Any document code which occurs on all or a reasonable proportion of the cards selected by the user is likely to identify a document which is relevant to his enquiry, and the retrieval and closer scrutiny of the documents themselves can proceed from this point. For example, enquiries for works on (a)'Fatigue of bearings in engines', (b)'Bearings in aircraft', or (c)'Fatigue in aircraft', could all lead to the production of document AM747, and in varying degrees this document has relevance for them all.

It is clear that keyword systems have certain obvious advantages when compared with a traditional classification. For one thing these systems are free to employ an open-ended vocabulary, which means that new terms can be introduced into the system at any time without taking account of the effect which this might have on the printed schedules of a classification scheme. The contents of a document can also be expressed at a level which goes beyond the practical limits of any shelforder classification, and in certain technical fields it is not uncommon for eighty or more terms to be assigned to a single report or specification. Not least important from the user's point of view is the fact that in certain circumstances keywords are an effective way of relating enquiries to particular documents. This has been established beyond a reasonable doubt by a series of tests first conducted in England under strict laboratory conditions and since repeated elsewhere. These experiments showed that the use of keywords without any kind of classificatory structure or term coordination is capable of a retrieval performance at least equal to that of a detailed faceted classification scheme covering the same subject field. The suggestion has therefore been made that classification is an unnecessary extravagance which has no part to play in matching the contents of documents against the needs of their users.

It is difficult to reconcile this attitude to classification with the view held by Körner[2]), of the Department of Philosophy in Bristol University, who maintains that the very act of thinking must involve an act of classification. The fact is, of course, that the use of a keyword index does involve classification, if by this we mean the recognition of certain inter-concept relationships which give meaning to a subject statement. In a keyword system, however, this act has to be performed by the user of the index, possibly at an intuitive level, and no attempt is made to display these relationships overtly at some intermediate stage between the document and its reader.

Before we abandon the idea of explicit classification altogether, however, it is worth noting that keywords have achieved their most obvious success in the indexing of specialised collections of documents in fields such as engineering and metallurgy. These are disciplines which clearly fall within the 'hard' sciences, which I would define as those fields of study in which (a)terminology is reasonably precise, so that a concept is normally expressed by an accepted term which is known and recognised by authors and readers alike, and (b)in which the name of a thing also frequently suggests the grammatical role it is most likely to adopt in the majority of subjects being indexed. These are special circumstances which reduce the need for an overt act of pre-classification, and since authors and their readers share common frames of reference and a common fund of well established terminology we should reasonably expect the user of a keyword index to be able to deduce a correctly formed declarative statement from its unrelated terms. The frames of reference which hold within the field of engineering, for example, are such that it would be difficult to misconstrue the subject which is implicit in a set of unrelated words such as 'Bearings', 'Engines', 'Aircraft' and 'Fatigue'. Even if these words were presented to the user in an order which suggested, for example, that 'Fatigue' is a part or kind of 'Engine', the logical model which regulates information transfer within this discipline would quickly expose the nonsense of these interpretations.

From the indexer's point of view these are very favourable circumstances. They relieve him of the most difficult stage in the indexing process: that of judging the classificatory role of each of the terms which together constitute a subject statement. Unfortunately these circumstances do not hold in every situation. Outside a certain range of special subject fields users of these indexes can no longer feel confident that the subject suggested by the words is the one intended by the author, even when they are able to reconstruct a coherent and meaningful subject statement out of its component parts. There are two reasons for this, one related to the extent of the subject coverage of the collection being indexed, the other to the relative hardness or softness of the subject field when indexing a special collection. From an indexer's point of view the greatest difficulties are, without doubt, encountered when we move away from these 'firmer' areas of knowledge towards the soft end of the subject spectrum. In fields such as politics and sociology it is a commonplace that the meaning of a term frequently changes with the viewpoint of its user, and the same idea can be represented by a variety of different words, each reflecting its user's attitudes. Many of the difficulties which occur in the indexing of the soft sciences can be ascribed to a lack of certainty concerning the correct interpretation of those components of a subject statement which convey the idea of an action: that is, verbs and other terms which imply that an act of some kind is taking place. In the hard sciences these are the terms which usually offer the safest clues to the most likely grammatical roles of the other components in a propositional statement. Actions in these fields tend to be 'one-way', and verbs are consequently capable of only one most likely interpretation, so that it becomes a relatively simple matter to deduce which of the remaining terms should be cast in the role of goal or patient to the action and which should be considered as the agent. For example, if a subject is represented by words like 'Haulage', 'Chains', 'Lumber' and 'Tractors', it is reasonably clear that the tractor is the agent engaged in the hauling of lumber through the medium of a chain. The obviousness of this interpretation arises from the presence of the action term 'Haulage' and our common knowledge of the fact that whereas tractors can engage in the hauling of lumber, lumber cannot conceivably haul tractors. However, we should not be able to deduce a subject with such confidende from a set of unrelated terms such as 'Behaviour', 'Modification', 'Students' and 'Teachers'. It is then no longer clear whether the teacher is modifying the behaviour of the students or *vice versa;* each of these human beings is equally capable of assuming the active or the passive role.

I have come to suspect the existence of a continuum of subjects which ranges from hard to soft and is related to the extent to which those entities which are the concern of a particular discipline can engage in complex or two-way interactions, so that they are capable of assuming a variety of grammatical roles. As we traverse from the hard end of the subject spectrum, in fields like mechanics, to social psychological studies at the soft end, we face an increasing need for the pre-coordination of terms in an index statement, and something more closely resembling the structure of natural language becomes necessary. Without this imposition of structure on a subject statement it is frequently difficult to deduce with confidence 'Who does what to whom', and the chances of a false coordination of terms, leading to the retrieval of irrelevant documents, is correspondingly greater. A typical example of this kind of irrelevancy is quoted by Dr. John Schneider (3), of the National Cancer Institute in Bethesda. Writing about the coding and indexing of the subject 'Production and properties of antibodies against tumour antigens', he points out that:

> "If the descriptor coding method [i.e. keywords] were used, this information would be dissected into three independent words: antibodies, tumour and antigens. Non-specific recombination of these three words describes almost every type of information about the whole field of tumour immunology rather than the limited aspect desired".

If pre-coordination of terms is necessary in the indexing of special fields outside the harder sciences, it follows that the need must be that much greater when we face the indexing of general collections covering all subjects in a single file. Not only are the particular problems which characterise the softer sciences endemic in this situation, but the increased breadth of subject coverage introduces its own problems, of which the lack of common frames of reference and a greater likelihood of homonyms in a verbal subject index are obvious examples. These are circumstances which appear to demand the systematic organisation of the subject file, and yet, as noted aerlier, this is exactly the area in which existing general classification schemes, based on main classes and intended for library organisation, have proved inadequate as citation retrieval systems, especially when considered from the viewpoint of machine filing and searching. The use of notation to represent particular concepts may overcome one kind of homonym problem, insofar as the different ideas which can be conveyed by the same word are coded in different ways, but we face a kind of inverse homonym problem in any scheme based on main classes, since the same idea can be represented by various different codes depending on the class to which it has been assigned.

The two kinds of subject retrieval mechanism which have been considered so far – that is, classifications based on divisions of the universe of knowledge and keyword indexes which are altogether lacking in explicit classificatory structure – probably represent the two ends of the classificatory continuum so far as library science is concerned. Each possesses certain advantages and disadvantages not shared by the other. In a library classification scheme, for example, ideas are expressed in the form of notation, which means that concepts can be organised systematically regardless of the words which convey them in natural language, and they are displayed in such a way that certain kinds of relationship between them can be readily exploited. At the same time complex subjects are identified as entities in their own right, so that the chances of an ambiguous interpretation of inter-concept relationship is reduced. On the debit side, however traditional classification tends to be restrictive, in the sense that the classifier has to assign a concept to some main class or other even though this kind of limitation may distort the author's intention, and these schemes also inhibit the classifier's ability to represent subjects freely and specifically as they are encountered. Keyword systems cannot be called inhibiting in this sense, since the constituent terms which together make up a complex subject can be entered as separate access points on a file as soon as they are encountered, but in abandoning systematic structure these systems have also surrendered meaningfulness to a large measure, at least so far as its deliberate display is concerned. This becomes increasingly serious as the subject field being indexed becomes softer, so that the frames of reference shared by the author and the user are no longer capable of suggesting a correct and unambiguous reconstruction of a subject from its unrelated parts.

These are the problems which have engaged the attention of the Classification Research Group in London for the past twenty years or so, and have led to a new approach to the whole question of concept organisation for subject storage and retrieval, especially where mechanised systems for use in a general or cross-discipline environment are concerned. The goal of this research is a computer-based system which permits the coding of a subject in such a way that:

(a) any subject can be constructed freely and to any depth as it is encountered, strings of concepts being organised consistently to a standard pattern by reference to rules which apply at both the input and the output ends of the system. Besides ensuring that common themes are collocated for scanning purposes, these rules would also help to promote optimum use of computer time by ensuring that enquiries put to the system match the subjects already held on the file.

(b) any selected component of a subject string can serve as an access point to the file, the subject being stated in full at each entry point and the string re-structured in such a way that the user can immediately establish the context in which his entry term has been considered without loss of meaning in the rest of the subject statement.

(c) specific concepts are represented by codes identify them on a once-and-for-all basis regardless of the subject in which they appear. These codes should not, however, be randomly assigned labels, but should express the fact that a concept is a member of one or more categories of semantically related ideas. A subject search might then be broadened or narrowed at the level of an individual component in a string, so adding a second or 'browsing' dimension to any machine-based search.

The development of a system capable of meeting these requirements demands an approach to the problems of concept organisation which differs in certain fundamental respects from that found in traditional library classifications. If the classifier is to have free access to specific concepts for the purpose of subject building he must be provided with a comprehensive and systematically organised source of ideas such as a general thesaurus, and since the intention is to organise these ideas into semantically related categories which are not generally conveyed by words, this thesaurus must be further equipped with a suitable index. The first task, therefore, consists in a systematic organisation of the universe of concepts, rather than of the universe of knowledge which has traditionally formed the starting point in the construction of a library classification.

The second difference is to be found at the level of the subject rather than of the concept. Because the traditional scheme functions principally as a tool for organising a collection of documents, the subjects set down in its schedules grow in complexity through the introduction of concepts in some order which reflects their relative significance as shelving factors, rather than in an order based on logical or predictable principles. Supposing, for example, that we are dealing with documents on 'Management of the purchasing departments of a hospital', and that the scheme we are using collocates works on management altogether on the shelves. This particular factor will then be emphasised in the sense that it establishes the class to which the document belongs, and the remaining concepts – that is, 'Purchasing departments' and 'Hospitals' – will be assigned to subordinate places within the class 'Management'. The fact that in this case an action concept such as management has been regarded as the principal factor does not, however, mean that we can recognise 'Action before thing' as a grammatical principle of division which holds throughout the scheme. In the class 'Music', for example, it is more than likely that the reverse situation will apply, and that an action concept such as 'Performance' will be subordinated to a class of instruments such as 'Voice' and 'Piano'. The needs of one class are often at variance with those of another in this respect, and I doubt whether we shall ever be able to establish general principles which can usefully regulate the order in which different kinds of idea should be introduced into the schedules of a shelf order classification.

However, there is no point in trying to impose factors of this kind on the design of machine-held files. In the kind of computer-based system which is envisaged it is intended that a basic string of concepts should be re-formatted or rotated by the computer in such a way that selected elements in the string come successively to the lead position, where each can serve in its turn as the user's access point to the file. At each stage in this process a new term will therefore assume the role of 'principal concept' in that particular version of the string, and it then becomes more important to ensure that strings are organised in such a way that the relationships between concepts are preserved throughout this reformatting operation, rather than to impose some arbitrary decision based on shelving factors upon the ordering of terms.

Two essential working tools will have to be created before this system can become a practical proposition. Besides the general thesaurus which has already been mentioned as the source of terms or ideas, we should also require a general decisionmaking model to which reference is made when the parts of a compound subject are being organised to a consistent pattern. I should hasten to add that neither of these tools exists at present, though tentative steps towards their creation have now been taken. In the sense that the creation of these tools involves some awareness of fundamental relationships between ideas, it is here that the interests of the classificationist begin to approach those of the analytical philosopher and linguist. Insofar as the construction of a general thesaurus involves the organisation of terms into categorial systems it is at least in part concerned with those *a priori* relationships between basic units of language which are of interest to linguistic philosophers, while the creation of a decision-making model for string writing clearly amounts to the formulation of an artificial syntax, which must, if it is to be effective, bear at least a superficial resemblence to the syntax of natural language.

Without doubt the classificationist has the easier task, since he is developing a tool, not explaining phenomena. This tool may be either a paralanguage – that is, a formal indexing language using a controlled terminology and its own kind of syntax – or else a metalanguage, in which specific ideas are represented by codes which are neutral with respect to the terms of any one natural language. In developing these tools the classificationist is in the fortunate position of being able to create his own rules as he goes along, and he can change the rules or modify his product until he is satisfied that the system he is developing performs in an acceptable manner. Linguists and philosophers, on the other hand, have the far more exacting task of analysing and explaining the phenomena of language and meaning as they find them; they can neither ignore nor change any part of these phenomena on the grounds that they defy a present explanation.

Since it is a reasonably simple matter to distinguish between these two basic kinds of relationship – that is, semantic or categorial on the one hand and syntactic on the other – it is convenient to deal with them separately here. I have tried elsewhere to present a general review showing how these different relationships have been dealt with in various recently developed index systems[4]), and shall limit myself here to a summary account of the findings of the particular project in which I have been directly involved, beginning with an account of current research into the problems of establishing a general syntactical model for use in the production of an alphabetical subject index from mechine-held files. Insofar as this project demonstrates many of the problems which have to be faced in the computerisation of index systems it can be taken as representative of much of the work which is now being carried out in this field. Although this research is limited in scope to alphabetical indexing I shall later consider its implications for a general system of data exchange and file searching using elements of notation rather than words to express the subjects of documents.

Research into this matter is at present being carried out in London as part of an indexing project financed by the Office for Scientific and Technical Information and accommodated at the *British National Bibliography*. This project, in its turn, grew out of a larger enquiry into the possibility that the serial parts and cumulations of the *British National Bibliography* could be printed directly from the magnetic tapes which this organisation produces in connection with the international MARC Project. These are tapes which carry catalogue data organised to an international standard and recording the national output of monograph literature published in Britain. They are dispatched regularly to a number of institutions including national centres in other countries. The *Library of Congress* in Washington, which is where the MARC Project originated, produces similar tapes recording the output of American publishing houses, and it is likely that further European countries and Canada will shortly join this cooperative international network. Throughout the research we have been constantly aware of this background, especially insofar as we could anticipate the ultimate need for a system capable of dealing with the exchange of subject information at the international and therefore trans-language level.

Our involvement in this project raised the immediate question of whether the national bibliography itself, including its subject index, could be produced directly from machine-readable records. We first, and naturally, tried to computerise the 'chain index' we were already using, but this proved to be more difficult than had been expected, and the product left much to be desired. A few words on these early stages of the project may be of general interest since they demonstrate two aspects of computerisation which are, I believe, common experience. In the first place the adoption of mechanised processing forces one into a careful analysis of existing procedures and usually reveals any inherent flaws in their logic. Secondly, the ability of these machines to handle large amounts of routine work quickly and accurately often promotes a higher level of expectation concerning the product.

The production of a chain index is geared directly to the schedules of whichever classification scheme is used to arrange document citations in the catalogue. After a subject had been assigned to its place in the schedules and its class number has been allocated, a 'chain' of entries is forged link by link, the first link being the name of the class to which the document has been assigned. As new principles of division are brought into play further terms are added to the front of the index statement, which continues to grow until the chain is complete. For example, the entries relating to the subject 'Women teachers as a profession' would be constructed in the following way:

Decimal Classification (14th edition)	*Chain index entries*
330	Economics
331	Labour: Economics
331.4	Women: Labour: Economics
331.48137	Teachers: Women: Labour: Economics

Although this procedure ensures that significant terms such as 'Women' function as access points to the index and are presented as part of a pre-coordinated subject statement, it can be seen that only one of these entries – that which is offered to the user under the last term in the chain – is in any sense co-extensive with the subject of the document. All other entries are less specific, and are therefore liable to mislead the user into supposing that a document dealing with one of these more general topics can be traced in the bibliography even when this is not the case. This particular aspect of the chain index has often been criticised, but the fault is inherent in the system. It should also be noted that the order in which terms are added to an entry exactly reflects the order of the classification tables. This means that consistency in the index depends entirely upon the extent to which principles of division are consistently applied in the scheme we are using. We have already seen that rigorous consistency of this kind has no part to play in the design of a shelf order classification, and we have to accept that the inconsistent phrasing of subjects which are basically similar is also an inherent feature of the system. As a final point we should note that the order of terms in the entries of a chain index, since it is based upon some conception of relative significance, is not necessarily related to the way in which words are used in natural language, and a correct and ready interpretation of entries often depends upon a working knowledge of the classification scheme in use. The human indexer can do much to ameliorate these troubles through a careful choise of terms, but this sort of intellectual discrimination is quite beyond the capacity of a computer.

In our search for a suitable system we subjected various other indexing methods to scrutiny of this kind, especially those which had already been successfully handled in the computer, but none appeared to meet our special requirements. For reasons which were considered earlier we felt committed to a pre-coordinated index rather than a keyword system. Documents covering the whole range of knowledge are handled on a day-to-day basis as they are received from publishers, and the next book to hand might equally be a pre-school reader, a manual on the readiological testing of welds or a review of the economic variables which influence academic achievement in slow learning children. Besides these subject factors we also had to take the physical form of the index into account. We produce a printed-page index, which means that our readers cannot use certain searching techniques, such as optical coincidence and hole punching, which are available to the user of a keyword index on cards. Our basic requirements amounted, therefore, to a fully pre-coordinated index which is capable of producing meaningful entries from machine-held files. To ensure that works on the same theme are consistently indexed under the same term or terms we wished to retain full control over the index vocabulary, just as we had with the chain index. In one respect, however, we raised our expectations, and proposed to offer the user an entry which is co-extensive with the subject under every significant term in an index string. From the viewpoint of machine economics it would also be an advantage if a full set of entries could be generated from a single string of terms, rather than store every entry on the file as a separate and discrete item.

Research into a system capable of meeting these particular requirements has led to the development of a new kind of subject index called PRECIS, (5,6,) or PREserved

Context Index System, and the results now appear regularly in the monthly index of the *British National Bibliography*. Perhaps not surprisingly, we found that we could achieve these aims only at the cost of surrendering some of our basic preconceptions concerning the content and layout of a subject entry; this is always a painful and difficult thing to do. To begin with we had to abandon the idea that it is necessary to order terms in an index entry according to some notion concerning their relative significance. We also found that the usual single-line entry of an alphabetical index inhibits the expression of certain kinds of inter-concept relationships, and we finally developed a two-line structure which of its own accord imparts a certain amount of relational information.

In the strings we now prepare for computer input no attempt is made to cite terms in an order which reflects their relative significance; this was, in any case, always a difficult thing to decide. Instead, terms are set down in such a way that one term establishes the context for the next one in the string, in the way, for example, that the name of a whole sets its parts into their context, or the name of a country serves as an obvious setting for its peoples and their interactions. Our interpretation of context-dependency has been derived from a study of various inter-concept relationships, such as whole-part and patient-action-agent. The basis of this order can be represented formally in the diagram:

A>B>C>D

which shows each term considered in the context of the term to its left. If this string is no looked at from the special viewpoint of one of its middle terms (the term 'C', for example) it will be seen that, regardless of the number of terms involved, the string breaks into three independent parts:

(A B) > (C) > (D)

Those terms which appear to the left of 'C' in the string can be considered altogether as establishing the wider context of 'C', while any term or terms to the right of 'C' ('D' in this example) are themselves context dependent on 'C'. This tripartite division of a string of terms forms the basis for the two line entry which is characteristic of the PRECIS index:

LEAD: Term 'C'	QUALIFIER. Terms B.A.
DISPLAY: Term D	

Three stages are involved in producing a set of entries structured to this pattern out of an initial string of terms:

(a) the indexer studies a document and writes an initial string which summarises its subject content and is organised according to the notion of context dependency

(b) he selects and marks those terms which are required as entry points in the index, i.e. those words which will appear in the lead position.

(c) machine instruction codes are then added to the string, which is keyboarded input to the computer.

The terms in the string are then 'shunted' by the computer through the three basic positions shown above, leading to a set of entries structured in the following way:

A
B.C.D

B. A
C.D

C. B.A
D

D. C.B.A.

It should be noted that as terms are moved across from the Lead position and into the Qualifier their original order is reversed, so that an elementary kind of transformation takes place. This means that the lead term is directly related to the terms which are adjacent in both lines of this entry, so that the user can immediately establish both the wider and the narrower terms which relate directly to his chosen entry word.

We have noticed that a string of terms organised on the basis of narrowing context bears some resemblence to the structure of the passive sentence in English, or at least corresponds to a reversal of the order of terms in the usual declarative statement. For example, a declarative statement such as 'Higher education of women in Canada', will be transformed during the initial stage of concept analysis into a string of substantives in the reverse of this order:

Canada / Women / Higher education

This string of terms would then serve as the raw material from which the computer would generate the multiple entries:

Canada
Women. Higher education

Women. Canada
Higher education

Higher education. Women. Canada

However, although the order of terms in the basic string may bear some resemblence to the order of terms in a passive sentence, this cannot be regarded as precise enough to serve as an instruction to indexers which would ensure that different indexers, or the same indexer on different occassions, will organise subjects consistently to the same pattern. English permits a variety of similar surface structures having the same semantic content, and the passive transformation is strictly valid only in those cases where a subject, verb and object are involved.

To ensure that entries are organised to a consistent pattern, therefore, the indexer uses a decision-making model developed from a study of syntactic roles which regulates the order in which terms should be cited in the input string. This model consists of a list of role operatores, the most important of which are shown below:

PRECIS. Role operators

(/) Quase-generic relationship
, Difference: *e.g. concept used adjectivally*

(p) Part
(q) Property

(0) Study region; Sample population
(1) Viewpoint, perspective
(2) Active concept
(3) Effect, action
(4) Key system
(6) Environment

When using this model the indexer first determines which concepts are necessary for an adequate summarisation of the subject content of a document, and he next considers the role which each of these concepts plays in the subject as a whole. These roles are then indicated by writing the appropriate operator in front of each term, which may consist of a single substantive element but can also include adjectives and prepositions. Finally, since each of these codes carries a filing value, terms are set down in the order prescribed by the codes, leading to a string of terms suitable for machine manipulation.

The technique is not so complex as this brief description would suggest. As a training aid we recently prepared a set of decision-making charts based on these operators, setting down in detail the steps involved in the content analysis of a document. We found, however, that indexers quickly tire of the 'decision-making game', and they soon work efficiently at a far more intuitive level. The kinds of decision involved can be demonstrated most simply by considering once more the straightforward subject of 'Higher education of women in Canada'. As a first step this is broken down into its three substantive elements: 'Women'; 'Higher education'; 'Canada'. A rule of the system states that every string must include a concept coded (3), which represents an action, and/or a key system, coded (4), much as every sentence must include at least a verb and an object. In this instance we can recognise 'Higher education' as an activity concept and 'Women' as the key system on which the action is being performed. Since these numbered operators are filed retroactively we can start the process of subject building by writing the string:

(4) Women (3) Higher education

This leaves the locality 'Canada' to be considered, and at this point it is necessary to examine the document more closely to determine the exact role of this concept. If the work is a review of this subject from a specifically Canadian viewpoint this locality is regarded as the 'Environment' and is coded (6), giving us the final version of the string:

(6) Canada (4) Women (3) Higher education

This is, of course, the order of terms which was established earlier when this subject was considered from the viewpoint of sentence structure. If, however, the author states that he is examining a general problem or propounding a general thesis but happened to use Canadian examples to make his point, then Canada is coded (0) to indicate a study region. This code declares that we are dealing with a different kind of relationship and calls for a change of typeface and a different layout in the printed index. Examples of this can be seen in the Appendix.

Once these role operators have been used to organise the parts of a subject into a string of terms they have, so far as indexing is concerned, finished their job and drop out of the picture. The indexer ticks those words which he wishes to appear in the lead position, and the string is next passed to an assistant who replaces the operators with machine instruction codes, producing a 'manipulation string' which is keyboarded ready for the computer. These machine instructions generate entries under the words selected by the indexer and they also control their layout and typography, but they are not concerned with intellectual content.

It can be seen that these role operators do more than organise the parts of a subject statement which already exists as some abstract conception in the mind of the indexer. By laying down certain conditions which have to be met and causing certain questions to be asked they also help the indexer to formulate the subject in his mind during the examination of a document. In the example shown above concepts fall into a straightforward sequence based on their one-to-one relationships, so that no special problems are encountered, but the system of operators is capable of dealing with far more complicated situations than this, including compound subjects in which two or more concepts share a mutual or coordinate relationship to some other component of the string.

The principal function of these operators is to ensure that terms are organised to a standard pattern which allows us to use the computer for what has then become the clerical task of producing a set of consistently structured entries out of a single string of terms. We have also found that the organisation of entries according to the roles of their conceptual elements, rather than some notion of relative significance, helps us to display homologies between different disciplines which have been obscured by the structure of a shelf order classification. In addition, we are naturally concerned with the meaningfulness of the entries we produce, though I would hesitate to go further than this and suggest that the system of operators we are developing and the notion of context dependency on which they are based may have wider relevance for a study of the way in which meaning is conveyed in natural language. Hutchins[7]) points out that:

> "For the purposes of searching index files there need be, in theory, no parallelism between the classificatory structure of index languages and the structure of natural language".

We have, nevertheless, tried to take account of the relationship between meaning and structure as it occurs in natural language in our search for an order of terms which by itself is capable of suggesting a correct interpretation of the subject, even though the index language is controlled, formal, and lacks many features of natural language such as articles and prepositions. We would go so far as to suggest 'invisibility' as one of the more desirable characteristics of an index entry, meaning that a user should be able to grasp the intellectual content of the entry without being made unduly aware of the mechanics of the information transfer process. Macdonald[8]) has shown that a successful paralanguage should have:

> "Many characteristics in common with natural language, utilising some portion of its vocabulary and its structure. Such a system must parallel the original language, fully or sketchily, if it is to fulfill its function as an indicator of the content of the original language. . . The problem of information storage and retrieval is one of translation between language and paralanguage".

Reactions from users of PRECIS have generally confirmed our expectation that the meaning of a string of terms, even when stripped of certain adjuncts of natural language, can nevertheless be perceived correctly and as a whole providing that terms are set down is such a way that one establishes the context for the next. Entries relating to complex subjects have been interpreted accurately and immediately by users who had received no explanation of either our entry structure or of our methodology, and it would seem that we are all, in reading as in many other things, particularly sensitive to the part which context plays in conveying a sense of correctness. If we suspect that a straightforward juxtaposing of terms will not of its own accord be sufficient to establish a readily understood context, or when the indexer considers that the entry which will be produced under some particular term is liable to an ambiguous reading, we can bring a further mechanism into play and introduce prepositional links between adjacent terms. In these cases the computer constructs prepositional phrases out of elements coded in certain ways in the manipulation string. It then produces different versions of the phrase in different parts of the entry depending on which word is in the lead position, so that a second kind of simple transformation is performed. Examples of this are also shown in the Appendix.

There is one other category of terms, apart from linking words such as articles and prepositions, which is generally dropped from a string when it is prepared for input. A policy of exclusion applies whenever two adjacent terms are related in such a way that one establishes the generally recognised category to which the next belongs, rather than its context. Except when the lower term is a homonym and the correct meaning can be indicated quite simply by naming the class of which it is a member, the higher of two terms in a class-to-member or categorial relationship is usually dropped from a string.

This is a characteristic which all pre-coordinated indexes share with natural language. In writing or speaking we would not normally interrupt the flow of communication by defining each term as we use it through reference to the higher class of which it is a member. If, for example, 'Rats' is an item of discourse, there would be no point in overloading a sentence with words explaining that we are referring to a kind of 'Rodent' which is itself a member of the class 'Animals'.

These relationships are already present by implication.

An index system, however, (unlike natural language) must make provision for the user who enters the file under the higher term without knowing that documents which may be relevant to his enquiry have been indexed under the lower. In PRECIS the link between these category-establishing terms is therefore handled by a separate system of *'See also'* references from higher to lower terms, and in addition the reader is directed via *'See'* references from unused forms of synonyms to the form which is used in entries. These references are generated from random access disc files by the simple procedure of quoting, as part of the PRECIS input, a numbered code which identifies the address on the disc of any term which appears in the lead position in the index. This address is then located by the computer, which reads off the name of the term to which a reference should be made together with further codes which establish (a) the kind of reference which has to be made to this term, whether *'See'* or *'See also'*, and (b) the address of the term *from* which the reference should be made. This second term is then located and the correct reference is automatically reproduced. Except in the case of synonyms this second address may itself include directions to print a yet higher level of reference, and the procedure is repeated until an entire hierarchy of references has been set up. For example, if the term 'Rats' occurs in a string and is needed as a lead term in the index, quoting a single code will automatically produce from the file both the first leven reference:

Rodents
See also **Rats**

and also any second or subsequent levels of reference such as:

Animals
See also **Rodents**

the machine continuing to extract these references until the hierarchy is exhausted.

In turning to consider this system of references held in a thesaurus it is clear that we have moved away from what might be called the syntactic aspects of an index language and on to the semantic or categorial features. The distinction between these different kinds of inter-concept relationship is also recognised by the analytical philosopher. Strawson [9]), during an interview at Oxford, discussed his views on language and remarked on a need "for closing the explanation gap between semantico logical features on the one hand and syntactic features on the other". So far as setting up an index system is concerned, however, we have found it more convenient to regard these two kinds of inter-concept relationship as sufficiently distinct to form the basis for separate stages in the indexing process. Each of these fundamental relationships is handled by a different team of people, their products are stored on different files, and the computer programs involved are also entirely different.

After a subject statement has been organised into a string of terms and operators it is examined by the team responsible for thesaurus construction. Inevitably there is interaction between this team and the string writers, insofar as certain terms are sometimes more appropriate in certain syntactical structures, but as a general rule the thesaurus team is less concerned with the logical structure of the string than with establishing any references which could appropriately be made to its component terms. Each term is therefore considered as an entity outside the context of the subject in hand, and is checked against various authorities, such as general and special dictionaries and thesauri and the schedules of classification schemes.

Two conditions have to be satisfied when these general to special references are being established:

(a) References should be made from higher terms which would function in their own right as access points in the index.

(b) A reference should remain valid in any of the different contexts in which a term is likely to appear, and should not be limited in its use to the subject in hand.

This second point is of some importance. The operators used in PRECIS can conveniently be regarded as coded versions of what Hutchins 7) calls the 'edges' between units of thought, usually conveyed by sememes. Each operator introduces a syntactic category, such as 'Active system', which is known only *a posteriori* and might be occupied by a wide range of different but equally appropriate concepts, and a particular concept may rightly assume a variety of roles in different kinds of subject. For the sake of a smoothly working system we need to establish references which are entirely neutral of context. We need the ability to quote the same code and generate the same references confident that these will continue to be useful and valid regardless of the role which a concept has assumed in the subject immediately to hand.

The criteria used in meeting these two conditions have tended to change as the indexing experiment has proceeded. At first we tried to establish a tight and logical control over the references admitted into the thesaurus, basing this on a severely limited interpretation of the class-to-member relationship. A relatively simple procedure was used to test for this relationship, a condition of 'Some-ness' being sought when reading from the higher to the lower term (e.g. '*Some* rodents are rats') and a condition of 'All-ness' when reading from the lower term to the higher ('*All* rats are rodents'). Both conditions had to be satisfied before a reference was created, and the test enabled us to distinguish between those generic links which hold good in any context, such as 'Rodents/Rats', and those which are, in fact, context dependent, such as 'Pets/Rats', in which case the upward reading 'All' condition cannot be satisfied. In the latter case both terms are cited in the concept string, a special role operator dealing with this 'Quasi-generic' relationship.

This test holds good in an obvious way when it is applied to physical objects in fields such as biology and chemistry which already possess well established taxonomies. It can, however, also be used to determine the existence of similar taxonomic groupings quite outside these fields, including relationships between actions such as:

Motion – Rotation – Spin

and abstract entities such as:

Shapes – Rectangles – Squares

It soon became evident,.however, that on commonsense ground we had to make certain references which would not pass this severely limited test. A string containing a term like 'Brain', for example, would be absurdly overloaded if, on purely logical grounds, we had to include other terms such as 'Central nervous system' to explain its presence. Altough this represents a whole-to-part rather than a class-to-member relationship, it is still better handled by means of a once-and-for-all reference which can be used whenever 'Brain' appears as part of any subject statement. The basis for reference construction has therefore been widened, and terms are now recognised as 'higher' if they are essential factors in a definition of the lower term or are clearly present by implication whenever the lower term is used. In the final resort this means that we now rely largely on our own judgement, backed by a careful scrutiny of the documents we handle, to tell us that the user who enters the index at some particular term is likely to find works on a 'narrower' concept still relevant to the subject of his enquiry.

It would seem at first glance that we have chosen to act unscientifically in setting aside a 'logically' testable condition and allowing subjective judgement to enter the system. In defence of our action, however, I would question how far this is really the case. I suspect that the grounds on which we all regard 'Rats' as belonging intrinsically and *a priori* to the class 'Rodents', but not to the classes 'Pests' and 'Pets', are in some respects also subjective, though the level of subjectivity has been raised from the personal to that of the culture. Körner[2]) makes the point that:

> "The manner in which a person classifies the objects of his experience into highest classes or categories, the standards of intelligibility which he applies, and the metaphysical beliefs which he holds are intimately related".

This being so, we can hardly avoid seeming to declare a particular standpoint whenever we assign a term to this or that category.

It could be argued that the category 'Rodents' was selected on what appear to be 'scientific' grounds, insofar as 'Rodents' is a grouping established by scientists which we are taught to regard as valid in any of the contexts in which the term 'Rats' appears. In that case 'Rodents' is obviously the class which best suits our purpose in setting up a general thesaurus, since this is the category which least inhibits our use of whatever reference is entered on the file. But we should have tried in just the same way to establish a 'most neutral' category if we had been working within a framework belonging to an entirely different culture; if, for example, we lived in a world which accepted as both obvious and scientific the delightfully fantastic categorisation of animals reported by Foucault[10]) as "appearing in a certain Chinese encyclopaedia". In this it was written that animals are classed as:

(a) belonging to the Emperor
(b) embalmed
(c) tame
(d) sucking pigs
(e) sirens
(f) fabulous
(g) stray dogs
(h) included in the present classification
(i) frenzied
(j) innumerable
(k) drawn with a very fine camel hair brush
(l) *et cetera*
(m) having just broken the water pitcher
(n) that from a long way off look like flies

Whether this taxonomy is apocryphal or not is beside the point; it is enough to demonstrate the role of the classificationist as one who selects, from among the categories available, the one most likely to hold good in the majority of subjects being indexed.

If even the categorisation of animals is subject to changing cultural fashion in this way it is obvious that we cannot hope to make a classification covering the entire subject spectrum, including soft and intrinsically partisan concepts such as 'Justice' and 'Democracy', and expect it to stand for all time; the best we can hope to achieve is a system which works for the present. We have, in fact, incorporated a correction mechanism which allows us to re-allocate concepts as and when necessary by the simple procedure of changing their address codes on the machine-held file. This concession to uncertainty may itself be symptomatic of our times. It should perhaps be seen in contrast to the confidence with which the great nineteenth-century classificationists established their systems of main classes.

In one respect this is an unusual approach to the construction of a thesaurus. As a general rule a fairly extensive and representative sample of terms belonging to a given field is gathered together and reviewed as a whole before any attempt is made to sort terms into categories. The classificationist then has an opportunity to step back and, as it were, survey the field from a distance, mapping out its frontiers and recording its internal boundaries before he begins to assign specific terms to their appropriate regions. This resembles in some ways the approach which the traditional classificationist brought to the universe of knowledge, in that principles of division first have to be established and then applied to the field as a whole, causing its breakdown into smaller and more manageable areas before the finer details are considered. Obviously we need quite different principles of division when facing a universe of concepts, and some of these principles and the outline categories they produce have now been tentatively established by the Classification Research Group in London, following work done by Farradane and reported elsewhere at this Conference.

In the PRECIS Project the approach to a system of categories has, of necessity been entirely different. In the course of day-to-day work we have to face each term as it comes, and we construct our hierarchies from the term upward than from the category downward, so that at present we have no clear picture of the kind of map we are making. Later in the project, however, we intend to print out this disc-held thesaurus and organise its contents on a systematic basis. When this time comes it will be interesting, to say the least, to see how closely the categories which have evolved in this working environment match those which were postulated on entirely theoretical grounds during the classification research.

There can be little doubt that when we reach this point and begin the task of systematic organisation we shall have to turn our attention to the most powerful weapon in the armoury of the classificationist: that is, a scheme of notation. A carefully chosen notation would not only serve to label each concept on a once-and-for-all basis but it could also be used to indicate the category or categories to which each concept belongs. Instances occur in all languages of closely related ideas which are represented by entirely dissimilar words. A scheme of notation can draw these together in such a way that their familial relationship is made explicit regardless of the words which represent them.

The introduction of notation into the scheme of things completely alters the horizons within which we work, taking us immediately out of the world of the alphabetical index and into the more abstruse and conjectural world of the metalanguage. Not that the boundary between them is entirely distinct. PRECIS evolved directly out of research into the theoretical basis for a new general classification which was conducted by the Classification Research Group during the 1960's, and many features developed during this research, including the role operators and the idea of categories neutral with respect to context, have been incorporated into the alphabetical index.

Remembering this classificatory background to the indexing project, and bearing in mind the fact that we work in an organisation which is closely concerned with the problems of data exchange at both the national and international levels, I should like to outline the reasons why I think we need to go on to develop the metalanguage, then briefly consider some of the problems involved.

Essentially we are trying to develop a tool which is capable of penetrating certain barriers to communication which have arisen through historical accident. We are faced by two kinds of barrier, one inter-disciplinary and the other interlingual. It has frequently been reported that scientists learn of relevant work in fields outside their own mainly by chance and usually too late to be of service. Much of this loss can be ascribed to the fact that workers in different fields use different sources of information, such as abstracting and indexing services, which never form part of a common data base. Some is due to the fact that different disciplines tend to use different words to describe what are essentially the same or closely related phenomena, so that even if a common data base did exist the loss might still occur. On a more obvious level work which is reported in one language is frequently missed by those who use another. The days of Latin as the *lingua franca* of science are long past, and no one language in current use seems likely to take its place.

These are not new problems. Duplication of effort has been a recurring theme in the field of scientific communication. This may, however, be an appropriate time to reconsider these problems and their possible solution in the new terms of common data banks and the international exchange of information in machine-readable form. I mentioned earlier that the MARC Project is at present limited in scope to the exchange of data concerning (a) monograph publications and (b) between two national bibliographic centres, Washington and London. There is no reason to suppose that these two limitations will continue indefinitely. In England a call has already been heard for MARC or some similar service to extend its coverage to report literature and ultimately to journal articles, and there is little doubt that non-English speaking countries will shortly join the MARC network. Hardware and software capable of dealing with information exchange on this scale already exist, and both are subject to constant improvement. An agreed international standard to deal with descriptive cataloguing is now being formulated. Yet we still lack a general technique which is capable of describing the subject content of a document in an entirely adequate way, and which is also designed to exploit the potential of machine-held files.

It is against this background that the concept of a neutral system assumes real importance, not so much as a means for producing printed indexes but as a basis for data exchange and file searching. In the context of an international exchange network the representation of subjects as strings of notated elements plus codes which identify the exact relationships between them possesses the obvious advantage of equal opportunity of access regardless of the language of the user. Notation also allows the classificationist to work at the level of the idea rather than the word. This does more than resolve certain problems such as homonyms which are inherent in any word-based system. It also adds an extra dimension to retrieval by drawing together in an exploitable way terms which are closely relaty in semantic content but are separated in an alphabetical file. It is not suggested that exactly the same code should be allocated to different words where valid distinctions between them are more than a matter of context. But we should at least be able, by the relatively simple procedure of changing the length or the final digit of a symbol, to draw the attention of a worker in one field to related studies in another; so that, for example, the sociologist concerned with formulae for measuring the spread of innovation can be made aware of relevant techniques used by biologists to study the growth rate of bacterial cultures.

The arrangement of concepts in categories and their representation in the form of neutral codes amounts, of course, to only half a subject retrieval language. We still have to face the problem of joining these separate pieces of notation together to represent compound subjects. The basic criteria are much the same as those which had to be taken into account when the scheme of operators was developed for PRECIS, in that (a) different indexers will be expected to achieve consistent results in the material they contribute to a common data base, and (b) the strings which are produced should be capable of interpretation by various users without distortion or ambiguity. In the case of a common retrieval language serving an international network the situation is made more complicated by the fact indexers and users in different national centres will bring to this work a variety of local language backgrounds. Again there is clearly a need for a neutral system, this time syntactical rather than categorial.

With this need in mind we have conducted some small-scale experiments in translating complex PRECIS strings term by term into various other languages, wondering how far a string of terms organised entirely on the basis of context dependency might still be capable of producing meaningful entries. The kind of result we obtained can be seen in the Appendix, which shows entries in English and French produced from a common manipulation string to demonstrate the system for the National Library in Ottawa. Similar tests have been carried out in German, Russian, Czech and Sinhalese, and the results have been very encouraging, especially since they revealed an unexpected relationship between the three different parts of a PRECIS entry and certain changes of case. This relationship could not have been detected so readily in English, which has generally shed its inflections where case is concerned. We are now considering an additional subroutine to the program which would allow us to alter the form of a word by selecting the appropriate prefix or suffix as the word is shunted through the three different positions in an entry.

These experiments are reported only to show the kind of enquiry which needs to be done, not to suggest that the syntax we use for PRECIS will serve for all purposes. This is a matter requiring more careful investigation than we have had time or resources to give. But at least a start has been made, and perhaps at this stage it is something to know that the aim of a general syntactical model for data exchange at the trans-language level is probably not beyond our grasp. The ultimate feasibility of this aim is further reinforced by recent developments in the field of linguistics, notably, of course, the work done by Chomsky and his followers in transformational grammar, with its underlying notion of a deep structure which is common to all languages.

I would set as an immediate aim a computer-held file which is capable of the following performance:

(a) it should allow a user to interrogate the file under any term in his enquiry, permitting the identification of subjects in which specific concepts or combinations of concepts have occured.

(b) the user should be able to discriminate between different combinations of concepts in terms of their varying inter-concept relationships.

(c) it should allow the user to explore round the subject of his enquiry, first by allowing him to broaden or narrow the extension of any of its components, and secondly by indicating on demand subjects in which different concepts have occurred in some specified role.

(d) finally, since both storage and retrieval would be conducted in the medium of codes or notation, the system should be capable of interrogation regardless of the language of the questioner.

It might, as an interim measure, be necessary to conduct this discourse between an information file and its users through the intermediary of a specialist staff who translate enquiries into their coded equivalents before they are matched against files. Ultimately, however, I should expect even this to become amenable to machine handling, so that the user can interrogate the file directly and in his own language without ever being aware of the notation which operates inside the processing unit.

It hardly needs saying that a system which is capable of this performance is at present a matter of speculation. There are those who maintain that the search for a single conceptual scheme which is capable of dealing with information across the subject spectrum is bound to be fruitless; who are convinced that the methods and concepts belonging to each discipline are so completely distinct that a general system must be beyond our reach. I see no point in accepting these view until at least some attempt has been made to set up a working situation capable of testing the principles outlined here. The machine side is less in doubt, and computers are already being used for the interactive searching of files of subject data held in a distant centre.

It is a challenging prospect. I am reminded of a Chinese saying which was frequently quoted by a colleague in the early days of the PRECIS Project: 'May you live in exciting times'. Any documentalist will realise that this was intended as a curse.

References

1) Conversations with philosophers: Sir Karl Popper talks about some of his basic ideas. *Listener,* January 7 1971, p 8–12
2) Körner, S. Categorial frameworks. Blackwell (Oxford), 1970
3) Schneider, J. H. Proposed categories for classifying the subject content of research grants. Division of Research Grants. U.S. National Institutes of Health. 1964
4) Palmer, B.I. & Austin, D. Itself an education. 2nd edition. Library Association (London), 1971
5) Austin, D. & Butcher, P. PRECIS: A rotated subject index. Council of the British National Bibliography Ltd. (London), 1969
6) Austin, D. An information retrieval language for MARC. *Aslib Proceedings,* 22 (10), October 1970, p. 481–491
7) Hutchins, W. J. Linguistic processes in the indexing and retrieval of documents. *Linguistics,* 61, September 1970, p. 29–64
8) Macdonald, R.R. Linguistic structure. *In* Information systems compatibility, Spartan (N.Y.); Macmillan (London), 1965
9) Conversations with philosophers – Peter Strawson. *Listener,* January 28 1971, p. 107–111
10) Foucault, M. The order of things: an archaeology of the human sciences. Tavistock (London), 1970

Appendix

Examples of PRECIS indexing. Entries in French and English produced by computer from a common string of terms. Concept analyses in English. Term equivalents taken from 'Liste commune de descripteurs', O.E.C.D., 1969.

Example 1

Concept analysis

(6) Canada
(4) Rural communities
(3) Social change

Entries in French

Canada
Communautés rurales. Changement social
Communautés rurales. Canada
Changement social
Changement social. Communautés rurales. Canada

Entries in English

Canada
Rural communities. Social change
Rural communities. Canada
Social change
Social Change. Rural communities. Canada

Example 2

Concept analysis

(6) Ontario
(4) Rural communities
(3) Adult education
(3) Role – of – in
(2) University of Ottawa

Entries in French

Ontario
Communautés rurales. Education des adultes. Rôle de l'Université d'Ottawa
Communautés rurales. Ontario
Education des adultes. Rôle de l'Université d'Ottawa
Education des adultes. Communautés rurales. Ontario
Rôle de l'Université d'Ottawa
Université d'Ottawa. Rôle en l'education des adultes. Communautés rurales. Ontario

Entries in English

Ontario
Rural communities. Adult education. Role of University of Ottawa
Rural communities. Ontario
Adult education. Role of University of Ottawa
Adult education. Rural communities. Ontario
Role of University of Ottawa
University of Ottawa. Role in adult education. Rural communities. Ontario

Example 3

Concept analysis

(6) Canada
(4) Industries, pulp
(w) Subsidies – from – to
(2) State governments
(0) Study region:
(/) Ontario

Entries in French

Canada
Industries de pâte. Subventions par les gouvernements d'un état – *Exemple: Ontario*

Pâte. Canada
Industries de pâte. Subventions par les gouvernements d'un état – *Exemple: Ontario*

Industries de pâte. Canada
Subventions par les gouvernements d'un état – *Example: Ontario*

Subventions aux industries de pâte. Canada
Par les gouvernements d'un état – *Exemple: Ontario*

Gouvernements d'un état. Canada
Subventions aux industries de pâte – *Exemple: Ontario*

Ontario. *Exemple*
Canada. Industries de pâte. Subvention par les gouvernements d'un état

Entries in English

Canada
Pulp industries. Subsidies from state governments – *Study region: Ontario*

Industries. Canada
Pulp industries. Subsidies from state governments – *Study region: Ontario*

Pulp industries. Canada
Subsidies from state governments – *Study region: Ontario*

Subsidies to pulp industries. Canada
From state governments – *Study region: Ontario*

State governments. Canada
Subsidies to pulp industries – *Study region: Ontario*

Ontario. *Study region*
Canada. Pulp industries. Subsidies from state governments

J. M. Perreault (Huntsville, USA)

Commentary

Suppe asked what a good many have dared not, what has philosophy to do with the broad theory of classification? I want do deal with that in the context of Austin's paper and with Austin's paper in that context, so I intend to expect your indulgence for the inevitable superficiality of the two sides of the coin, but shall read fast enough to keep you from sinking into awareness of that superficiality (what goes fast enough, remember, goes in a straight line).

Is classification the same for the philosopher and the librarian (the last taken as the type of the broad-theory-of-classification-ist)? First we must see what it is to each. To the librarian it is a layered complexity: layer 1_B is the construction of a system or of the rules for constructing the various bits of a system, dependent implicitly or explicitly on a theoretic layer 1_A. The system or its constructive rules either actually or virtually include all the possible "messages" of which it is capable; it is a language, but is unlike a natural language, in that the total number of its resultant messages is predictable –, unless of course it *is* only natural language. Layer 2_B is classification in a wholly different sense, in the sense of the mapping of documents into the bits of the system and thus in the promulgation of actual messages; this in turn is dependent upon layer 2_A, the analysis of the content of the documents and the subsequent translation of this analysis into the system-bits. (Suppe, fn. 30a, gives out that this part of the biological taxonomist's effort is easier than is the definition part analogous to layer A_2, but Polanyi* says just the opposite: the student needs to be trained to make discriminations unrecognizable from a verbal or perhaps even a pictorial description.)

Then, in contrast, how is classification defined by common sense? How about Putting like with like–? This goes well enough till at least a few apples are so equally green *and* red that one sees the need to start a new pile and go back to reexamine previous misassignments.

For the philosopher, I submit, it's not all that crude, though operationally it gets not much farther, despite its relying more on *distinction* than on plebeian likeness (Hooker, p.1). What we are indebted to philosophers for is primarily the main classes of the system as a whole. But for m o s t librarians (surely for Austin) the primary classificatory concern is not with this super-structure but with the infrastructure of the individual message.

* Michael Polanyi, *Knowing and Being: Essays* (Univ. of Chicago Press, 1969), p. 123; the whole essay "Knowing and Being", pp. 123–137, is worth reading in the context of this Conference.

Yet we have had some general philosophical or logical points from our philosopher-conferees that go below the super-structure: Ashworth and Heelan point out that x<y means that x is less well defined than y and thus is more, not less, general (intensional definition is the foundation of extensional). Fairthorne (our resident cautionary philosophico-mathematical guru) shows that the narrower the topic the more nearly mention is equivalent to 'aboutness'. Suppe shows that a variety of hierarchies is possible above the level of *infimae species,* and that several if not all of them may be natural.

Look for a moment at another difference, which will lead to the admission that a library classification must be artificial in at least some part of its intermediate structure: the entities classified by a music store are all of one or other pitch, and of one or other sound-producing mechanism; the scores are all for some medium and in some form or other (even if some of these last cannot readily be named). But the secondary literature on music may concern only a few isolates: "Strauss' *Don Quixote* as example of late-romantic variation-technique" –whereas the score is ineluctably also in D, for such and such a *Besetzung,* including the violoncello- and viola-concerto aspect and the presence of a wind-machine, is program-music with an extant and known allusion, etc. In libraries, in a word, much of the intermediate structure is left out in promulgating infra-structure messages, but it never (ideally) can be for 'real' things. This I see as one need for main classes on which to hang these tenuous rags of meaning.

Austin decries the attitude of dependence on philosophy, but in the same sense that a man's epistemology is a justification of his (often unconscious) metaphysics, so there is an underlying philosophy to each system or group of system for information retrieval, however trivial philosophically. (De Grolier (p. 73) complains that three attempts at theorizing about classification do no more than justify the preferred system, but I would counter that investigation of a language to determine its grammar is of precisely the same sort, and is not to be scorned. The three are Ranganathan, Perreault, and Soergel.)

So, to abandon the super-structure and deal with the infra-structure of the individual messages, I see three possible families of infra-structures:

a: categories of terms having no imposed order
b: categories of terms having a single allowed order
c: categories of terms having an allowed order but allowed to recycle or "nest". The *a* structure is close to LCSH, the type of the *b* structure is Kaiser's theory or Vickery's "standard" citation order, the *c* structure is available only in PMEST with rounds and levels, so far as I know. To which does PRECIS belong? Surely not to the first, since "The principal function of these operators [and of the imposed order between them] is to ensure that terms are organized to a standard pattern" (p. 23). We shall then place it in *b* or *c*; after discussing examples such as 6> 4> 3> 2> 0/ Austin says that "the system of operators is capable of dealing with far more complicated situations" (p. 23); but we are

shown none, so must conclude that despite the availability of further subdivisions by *p, q,* and the comma, we have a *b*-structure with its inherent inflexibility.

The universe Austin would deal with is that of concepts, not of knowledge (pp.2, 13); this is also put in terms of the shelf-organizational need for "subjects" (disciplines), whereas the searcher needs "concepts" (p.5.). The choice of terms here seems ill-advised, since we have at least as much knowledge of argon as we do of chemistry, if by knowing chemistry is meant grasping: who can grasp it, alone; and: is the "grasp" of chemistry by all chemists taken together a real grasp? But operationally, Austin is upset at distributed relatives the same way as Farradane is: "Since the whole field of knowledge is fragmented down in a series of steps to the point at which a particular concept is named within the context of its class, it follows that the same idea is bound to be identified by various different class numbers which are scattered throughout the classification." Since turning the process on its head probably results in distributing the disciplines instead of the objects, a perfect classification *that would offend no one* would have to apportion objects to quasi-disciplines such that no discipline contained any object contained in another. This may be impossible, but a negative experimental proof would be sadly welcomed as condemnation to the status quo.

Austin sees "The principal function of a traditional library classification [as] the organization of large collections of documents regarded as so many physical entities." (p.3) This implies a misreading of the introductory matter to every live library classification, even LC – all are equally or primarily intended for use in a classed catalog. (Of course what Austin may mean is that he sees through that subterfuge to the hidden intent behind it.) He argues that main classes have the advantage of short codes, " [b]ut it must then follow that these systems can only occasionally identify as independently retrievable elements the separate parts of a composite subject" (p.5); but while this may not be true, it is surely true that it *is* true of the converse point in the converse way, e.g., when large clusters of objects (seen in traditional library classifications as defining a discipline) are searched for, if there is no simultaneous deductive subdivision of main-classes. Queries, we are told, are just terms, not parts of classes; to this there are two replies: *1:* to speak of each term as member of a category is to see it as part of something somewhat very like a class, and perhaps to speak of categories is to speak of a total integration of categories into a grammar, if not into an ontological order – the process of integration, once begun, as Kant showed, proceeds ineluctably to the universal integration–; *2:* many terms are indeed the names of an object as the object of that and only that discipline, though this normally is the case only with object-definable disciplines.

But anyway it is the healing of the wound of distributed relative that seems to motivate, to me; and of course this has been tried before, first of all by Dewey's relative index. While Austin's solution may indeed be superior to the relative index, or to the chain index, its primary sense is as a subject-heading system -- and who does not welcome escape from LCSH with its vagaries? But subject-heading systems are analyzable into classifications by the application of the

A sa B = A>B
A xx B = A<B
A sa, xx B = A, B<a

test. We need not wait for the whole thing to be finished to do such an analysis, though to wait might avoid progressive reinterpretation brought on by changes of the earlier structure of *sa*'s and *xx*'s.

The hidden meaning of classification as shelf-order-oriented is that the infra-structure depends on "relative significance [of component concepts] as shelving factors without taking account of the effect which this might have on the meaningfulness of a subject string" (p.6); first of all, this is a chcracteristic of *b*-infra-structure only (wo which PRECIS seems to belong); the point is modified a bit on p.14: "relative significance [of the component concepts] as shelving factors, rather than in an order based on logical or predictable principles." I submit that CC was in fact designed for the classed catalog, that it was the first to see the problem of infra-structure as one solvable only by recourse to a categorial order, and that it is more flexible than PRECIS even if it bears the stigma of not having dumped main classes.

To charge that *some* classifications are "restrictive, in the sense that the classifier has to assign a concept to some main class or other even though this kind of limitation may distort the author's intention" (p.12) is perfectly true, but only shows a need for a careful choice and for improvements wherever called for; Austin then immediately admits that "in abandoning systematic structure [keyword] systems have surrendered meaningfulness to a large measure." The infra-structure is to save this meaning; the super-structure has become superfluous. What is the character of this infra-structure? It is, in a word, context dependency (we ignore the supposed passive-English-sentence origin). What strikes one as odd is the idea that not the leading term in A>B>C>D is the real subject, but that the leading term is a context – as if the object itself could not be expected to stand there all naked and afraid. In this I detect a crypto-classificationism. Again, after the polemic against "relative significance", we see that the numeration of the categories in effect produces a weighing factor in favor of environment, rather than of object studied. And yet again, how are environment and study region to be distinguished? e.g.,

Canada. Women. Higher education
from Women. Higher education. Canada. --?

And if the central term is not the leading term, then when the entry is rotated A>B>C>D must become B<A>C>D
C<B<A>D
and D<C<B<A;

but in fact there is no way to reverse the direction of the context signs except by reintroduction of the prepositional equivalents of the category signs; why take them out to begin with? If they are removed, the indexer must at least mentally rotate each entry to see if he needs to insert them somewhere to prevent ambiguity.

The impression this gives may be harsh, and there are several other technical points I should attack if time were freely available. But let me assure you that I do indeed support such an effort; I have an unpublished MS from 1966 or 1967 in which much the same point is adumbrated, and Mr. Bhattacharya has even published something along the same line, though he feels it needs much more work to be a clue to a viable system. The central desideratum is (to me) to allow controlled terms to be plugged into a flexible categorial-order grammar, rather than being translated into phraselike subject headings or into class codes. But if we seek to escape from classification in its broad sense we are fooling ourselves; this bit-by-bit approach constitutes a classification whether we will or no, so we need to see the super-structure we are creating all along rather than merely when our day of rest after creation comes.

Robert A. Fairthorne (North Farnborough, U.K.)

Temporal structure in bibliographical classification

Neither in theory nor in practice does contemporary classification ignore temporal sequence in the sense of history. But it is regarded in a rather static way, as if classification could be reduced to some unique landscape to be viewed by the classifier as from a balloon. Ahead of him the future landscape is in shadow, but is assumed to exist, even if currently obscured. Vision backwards of the past is hampered by ignorance, but ideally could be observed in detail. To each side the landscape of subject interests is visible in as much detail as you care to take the trouble to observe. Observers looking at the same field from different viewpoints will map the landscape rather differently, the differences being regarded as a kind of perspective distortion of an unique fixed landscape.

According to this notion of classification, as the observer's balloons rise higher the perspective effects disappear, the view becomes the same for all observers and can be studied in as much detail as time and facilities permit. Moreover, provided the observers work fast enough to produce a synoptic map, this can be useful to all contemporary readers in all topics who, of necessity, are grounded at the boundary between present and future.

In short, according to this way of thinking the classificatory landscape is unique and exists independent of observers or classifiers. It could be described usefully in indefinite detail by an ideal omniscient observer working infinitely rapid, but this is regarded as a problem of practice, not of theory. Indeed, observers and agents play little part in contemporary classificatory theory which takes their necessarily finite powers and numbers as defects in application, not as essential elements of the classificatory structure which, after all, is created by them.

I admit that this sketch of conceptual background is far from fair to the best of explicit classificatory thinking. I doubt whether it is unfair to the implicit conceptual background of classificatory acting. Unless one takes time off to think about the assumptions behind our actions, there lurks undisturbed a background of belief in a landscape of topics, to which books and other recorded discourse can be assigned, and the belief that this ladscape is unique, independent of classifiers, and can be described in as much detail as you like by an appropriately omniscient and omnipotent observer. A less naive belief is that differences arising from different points of view can be dealt with by the classificatory equivalents to rules of perspective.

For many day to day operations these beliefs are good enough. Their inadequacies tend to accumulate over a number of days and a number of operations. For even if the classificatory landscape existed and were unique to an omniscient observer, we have no omniscient observer, only a collection of finite ones. These can approximate to an omniscient observer only by collaboration. Collaboration demands discourse and the management of discourse. Both of these entail language and delay. So to approximate to an omniscient and ubiquitous observer, the collection of finite observers would themselves have to act as authors and readers and communicators, utterers and listeners. Thus their discourse, which is about the patterns of discourse, would – and, indeed does – itself require classificatory aids, compiled by an omniscient observer. If we decide, as we must, that this omniscient observer can be approximated by a collection of finite observers of observers, another observer is needed for the classification of their hyper-discourse about discourse. And so on. If we call a halt somewhere, as we must, we might as well call a halt at the first stage, and accept that a finite set of finite observers, communicating between themselves in the same mode and in the same physical universe as those they observe, cannot be equivalent to a single omniscient and ubiquitous observer. If an omnipotent observer cannot be approximated by finite means, he, she, or it, must be dismissed from classificatory theory either as an explicit element, or as an implicit ghost haunting the conceptual background. This entails the dismissal of the concept of a classificatory landscape that can be viewed all at once without delay. This raises an important question, what delay is inevitable in the synoptic view of a given classification range?

Consider now some special range of discourse, viewed by observers who have different interests. That is they act on behalf of readers who share in common some of the matters mentioned in the discourse, but not the topics the discourse is about. Put otherwise, the authors that interest them speak *of* the same matters, but do not speak *on* the same topics. For instance, many authors may mention, – i.e. speak of – the entities called in English, "rabbits". Their discourse considered as a whole may be on or about pets or pests or provender. Similary, "rabbits' feet" may be mentioned or spoken of to illustrate discourse on anatomy, animal locomotion, or amulets.

This distinction between what discourse mentions and what the discourse is about is fundamental to the relativistic nature of bibliographical classification. Some further examples may make the matter more clear. With some extreme types of discourse it is impossible to find out what the discourse is about from what is mentioned. One must have in addition some meta-discursive aids, such as a title or introduction or instructions for use and interpretation. A familiar example is a table of numerical values of some mathematical function. Without external knowledge, the most one can say about the displayed values and arguments is that they represent the values of certain types of polynomial for these arguments. We cannot tell what function they represent over these arguments, because there are an infinite number of functions that have the same values, just as there are an infinite number of curves that can pass through a finite number of given points. Indeed the more knowledgeable often used tables labelled "Trigonometric Functions" rather than the rarer tables labelled "Bessel Functions of High Order" because the numerical contents were the same when considered piecemeal instead of as a whole from the point of view of the author.

To say that we say only that the displayed values and arguments represent the values of a certain polynominal (Weierstrass' Theorem) is no more and no less than to say that we can reconstruct the table from an algorithm. We can describe its numerical contents exactly, but cannot reconstruct the intentions of the author; that is, we cannot tell what this collection of digits is about. For this we must look at the page or chapter heading, the introduction, on the title on the title page, the label on the shelf we took the book from, or other metatextual aid.

Mathematical tables are rather pathological forms of discourse which, when used solely for look-up or 'data-retrieval', may not be discourse at all, but just compilations of data. That is, they degenerate into a collection of distinct expressions each used independently of the others. It should be noted that whether the collection be rated as data or discourse depends upon how it is used, not upon the form or interpretation of the expressions. For instance it is held by some that expressions containing number names in digital form are necessarily 'data', and conversely. According to this a well known piece of discourse would become data if its opening words were replaced by '87 years ago. . ." Also it overlooks the widespread practice of using poetic, prophetic, and totally non-numerical discourse as sources of data, usually in the form of citations torn out of context, which is precisely what differentiates data from discourse. Indeed, certain works are so commonly used in this way that they are produced in formats that facilitate it at the expense of ordinary sequential reading; e.g. isolated and numbered lines and verses in Skakespeare and the Bible.

What discourse speaks of, – that is, what it mentions by name or description –, are amongst its extensional properties. What discourse speaks on, – that is, what it is about –, is amongst its intensional properties. This, its topic, cannot be determined solely from what it mentions. For this one must take into account extra-textual considerations, such as who is using it for what purpose, what purpose the author intended it to be used for, and for whom and for what the librarian, or other manager of messages, acquired it. The last is important because a specialist using a specialist library or libary service assumes that items he gets from it are on that special topic, else they would not be there. Thus he often is justified in using extensional properties to differentiate sub-topics which, if within the main topic, are often defined in the first instance by what is mentioned. The narrower the sub topic, the more exact is the correspondence between what is mentioned and what the discourse is about.

This does not deny the distinction between what is spoken of and what is spoken on, because it assumes that this distinction has been made earlier when the document was acquired or disseminated. It assumes also that this distinction was made correctly. If, to take a standard example, the librarian or classifier, were to assume that a volume with the title "Cricket on the Hearth" had the topic of Indoor Sports, no study of that novel is likely to throw much light on that topic.

It may do so because anything may throw light on anything else, if directed to it by a sufficiently enlightened and original mind. But this lies outside the powers of classification which of necessity can deal only with knowledge that is already known and about which there is some public agreement as to topics and as to how they are talked about. By any useful definition a topic involves public behaviour in terms of language and discourse. It is this kind of behaviour that a classifier must know about, and this is not the same as what those who actually discourse must know about. The function of the classifier is to detect and identify topics of discourse, not to practise them.

What must a classifier know in order to detect, distinguish, and identify the topics of specialist discourse? Anybody, of course, should know as much about as much as he can, whether he is a classifier or anything else. But people, and machines, are finite and therefore must determine what is certainly necessary for a given end, even if in real life one can never guarantee that this will be sufficient. The general problem of what are the necessary qualifications of classifiers is as controversial as it is complicated and profound. Fortunately it is outside the scope of this Conference. However, we will discuss a special part of this problem, what is needed to distinguish the topics of discourse that mentions the same things. This is special enough to exclude many irrelevancies. On the other hand it is far from trivial in either theory or practice, because it raises some issues common to all classification of discourse. So what one needs to solve it one may need to solve wider problems, though certainly one will need other things as well.

Note first that if two writings mention only one thing in common, then they are unlikely to be on the same topic. For instance, mention of a handkerchief is not by itself enough to justify classifying discourse with "Othello". Conversely, to establish their difference we must have only a relatively low order of bibliographical skill, just enough to distinguish what are mentioned by name or description. If one language only is involved, still lower order skills are enough. Often automatic language analysis, statistical techniques, and background knowledge of current modes of expression are sufficient.

Sometimes they also are sufficient to differentiate and identify specialist topics. But this required, as mentioned above, that the discourse has already been classified to some extent, the topics being implicit in the authors, journals, or other circumstances of production and supply. Thus the many 'key-word' and other vocabulary based classifying techniques are usually techniques for refining classification made by other means that demand higher skills and wider extra-textual knowledge. An interesting, long established, and apparent exception is the use of proper names by press-clipping agencies. Here recognition of a proper name, usually the name of the person employing the agency, is enough to guarantee that the clipping will interest the owner of the name. But his, or her, interest is not in fact in a topic, but in the occurence of the name in the text of discourse upon any topic.

But topics are not the properties of text marks as such, but of discourse. Discourse entails interests common to people, not to one person, as well as a common language. Clearly what is said or written is, in the literally literal sense, the indispensable starting point. A book of blank pages is unsatisfactory as a representation of a message, or as a source of clues to its topic. But to create or assign topics to a text we must consider it in the wider context of what kind of person uses it for what, what other texts are used, and in what ways do these texts depend upon each other. Because a text can be influenced only be texts that existed before it did, interdependence of texts, and therefore the nature of topics, implies some kind of temporal sequence. Often the importance of the temporal sequence is masked by the fact that a classifier's decisions are made, and rightly made, and checked from a wealth of evidence and experience, not from the logically necessary minimum. Also custom and convenience usually require that the tools and results of classification be presented in 'snapshot' form from which most traces of temporal structure have evaporated. As in a motion picture film, this structure is replaced by a succession of static revisions. However convenient this may be for the user, the classifier must not let himself confuse this form of presentation with the actual structure of the emergence and growth of topics. The temporal structure is an essential element of the practice as well as the theory of topic analysis.

Consider a group of classifiers, each having a different point of view; that is, serving different reader interests. They have before them some documents, each of which mentions the same things and only these same things. How can any of the classifiers decide which belongs to or does not belong to his particular interest? Clearly he cannot do this from what the text mentions alone. Some extra-textual criteria are needed.

In practice one often uses criteria that implicity demand very high level skills, such as understanding of the subject matter itself. But this should not be demanded, because the classification of discourse about, say, physics is most certainly not a part of physics. The criteria must demand only what one may call bibliographic skills, those that demand knowledge and understanding of discourse and of those who take part in it, not of what the discourse may be about. This keeps the problem and its solution within one discipline in which it is reasonable, and necessary, for the classifier to be more knowledgeable and adept than those he serves. Also, within this discipline, we should prefer methods that do not appeal to class membership in order to determine membership of a class, the class being here the topic of discourse.

Bibliographic methods are of course used widely, but often embedded in the "understanding of the subject matter" used as a supposed short cut. Too often it is a detour through irrelevancies, leading to individual conclusions not reached by others. The purely bibliographic methods, necessarily more consistent, include consideration of authors' names, citations, circumstances of production, vocabulary and style. The last two are used, as are the others, extra-textually for relation to and comparison with the corpus of discourse.

Whatever characteristics may be used, they are used for one purpose only; to link documents (or records of any kind of any kind of discourse) with documents previously studied. For documents that mention the same things, but are on different topics, will have different ancestors, in the sense of preceding documents to which they are linked by various bibliographic characteristics. Similarly they will have different descendents. So the problem of classifying documents that mention exactly the same things is to be solved by classifying, not the documents, but the sets of documents that form their histories, bibliographical 'world lines'.

Take, as a simple example, the very strong form of linkage by direct citation. Texts that mention the same things may cite different earlier documents. These may cite others, earlier than they, and so on. This produces 'fans' or 'spreads' (in the Brouwerian sense) of ancestors that have converged on the documents under consideration. If these fans differ enough in the documents that form them, they indicate that their elements are on different topics though, of course, this is not enough to identify them. Identification of topics can be done only on a higher level, with higher level background knowledge and rules of inference. However high the level, these last do not demand knowledge, other than 'book knowledge' in the literal sense of the subject matter. Identification of the topics discussed in physics, say, is not a part of the science of physics but of physicists, with respect to their habits and habitats of discourse.

The importance of temporal structure is shown very clearly when considering the situation dual to that just described. Here, instead of having documents mentioning the same things but with different ancestors and descendents, we have documents with at leat one common ancestor and at least one common descendent. For instance each may cite – amongst others – the same document, and all may be cited in a subsequent document. Here there is little doubt that they all have a topic in common, but what topic it is has to be determined on a higher level.

These simplified examples show that a rather attenuated form of history is essential to the classification of discourse. So far as history is concerned, this comment is hardly novel. Historical considetations are and have been used explicitly and implicitly by all competent classifiers. The novelty lies in the attenuation of the history which, to be effective, demands only that one can determine, within the documentary nexus – not with reference to the physical universe – that one document is or is not 'descended' from another.

As mentioned above there are many characteristics that can be used for this from strong, such as explicit citations, to weak, such as vocabulary and style. The important point is that though these differ in efficiency and precision, they are self consistent, converging on the same temporal order in the sense of bibliographic sequence.

Thus the documentary corpus has a loose but irrevocable order, bibliographic time, generated by the fact of continual growth. This must not be identified with physical time, though clearly bibliographic time does not contradict it when the two can be compared. New documents may come to the attention of different observers, authors, and readers in different orders and with but a weak relation to their dates of publication in the outside world. Nor is 'publication' a precise physical, social, or legal, event. No observer, reader or author can be aware of all documents at any time or, for that matter, retrospectively. Thus we must not assume bibliographic history of the type we are discussing to be embedded in some physical space-time framework. Nor can we assume simultaneity of bibliographic events, because it is impossible to establish this without instantaneous communication between finite observers. Communication entails delay.

In short, the classificatory landscape is a relativistic landscape, and therefore not a landscape at all, to be viewed as a simultaneous whole. True, it is the product of people who work in a world that, for everyday physical activities, it not significantly affected by relativistic phenomena. This is because physical signals can be so fast that the time take to travel is very much less than the duration of the physical events they influence. However, as is only too obvious, the time taken to communicate socially effective messages is often far longer than the time taken to perform the actions the communications are supposed to influence. Hence social events, such as discourse, should be expected to be and in fact are relativistic. That is, any strict laws and measures appropriate to them are in general only local and contemporary.

This still leaves open the possibility of using topological methods and measures to deal with the formal aspects of some social sciences. All that is needed for that is a situation with elements whose formal properties correspond to the basic topological concept of 'neighbourhood'.

Unfortunately a purely static view of discourse does not permit the concept of an invariant topical neighbourhood. In it no pair of documents might not be 'close' to each other from one point of view, and 'widely separated' from another. That these points of view may be artifical or implausible is beside the point, which is that the property of being almost about the same topic is not intrinsic to the documents as such. This causes ambiguities when one tries to design criteria for 'clustering' and the like in computational taxonomy.

As we have seen, the ambiguity disappears – in theory, anyway – when we consider, not documents, but the sets of documents that make up their history, in the sense of ordered bibliographical linkage. These histories generate a local 'time' and also a local 'space' corresponding to topics of discourse. Once more we must emphasize that the space-time is self-subsistent, and logically need not be embedded in a wider framework. Empirically we, being outside this growing bibliographic universe, look upon it as an abstraction from social activities carried out in our own physical space-time. This abstraction ignores all phenomena irrelevant to the specific ussue at hand, the taxonomy of discourse, though of course it neither denies nor affirms the existence of irrelevant phenomena.

The logical justification for an derivation of bibliographic space-time was outlined by myself some years ago (Fairthorne, 1961). They need not be repeated here because, though not necessarily widely known, they are accessible to anyone interested. It was shown how this rather sparse structure threw light on such problems as the distinction between document relevance and document description, and the relation between translation and classification of discourse. For instance, to study the last we consider re-translation of a translation back into its original language. Even if the translation and the re-translation were perfect, resulting in a word-for-word replica of the original, it would not be the original, but a descendent of it. The descent here is not along a direct purely temporal route, but 'sideways' as well as forwards. This shows that, in bibliographical space-time, linguistic translations are kinematical translations also. It illustrates also another fundamental, that double negation – here represented by translation followed by re-translation – is not equivalent to identity. Hence the algebras appropriate to topics cannot be stronger than those of Brouwer or Heyting, in which triple negation is equivalent to single negation.

Up to now we have considered only the structure as observed by classifiers. So the word 'document' has been used without definition, because there could be no ambiguity, in the sense of what librarians would call a 'title'; i.e. a unique element of discourse of which a multiplicity of physical documents may be the physical representations. It is one of these physical representations that a reader must use to know the text or 'title' in which the author expresses his message. To preserve this distinction I will refer to the physical representation as a 'copy'.

To fit readers explicitly into bibliographic space-time, we must attach to each document ('title') a fan into the future corresponding to its copies. A reader is then adequately represented, for the purpose of topic analysis as the aggregate of copies he has read, connected in the order he has read them. More precisely, this represents his reading history which spreads like a wake behind him into the past from the latest copy he is reading. This wake of copies generates a wake of documents of which they are copies. The temporal order of the document wake corresponding to a reader is not exactly the same as the temporal order of the documents between themselves. An observed reader history is always out of date. The reader himself is the first and only person to know what he is currently reading. Observers, classifiers, and librarians are necessarily in his wake.

Accepting that observations must be retrospective, we see that the overlap of reading histories can be used to detect common interests between readers. Also of authors, whose reading histories differ from those of non-authors only in that they pass through the titles of the titles they write, rather than through the copies. The wider the range of topics observed the further into the past of the reading histories we must go in order to observe an overlap of wakes wider enough to correspond with the range. One can classify readers according to their interests only by studying how their wakes overlap. The more readers and the more their interests, the further in the past this will be.

In general the wider the scope of classification, the earlier must be the observations, and the more out of date the topics classified. This is inevitable, but the space-time structure outlined in this and the earlier paper should provide estimates, in bibliographical terms, of how much delay must occur in some well defined situations.

Even for an observer with infinite memory, a 'general' classification must be indefinitely out of date. Otherwise it must be incomplete. We can of course have a general classification of a universe of discourse that is dead.

What happens if we have a finite number of finite observers working together to compile some sort of 'general' classification? Because they must communicate together by means of discourse, about the classification of discourse, they themselves are readers and authors in bibliographic space-time. Because of the time structure this does not create the paradoxes of self-reference that arise in the purely static view. They are of course subject to inevitable delays as are others observing special topics. But the number of classifiers and the quantity of their discourse is, or should be, much less than the numbers of readers and titles they discourse about. Therefore the delay added by such cooperation should be less than the delay added if there were only one classifier to scan and consider, even though the single classifier does not have to communicate with others.

Although the vague and naive idea of a 'general' classification must be abandonned, we can design attainable and useful compromises. These will balance delay and effort against range of topics and number of readers served. The study of the temporal structure of bibliographical classification destroys a mirage, to replace it with smaller and accessible targets, together with some hints as to how to hit them.

Reference

FAIRTHORNE, R.A., Basic Postulates and Common Syntax. In: Information Retrieval and Mechanical Translation, vol 2, edited A. Kent. Interscience Publishers, 1961, reprinted in Fairthorne, R.A., *Towards Information Retrieval,* Butterworths 1961, Archon Books, 1968.

Phyllis A. Richmond (Mayfield Village, Ohio)

Review of : R. A. Fairthorne. 'Temporal structure in bibliographical classification"

This is one of the most interesting papers to come along in many a moon. Robert Fairthorne begins by discussing "the classificatory landscape", a grand overview supposed to exist independently of observers, something that could be seen as a whole from afar. It is very common to find students vehemently proclaiming that a universal classification system exists in nature and it is up to the human element to find out what it is. All present descriptions are recognized as partial, incomplete and inaccurate because the classificationists have not yet seen the landscape clearly enough.

Fairthorne points out that actually we can *approximate* the landscape as it would appear "to an omniscient observer only by collaboration. Collaboration demands discourse and management of discourse. Both of these entail language and delay." After discussion of the implications of this, he concludes that it would be difficult if not impossible to approximate the omniscient observer and the idea of a view of a classificatory landscape to be seen all at once without delay is not productive.

We are reminded that what discourse mentions and what it is about are two different things. The mentioning uses a name or description; the aboutness indicates which meaning is involved. The problem of aboutness requires a broad background of knowledge for the classifier, usually more than that possessed by any one person. A great many systems assume a classification already made. "Thus the many 'keyword' and other vocabulary based classifying techniques are usually techniques for refining classification made by other means that demand higher skills and wider extra-textual knowledge."

Fairthorne then goes on to point out that no text appears, functions or is classified in a vacuum. Each text is related to earlier texts and will be related to later ones. Furthermore, the experienced classifier can recall related texts in the same area of discourse in making his classification assignment. The "tools and results of classification" are maintained in "snapshot form from which most traces of temporal structure have evaporated. As in a motion picture film, this structure is replaced by a succession of static revisions." However, in linking older and new documents, knowledge of bibliographic method takes precedence over subject knowledge in certain areas: author's names, citations, imprint, vocabulary, style. The characteristics are used for linking documents. "So the problem of classifying documents that mention exactly the same things is to be solved by classifying, not the documents, but the sets of documents that form their histories, bibliographical 'world lines'."

The process of tracing backwards provides a "fan" of ancestors for each document, a fan-history.

Parenthetically one might comment at this point that this applies not only to external features of the document, including its citations, but also to internal literary features which would include borrowed matter not cited. Sometimes this latter is easy to see, but usually it is so well disguised that the author himself does not recognize it. For example, some years back it was suggested that every historian insert an introductory essay at the beginning of his work indicating his biases. Very few do this and then only conscious biases. The identification of biases is most apparent in cross-cultural criticism, and least apparent among members of "ingroups", where the critics are also members of the group.

Fairthorne also notes a fan of descendents of a document. One can tell from internal characteristic factors whether one document is descended from another. A documentary corpus has a distinct order which is in *bibliographic time,* not necessarily the same as physical time. Bibliographic time is the time in which the documents presented themselves to the reader or author. This time has no relation to time of publication. I may, for instance, "discover" a paper by George Sarton which is meaningful for my current purposes even though it was written thirty years ago and he has been dead over ten years. It may get cited in a review article published in 1972 and will be having a whole fan of antecendents and, hopefully, some descendents. The communication delay is obvious. It is this delay, in fact, that keeps a good number of presently unpopular writers writing, because we have ample evidence of post-mortem acceptance.

Fairthorne's fan-history is perhaps more accurately a genealogical map, but not one in which the descendent is limited to two parents. Conceivably, at a given moment, one might have 50 or 100 parents, depending on how much one reads, and similarly the antecendents' fan histories would depend on their volume of reading. The "bibliographic world line" would look more like a cloud.

Because the classificatory landscape is essentially relativistic, the author suggests developing a bibliographic space-time view, to be dealt with according to the basic topological concept of "neighbourhood" applied to historical sets of linked documents and using algebras of the Intutitionist School, of Brouwer and Heyting. Says he, "To fit readers explicitly into bibliographic space-time, we must attach to each document ('title' [in the librarian's sense, Lubetsky's "work"]) a fan into the future corresponding to its copies. A reader is then adequately represented, for the purpose of topic analysis, as the aggregate of copies he has read, connected in the order he has read them. More precisely, this represents his reading history which spreads out like a wake behind him into the past from the latest copy he is reading. This wake of copies generates a wake of documents of which they are copies." He points out that everybody else is always behind the reader because only he knows what he is currently reading.

Fairthorne would look at the overlap in wakes of readers to detect common interests. He applies the same idea to authors, substituting titles written as source of reading histories. He would classify readers by what they have read to discern their interests, using the wakes developed in bibliographic space-time.

One might comment here on the assumption that an educated person is the sum of all his reading, and a specialist of his reading in things which he considers related to his specialty or which he *makes* related to it. Fairthorne's discussion is limited to bibliographic classification of that which comes to a person via the eye. One could also assume that information received through other sense organs, especially the ear, gets marked and parked in the brain and joins the "bibliographic world line" or fan-history. In addition, there is the possibility of mis-information to be considered, bad connections, erroneoneous connections, imperfect recall and so on. A good many factors are involved. One cannot safely assume that the reader took in everything to which he was exposed. The ground is surer when limited to written bibliographic forms, traced back directly through citations and indirectly through various methods of style analysis or word and idea patterns.

Last but not least, in a discussion of "general" classification meaning, I assume, a universal one, he inclines to the view that the intersection of a multitude of classifiers with intersecting and oberlapping fan-histories, though involving a communication delay, would produce a better system, although no such thing as a once-and-for-all general classification is possible. The better system would be a mass of workable compromises.

The fan-history idea introduced by Fairthorne is a most useful theoretical underpinning for the process of searching for a core literature through multiple citation indexing. How a librarian or information scientist would pry a reading history out of a patron remains to be found, although there is little reason to think that this would not produce much better matches between readers and documents. Technically no two people are alike or ever could be, so that technically no two landscapes could ever be exactly alike. There is, however, enough consensus to make communication possible and it is in this area that application of the theory may be possible. With the emphasis on written reading and transmission, the importance of oral communication of ideas should not be overlooked. Perhaps the inquiry for a fan-history should also include the question, "To whom have you been talking lately? " The fan-history also includes a fan of oral communication and this is especially likely to turn up in a request, or in authorship without any citation.

Ejnar Wåhlin (Stockholm)

The AR-Complex – Adapted Systems Used in Combination with a Common Reference System

Summary

In this report a coherent complex of classification systems is proposed which should be constructed in such a way that every component in this complex is built with consideration to the other components.

The complex contains an unlimited number of *adapted systems* (A) and one common *reference system* (R), which explains the denomination "the AR-Complex".

The R-system has the character of a universal system, containing all basic concepts.

The A-systems should not go deep as all necessary specifiactions can be delivered by the R-System.

A composed AR-code has generally the following structure:

N – a:b or N – a:b:c

Here N represents an A-system, a and b the R-system. N – a:b may alternatively be written N – b:a, which demonstrates the flexibility of the R-system.

Certain A-systems are adapted to a general category of objects, e.g. products (commodities), occupations, economic activities, etc. Other A-systems are adapted to a certain branch or a certain institution. If one has a system in use that one does not want to give up this could be kept as a A-system. By using the R-system for specification one can consequently obtain the necessary specifications without enforcing alterations in the traditional routines concerning the order on higher levels.

The principles of using two systems in combination will in different ways reduce the tension between on the one hand scientific connections and on the other hand the adaption to practical conditions.

Two A-systems are specially studied, a general system for products, PC, (Product Classification) and a document system for the building trade, AZ. Both these systems have been tested on a rather big material and the proposal is thus not only a theoretical outline, but a method with a rather firm foundation.

The AR-complex can solve many problems and conflicts and give a consistent approach for the development among the classification systems of different kinds that are now going on without uniform guidance. The conflict between special and universal systems that now is experienced in many branches can be substituted by peaceful co-existence. Systems for documentations, for statistics and many other purposes can be united.

A Introduction

With the ideas presented here I shall try to introduce a new theory of classification systems with the aim of solving two different conflicts which in different ways obstruct the sound development of classification theory and thereby prevent world-wide solutions.

The first of these conflicts is experienced by anyone who attempts to construct a classification system. Two different guiding principles are then in opposition: the desire for a logical ordering of concepts (subjects) and the desire for the system to correspond to relations and conditions in a certain field of interest.

The second conflict can arise when an institution or a branche has to decide what kind of classification system one should use, a universal or a specialized system. The *universal* system has the advantage of representing the whole field of knowledge and all kinds of concepts, the *specialized* system can more easily be adapted to the user's need. These two conflicts are somewhat similar as to logical order and practical adaptation.

A way to settle this problem seems to be to use *two systems in combination,* one representing logical order, the other practical adaptation. All concepts are arranged in a universal system or a general *"reference system",* the structure of which is determined by logical-semantical rules only, without any special inclination to adaptation. The user's requirement could be satisfied by a separate *adapted system* superimposed on the reference system. The adapted systems should not go very deeply into detail, since the necessary specifications are made with the help of the reference system.

We should thus have an unlimited set of adapted systems all using a common reference system for specification. This is called *the AR-complex* (A for Adaptation, R for Reference).

B The Reference System

The classification system here applied in the role of a reference system is called US–2b, where US stands for Universal system. Alltogether four US– alternatives are elaborated, US–1, US–2, US–3 and US–4. Only US–1 and US–4 are presented here.

This idea has been studied thoroughly in two different applications, one for product (commodity) classification, and the other for document classification in the building field.

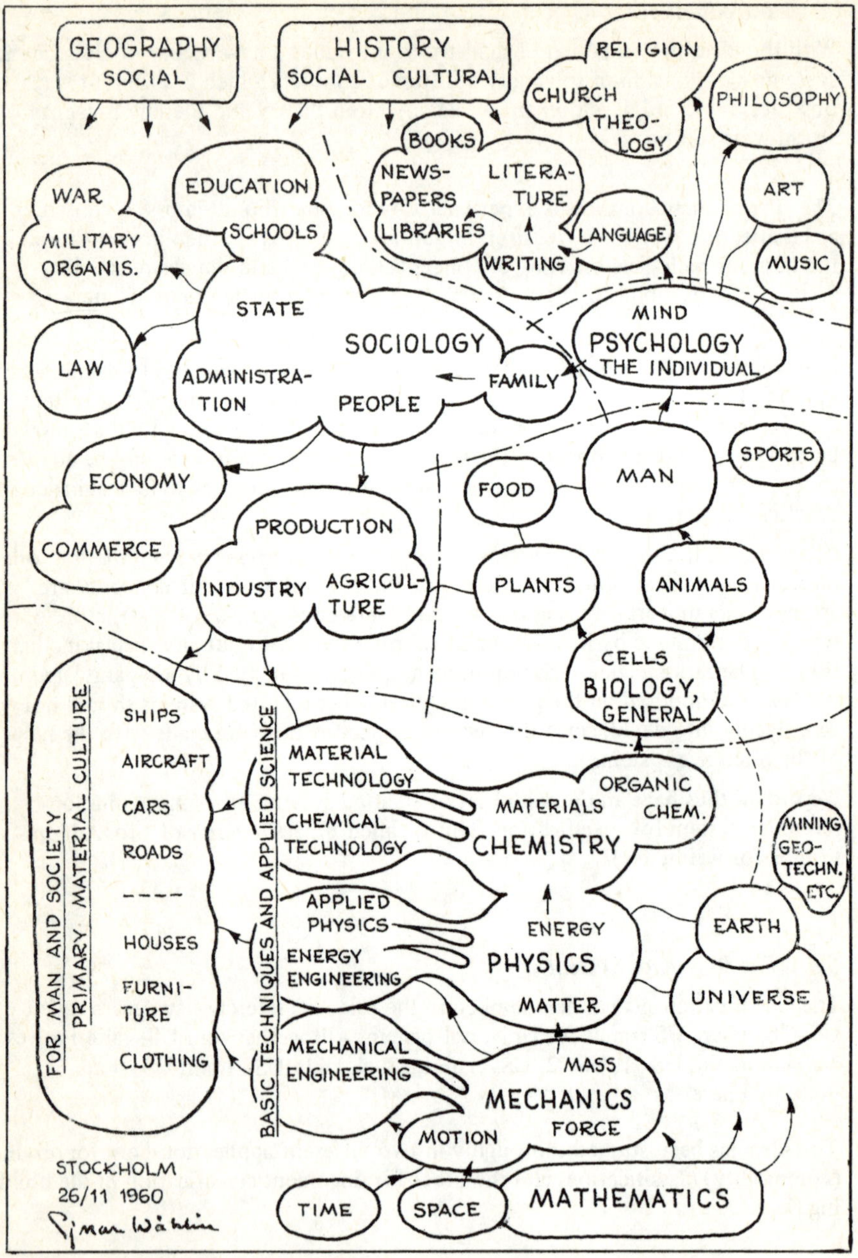

Fig. 1. Diagram showing the most important fields of knowledge and their connections.

US–1 was earlier called TUS (T for Technology) as this system has a special feature useful for technology, applied natural sciences being brought together with the corresponding pure sciences. US–2 and US–4 do not have this feature and the technical class here contains also applied science. US–4 differs from US–2 in that respect that the technical class in US–4 is given a structure corresponding rather well with the proposed classification for products, PC. See further section C. Another difference concerns the position of technology, which in US–2 is class 4, in US–4 class 8.

The fundamental idea behind TUS (now US–1) is visualized in Fig. 1, earlier presented in an article in Journal of Documentation 1963. This structure which is independent of any coding, can in broad lines represent the common idea for all the US– alternatives.

The two digit structure of US–1 and US–4 is shown in Appendix 2 and 3. Appendix 3 is an extract from US–4 demonstrating the use of colon combinations.

All the US– numbers in this articles refer to US–4.

Combinations of concepts within a universal system can be illustrated graphically by a square in which both axes represent the universal system.

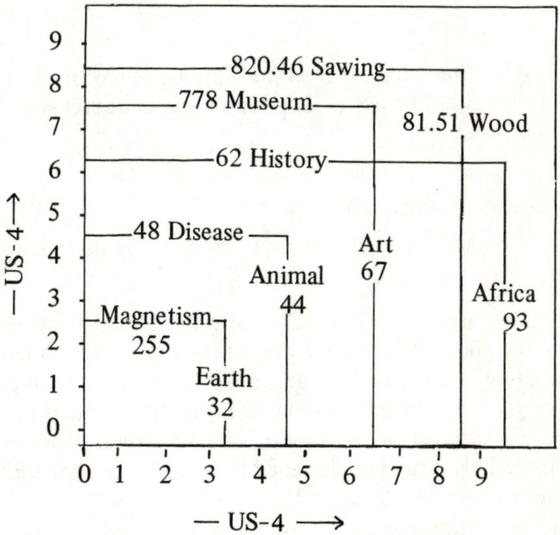

Fig. 2. US–4 in two-dimensional display. If the order of sequence within the combinations is reversed (b:a instead of a:b), one obtains other points located symmetrically with respect to the diagonal, which is not shown here for the sake of clarity.

Other examples of combinations of concepts are shown below with the corresponding solid UDC code on the right.

	US–4	UDC
Taxes for dogs	734.5 : 44.931	336.216.33
International law for polar regions	722 : 95.8/9	341.228
Offences against public authority	725 : 71	343.343
Criminal psychology	47 : 728	343.95
Criminal pathology	483 : 728	343.96
Regulations for theatres	724 : 692	351.758
Insurance for maritime transport	739 : 766.5	368.23
Acoustics, measuring methods	22 : 13	534.6
Farm management	436 : 751	631.1
Agricultural tools and machinery	821 : 436	631.3
Bridges, steel	886 : 81.22	624.21.013.4
Bridges girder	886 : 212.41	624.27

See also Appendix 3. In principle, the combinations are reversible, but when they are reversed it may often be desirable to complete or abbreviate one or the other of the two numbers.

C Product Classification, Main structure

PC signifies Product Classification. This is a system specially intended for classification of products (goods, commodities).

The reason for using a special product system for product classification is that a classification made for documentation purposes does not give sufficient guidance for unambigous and logical placement of all products, and even if all products could be placed, they would not fall together in groups convenient for use in industry and trade. If a clear survey of all products is wanted, a special product system must be made. This should be designed for classifying *the products themselves,* not knowledge of products.

Generally PC in the four first classes only goes down two, three or four steps in the hierarchy (2–4 digits). The necessary specifications within the product classes according to origin in nature, material, form, function, etc., however, demand a general reference system from which these specifications can be taken. US–4 is used for this purpose.

A suitable principle for ordering the main classes seems to be to follow the flow from production to consumption. The following main classes are proposed:

Main classes in PC: (Cf. Table I, page 424–425)

PC 1, Natural products (NP), contains all products (goods) taken directly from nature, including materials which have been lightly treated, i.e. undergone treatment which does not essentially change the character of the product (e.g. cleaning).

PC 2, Intermediate products (IP), (industry intermediates), includes products which have been subjected to certain radical treatment but are not ready to be used for direct manufacture of articles or for consumption. This group represents raw materials for further technical processing.

PC 3, Structural products (SP), represents fully processed raw material for manufacturing of articles, machines, buildings etc. To a large extent these products have a certain geometric form (tiles, pipes etc.), but the group also includes certain formless products (e.g. paints, glue).

PC 4, Consumption products, (CP) contains products ready for consumption, serving their purpose simply by being consumed (e.g. fuels, foodstuffs).

Energy, can be taken up here, as energy (electricity, heat) is a market product.

PC 5, Finished objects (FP), includes all objects with a design especially adapted to a given function. Contrary to group 4, a certain durability is desirable here, even if the lifetime be short.

This structure is based on the fact that the aim of industry to a great extent is to produce new and more complicated products from elementary materials. The substance of a ready made product has usually passed through several stages of processing. In certain cases, however, a product may go unchanged from nature to consumption (e.g. vegetables, fruits) and in this case it will be defined by double numbers, the numbers to be used depending on the purpose. In the retail trade the consumption number is used, in agriculture the number from the natural products group.

Subdivision of the primary groups (see table I)

In PC 1–3 the material (substance) is the most obvious determinant of subdivision. A certain unifomity regarding the order of the secondary groups has been attained, and thus the sub-groups represent mainly the following original materials:

–1	Of rock and soil	–4	Fossil [2]
–2	Metallic	–5	Vegetable
–3	Chemical [1]	–6	Animal

In PC 4 and PC 5, where the division is made according to the function of the products, this principle can not be continued, but it does emerge occasionally for the consumption products (in 43–45).

Number –7 in this series is used in PC 3 for textiles (37) and number –9 in all classes for "other products". In PC 4 the numbers 41, 42 and 48 are unoccupied. In PC 5 all numbers are used (51–59). We can thus illustrate PC with the following figure:

Origin		NP	IP	SP	CP	FP
		1	2	3	4	5
Rock. . .	–1	11	21	31	–	51
Metall.	–2	12	22	32	–	52
Chem.	–3	13	23	33	43	53
Fossil	–4	14	24	34	44	54
Veget.	–5	15	25	35	45	55
Anim.	–6	16	26	36	46	56
–	–7	–	–	37	47	57
–	–8	–	–	–	–	58
Other	–9	19	29	39	49	59

Fig. 3a. The PC square

In other publications the following material series is proposed: –1 Vegetable, –2 Animal, –3 Fossil, –4 Metallic, –5 of rock and soil, –6 Chemical.

A more detailed picture gives table I, with examples and references to US- and BTN-classes (roman numerals). The structure here, esp. in PC–5 is determined with help of the investigations reported in section E.

1) Preferably inorganic

2) In some cases also other organic chemicals

Other arrangements of PC are of course possible, with retaining the main idea of one dimension characterized by production stage and another dimension – through the first stages – characterized by material.

Some possibilities for arranging the classes are shown here:

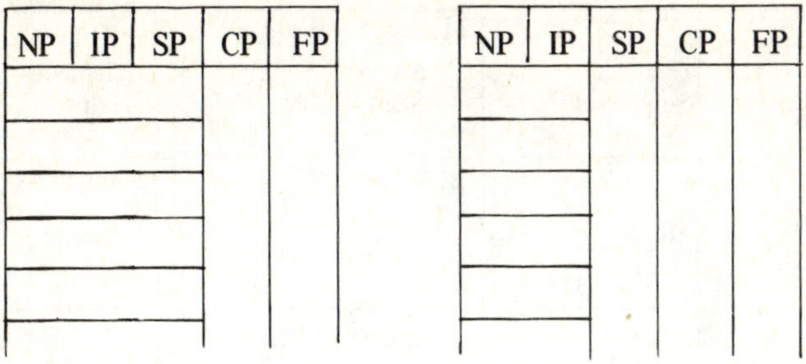

Fig. 3b. Other ways of arranging the first classes in PC

If the trade should prefer the structure in Table I and the industry a rearranged structure, these both parties can have different systems based on the same network that admits simple and exact connections by the aid of a translation key.

D *Product classification. Principles of specification*

We want a system by which it is possible to give every product a logically determined position, including products not considered when the system was constructed.

A designation is generally composed of one PC-number (N) and one or more US-numbers *(a, b etc).*

N – a:b (or *N – a:b:c . . .)*

The numbers *a* and *b* are separated by a colon. In the US-number there is generally a decimal point after three digits, but in some cases we have the point after two digits. This is the rule for chemical elements (in 28), species of plants (43) and animals (44), languages (61), materials (81) and nations (91/95).

Example:

Apple (fruit), as a horticulture product;

153 – 432.7 : 43.5761

Vegetables, fruits, berries (153) — fruit (432.7) — apple tree (species) (43.5761)

Here 432 is "Parts of plants" and 432.7 is "Fruit". Plant species is designated 43. (note the point), and 43.5761 is a supposed botanical number for the species apple tree. If all products in PC 153 are classified first by plant part and then by plant species, the numbers 432 and 43. can be eliminated and there will result a

Table I. PC – Product Classification

		0		1		2
		Origin		Natural products NP		Intermediate products IP
1	Of rock and soil	Quarries Soil layers 332	→	11. Stone and soil NP Stone Sand – Clay V 81.1	→	21. Stone and soil IP Cement Lime V 81.1
2	Metallic	Mines 335	→	12. Metallic NP Ores V 81.2	→	22. Metallic IP Raw metals XV 81.2
3	Chemical	Chemical deposits 336	→	13. Chemical NP Salt Phosphate V 81.3	→	23. Inorganic chem. IP Cuastic Soda Sulphuric acid VI 81.3
4	Fossil	Coal deposits Oil deposits 337/338	→	14. Fossil NP Coal Raw petroleum V 81.4	→ → 44	24. Fossil - organic IP Petrochemicals Ethylene V 81.4
5	Vegetable	Plants 43	→	15. Vegetable NP Cereals. Fruit Timber. Cotton II, IX··, XI·· 81.5	→ → 45	25. Vegetable IP IP for food " " textiles Pulp III, IV··, X··, XI·· 81.5
6	Animal	Animals 44	→	16. Animal NP Carcasses. Fish Skins. Hides. Wool I, VIII··, XI·· 81.6	→ → 45	26. Animal IP III etc. 81.6
7		43/44				
8		43/44				
9	Other	Lakes Rivers 35	→	19. Other NP Water 81.9		29. Other IP

Table I. Contin. Roman digits refer to BTN
Arabic digits refer to US-4

	3		4	5
	Structural products SP		Consumption products CP	Finished objects FP
→	31. Stone and soil SP Concrete tiles Brick. Glass XIII 81.1			51. 81
→	32. Metallic SP Metal sheets " pipes XV.. 81.2			52. Mechanical group Units. Tools. Working machines XVI 82
→ →	33. Chemical SP Paints. Glue VI 81.3	23 → 24	43. Chemical CP Cosmetics Cleaning agents VI 81.3	53. Energy group Engines Electric machines XVI 83
→	34. Plastics. Rubber Plastic sheets " foils VII 81.4	14 →	44. Fuels Coke. Wood. Gas. Fuel oil V·· 81.4	54. Transport group Cars. Ships Aircrafts XVII 84
15 → →	35. Vegetable SP Wood excl.37 Paper IX, X, XI 81.5	15 → 16	45. Foodstuffs II, III, IV·· 81.8	55. Data- and transmission group. Instruments XVIII 85
16 →	36. Animal SP excl.37 Leather VIII 81.6	15 →	46. Beverages. Tobacco IV·· 81.88	56. Graphical group Writing devices Documents X·· 86
24 → 5,26	37. Textile SP Yarn Fabrics XI·· 81.7	23 → 24	47. Medicines. Drugs VI·· 492 + 81.36	57. Personal group Clothing. Furniture Cultural objects XI, XII, IX·· 55··
				58. Buildings. Installations XVI·· 88
	39. Others SP	19 →	49. Water. Energy	59. Other objects 89

shorter "practical number" (in which the colon is replaced by a point):

Apple = 153 – 7.5761 Maize grain = 154 – 8.331'31

In order not to confuse a decimal point in US with the point that here separates the two US-codes, we here have an apostrophe (333'31). Four digits (e.g. 5761) may be written without any sign. 154 is Cereals in PC.

By mechanized methods it is possible to extract all products containing a certain number within colons or points. Grain, meal and flakes of maize all contain the number 33131 (= maize) within points and thus all maize products (agricultural products, foodstuffs in retail trade etc.) can be collected together.

In the field of chemistry the apostrophe has a quite special function, which is to differentiate between an element and a compound containing this element in the negative ion. Example:

The mineral Magnesium Silicate is written

132 – 28'42 : 28.22

Inorganic minerals (132) — Silicate (28'42) — Magnesium (28.22)

28'42 = Silicate 28.42 = Silicon 28.22 = Magnesium

If all minerals in group 16 are given a chemical designation, the number 28 (= chemistry) can be eliminated and we obtain the "practical number"

132 – 42.22

From this example and from the botanic example it appears that the main numbers in US (e.g. 43 and 28) disappear when we come to the practical number. The principle of using two systems in combination does not here result in longer numbers.

The above-mentioned designations apply to minerals consisting of two components. In more complicated cases the designational principle may be modified. It may also be necessary to differentiate between two separate forms of the same chemical compound. An example is Magnesium Silicate, which exists both as talc and as asbestos. These can be separated by supplementary numbers, e.g. with two digits. Vacancies make it possible to introduce new products. Example:

132–42.22 .01 = talc
132–42.22 .05 = asbestos

In the retail trade the assortment is small compared with that proffered by nature or industry. Then short "trade-numbers" or retail numbers may be substituted for the US-numbers, e.g. for food-stuffs. Thanks to vacancies a new product generally can find a position of its own. This is the same principle as in the matter of talc – asbestos above.

Other examples of codings are given in Fig. 4.

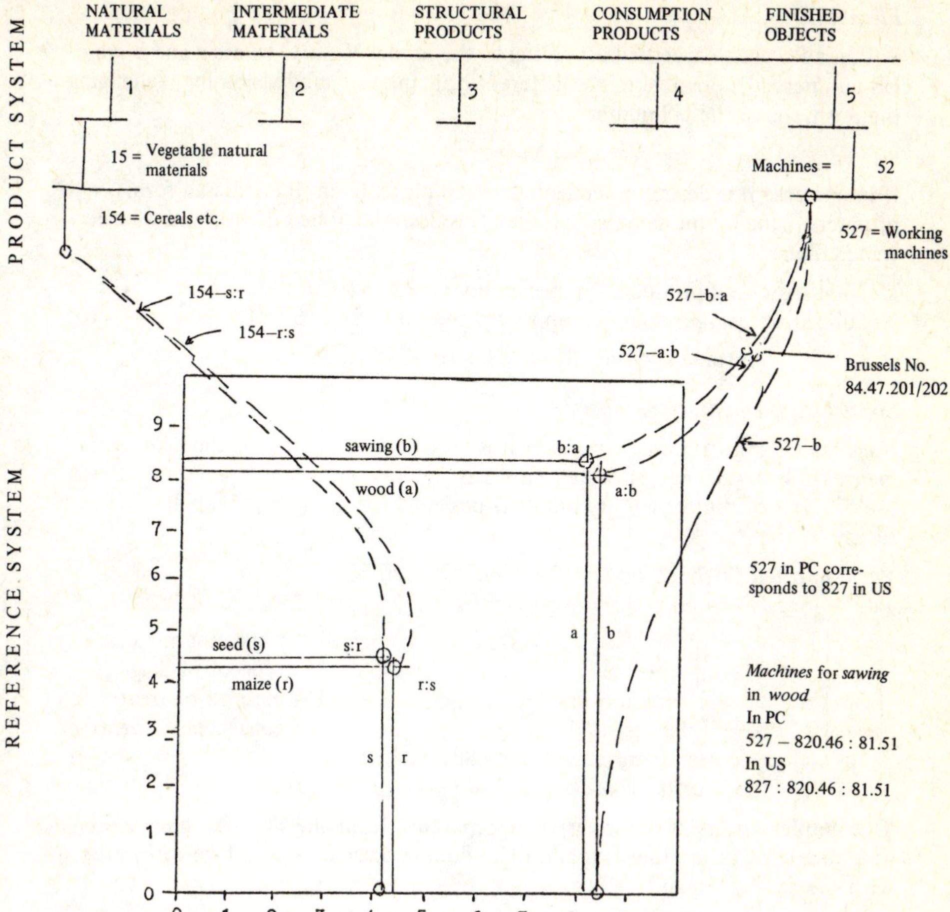

Numbers according to US–4

a	=	wood	=	81.51
b	=	sawing	=	820.46
		machine	=	827
r	=	maize	=	43.331'31
s	=	seed	=	432.8

Document indexing

Machines for *sawing* in *wood*

827	:	820.46	:	81.51
"	:	81.51	:	820.46
820.46	:	81.51	:	827
"	:	827	:	81.51
81.51	:	827	:	820.46
"	:	820,46	:	827

Fig. 4. Examples of product classification and thereby coordinated document classification

Flexibility

Within a PC-group specified according to the model $N\text{–}\ a{:}b$, where a and b are US-numbers it is possible to use different principles of subdivision by exchanging the positions of the US-numbers:

$$N\text{–}\ a{:}b = N\text{–}\ b{:}a$$

If in one case it is desired to collect all metal pipes, in another case all form products made by the same metal, then this desire may be satisfied and we can write either:

Metallic structural products – pipes: copper = 32–810.76 : 81.216 or
Metallic structural products – copper : pipes = 32–81.216 : 810.76

Shortened coding: 32–76.216 or 32–216.76

Special supplementary numbers

For detailed registration of products it is necessary to express methods of treatment, kinds of package, sizes etc. This may be done in different ways with the use of series of supplementary numbers designed for the special branches. Example:

Vacuum-dried finely ground coffee in plastic bags

```
453–887 = 152 .28 .35 .46
 |    |     |    |   |  L___plastic ........ (4 = packaging material)
 |    |     |    |   L__bag ................. (3 = type of packaging)
 |    |     |    L__vacuum dried ............ (2 = method of treatment)
 |    |     L____finely ground .............. (1 = consistency, form)
 |    L_________coffee (shortening of 81.887 in US)
 L______________food-stuffs, taste-stuffs, additives (PC number)
```

The supplementary numbers are here separated from the PC–US–numbers by a double dash or another symbol. (The double dash should not be mixed up with the sign of equality.)

If the positions are determined, the number can be written

453–887 = 52.8.5.6

Checking of PC

The system has been checked by inserting all products in the Brussels Nomenclature (BTN) into the two-digit structure of PC. The five PC-classes have received fairly equally sized parts of BTN. Within all classes detailed studies have been made with products from schemes of different manufacturing companies and sectors of industry and trade. Vegetable materials and foodstuffs have been specially investigated. The experience gained so far indicates that construction of a complete system for products in this way is possible.

E Comparative study of product classification in different systems

In a universal classification intended for use in industry and technology it is important to have *materials, products, processing methods* and *fields of industry* arranged in a logical way.

It seems that it could be of interest to investigate different classifications used in technology and industry, with the aim to find their characteristic features with regard to the above-mentioned categories. These investigations should also include comparisons with proposals of new classifications.

The study described here should merely be regarded as a first brief attempt. Time has not permitted a more careful study of the systems question nor discussions with those who are experts in applying these systems. The tables and figures should be studied as to the principals only.

1 Materials and Products in UDC

Let us first investigate how the most widely known international classification, the *Universal Decimal Clssification,* UDC, accounts for materials and products in comparison with US–4.

UDC has no complete and logically constructed material sequence, order according to the relationship of materials. The most important materials belong to Classes 66 and 67, which together may be said to include the industries based on the processing or production of materials. The headings here at times indicate materials and at other times industrial branches.

The two most important UDC-classes for materials and products[1])

UDC 66 Chemical and allied industries		US–4
661	Chemicals (fine, heavy, etc.)	81.3
662	Explosives. Fuels	81.3/4
663	Technical microbiology. Fermentation industries. Beverages Stimulants, narcotics.	81.8
664	Preparation and preservation of foodstuffs	81.8
665	Oils, Fats, Waxes.	81.3/4
666	Glass. Enamels. Ceramic (Clay) Industries. Plasters and compositions. Cements and concrete.	81.1
667	Colour industries	81.3/4
668	Various organic chemical industries (Soap. Gums. Perfumery. Cosmetics. Tar. Syntetic organic duestuffs)	81.3/4
669	Metallurgy (ferrous and non-ferrous)	81.2

1) UDC-headings according to the trilingual abridged edition 1958.

UDC 67 Various industries, manufactures and crafts. Mechanic technology

(671	Precious metal and gem industries. Jewellery. Trinkets)	–
(672	Iron and steel goods generally)	–
(673	Non-ferrous metal goods other than precious metals)	–
674	Woodwork and timber industry	81.5
675	Leather industry. Skin. Hides	81.6
676	Paper industry pulp and board	81.5
677	Textile industries. Cordage	81.7
678	Industries based on macromolecular materials. Rubbers. Plastics	81.4
679	Industries based on various processable materials	

As a rule, within these classes we also find what could be called "simple goods" i.e. both formless products and products of a simple form consisting of a single material.

The extraction of materials from natural sources comes under Classes 622 (inorganic) and 63 (organic), which classes also contain the extracted materials, thus leading to a certain amount of overlapping of two classes. Overlapping is also caused by certain products appearing both as industrial products in 66 or 67 and as material for construction or manufacturing. This applies to all building materials, among others.

Overlapping of certain concepts may be considered as an admitted and perhaps necessary phenomenon but it is essential that clear rules are formulated and cross-references given.

More complicated products are usually found associated with the function of the products, e.g. machines in 621; military materials in 623; vehicles in 629; clothes in 646 (and 687); instruments and certain special machinery in 681 and so on. Class 7 also contains products of a "light" kind, such as works of art, photographic materials, musical instruments, sporting gear, and equipment for games.

Fig. 7 (s. page 436) demonstrates how the product containing classes in UDC correspond to the PC-classes.

2 *The Brussels Tariff Nomenclature (BTN) and SITC. Relation BTN-PC*

A well known example of a general product classification (containing products only) is the Brussels Tariff Nomenclature, BTN, which is used for statistical purpose and custom tariffs. This classification is based on one clear principle: it begins with natural products and ends with finished products. This seems to be the most suitable arrangement, as we start with simple products and proceed to more and more complicated ones. This idea has in fact been an important factor for the creation of PC.

An other international product (commodity) classification is SITC, Standard International Trade Classification, published by the United Nations. This classification has many features in common with BTN.

BTN has 99 Chapters brought together within 21 sections, coded with roman numbers. SITC has 10 main sections (0–9) subdivided decimally.

Neither BTN nor SITC makes use of combinations for their coding. Pipes of steel is 73.18 in BTN and here 73 is products of iron and steel, but 13 is not a code for pipes; it is an ordinary running number. One has thus not the possibility to rearrange the items and e.g. collect pipes of all metals. BTN is to a great extent characterized by its original function, to regulate customs tariffs.

SITC has the same basic units as BTN and this makes an exact translation possible. The main structure of SITC is easier to survey than that of BTN.

A comparison of BTN and PC is shown in Fig. 6 b.

3 ISIC and its relation to UDC, US and PC

Sectors of industry are arranged in an international classification used for statistical purposes, ISIC, the International Standard Classification of All Economic Activities, published, as SITC, by the United Nations. Here we find also fields outside industry, such as transportations, economic activities etc. Again there is the same material flow as in the Brussels system, from natural to finished products. The relationship between ISIC and US – in very broad lines – is shown in the following table, in which UDC appears as well.

ISIC		*UDC*	*US–4*
1	Agriculture, Hunting, Forestry, and Fishing	63/64	4··
2	Mining and Quarrying	622	3·· + 88
3	Manufacturing of		
31	Food, Beverage and Tobacco	663/664	81.8
32	Textile, Wearing Apparel and Leather	675, 677	81.6/7 + 55
33	Wood and Wood Products including Furniture	674, 645	81.5 + 55
34	Paper and Paper Products. Printing and Publishing	676, (655)	81.5 + 55
35	Chemicals and Chemical, Petroleum, Coal, Rubber and Plastic Products	661, 662, 665 667, 668, 678	81.3/.4
36	Non-Metallic Mineral Products (except Products of Petroleum and coal)	666	81.1
37	Basic Metal Industries	669	81.2

ISIC		*UDC*	*US–4*
38	Manufacturing of Fabricated Metal Products, Machinery and Equipment	621	82/85
4	Electricity, Gas and Water	628.1 etc	---
5	Construction	69	88
6	Wholesale and Retail Trade and Restaurants and Hotels	658 etc.	7..
7	Transport, Storage and Communication	654, 656	84
8	Financing, Insurance, Real Estate and Business Services	3. . .	7..
9	Community, Social and Personal Services	3. . .	7..

ISIC 1–5 represents the producing and manufacturing activities or – more precisely – the complex of activities and equipment that constitute an industrial sector. We can illustrate the relation between the activities of ISIC 1–5 and the product classes of PC in the following way:

(PC 0) Natural resources
Exploitation
ISIC 1–2

PC 1 Natural products
Processing of materials
ISIC 31–37 + ISIC 4

PC 2 Industry products — *PC 4 Consumption products*
Processing of materials
ISIC 31–37

PC 3 Structural products
Manufacturing of objects
ISIC 38 (+ parts of 31–37) + ISIC 5

PC 5 Finished objects
Transportation ISIC 7
Trade ISIC 6

Fig. 5 The material flow in industry expressed in ISIC and PC classes.

The relation ISIC-PC is demonstrated in Fig. 6d.

4 *The Relation between PC and US regarding materials and products*

The concept material (=substance) is essential in a universal system (document system). We have to classify documents on e.g. properties of steel, glass etc. regardless what kind of product. In a product system, however, we cannot locate a material; here we have to consider products made of different materials. The vertical division of PC, according to product type, seems not to be suitable in a document system, where a horizontal structure, with a division according to material seems more appropriate and this structure is generally applied in document systems. US follows also that way of grouping.

In spite of the quite natural and necessary differences between a document system and a product system, there are several reasons for getting a certain correspondence. Firstly it is an easier task to get a clear and consistent subdivision of products, which are concrete objects, than of "subjects" or "concepts" of all kind, which we have to locate in a document system. A good product classification can therefore give a valuable guidance for designing the technical part in US, as technology in a high degree is linked to products. It will also be an advantage from many different aspects if a system used for statistics and other administrative purposes can be easily translated to a system for documentation.

Behind PC, as shown in table I, behind the three-digit structure of PC here not shown and behind Class 8 of US–4 (Appendix 2) lies a work in several stages for mutual adapting of US to PC and PC to US in those parts containing materials and products.

The adjustment of US 81 has led to good compatability with the first classes in PC which appear here:

US		*PC*	
81.1	Rock and earth materials	11, 21, 31	Rock and soil NP + IP + SP
81.2	Ores and metals	12, 22, 32	Metallic NP + IP + SP
81.3	Chemicals	13, 23, 33, 43	Chemical (unorg.) NP + IP + SP
81.4	Fossil materials. Fuels. Plastics	14, 24, 34, 44	Fossil and fossil-chemical NP + IP + SP + CP
81.5	Vegetable materials[1])	15, 25, 35[2])	Vegetable NP + IP + SP
81.6	Animal materials[1])	16, 26, 36	Animal NP + IP + SP
81.7	Textiles	– – 37	Textile SP
81.8	Foodstuffs	– – – 45	Foodstuffs CP
81.9	Various materials	19 – – 49	Other products

1) Excluding textile materials and foodstuffs
2) Excluding textiles

While perhaps the half volume of PC corresponds to one subclass in class 8 of US–2b, the whole class 8 corresponds rather well on the two-digit level to one subclass in PC, finished objects PC5, as shown both in table I and in Fig. 6c.

5 *PC as a basic scheme for analysis of other systems*

In Fig. 6 and 7 the PC– network is a reference scheme for comparisons of the systems BTN, US, ISIC, and UDC regarding the way materials, products, and industrial sectors are arranged.

What general conclusions can be drawn from these figures. First it is evident that both the BTN and the ISIC classes have rather easily been fitted into the PC-network. One reason for this is obviously the two-dimensional pattern of PC, which allows extending the classes in two directions, representing production stage and material or function. But this also can be interpreted as a sign that all these classifications mirror the same reality, i.e. the material flow in the industry and our stock of products, this reality, however, being looked upon from somewhat different aspects.

The vertical borderlines in PC protrude most evidently in BTN, which – like PC – is a product system. Here some of the structural products (PC 3) are separated from the corresponding industrial products in contrast to ISIC, but infiltration both to the left (e.g. metals XV) and to the right (e.g. products of textiles) occurs.

Consumption products (PC 4) do not appear as independent categories in BTN and neither in ISIC, depending on the fact that these products are very close to the natural or industrial products.

Finished objects (PC 5) is in all the systems a relatively distinct category. Here a certain correspondence PC – BTN appears.

The comparison between PC and UDC (Fig. 7, see page 436) indicates that their relationship is rather complicated.

As a general conclusion it may be said that detailed investigations of the material flow in industry are necessary for arriving at the best classification of products, processes and industrial branches. Such industrial classifications are also necessary as a foundation for universal classification, which ought to be the result of theoretical analyses in combination with studies of all types of classifications designed by experts for practical application.

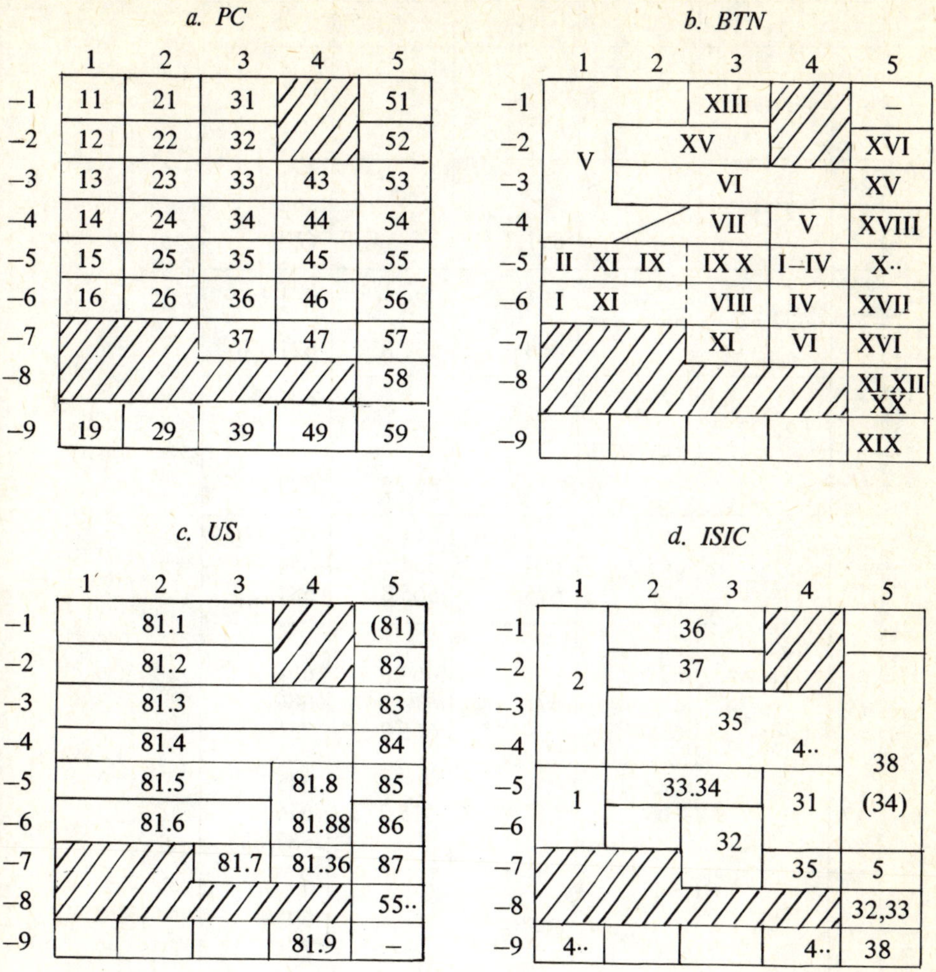

Fig. 6 a–d. BTN, US and ISIC introduced into the PS-square

	1	2	3	4	5
–1	622.35	666	691.2/6	–	–
–2	622.34	669	691.7	–	621.8/7 672 683
–3	622.36	661. .	667 668.3	662.1/4 631.8	621.1/36 621.4/6
–4	622.32 .33	661.7	678	662.6/9 665.5	625 629.1
–5	633 635 634	664. . 665.3 676.1	674 676. .	641 664. . 665.3	621.37/39 654 681
–6	636 637. . 638 639	637. . 665.2	675	664.9 665.2	65. . 686 73/79
–7		23 06	677	615.777	645 646 687/688
–8					69. . 72
–9	628.1			628.1	623 - - - - - - - - -

Fig. 7 The UDC-classes in the PC-network

F Classification of documents in a special technical field

While in the case of product classification the background for introducing the AR-idea (PC–US) was the desire to combine a classification adapted to products with access to a universal set of concepts for specification of the products, the actual motive in the building trade was to reconcile the differing opinions of whether a universal or a special system should be used for registration of building research projects and also statistics for research costs.

The method of using two systems in combination – one special and one universal – was presented in a report to the CIB[1]) Working Conference, Dec. 1969. Section F and G is mainly a citation from this report:

When the classification of documents in the building trade is discussed, the question often arises whether to use either a universal system or a special system (a building system), or to have one system of each kind for separate use (a universal system in libraries, a special system in offices). The documents relating to discussion within CIB seem to suggest a preference for a special system.

Building research, however, relates to numerous problems that lie outside the building sector proper and a universal system offers the advantage, in principle, of providing room for everything. A universal system is divided according to an overall view, which is determined in some way by the relatedness of different fields of knowledge. Naturally, in this way one does not obtain a sequence that relates directly to the building process.

Thus there are strong arguments in favour of both a building system designed with. consideration to the building process and a universal system. It seems that the problem could be solved *by using a building system at the highest level and a universal system below that* for the necessary specification.

There are indications that a combination of two systems, *one directed towards the user's field of activity and the other directed towards the store of information,* is a method that may have some prospect of success.

A need for two different systems is also indicated by the interest expressed among certain circles within the building industry in using a special system (SfB) in combination with a universal system (UDC) ("Principle of complementarity"). The development of this idea would be promoted if the question of both the structures of the systems and the manner of their combination were made the subject of fundamental studies.

The building system that I proposed, called AZ (as I here have letters instead of numerals for the main classes) has here been subjected to some minor alterations in relation to the structure given in the CIB report. Here is the current main structure:

---

1) CIB = International Council for Building Research

The AZ-Building System

A. General
B. Materials and products
C. Protection. Hygiene
D. Structure. Building elements
E. Safety. Installations
F. Construction. Maintenance.
G. Building work (labour)
H. Rooms and buildings
I. Constructions (Civil engineering plants)
K. Community planning
M. The building trade and society
Z. Other

Example of subdivision:

C. Safety. Protection. Hygiene

1 Statics. Strength of materials
2 Geostatics. Geotechnics
3 Acoustics
4 Climate
5 Indoor climate. Heat insulation
6 Lighting
7 Fire protection
8 Damages, general
9 Other

For a document "Sound insulation of doors" may in a smaller office the code C3 be sufficient, but as the number of documents within C3 grows, US numbers can be introduced for specificational purposes:

C3 – 224 : 882.75, or reversed: C3 – 882.75 : 224
    └ door (882.75)
    └ sound insulation (224)

"Sound insulation of doors in hospitals" can be designated as:

C3 – 224 : 882.75 : 884.41 (= C3 – a:b:c)
    └ hospitals (884.41)

When registering documents systematically, one can, on the lowest level, use key words ordered alphabetically in order to avoid building out the hierarchy of the system on the lowest stratum. (5)

G Conformity between adapted systems and the reference system

A certain conformity can be established between the adapted (special) systems and the reference system. If we presuppose that the adapted systems contain concept series or "facets" designed by thorough studies, these facets can often be used on lower levels in the reference system. As an example we can take the facets for building elements (D) and building types (H) in AZ. These can be found in US in the appropriate places. Thus doors in AZ is D 75, and in US 882.75, where 882 is building components.

We can also have internal conformity between different special systems. A system for traffic may constitute one class in a system for community planning, and this class may be identical with the class Community Planning in US and also constitute the class Community Planning in AZ. (See Fig. 8, page 440).

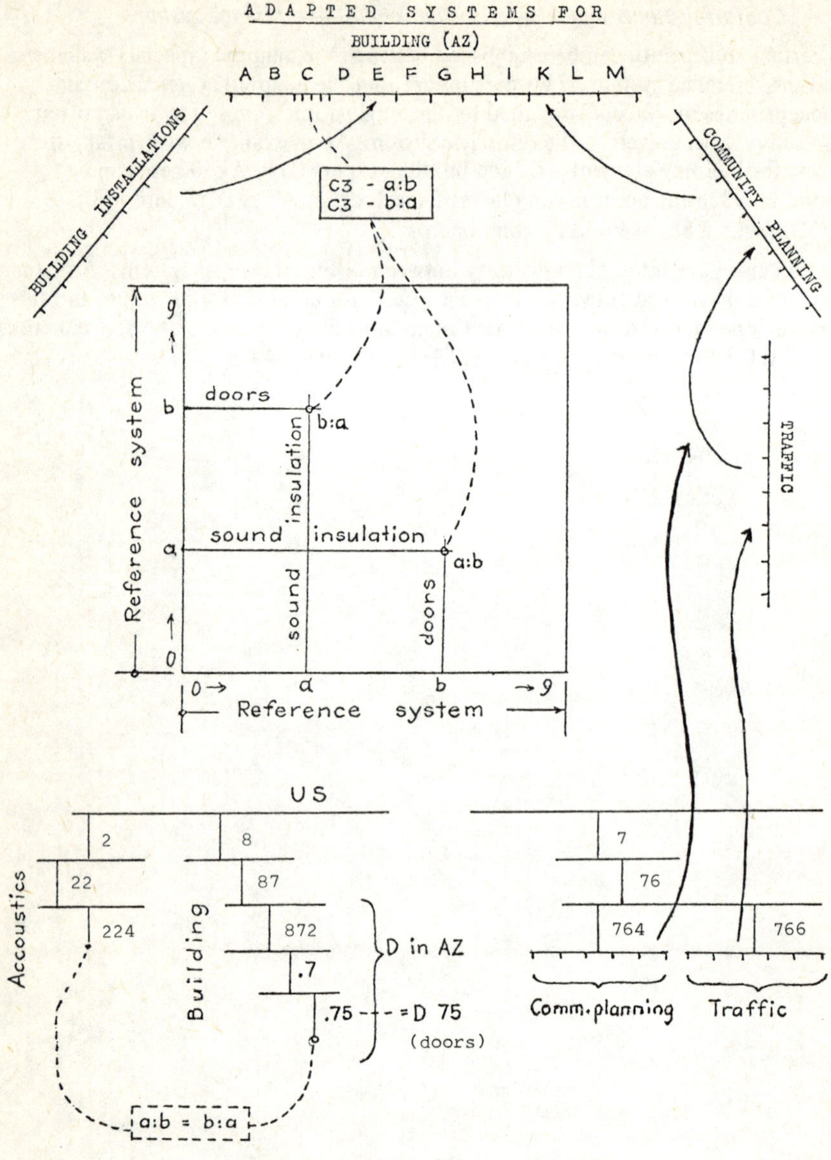

Fig. 8. Connection between different adapted systems and between an adapted system and the reference systems.

H *Classification of occupations*

The problem of fitting together an employer's desire for qualifications for a certain appointment with the qualifications of an applicant seems to be closely related to the general documentation problem, implying a person's need of knowledge for a special occasion being matched with the knowledge contained in the stored documents.

If we examine the "Nordic Classification of Occupations" (here called N.Y.K.), which is based on the International Standard Classification of Occupations, ISCO, we find that occupations are characterized by the same kind of terms as occur in a general classification, e.g.

> House, town, drawing, work, control, studies, planning, car, boat, measuring, mines, maps, chemistry, physics, laboratory, geology, genetics, fishes, drugs, agriculture, gardening, wood, hunting, hospital, medical care, school, music, textiles, etc.

These concepts reappear in many connections. The problem here is analogous to the problem of product classification. It seems to be possible to separate a coarse classification of occupations, representing the general character of the occupations, from the specified objects of the occupation, e.g.

A-system	*R-system*
Teacher	Mathematics
Consultant	Measurement
Labourer	Chemistry
Researcher	Mining
	Agriculture
	Law
	History

As an example we can take occupations involving the sawing of wood. We have in N.Y.K.

77	*Wood Work*
772.10	Frame sawyer
.15	Circular log sawyer
.20	Cross cut sawyer

These occupations could be expressed as:

Labourer – Wood : Sawing (or sawing : wood)

and the analogy with product classification can be illustrated by:

Machine – Wood : Sawing (or sawing : wood)

See also Table II, page 445.

I An integrated complex

In Fig. 8 (page 440) the *reference system* with its two-dimensional structure is illustrated by a square in which the same system – as in Fig. 4 – lies in both horizontal and vertical directions. A combination of two concepts as Wood : Sawing or Sawing : Wood represents two different points in the square, but only one point is shown here.

Along the circle we have the *adapted systems,* all making use of the same reference system. To the left along the circle we find systems adapted for special trades or branches, building, traffic etc. To the right we have adapted systems of "general type", that is systems which are complete in their kind, adapted not for a special branch but for a certain general category of objects, e.g. products and occupations. Perhaps patents could be included here.

The combination *wood : sawing* can correspond in product classification to a machine, in occupational classification to a worker, or in a universal classification to a heading Material Treatment or the like in the technological class. The combination *pig : liver* corresponds to the category Food in the product classification, to Veterinary Medicine or Zoology in other connections.

A universal document system is also indicated. We have here an interesting problem which may be indicated. In the present debate on the possibility of obtaining a new universal system, there are some people who advocate that a structure based on traditional disciplines should be maintained, while other experts prefer a universal system based on concepts, perhaps with a faceted structure.

A discipline system as an A-system in combination with a reference system based on concepts is perhaps a solution that should be investigated.

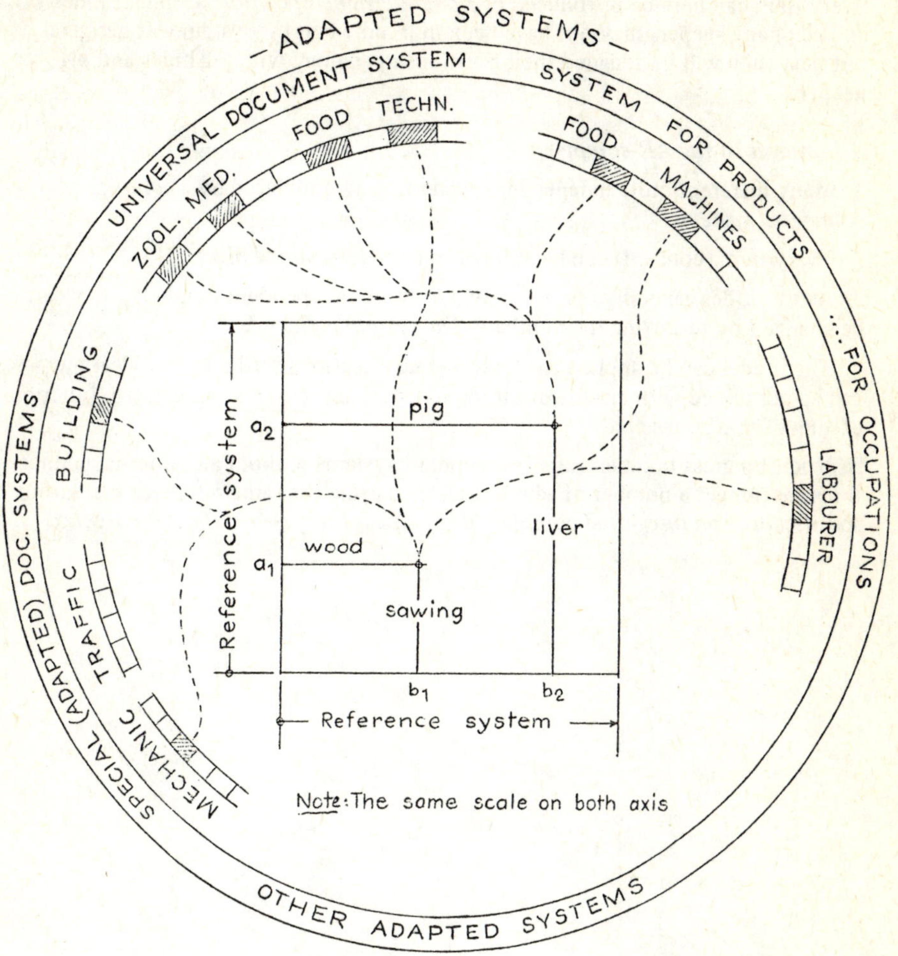

Fig. 9. Different adapted systems, along the circle, make use of the same reference system, the square. Within the reference system two-dimensional classification is applied.

Table II shows how codes for documents, products and occupations from three international systems, all with a certain official position *can be replaced by codes from the AR-complex.*

The table shows also how the product numbers can be written in shorter notations.

Every item can here be introduced *in a predetermined position:* elephant hides and elephant keepers may not have been in mind when the system was designed, but here they will be assigned their positions in analogy with pig hides and pig keepers.

Advantages of the AR-complex

1 Many different quite independent systems can be replaced by a *complex of connected systems.*

2 Every item (concept) can be entered in a *predetermined place.*

3. As the codes generally consist of two or more independent numbers, they can be changed by re-sorting the numbers. *The system is flexible.*

4. The model can be applied to all areas. Classifications for bibliographical purposes can be combined with classification for statistics and other administrative functions, editing of textbooks etc.

Instead of a great number quite independant systems without any internal connections we can get a number of adapted systems using the same reference system for specification, *an integrated complex of classification systems, the AR-complex.*

Table II. Comparison between the AR-complex and other systems

In order to obtain a certain uniformity in the shortened numbers the digit 0 sometimes is added.

SYSTEM FOR:	THE AR–COMPLEX		OTHER SYSTEMS	
DOCUMENTS	US – 2		UDC	
Pig, zool.	44.975		Pig, zool.	599.731.1
Pig breeding	446.4...		Pig, domestic animal	636.4...
Pig hide (= hide : pig)	81.62 : 44.975		Pig hide	675.034
Elephant hide(= hide : elephant)	81.62 : 44.951		Elephant hide	675.03:?
PRODUCTS	PC–US	Shorter No.	BRUSSELS	
Living pig = = Living animals – pig	161 – 44.975	161.00.975	Living pigs	01.03
Pig hide = = Animal struct. prod. – hide : pig	36 – 81.62 : 44.975	360.20.975	Hides of other animals	41.01.903
Elephant hide = = Animal struct. prod. – hide : elephant	36 – 81.62 : 44.951	360.20.951		
Wood sawing machine = = Working mach. – sawing : wood	527 – 820.46 : 81.51	527.46.51	Circular and frame sawing machines	84.47.201/202
OCCUPATIONS	. . . . – US		NORDIC CLASS. OF OCCUPATIONS	
Pig breeder	XX – 44.975		Pig breeder	412.40
Elephant keeper	XX – 44.951		–	–
Tanner = = Worker – chemical : hides	XX – 28 : 81.62		Tanner	854.30
Wood sawyer = = Worker – sawing : wood	XX – 820.46 : 81.51	XX – 46.51	Frame-, cross-, cut- and circular-sawyer	772.10/.20

Appendix 1

UNIVERSAL SYSTEM. VARIANT NO 1

Part A. Theory – Technology

0. GENERAL CONCEPTS

1. QUANTITIES – DATA

   11 Number. Mathematics
   12 Data. Data processing
   13 Measurement
   14 –
   15 Space. Time. Motion
   16 –
   17 Organization. Sciences

2. MATTER – MATERIAL

   20 Matter, general
   21 Physics of matter
   22 Physics of atoms
   23 Chemistry
   24 Stone and earth material
   25 Metals
   26 Organic materials
   27 Special materials and products
   28 Material processing
   29 Products, general. Form types

3. FORCE – ENERGY

   30 Energy, general
   31 Mechanics – Dynamics
   32 Statics
   33 Acoustics
   34 Heat. Heat technology
   35 Radiation. Light. Optics
   36 Electricity. Magnetism
   37 Electronics. Teletechnology
   38 Nuclear energy

4. GENERAL AND FINAL TECHNOLOGY

   40 Technology, general
   41 Tools, Machines, etc.
   42 Energy technology
   43 Manufacturing engineering
   44 Vehicles. Transport
   45 Building
   46 Individual equipment. Household
   47 Graphic technology
   48 Military technology. Weapons
   49 Other fields of technology

Part B. Nature – Society

5. EARTH – UNIVERSE

   51 Universe
   52 The earth
   53 Rock stratum. Geology
   54 Soil stratum. Ground
   55 Water stratum
   56 Air stratum

6. LIFE – MAN

   60 General biology
   61 Microorganisms
   62 Plants
   63 Animals
   64 Man
   65 Organs and functions of the body
   66 Nervous system. Perception
   67 Diseases. Injuries. Hygiene
   68 Food. Stimulants
   69 Psychology

7. CULTURE

   70 Culture, general
   71 Language
   72 Religion. Ethics
   73 Philosophy. Theory of sciences
   74 Literature
   75 Art
   76 Music
   77 Dance and spectacles
   78 Individual environment

8. SOCIETY

   80 Society, general
   81 The state
   82 Law. Justice. Regulations
   83 Economy
   84 People
   85 Work. Industry
   86 Town and country. Communications
   87 Education. Information
   88 International relations
   89 Other social matters

9. GEOGRAPHY–HISTORY

   90 Geography, general
   91 Europe
   92 Asia
   93 Africa
   94 America
   95 Australia. Oceania
   96 History
   97 Individuals

Appendix 2

UNIVERSAL SYSTEM – VARIANT 4

1. EXACT BASIC SCIENCE
   - 11 Number. Mathematics
   - 12 Data. Data processing
   - 13 Measurement
   - 14 –
   - 15 Space. Time. Motion
   - 16 –
   - 17 Organizational sciences

2. PHYSICS. CHEMISTRY
   - 20 Physics, general
   - 21 Mechanics
   - 22 Acoustics
   - 23 Heat. Thermodynamics
   - 24 Radiation. Optics
   - 25 Electricity. Magnetism
   - 26 Atoms
   - 27 Physics of matter
   - 28 Chemistry

3. UNIVERSE. EARTH
   - 31 Astronomy
   - 32 The earth. Physical geography
   - 33 Rock stratum
   - 34 Soil stratum
   - 35 Water stratum
   - 36 Air stratum

4. LIFE
   - 41 General biology
   - 42 Microorganism
   - 43 Plants. Agriculture
   - 44 Animals
   - 45 The animal and human body
   - 46 Nervous system. Senses. Life
   - 47 Psychology
   - 48 Diseases
   - 49 Hygiene. Medical care

5. THE INDIVIDUAL
   - 51 The individual in society
   - 52 Family. Sexual life
   - 53 Daily life. Customs
   - 54 Special cultures. Ethnography
   - 55 Food. Clothing. Home
   - 56 Sport

6. LANGUAGE. HISTORY. SPIRITUAL CULTURE
   - 60 Culture, general
   - 61 Language
   - 62 History
   - 63 Biography
   - 64 Religion
   - 65 Philosophy. Science. Ethics
   - 66 Literature
   - 67 Art
   - 68 Music
   - 69 Dance. Spectacles

7. SOCIETY
   - 70 Society, general
   - 71 The state
   - 72 Law. Justice. Crime
   - 73 Economy
   - 74 Demographics. Social matters
   - 75 Production. Industries. Trade
   - 76 Town and country. Communications
   - 77 Education. Research. Information
   - 78 International relations. Defence. War
   - 79 Other social matters

8. TECHNOLOGY
   - 80 Technology, general
   - 81 Materials
   - 82 Industrial work. Tools. Machinery
   - 83 Energy engineering
   - 84 Transportation. Vehicles
   - 85 Teletechn. Calculation. Regulation
   - 86 Graphic technology
   - 87 Ground and terrain work.
   - 88 Buildings and Constructions
   - 89 Other fields of technology

9. THE NATIONS. SURVEYS
   - 90 International organizations
   - 91 Europe
   - 92 Asia
   - 93 Africa
   - 94 America
   - 95 Australia. Oceania. Arctic regions
   - 96 –
   - 97 –
   - 98 –
   - 99 General documents

Appendix 3

	Examples from US–4	Reference to US	SAB[1]	UDC	
3	UNIVERSE, EARTH				
31	ASTRONOMY		Ua	52	
:152	Astronomical time calculations		Ua-f		
:20	Astro-physics, general			523.0	
:24	Astro-radiation			.035	
:25	Astro-electricity			.037	
:255	Astro-magnetism			.038	
:4	Astro-biology			.07	
310	Astronomical science, general	(= 31:656)	–e	522	
311	Astro-mechanics. Motion. Orbits	(= 31:21)		521	
312	Universe. World systems		–a		
.1	Primitive systems			523.18	
.2	Astrology				
.3	Geocentric			133.52	
.4	Heliocentric				
.5	Modern system. Cosmogony			523.1..	
313	Galaxies and stars		–e	.8	
314	Sun. Solar system		–ba	.7	
315	Planets		–bb	.4	
316	Moon		–be	.5	
317	Comets. Meteors		–bf/bg	.5/6	
318	Space exploitation (Space vehicles 847)				
32	THE EARTH. PHYSICAL GEOGRAPHY		Ud	52.. + 55..	
320	Geophysical and geolocial science, general		Ud	550.3	
321	The globe	(= 32:656)	Ud		
:13	Measurement	Geodesy 137		528.2	
:151	Form			525.14	
:152	Time			.5	
:153	Motion			.3	
:211.1	Mass			.11	
:211.4	Gravitation			.11	
:23	Heat			.21+550.36	
:24	Radiation			.22	.35
:25	Electricity			.23	.37
:255	Magnetism		Ub–d	.24	.38
:28	Chemistry				.4
.1	Core			551.14	
.2	– –				
.3	Crust			.16	
322	Evolution of the earth		Ud–ag	551.7	
.1	Origin of the earth				
.2	Archean time			.71	
.3	Eeozoic, Algonkian time			.72	
.4	Paleozoic time			.73	
.6	Mesozoic time			.74	
.7	Tertiary time			.76	
.8	Quarternary time			.78	
.9	Future of the earth			.79	

1) SAB = The general Swedish library classification.

References to reports and articles of the author

1) E. WÅHLIN, "Basic principles and outline of a new universal system of classification". Conference in Building Documentation, Geneva 1949.

2) E. WÅHLIN, "Classification systems and their subjects". American Documentation, 1966: Oct. Also introduced in "Introduction to Information Science". R.R. Bowker Co., New York, 1970.

3) E. WÅHLIN, "Klassificeringssystemens roll i samhället" (The role of classification systems in society). Teknisk Tidskrift, Stockholm 1967: 36.

4) E. WÅHLIN, "Dokumentationens dilemma" (The dilemma of documentation). Teknisk Tidskrift, Stockholm 1968: 28.

5) E. WÅHLIN, "Klassificering av materialteknisk litteratur" (Classification of documents in material technology). Report to SINFDOK (Swedish council for Scientific Information and Documentation) Stockholm Jan. 1969.

6) E. WÅHLIN, "The structuring of information on research projects". CIB, W 52 Conference Dec. 1969.

7) E. WÅHLIN, "Klassificeringssystem". Teknisk Tidskrift, Stockholm 1970: 18.

8) E. WÅHLIN, "Classification Systems". Report to SINFDOK. March 1974 (in English).

Ingetraut Dahlberg (Frankfurt/Main)

Principles for the construction of a universal classification system. A proposal

1. Introduction

1.1 A "useful compromise"

In presenting a proposal for a structure of a universal classification system it should be stated in advance that we ought to be well aware of whatever could be achieved by such a system can necessarily only be some kind of a compromise between reality and human insight; a compromise which, however, as R. Fairthorne has put it in his paper at this conference[1]), will hopefully be a "useful" one, meeting a great number of needs which will not be emphasized here. Some of these needs have been listed earlier[2]). Regarding the attempts for a structure in former or existing classification systems the paper presupposes that such are known. Most of them are well described in the work of Šamurin[3]).

1.2 Definitions

For the specific terms of this paper the following definitions are offered:

Regarding the definitions of "class", "classification" and "system" I would like to refer to the results of discussions conducted by members of the Philosophical Institute, University of Düsseldorf at a meeting in April 1968 and published in[4]). They defined

- a *class* as a unifying characteristic (attribute, feature, etc.);
- a *classification* as a form of order which puts together a body of (a whole of) any kind of elements (things, ideas, sets, also classifications and systems) according to the characteristics which they have in common. These are called 'classes' to which elements are adjoined according to their existence (Maßgabe); the whole of which just as often also is called a classification;
- a *system* as a form of order that puts together a whole of elements into a structural connection which in itself is often also called a system.

A *'classification system'* (thus) is a form of order that puts together into a structural connection a body of elements according to the characteristics which they have in common.

'Characteristics' can be defined as intrinsic or additional attributes of things or any kind of attributes of relations existing between objects (large sense – Gegenstände).

The principle underlying the structural connection of a system is called the *systematifier.*

By *'universal'* as the attribute for the proposed system is meant

- a system that can be used by everybody,
- for every application purpose (e.g. in archives, libraries, documentation and information services, in education, administration, editorial organizations)
- in order to survey from a universal point of view
- all the knowledge fields existing.

Knowledge means to have experiences, cognitions or notions, subjectively and objectively certain and from which judgements and inferences may be formed, which also seem sure enough to be recognized as knowledge.

Knowledge elements are single experiences, cognitions or notions of objects (large sense – Gegenstände) (things or phenomena or aspects under which these can be seen) which can be comprehended by concepts and which can be made communicable by names or (better) 'descriptors'.

Knowledge fields consist of clusters and concentration of knowledge elements. There are different sizes of knowledge fields as well as different stages of development; also, according to the different purposes of knowledge fields subject – and functional fields may be distinguished.

A *scientific discipline* represents the last stage of a knowledge field which is only reached when an assemblage of knowledge elements referring to the same object area is brought together into a structured, interdependent connection (Begründungszusammenhang).

The term *"objects"* has here been used in its larger sense. In the paper it will, however, be always used in its narrower sense, denoting things, mostly material; *"subjects"* is used for topics, e.g. topics of documents; *"facets"* are the specific applications of terms of general categories (e.g. objects, processes, characteristics, relation, determination) in one knowledge field, organized in separate hierarchies.

2. Premises and postulates

2.1 Premises

The following premises will be used for the conception of the system:

(1) The universe of knowledge is a universe of knowledge elements which can be comprehended as a universe of concepts represented by single, combined or complex linguistic or other symbolic means, descriptors;

(2) The universe of concepts expressed by descriptors can be grouped into classes applying class characteristics or class relations;

(3) The universe of classed descriptors can be arranged according to a predetermined structure into a universal classification system using a systematizing principle, a systematifier.

4) The element positions of the universal classification system, determined by the systematifier, can be represented by a notation, the symbols of which have mnemonic value (signifying the structures of the concept system) which can be used to express concept combinations within the paradigm of the system and for the syntagmatic organization in formulating propositional statements.

2.2 Postulates

The following postulates should be observed in constructing the proposed system:

Concerning arrangement

1) The system, in order that it may be handled universally, must be organized distinctly and intelligibly;
2) The elements of the system ought to be meaningfully and logically grouped and arranged;
3) Except for the preferred hierarchy of aspect fields the system should have an overall faceted structure, allowing for additional hierarchies according to objects, methods and processes, general and special characteristics, time and space categories and further aspects;

Concerning contents

4) The system should allow for determining past, present and future subjects;
5) The system should allow for continuous merging of fields, creation of new fields, be it in the object – or the aspect fields area; it should provide insight into the respective contents of existing or prospective fields;
6) The system should contain all the necessary concepts for
   a) description of subjects in discourse (the subjects may be e.g. objects or methods, theories or aspects)
   b) labelling of objects (e.g. in primary data documentation).

Concerning relations

7) The relations between the classes should be made easily recognizable or they should be stated expressedly;
8) Although the system will include one preferred hierarchy of fields it also ought to contain indications of additional hierarchical relations;
9) In addition to the relations given by the structure of the system (paradigmatic organization) semantic relationships between descriptors can be indicated within the main systematic part (by partial application of the thesaurus-method, including the additional hierarchical relations from 8);

Concerning notation

10) The notation for the descriptors of the system should consist of decimal numbers to be subdivided by direct or interrupted subdivision (the interruption being expressed by a period between the numbers);

Concerning presentation

11) Any presentation of the system should consider the psychology of its users and should therefore be adapted to the expectations of different user groups by edition of different versions of the system;

12) The system should contain in each case of presentation at least the following 5 different parts:
   a) Introduction into and rules for the use of the system together with highest level outline of the whole system showing also the access to the different levels of the structure of the system;
   b) a systematic index to the whole system (containing notation and descriptors only, without additional relations);
   c) systematic main part (including additional relationship indications);
   d) systematic part of precombined concepts;
   e) alphabetical index (referring to notations of systematic parts c and d; descriptors of precombined concepts should be entered in natural and inverted order.)

3. The structure of the system

3.1 The problem of the 'main class'

In earlier systems it was tried to group a universal system according to subjects fields or disciplines (called 'main class' by librarians) with insertions of the specific objects of such disciplines wherever they were deemed correctly localized; by this procedure however, difficulties were experienced whenever the same objects were treated by different disciplines; this leads to numerous parallel divisions and an irritating abundance of object-arrays seen under different aspects. In newer systems therefore it was decided to omit such disciplines as 'main class'; it was tried to elaborate special classification systems and likewise rules for universal systems based completely on terms for entities and attributes or a number of facets only [5], [6], [7]. In the latter case the descriptors denoting fields or disciplines could – if necessary – be inferred from the contents of the entities[8].

Neither of these two standpoints seems to be realistic.

3.2 Five fundamental categories

The reasoning for these categories cannot be elaborated here. A paper containing it is in preparation.

It is proposed to separate the main objects and their aspect fields (subject fields, disciplines) from each other establishing thus the first two fundamental categories of the proposed system:

1– category for object descriptors (objects relating to one area of aspect fields only),

2– category for descriptors from aspect fields.

Three further fundamental categories will be needed as basis for the determination of concepts in discourse; i.e.

3– category of descriptors for time indications,

4– category of descriptors for space indications,

5– category of general descriptors.

Each of these categories could be made the 'main class' or starting point for a propositional statement, to be used e.g. in card files or computer retrieval systems; however, if arrangement of books on shelves is concerned the order of these categories might be fixed, starting in any case with category 2 as preferred hierarchy in order that the user may browse by broad subject areas and specific aspect fields. As an additional help the catalog or index must lead the user to the descriptors of the facets of aspect fields or to the other categories desired.

Later on it will be shown that all of these fundamental categories could be joined together into one single frame. For reasons of easier syntagmatic organization of propositions (including the question of short notations) it seems however useful to separate the five categories in the manner proposed.

In the following we will only be concerned with the first two categories, leaving the latter three to a later elaboration of their respective main structures. As far as the category of the "general descriptors" is concerned it should be noted that for these a proposed classification could be used which A. Diemer submitted in 1968 to the members of the Committee on Thesaurus Research of the German Documentation Society[9]). So far for this category some 4000 terms have been collected. Besides the most abstract terms for objects, processes and attributes this category will also contain relators as proposed by J. Perreault[10]) and might well be identical with the idea of a universal scheme of role indicators as proposed by D. Soergel[11]).

3.3 The category of object descriptors

According to the principle of evolution of entities and beings, adding to them the products of the activities of beings the following array of object areas is proposed taking into account the fact that "before" matter could be or come into existence, ideas and principles underlying the construction of matter must have "existed". (As a mnemonic help the numbers for objects are always preceeded by an 0.)

01 formal entities (numbers, concepts, principles, etc.)
02 material entities (atoms, molecules, compounds, etc.)
03 cosmological entities (earth, planets, sun, stars, etc.)
04 biological beings (plants, animals)
05 human beings (persons according to age, sex, race, etc.)
06 sociological beings (organizations, populations, etc.)
07 material products (industrial and trade products, etc.)
08 intellectual products (works presented in documents)
09 mental and spiritual products (e.g. pieces of art – music, literature, paintings; metaphysical works)

The object areas can be developed into a classification by subdivisions according to the following subsumptive relations:

a) genus-species relation (or kind-type relation)
b) whole-part relation
c) subject-property relation,

where relation a) is meant to be considered the constant and b) and c) the variables, the "analytical" subdivisions, so that to each of the subdivisions of a) any subdivision of b) and/or of c) could be attached, using the principle of interrupted subdivision (see[12]) by using the period for designation of the additional concept of b) and/or of c). It might be argued that the subdivision of a) already takes place by considering the characteristics (common attributes constituting a specific class) and that the subject-property relation then would become a redundant subdivision. This, however, should not occur, if within a) only intrinsic, substantial attributes and within c) only additional, varying attributes of certain objects are used.

An example of the subdivision of an object area (08) will be given later (see 4.2).

3.4 Aspect fields

Corresponding to the evolutionary array of object areas a range of aspect fields can be seen to exist so that an object area may serve as a sort of "specific object" for such an aspect area. The name "aspect fields" was chosen since each of these fields considers objects of its area according to a specific aspect. The following aspect areas with some fields meant as examples are proposed:

1 Form and structural area (logic, mathematics, statistics, etc.)
2 matter-handling area (physics, chemistry and their technology)
3 geo- and cosmo-area (geosciences, astronomy, etc.)
4 bio-area (biology, veterinary medicine, agriculture etc.)
5 anthropological area (human biology, medicine, psychology, language, etc.)
6 sociological area (politics, administration, social psychology, history, law)
7 production area (economics, industry, trade, traffic, etc.)
8 information area (archivistics, documentation, information science, journalism, book-trade)
9 humanities area (Geisteswissenschaften) philosophy, art, religion)

The fields of these areas can only be subdivided into a classification by using the following two relations:

a) whole-part relation (if a number of subfields make up one field) and
b) appurtenance or association relation (for the relation between the concept of a field and its specific facets).

It must be kept in mind that by direct subdivision of a field using the subsumptive relations (which can be applied with objects as stated above) it will never be possible to "get" to the specific facets of a field which are either theoretical problems, specific objects, methods and processes, or specific aspects etc., since these are elements of their own categorial hierarchies, applied to the aspect fields. Thus the facets of a field must always be "attached" to the descriptor of the field by interrupted subdivision. They are thus the variables of the field designations and can be used as a single interrupted subdivision or in sequential form, adding one facet-division after another according to the proposition of the subject.

The following array of facets is proposed for the subdivision of aspect fields:

.1 Theory, science of a field, general problems
.2 Specific objects of a field (parts, components, systems etc.)
.3 Processes (methods, techniques, actions)
.4 Attributes (properties genuine to a field, characteristic features, etc.)
.5 Administration facet (management of center of a field)
.6 Organization facet (organization of field in a larger frame; orientation)
.7 Relation facet (not as yet exemplified)
.8 Determination facet (use, function, influence, effect of field)
.9 Evaluation facet (application of normative principles, rating, tests, etc.)

In elaborating a number of aspect fields it has been discovered that it is not always possible to find for each of these facets specific descriptors in a given field. It probably depends very much on the scientific status of a field whether or not these facets are "occupied".

Regarding the facet .2 –'specific objects of a fields' it should be pointed out that these are different from the objects to which the whole area of a field is related.

4. Examples for use of proposed classificatory principles

4.1 Two schemes elaborated

For two practical cases the proposed principles for the subdivision of objects and its particular object fields (defined as the knowledge of the kinds, parts and properties of an object) and the corresponding fields of an aspect area could be applied in the elaboration of

1) Proposal for a classification of kinds of documents and documentology[13]), to be called scheme 1 and
2) Design for a descriptor system for the information sciences[14]), to be called scheme 2.

In both cases the principles outlined above were followed; they have proved to be valid ones. For some reasons, however, the two schemes differ considerably in contents volume and outline. Scheme 1 was elaborated for a general purpose on an international level; scheme 2 for the indexing work in an operating documentation center.

4.2 Description of scheme for kinds of documents (scheme 1)

In order to follow a decision of the Central Classification Committee (CCC) of the Fédération Internationale de Documentation (FID) to revise Class 0 of the UDC entirely and to summarize all the notations meaning concepts on documents or documents as such within UDC 01/09 and within UDC 1/9 into the notation 03 (to receive then a new meaning), a revision proposal was elaborated and presented in scheme 1 which contains 1160 descriptors for all kinds of documents as well as parts and properties of documents using only the three subsumptive relations for objects as mentioned in 3.3 except for the subdivision of .0, where some general aspects are included. It cannot be said that this scheme – the outline of which is presented below – was the result of the two years' work of a special CCC Working Group, it must be stated however, that much of the contributions of the members of the Working Group went into the scheme. The proposal was not accepted by the CCC in its structure, but partly so in its contents. The outline presented below is taken from the proposed scheme itself and therefore contains also the UDC notations – 03 instead of 08 if it was the object area of our proposed object array and .01/.09 instead of .1/.9 for the analytical subdivisions which in the UDC are always introduced by a .0. (In our proposal the period between decimal numbers receives a meaning and can therefore be used for the interrupted subdivision.)

From the presentation of the scheme it cannot be recognized that the subdivisions of .03 'Form and shape of documents' until .09 'Evaluation of documents' are actually all properties of documents; they are, however, seen from the same viewpoints or as results of actions implied in the facets which are outlined under 3.4 as the specific facets of aspect fields.

Outline of scheme 1: Example of subdivisions of an object area

A SURVEY OVER MAIN DIVISIONS

03 DOCUMENTS

031 ENCYCLOPAEDIC TYPE WORKS
. 1 Encyclopaedias
. 2 Lexicons
. 3 Dictionaries
. 4 Classification systems, documentation thesauri
. 5 free
. 6 Handbooks
. 7 Advances, Annual reviews
. 8 Digests, Surveys, State-of-the-art reports
. 9 free

032 LISTS OF ALL KINDS
. 1 Data lists, Tables
. 2 Lists of materials
. 3 Lists of plants and animals
. 4 free
. 5 Lists of persons
. 6 Lists of Institutions, organizations
. 7 Lists of products, services
. 8 Indexes to documents, Concordances
. 9 Lists of literary citations, Lists of abbreviations

033 BIBLIOGRAPHIES AND CATALOGS OF DOCUMENTS
. 1 Bibliographies in general
. 2 B. according to chronological issuance
. 3 B. according to geographical and language origin
. 4 B. of certain authors or author groups
. 5 Abstracts services
. 6 Catalogs of document collections
. 7 Union catalogs
. 8 Catalogs of the book trade
. 9 free

034 MACRODOCUMENTS
. 1 Document collections
. 2 Collected works; polygraphies
. 3 Journals
. 4 Newspapers
. 5 Newsbulletins
. 6 free
. 7 free
. 8 free
. 9 free

035 MONOGRAPHIC TYPE DOCUMENTS
. 1 Monographs
. 2 Papers
. 3 Journal articles
. 4 Reports
. 5 Dissertations
. 6 Patentletters
. 7 Standards
. 8 Letters, Circulars
. 9 free

036 DOCUMENTS WITH INTENDED FUNCTIONS
. 1 D. with announcement and advertising functions
. 2 D. with an advisory or regulatory function
. 3 D. with evidence function
. 4 D. with emotional features
. 5 D. for educational purposes
. 6 D. meant to be filled out and filled-out d.
. 7 Documents as models
. 8 free
. 9 free

037 DOCUMENTS RELATED TO SUBJECT FIELDS
. 1 free
. 2 Religious documents
. 3 Legal documents
. 4 Parliamentary documents
. 5 Business, commercial and banking documents
. 6 free
. 7 Music documents
. 8 Literary documents
. 9 D. of historial value and historical evidence

038 GRAPHICAL, PICTORIAL AND THREEDIMENSIONAL D.
. 1 free
. 2 Graphical presentations
. 3 Pictorial presentations
. 4 Geo- and astrographical representations
. 5 Moving pictures
. 6 free
. 7 Threedimensional documents
. 8 free
. 9 free

039 FREE

(scheme 1 continued)

B SURVEY OVER SPECIAL FACETS

03.01 DOCUMENTOLOGY

.01 THEORY AND GENERAL ASPECTS OF DOCUMENTS
.011 Science of a document
.012 Analysis of documents
.013 Importance of documents
.014 Problems of documents
.015 Developments, trends of documents
.016 Innovations on documents
.017 Reforms for documents
.018 Rules, conventions for documents
.019 free

.02 ELEMENTS AND PARTS OF DOCUMENTS
.021 Peripheral parts of documents
.022 Identification elements of documents
.023 Text parts of a document
.024 Non-text parts of documents
.025 Additional elements to the text
.026 Contents indicating and opening parts
.027 Other parts
.028 free
.029 free

.03 FORM AND SHAPE OF DOCUMENTS
.031 free
.032 Kind of information entrance
.033 Documents being reproductions
.034 Shape of data carrier
.035 free
.036 Issuing of documents
.037 free
.038 free
.039 free

.04 EXTERNAL AND INTERNAL CHARACTERISTICS
.041 free
.042 Characteristics of material
.043 Physical attributes of document material
.044 Volume, size, weight and age of documents
.045 free
.046 Attributes of formal arrangement of contents
.047 Stylistic and presentation form of contents
.048 Languages and translations of documents
.049 free

.05 STATUS FORMS OF DOCUMENTS
.051 Original documents, primary documents
.052 Secondary documents, adapted documents
.053 Tertiary documents
.054 Cumulations of documents
.055 Conversion of documents
.056 free
.057 Different editions of documents
.058 Documents according to value judgments
.059 free

.06 ORIGIN AND LEGAL SITUATION OF DOCUMENTS
.061 Institutions as authors and publishers
.062 Author team as authors or publishers
.063 Single persons as author or editor
.064 Documents published under a pseudonym
.065 Documents published anonymously
.066 free
.067 Availability of documents
.068 Legal questions related to documents
.069 free

.07 FREE

.08 DETERMINATION OF DOCUMENTS
.081 free
.082 D. for specific user groups
.083 D. for specific level of understanding
.084 D. meant for specific purposes
.085 D. for restricted user population
.086 Use of documents
.087 User feedback on documents
.088 Environmental influences on documents
.089 free

.09 EVALUATION OF DOCUMENTS
.091 free
.092 Evaluation of quality of external presentation
.093 Evaluation of intended level
.094 Evaluation of quality of contents
.095 Timeliness and originality of documents
.096 Evaluation of attainment of author's aims
.097 Free
.098 Evaluation of documents for documentation purposes
.099 free

4.3 Description of scheme for the information sciences (scheme 2)

In scheme 2 all of the 1255 subject headings are presented that are used in the form of single or precombined descriptors for the preliminary organization of the systematic catalog in the documentation center of the German Documentation Society, Frankfurt. With this scheme some 120 000 titles (from the period 1963–1970) have been arranged in the following order:

Kinds of documents, documentology

Information sciences, core fields

- Archivistics
- Museology
- Library Science
- Documentation (= information science)
- Information (= publicity and journalism)
- Communication science

Information sciences, fringe fields

- Linguistics, Terminology, Translation
- Classification (science)
- Computer technology
- Handsorted punched cards technology
- Office technology
- Reprography and microfilm technology

Influence and application fields

(without the core- and fringe fields, arranged in the sense as outlined in 3.4 of this paper)

Geographical subdivisions

Each of the fields has been subdivided by pertinent descriptors in the fixed facet array as outlined for aspect fields under 3.4; those used for the core fields of the information sciences are given below.

Whereas scheme 1 was done in German and English this one was only prepared in German, and its notation differs insofar from the proposed one as for reasons of brevity and disguise letters were used for the designation of fields in combination with decimal numbers for the subdivision into subfields and for the facets. And whereas scheme 1 contained all or nearly all of the concepts possible in relation to kinds of documents, scheme 2 contains only those descriptors that have actually been used. Therefore the scheme cannot at all be considered complete regarding the potential classes in this area. The scheme does contain, however, a great number of precombined descriptors which were needed for the subdivision of occasionally occuring large amounts of references to the same subject.

Subdivision of facets in the core fields of the information sciences

Facet 1:

.1 Theory and general
.10 Science of a field
.11 Theory of a field
.12 General questions (tasks, problems)
.13 Methodological problems
.14 Concept of a field
.15 Research in a field
.16 Politics in a field
.17 History of a field
.18 Development trends in a field
.19 Reforms in a field

Facet 2:

.2 Specific objects of a field

.21 Formal units
.22 Information units
.23 Bibliographical units
.24 Card files, collections
.25 Groups of persons in a field
.26 Institutions in a field
.27 Systems in a field
.28 Media and services

Facet 3:

.3 Methods and processes

.31 Originating of information and documents
.32 Acquisition
.33 Analysis and cataloguing
.34 Abstracting, indexing, and classifying
.35 Storage and retrieval
.36 Secondary preparation of documents
.37 Dissemination
.38 Interchange of inform. and docs.
.39 Reading

Facet 4: *)

.4 Special characteristics

Access to information and docs.
Compatibility of (systems)
Contralization (in IS)
Redundancy (of information)
Selectivity
Backlog
Critics
Humor

Facet 5:

.5 Administration (of centers)

.51 Management
.52 Buildings, equipment
.53 Personnel
.54 Work studies
.55 Accounting
.56 Control
.57 Marketing
.58 Taxes and insurance

Facet 6:

.6 Organization of IS institutions

.61 Planning (national, regional)
.62 Systems design
.63 Systems construction
.64 Organization of systems on natl. or internatl. basis
.65 Organization of systems or centers
.66 Information flow
.67 Cooperation
.68 Relations between institutions

Facet 7:

free

Facet 8:

.8 Function and use of IS installations
.81 General use
.82 Circulation
.83 References and advice
.84 Information needs
.85 User psychology
.86 User sociology
.87 User barriers

Facet 9:

.9 Evaluation

.93 Evaluation, relevance and recall
.94 Evaluation, effectiveness and efficiency
.95 Evaluation, economic situation

---

*) here only some examples can be given, the notation of these Descriptors would have 3 digits; it would lead too far to derive them in each case.

Another observation must be mentioned regarding the contents of scheme 2: in the area of the information sciences there are a number of scientific fields, here called 'core fields', e.g., archivistics, library science etc. and there are some technical fields, e.g., computer science, reprography etc. which here are listed together with the two disciplines (linguistics and classification science) of heavy influence on the information sciences, as 'fringe fields'. Then, there are other influence fields, e.g.: science of science, logic, methodology, organisation science, cybernetics, information theory, standardization, etc. and in addition the great number of application fields, for e.g. 'documentation of chemical literature', 'information systems in medicine' etc. And finally there are fields for which the information sciences are meant as application fields, as e.g. education (for information science), legal questions (in information science) etc. Within a special classification all of these fields, even with a number of subdivisions each, must be elaborated in the descriptor system, as well as, of course, all the necessary geographical subdivisions equally. This would not apply if special classifications were derived from a universal scheme.

As can also be seen above, descriptors from the object area corresponding to the information sciences, 'Kinds of documents, documentology' have been incorporated into scheme 2 also; however, from the total number of 1160 descriptors only some 260 were used. This is partially due to the fact that there is not that much literature on documents; secondly it can be explained by the huge amount of properties of documents listed, which do not apply very much to the purpose of this documentation center; thirdly, many kinds of documents are listed in scheme 1 which refer to other fields, especially also to the arts; they should not have been listed there at all. On the other hand, among those aforementioned 260 descriptors for 'documents' many have been used in the second possible way descriptors for objects may be used, namely as labels, saying e.g.: "this is a textbook, a bibliography, a journal". With the corresponding notation:

0831 bibliography   0814 classification system
0831:0814 bibliography on classification systems

(here the newly proposed notation is used). Such a combination implies an althogether different approach to the former socalled form-divisions, the reasons for this were given in the introductory remarks to scheme 1[13]).

With regard to the descriptors used in the facets of the core fields of the information sciences (page 459) it should be pointed out that the underlying principles for their subdivision are found roughly outlined in the preface to scheme 2[14]). It will be noted that the divisions of .1 and .5 are quite general; they might occur in every field.

At this time experiences with these facets are collected, the system being in use with these notations since January 1971. The descriptors were established in 1963 and have since been updated or supplemented.

It would surely be interesting to review the descriptors in other fields outside the information sciences and to learn whether and how the same facets might be applied.

4.4 Conclusions from the contents of the two schemes

It can be concluded from the different amount and contents of the descriptors in the two schemes that apparently for every specific application purpose, that is, for every special library or information center or any other facility, two schemes are necessary for indexing work:

- one scheme containing all the concepts existing, with provision for extensive combination of descriptors, and
- another scheme containing only the descriptors in actual use.

Both schemes should be based on the same structure (referring to the same notation) so that, in a further development of a field, descriptors from the "background" system would only have to be moved into the operational sphere.

5. A survey on existing knowledge fields

5.1 The collection of "field-terms"

In his article on recent universal classification systems, E. de Grolier[15] had proposed – as a first attempt for a systematic collection of newer knowledge areas – to collect from the indexes of reference books terms denoting fields of activity to be incorporated into a possible new universal classification scheme. I was not aware of this suggestion when in the spring and summer of this year I started to collect the index terms from the following reference books

1) Directory of German documentation centers[16]
2) Directory of German special libraries[17]
3) Handbook on German scientific societies[18]
4) Bibliography of lexicons and dictionaries[19]
5) Bibliography of German abstracting services[20]

and established from these a cardfile of about 3000 entries, the terms being derived either from the headings of the subdivisions or from the stated subject interests of the centers. Each card contains the following information

- name of field,
- definition of field,
- indication of number and location of sources,

the latter being also of statistical value in that they show the amount of "activities" in a given field.

So far all the cards are still arranged alphabetically; they will soon be sorted into an order according to the breakdown of the universe of knowledge into areas as shown in chapter 3.

5.2 *Construction of a universal thesaurus of field-names*

With the help of the terms in the above mentioned cardfile a thesaurus of "field-names" might be constructed to serve also as a basis for the construction of a universal classification system according to the ideas outlined above. Field-names alone are of course not sufficient for the identification of an actually existing field; also, the suffixes, like

– logy, – nomy, – graphy, –sophy, –metry,
–ic, –istic, or –science, –technology, etc.

do not signify the contents, size or state of development of a given field. Theories ought to be found that help in recognizing a field by the contents of concepts which are connected with it. This would also hold for the merger fields between fields of the object area and the aspect area or in between the latter. Also the relations between the fields could be indicated by the thesaurus method, including of course also the identification of synonymous names of fields.

5.3 *Indicative indexing*

One application field of a universal thesaurus of the above shown kind would be its use for fast "indicative indexing" – as I would call it – versus "informative indexing". Such indicative indexing might meet a number of purposes, as e.g. arrangement of entries in reference works, arrangement of books in libraries, determination of information systems, information centers; it could also be used for administrative purposes in education, etc. Such a survey could conceivably be given to each author who might take from it the notation to index (just roughly) his own publication, e.g., for easier handling of national bibliographical entries. (A referee or editor might of course still check the author's assignments of field terms and add perhaps some more aspects to them, if applicable.)

6. The rôle of philosophy

Let us now think about the rôle which philosophy might play in the conception, design and development of a new universal classification system. There seem to be the following important tasks:

6.1 *Epistemologically valid arrangement of classes*

In historical attempts of philosophical classifications of knowledge very often dichotomic or triadic partitions have been used for the highest level division of knowledge fields. Thus Avicenna used a dichotomic division of "practical and theoretical sciences" which in the 19th century often was used as "abstract and concrete sciences", Ampère distinguished dichotomically between "cosmological and noological sciences", whereas triades were used much more often, e.g. by Aristotle, F. Bacon, Hegel, Wundt and Tillich. Such attempts of a meaningful interpretation of ontological states or human abilities should not be underestimated, they may serve as an intellectual tool for Man's understanding of his environment and world. One might perhaps apply such a view in the following case where the fields of

I 1–5 — what man discovered (nature)
1
2
3
4
5
II 5–9 — what man made (culture)
6
7
8
9

may be divided into two parts (1–55, 56–9) so that all the "areas" of the first part existed before man entered the world and became conscious of himself, and all the "areas" of the second part came into being as a consequence of man's existence.

Or, one might form triades as follows:

I 1, 2, 3 — unliving
II 4, 5, 6 — living
III 7, 8, 9 — creative,

by which different stages between the unliving, living and creative stage could be determined, or, alternatively, according to Hegel:

I 1 — idea in itself
II 2, 3, 4, 5, 6, 7, 8 — idea outside of itself
III 9 — idea returning to itself

which might also be pictured in a circle leaving open one segment to show the tension within the entire reality:

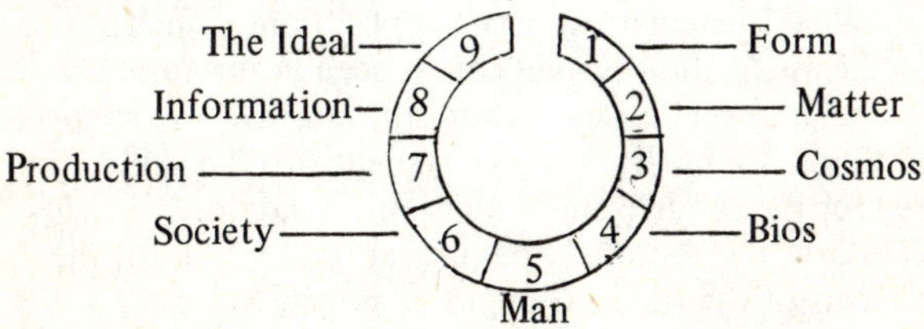

6.2 The rôle of philosophy of science

One of the tasks of philosophy of science (Wissenschaftstheorie) is definition theory. It could be extended to become an applied task in the sense Storer & Parsons have indicated[21]) when proposing to theoretically define fields to clarify their contents and by this to contribute to an overall theory of subject fields:

> "It may be questioned whether current difficulties in facilitating access to knowledge are due simply to the publication explosion . . . or to the more basic fact that the fields in which all this knowledge is being produced are themselves in states of relative disorganization. At base, however, the two factors are intimately related. An increase in the quantity of information relevant to a discipline makes for greater strain on the *integrative capacities of theory in that discipline,* (my underlining, I.D.), so that in the end, we may have a kind of Malthusian, or at least Parkinsonian, steadystate in each discipline, where the quantity of material added to the literature is limited by a discipline's ability to organize this information. As organizational or integrative ability increases, the output of information facilitated by it also increases, so that the degree of "strain" may remain relatively constant."

Storer even recognizes a mutual interdependence between the theoretical order in a field and the informational power of its concepts which of course means a further reason for the vast theoretical task of the "queen of the sciences"! Although this is no really new aspect in philosophy – since such philosophical "interpenetrations" of fields have always been the concern of philosophers, see e.g. 'philosophy of biology', 'philosophy of mathematics', (see also[22])) – it is getting more and more recognized as a specific task of applied philosophy of science; this has also been stated by the result of another colloquy at the University of Düsseldorf, Philosophical Institute, April 1968, published in[23]).

6.3 Development of a systems theory for a classification system

A classification system can be regarded as a form of a metalanguage by speaking in terms of descriptors about knowledge[24]). In introducing the structure of the fixed facet arrays as a common subdivision in each field, it is perhaps allowed to speak of a formal superstructure imposed on the metalanguage of the classification system, a metalanguage to the metalanguage. The usefulness of such a metalanguage for indexing, called "metalanguage of context", was shown already in 1963 by Goffman, Verhoeff and Belzer[24]).

In the definition part of this paper (see 1.2) the principle underlying the structural connection of a system was called 'systematifier'. The application of such a systematifier in the proposed classification system can be seen in the formally identical use of subsumptive relations with objects and subsumptive and appurtenance relations with aspect fields. We might now give thoughts to the subject of a systematifier (or a theory) for the subdivision of areas into fields and of fields into subfields. It is also necessary to elaborate a theoretical foundation for the subdivision of the specific facets of the fields. It might be possible to use e.g. for the subdivision of the object-facet .2 the 'integrative level theory' of Needham and later Feibleman[25]), the laws of which have been well explained by D.Austin[26]).

The problems mentioned do by no means exhaust all the open questions regarding a system theory for the proposed classification system. It seems, however, especially with regard to the valuable research work accomplished by the British Classification Research Group in recent years, described in[27]), that from now on a new system should only then be elaborated, if a predetermined systematifier can be applied in order that the structure principle may serve as a guide to the system and as a help for its consistent construction, development, correction and use.

7. Concluding remarks

7.1 The system as seen within one frame

It was stated earlier that the system, instead of being set up in five fundamental categories of

object-, aspect-, time-, space-, and general descriptors could also have been constructed within one single frame. This might have been achieved by moving the object areas "on top" of the aspect areas, so that the 0 (for objects) would become the zero class, the 'general descriptors' would find a place in area 1 and the 'time- and space-descriptors' one in area 3. Thus we would get:

01/09 Objects of general importance
1 Form and structure (incl. General descriptors)
2 Matter area
3 Cosmological area (incl. Time and Space descriptors)
4/9 (remains as before)

By such an arrangement one would perhaps become more conscious of the fact that the entire system of knowledge is a unity of its own. However, for the use of the system it seems to be more practical to think in terms of five categorical hierarchies, the descriptors of which can be combined with each other in any direction or form. With a presentation in the unique form, as shown above, perhaps even controversy might arise from the placings of 'objects' before 'form', or 'time and space' after 'matter', etc. since man with his intuitional power will anyhow regard the entire structure as an expression of historical development.

In order to get a composite view of the interconnectivity of the whole system one might perhaps arrange the descriptors of the five fundamental categories in a circular way, where the combinatorial lines form a pentagramm:

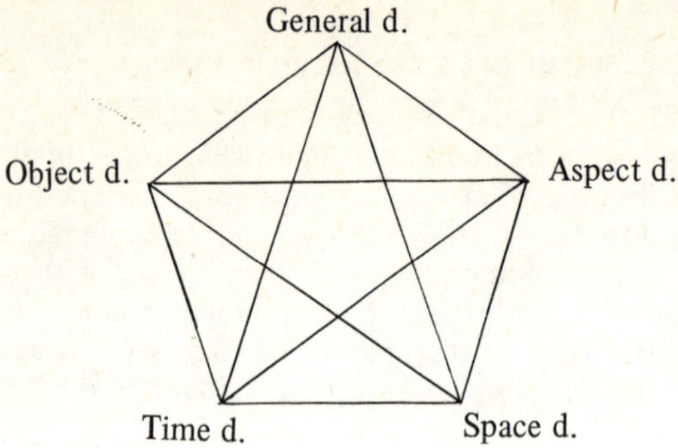

These categories constitute the second specific feature of the whole system besides its ontologically based object and aspect areas so that one might call it an ontologic-categorial system.

7.2 Conclusions

It is hard to summarize in such a short paper the essential features that have crystallized out of historic and recent research and which we would need to know as a basis for a new universal classification system. Many brilliant thinkers have tackled the problems involved; it does not seem fair not to even mention their contributions although most of our knowledge has been derived from them. To refer to just one of them, credit must be given to S.R. Ranganathan for his idea of facet analysis and his distinction of the idea-, the verbal- and the notational plane in classification.

Classification theory has only developed since about 100 years with M. Dewey starting to lay the foundation for concept combinations by notational assistance in 1876. It has reached a point today where we could use its general and special findings and build with them a truly theoretically based and conceptually and structurally formalized classification system. This paper was meant to show how one would have to proceed using such theories, even if any accomplishment of ours in this regard was nothing but a "useful compromise". We may not however, sit and wait, because the flood of information has become too voluminous to be controlled or because the "categorial frameworks"[28]) into which knowledge elements could be placed are too diverse to be usefully applied or be comprehended. We should design the maps for the countries, towns and roads which we need for intellectual housing and traffic of today, even considering the ongoing progress; for we all are in urgent need of a conceptually based classification system for the organization of our knowledge.

Literature:

1) FAIRTHORNE, R. A.: Temporal structure in bibliographical classification. – Paper presented at 1st Ottawa Conf. on the Conceptual Basis of the Classification of Knowledge, Oct. 1971

2) DAHLBERG, I.: The UDC and an ideal indexing language. Paper presented at Intern. Symposium on UDC in relation to other indexing languages, Herceg Novi, Yougoslavia, 28.6.–1.7.1971

3) ŠAMURIN, E.I.: Geschichte der bibliothekarisch-bibliographischen Klassifikation. Bd. I und II (1955–1959). München-Pullach: Verlag Dokumentation 1967. 404 + 781 p.

4) DIEMER, A. et al: System und Klassifikation. Versuch einer terminologischen und thematischen Klärung des Begriffsfeldes. In: Diemer, A. (Hrsg.): System und Klassifikation in Wissenschaft und Dokumentation. Meisenheim am Glan: Verl. A. Hain 1968. (Here p. 150–156)

5) AUSTIN, D.: Prospects for a new general classification. J. Librianship 1 (1969) No. 3, p. 149–169

6) FARRADANE, J.: A psychological basis for general classification. Paper presented at 1st Ottawa Conf. on Conceptual Basis of the Classification of Knowledge, Ottawa, Ont., Oct. 1–5, 1971

7) VICKERY, B. C.: Faceted classification. A guide to construction and use of special schemes. (Reprint with additional material 1968) London: Aslib 1968. 70 p.

8) Oral communication from J. Farradane, Ottawa, Ont., Oct. 2, 1971

9) DIEMER, A.: Versuch einer Systematik der 'allgemeinen Wörter'. Paper submitted at the meeting of the DGD-Committee on Thesaurus Research, 9 May 1969

10) PERREAULT, J.: Categories and relators: a new schema. In: Rev. Intern. Doc. 32 (1965) p. 136–144. Also in: Towards a theory for UDC. London : Bingley 1969. p. 119–148 (incl. emendations)

11) SOERGEL, D.: Dokumentation und Organisation des Wissens. Berlin: Duncker & Humblot 1971. 380 p. (here p. 337–338)

12) DAHLBERG, I.: Possibilities for a new universal classification. J. Doc. 27 (1971) No. 1, p. 18–36; in German: Nachr. Dok. 21 (1970) No. 4, p. 143–151

13) DAHLBERG. I.: Vorschlag für eine Klassifikation der Dokumentenarten und der Dokumentenkunde. Proposal for a classification of kinds of documents and documentology. Frankfurt: Sept. 1970. 37 p.

14) DAHLBERG, I.: Entwurf für ein Deskriptorsystem der Informationswissenschaften als Basis eines Bereichsthesaurus. Arbeitsunterlage des DGD-Komitees Thesaurusforschung. Frankfurt: Febr. 1971. 61 p.

15) GROLIER, E. de: Quelques travaux récents en matière de classification encyclopédique. In: Bull. Bibl. France 15 (1970) No. 3, p. 99–126

16) DEUTSCHER NORMENAUSSCHUSS (Hrsg.); Barlen, S. (Bearb.): Verzeichnis von Schrifttum-Auskunftsstellen. 6. erg. u. erw. Aufl. Berlin-Köln-Frankfurt: Beuth Vertr. 1968. 217 p.

17) ARBEITSGEMEINSCHAFT DER SPEZIALBIBLIOTHEKEN (Hrsg.); Meyen, F. (Bearb.): Verzeichnis der Spezialbibliotheken in der Bundesrepublik Deutschland einschließlich West-Berlin. 2. erw. Aufl. – Braunschweig: Fr. Vieweg 1970. 207 p.

18) DOMAY, F.: Handbuch der deutschen wissenschaftlichen Gesellschaften einschl. zahlr. Forschungsinstitute und Arbeitsgemeinschaften in der Bundesrepublik Deutschland. – Wiesbaden: Steiner 1965.

19) ZISCHKA, G. A.: Index lexicorum. Bibliographie der lexikalischen Nachschlagewerke. Wien: Verl. Br. Hollinek 1959. 290 p.

20) FORSCHUNGSINSTITUT FÜR RATIONALISIERUNG, TH Aachen (Bearb. u. Hrsg.): Deutsche Literaturdienste und Zeitschriften mit Literaturnachweisung. Berlin–Köln–Frankfurt: Beuth-Vertr. 1964, 530 p.

21) STORER, N. W., PARSONS, T.: The disciplines as a differentiating force. In: Montgomery, E. B. (Ed.): The foundations of access to knowledge. A symposium. Syracuse Univ. 1968. p. 101–121 (here p. 119)

22) DIEMER, A.: Grundriß der Philosophie. Bd. II: Die philosophischen Sonderdisziplinen. Meisenheim am Glan: Verl. A. Hain 1964. 861 p.

23) DIEMER, A. (Hrsg.): Der Wissenschaftsbegriff. Historische und systematische Untersuchungen. Meisenheim am Glan: Verl. A. Hain 1970. 277 p. (Pt. II contains the papers of a colloquy "IV. Wissenschaftstheoretisches Kolloquium der Universität Düsseldorf, April 1968" in which science problems in mathematics, biology, philosophy, psychology, and art have been discussed).

24) GOFFMANN, W., VERHOEFF, J., BELZER, J.: Use of meta-language in information retrieval systems. In: Amer. Doc. Inst., 26th Ann. Meeting, Chicago, Ill., Oct. 1963. Automation and Scientific Communication. Short papers, p. 181–182.

25) FEIBLEMAN, J. K.: The integrative levels in nature. Nachdruck in: Kyle, B. (Hrsg.): Focus on information. London: Aslib 1965. (here p. 27–41)

26) AUSTIN, D.: The theory of integrative levels reconsidered as the basis of a general classification. In: Classification Res. Group (Ed.): Classification and information control. London: The Libr. Assoc. 1969. 130 p.

27) FOSKETT, D. J.: Classification for a general indexing language. London: The Libr. Assoc. 1970. (Report on the studies of the Classification Research Group 1963/69)

28) KÖRNER, S.: Categorial frameworks. Oxford: Blackwell 1970

Frankfurt, 12. Oktober 1971

Б.М. Кедров /Москва/

КЛАССИФИКАЦИЯ НАУК. ЕЕ ПРИНЦИПЫ. ЕЕ ЦИКЛИЧЕСКАЯ ФОРМА,

1. ИСТОРИЧЕСКИЙ ВЗГЛЯД НА КЛАССИФИКАЦИЮ НАУК

1. Классификация наук составляет одну из центральных проблем общей теории научного знания, общей теории науки. Она может быть сформулирована как проблема раскрытия взаимной связи между всеми отдельными отраслями науки, следовательно интеграции всего научного знания, его синтеза, его систематизации, привидения к единству разобщенных между собой наук, научных дисциплин с их разделами и подразделениями. Для постановки и решения этой проблемы имеют особое значение два обстоятельства: первое - те принципы, которые лежат в основе предлагаемого решения, второе - конкретная форма выражения взаимосвязи между науками, то есть их взаимное схематическое расположение, в котором воплощаются соответствующие принципы. Первый момент можно рассматривать как содержание предлагаемого решения, второй - как его форму.

2. В истории науки данная проблема прошла три основные стадии, хотя они и не равноценны между собой по своему историческому и познавательному значению: первая - античность; вторая - с 17-го в. примерно до середины 19-го в.; третья - с середины 19-го в. по настоящее время. Каждой из этих стадий отвечал определенный, характерный для нее метод мышления ученых, на основе которого ставилась и решалась данная проблема. Исторический подход к вопросу позволяет понять изучаемый вопрос в его исторической связи и перспективе: поэтому и в данном случае мы будем его придерживаться.

3. Первая стадия /античность/ характеризуется господством натурфилософского взгляда на предмет познания. Древние греки и римляне не достигли еще ступени анализа в изучении природы и общества, равно как и нашей собственной

духовной деятельности, нашего мышления. Поэтому у них не было науки в современном понимании слова "наука", а существовало единое, нерасчлененное знание, развившееся под эгидой философии. В значительной мере это недифференцированное знание носило еще диффузный характер; первые наметки будущей классификации наук возникали лишь внутри этого единого знания, а не между разделенными достаточно четко его отдельными отраслями. Первые зародыши дифференциации наук возникли лишь в александрийский период, но они не получили тогда заметного развития.

4. Действительная дифференциация наук началась в эпоху Возрождения и развернулась полным ходом в 17-м в. В науке господствовал в это время односторонне аналитический метод исследования. Главную свою задачу ученые видели тогда в расчленении предмета исследования и всего мира на обособленные между собой области. Это нашло свое выражение в распространении таких наук, как анатомия растений и животных, как аналитическая химия и др. Классификация наук в это время принимает также односторонне аналитический характер. Фактически науки остаются изолированными между собой и лишь внешним образом прикладываются друг к другу, приводятся во внешнюю связь между собой. Подлинной интеграции наук еще не наблюдается; задача такой интеграции ограничивается лишь тем, чтобы не дать научному знанию совершенно распасться на обособленные участки при полном исчезновении какого-либо объединяющего их начала. Следовательно, дифференциация наук остается доминирующей тенденцией их развития, полностью подчиняя себе противоположную тенденцию к их интеграции.

5. Основной принцип классификации наук на этой второй стадии /17-й в.- середина 19-го в./ можно охарактеризовать как принцип сугубо формальный, который известен как принцип координации /соположения/ наук. Вначале /от Ф. Бэкона до Д'Аламбера/ он сочетался с чисто субъективным толкованием наук с точки зрения особенностей человеческого интеллекта. Науки располагались и группировались соответственно таким сторонам духовной деятельности человека как память /история/, воображение /искусство/, разум

/теоретические науки/. Сен-Симон, а вслед за ним Конт ввели объективный принцип: расположение наук соответственно усложнению самого предмета исследования - явлений природы и общества - в порядке перехода от более простого к более сложному. В противоположность Конту Спенсер, придерживаясь общего принципа координации, выдвинул принцип расположения наук в порядке перехода от более отвлеченных /и в этом смысле более абстрактных/ к менее отвлеченным /более конкретным/. Это было шагом вперед, однако все еще в рамках односторонне аналитического подхода, без того, чтобы подняться на уровень уже назревшего в середине 19-го в. теоретического синтеза всего научного знания.

6. Третья стадия, начавшаяся примерно в середине 19-го в., характеризуется тем, что она опирается на широкое проникновение в это время идеи всеобщей связи и развития в естествознании. Клеточная теория в биологии /конец 30-х гг./ и в особенности открытие закона сохранения и превращения энергии /40-е гг./ создали предпосылки для того, чтобы перенести идею связи и развития, которая осветила жизнь природы и общества, на самый процесс их познания, на науку, в том числе и на классификацию наук. Такую задачу выполнил Ф. Энгельс во второй половине 19-го в. В противоположность односторонне аналитическому воззрению на природу, которое господствовало в естествознании 18-го в. и лежало в основе принципа координации /внешнего сопоcostположения/ наук, Энгельс разработал принцип историзма, согласно которому науки следует не просто прилагать одну к другой, не просто сопологать их внешним образом, а развивать /выводить/ высшие из низших, более сложные /по своему предмету/ из более простых. Это был принцип субординации наук, т.е. их последовательного соподчинения; благодаря этому преодолевался формализм прежних, чисто аналитических классификаций.

7. Принцип историзма, принявший конкретную форму принципа субординации наук, вполне сочетался со строго объективным взглядом на содержание самих наук, каждая из которых считалась отражением определенной формы движения материи в природе, а также /в области социальных, истори-

ческих/ наук - развития общества. Принцип историзма, сочетающийся с принципом объективности, лежит в основе классификации наук, предложенной Энгельсом. Однако здесь уже был намечен и другой вариант принципа субординации, который распространял принцип историзма на самый процесс познания /в части перехода от математики к естествознанию/.

8. Одним из самых существенных факторов, обусловивших необходимость перехода от второй стадии в развитии классификации наук к ее третьей стадии, было появление первых междисциплинарных наук и научных отраслей во второй половине 19-го в. Первой из них была астрофизика /возникла в начале 60-х гг./. За ней последовала физическая химия: химическая термодинамика /70-е гг./, электрохимия /80-е гг./, биохимия /на рубеже 19-го и 20-го вв./ и др. Огромным толчком в том же направлении послужили два великих открытия однотипного характера: эволюционное учение Дарвина /для органических видов живой природы/ и периодический закон Менделеева /для химических элементов, т.е. основных видов вещества в неорганической природе/.

9. Конкретная схема классификации наук, как правило, носила линейный характер. В основном /при всем различии положенных в ее основу принципов классификации/ она совпадала у Конта и у Энгельса: математика; механика /включая астрономию как механику неба/; физика; химия; биология /включая и физиологию/; наука об обществе /включая в нее и социологию/. Но наряду с линейной схемой в первой половине и середине 19-го в. возникли разветвленные схемы, а также табличные схемы /у Ампера/, которые однако продолжали опираться на уже устаревший к тому времени, утративший силу принцип координации с его сугубо формальным характером.

10. На третьей стадии развития данной проблемы ясно выступило то обстоятельство, что тенденция к дифференциации наук, которая до тех пор доминировала в одностороннем порядке, начала отступать на второй план и все больше стала подчиняться противоположной ей тенденции к интеграции наук. Это проявилось уже во второй половине 19-го в. - в

начале 20-го в. и в еще большей степени в последние десятилетия 20-го в. Но теперь такое ударение на тенденцию к интеграции наук происходит не в порядке подавления или ослабления тенденции к их дифференциации, а как раз напротив - в порядке дальнейшего усиления этой последней. Действительно, никогда раньше дифференциация и специализация научного знания не достигала такой интенсивности, как в наше время. Однако сегодня самый процесс интеграции наук совершается не изолированно от процесса их дальнейшей дифференциации, а через его посредство и благодаря ему. Продолжающаяся дифференциация наук не только не препятствует их интеграции, как это было раньше, на второй стадии развития проблемы, но прямо ее вызывает и стимулирует.

11. Объясняется это тем, что вновь возникающие в ходе их дифференциации науки носят теперь особый характер: они возникают либо между ранее существовавшими изолированно между собой основными науками, т.е. как междисциплинарные науки и отрасли знания, либо проникают как бы внутрь сразу нескольких наук, подобно кибернетике. В итоге в первом случае новые науки выступают как цементирующие прежние науки, а во втором случае - как связывающие их подобно нити в бусах. В обоих случаях как раз и происходит то, что одна тенденция в развитии современных наук /к их интеграции/ осуществляется через посредство другой, прямо ей противоположной тенденции /к их дифференциации/.

12. Соответственно тому, как в настоящее время усложнилась общая картина взаимосвязей между науками, усложняются конкретные схемы классификации наук, а принципы ее требуют дальнейшего развития и уточнения. Современные схемы классификации, как правило, не поддаются выражению в линейной форме. Лишь искусственно, например, для утилитарных целей библиотечной классификации, возникает необходимость свести более сложную схему к линейной. Наряду с нею, кроме разветвленных и табличных форм, возникли замкнутые формы /циклические, треугольные и др./, а также сетеобразные, объемные и др.

13. Принципы классификации также подверглись дальнейшим модификациям и уточнениям. Особенно важное значение имеет вопрос о соотношении 1/ тех связей между науками, которые выражают объективные зависимости между самими предметами, которые изучаются этими науками, и 2/ тех зависимостей, которые возникают между науками в процессе познания этих предметов и которые по этой причине существуют в голове ученых лишь как субъективный образ внешнего мира. Если не учитывать подобного рода соотношений, нельзя правильно классифицировать различные отрасли отдельных фундаментальных наук, которые имеют своей задачей изучение одного и того же предмета, но с разных его сторон или с разной степенью глубины, как это имеет место при соотношении между экспериментальной /эмпирической/ частью той или иной науки и теоретической частью той же науки. То же касается наук, в основе которых лежат различные методы исследования, например, аналитическая химия и химия органического синтеза, или, соответственно, анатомия растений и физиология растений.

14. Мы не останавливаемся на деталях современных классификаций наук и их истории. Наша задача: во-1-х, сформулировать и разобрать выдвигаемый нами общий принцип классификации наук, объединяющий в себе оба момента - объективный и познавательный /субъективный/; во-2-х, на его основе рассмотреть предложенную нами еще в 1946 г. циклическую схему наук, в которой были бы учтены обе отмеченные выше черты происходящей ныне дифференциации наук, благодаря чему совершается их цементирование и их связывание между собой.

2. ОБЩИЙ ПРИНЦИП КЛАССИФИКАЦИИ СОВРЕМЕННЫХ НАУК

15. Мы исходим из тех принципов классификации наук, которые выдвинул Энгельс и на основе которых он построил свою линейную схему. В качестве фундаментального, носящего всеобщий характер, мы считаем принцип развития, принимающий конкретную форму принципа субординации наук. Точно также за основу мы принимаем объективный принцип соответствия последовательной связи наук /порядка их располо-

жения/, исторической последовательности ступеней и уровней развития самого материального объекта /природы и общества/. Наша задача найти способ, каким можно было бы объединить на его основе учет объективного момента с познавательным /субъективным/ моментом, возникающим в процессе развития самих наук. Отсюда должен вытекать принцип, могущий с единой точки зрения определить собой постановку и решение проблемы классификации наук в современных условиях.

16. Таким достаточно широким и общим принципом, в котором нашел бы свое адекватное выражение принцип историзма, мы считаем принцип восхождения от абстрактного к конкретному. Здесь понятия "абстрактное" и "конкретное" не совпадают с тем их содержанием, которое обычно в них вкладывается и которым оперировал, в частности, Спенсер. Сами эти понятия трактуются здесь в духе принципа историзма: абстрактное - это неразвитое, бедное содержанием, лишенное почти всех своих определений, которые присутствуют лишь в их зародыше, имплицитно; поэтому абстрактное совпадает с зачаточным, зародышевым, подобным понятию клеточки в биологии. Напротив, конкретное выступает здесь как развитое, богатое содержанием, успевшее уже развернуть присущие ему определения, которые потому и выступают у него в явном виде; поэтому конкретное совпадает с развитым телом в биологии.

17. Сказанное относится прежде всего к предметам внешнего мира, которые служат предметом изучения их различными науками. В этом случае процесс развития, совершающийся в природе или обществе, можно представить как движение от неразвитых, зачаточных форм предмета /его "клеточки"/ к развитым его формам /к "развитому телу"/. В таком случае речь идет о том, что движение совершается от чего-то низшего, более простого, каким всегда является неразвитая форма предмета, к чему-то высшему, более сложному, каким является его развитая форма. Поэтому можно говорить именно о "восхождении" в смысле перехода от одного, более низкого уровня процесса /или предмета/ к его более высокому уровню. В этом смысле в дальнейшем для краткости употребляется термин "восхождение".

18. Принцип "восхождения" может быть столь же успешно применен и к изучению самого процесса познания, в том числе в том его аспекте, который выступает в классификации наук. Здесь абстрактное выступает как более отвлеченное, как лишенное тех определений, которые присущи конкретному и которые придают конкретному характерную для его содержательность. Благодаря этому одними и теми же понятиями "абстрактное" и "конкретное" удается охватить две существенно отличные между собой области, представленные в любой классификации наук: объективную и познавательную /субъективную/. Обе эти области соотносятся между собой как предмет и его отражение, а потому они не отвергают одна другую, но в конечном счете совпадают одна с другой - субъективная с объективной. Однако по форме здесь имеются весьма существенные различия, которые нельзя игнорировать. Иначе неизбежна вульгаризация, отождествление одного с другим, субъекта с объектом.

19. Примером наук первого рода, которые основаны на учете объективного момента в развитии наук, могут служить естественные науки, начиная с физики и кончая биологией и стоящей на грани между естествознанием и историей общества - антропологией. При построении каждой из этих наук и всей их совокупности в целом действует тот принцип, что низшее, более простое /и в этом смысле абстрактное/ кладется в начало данной науки или их совокупности, а высшее, более сложное /и в этом смысле конкретное/ представлено как возникающее из этого начала /абстрактного/, как производное, как развившееся из него. В биологии - это "клеточка", в химии - химический элемент и т.д.

20. Примером наук второго рода, которые основаны на учете процесса познания /субъективного момента/ в развитии наук, могут служить логика /математическая, формальная/ и математика. Существенной их чертой является отвлечение от непосредственно чувственных данных, от наполнения всеобщих форм бытия каким-либо материальным содержанием. Движение /"восхождение"/ внутри этого класса наук можно представить себе как переход от одной области знания, лишенной содержательных определений, к другой или другим, где

такие определения последовательно возникают и включаются в рассмотрение, т.е. учитываются при последовательном движении нашей мысли от абстрактного к конкретному в пределах указанной области научного знания.

21. Сказанное можно пояснить движением мысли от области логики к наиболее абстрактным и общим разделам математики, затем - от них к более конкретным ее же разделам, в которых фигурируют понятия величины, числа, функции и т. д., как это видно, например, в теории чисел, анализе бесконечно малых, теории вероятности и т.д. Наконец, за дальнейшую конкретизацию можно было бы считать включение признака пространства /пространственных отношений/, благодаря чему осуществляется переход к таким разделам математики, как геометрия и топология. Мы не ставим в данном случае задачу дать классификацию математических наук, но ограничиваемся исключительно тем, чтобы пояснить принцип "восхождения" применительно к логико-математическим наукам.

22. Почти во всех прежних и современных классификациях наук математика стоит либо на первом месте, либо на одном из первых мест. За нею обычно следуют механика и физика, причем осуществляется фактически переход от той области науки, которая оперирует лишь понятиями и абстракциями, к той, которая /подобно физике/ оперирует в своем экспериментальном исследовании конкретными, чувственно осязаемыми вещами. Механика стоит на стыке между обеими областями науки, причем иногда она рассматривается как наиболее конкретная часть математики /так это было у Конта/ или же как наиболее абстрактная часть естествознания /так это было у Энгельса/. Однако "механизм" перехода от одной области наук к другой при этом не раскрывался.

23. Для того, чтобы раскрыть такой "механизм", рассмотрим простейший случай, каковым может служить при данных условиях переход от математики к классической механике. Этот случай соответствует состоянию научного знания 18-го в. - начала 19-го в. Последовательная конкретизация предмета научного изучения может быть представлена следующим

образом: сначала были введены определения /или признаки/, касающиеся величин, чисел и т.д.; после этого были введены определения, связанные с пространственными отношениями, что повлекло переход к геометрии и топологии. Но пространство есть одна из двух основных форм всякого бытия. Другой, столь же основной формой служит время. Пока временные отношения не включались в рассмотрение, все наши абстракции оставались в пределах математики. Но как только после пространства мы в порядке движения от абстрактного к конкретному включили в наше рассмотрение время, так сейчас же из области математики мы перешли в область классической механики.

24. Дальнейшее движение в том же направлении связано с включением еще одного важного определения, касающегося свойства массы: когда к пространству было подключено одно только время, мы попали в наиболее абстрактную часть механики - кинематику, которую именуют "геометрией движения". Когда же затем было подключено свойство массы, то есть когда механически движущееся тело стало рассматриваться, как наделенное массой, мы вышли из области кинематики и вступили в область динамики, как более конкретную область классической механики. Последующее включение момента механического равновесия приводит нас в еще более конкретный раздел механики - статику.

25. Последующее движение /"восхождение"/ все в том же направлении от абстрактного к конкретному приводит нас от механики к физике; это происходит тогда, когда наряду с чисто количественными определениями форм бытия и механических свойств тел /например их массы/ включается в рассмотрение качественная характеристика предмета изучения, от которой классическая механика отвлекается /абстрагируется/ по самому своему существу. С этого момента процесс дальнейшего "восхождения" становится зависимым уже не от того, какие определения и признаки мы включаем в наше рассмотрение, а от того, какие определения и признаки возникают в процессе объективного развития самих изучаемых науками предметов.

26. Мы обращаем внимание на то, что в ходе всего отрезка движения мысли от математики к механике и от механики к физике действовал один и тот же общий принцип "восхождения", поскольку во всех пунктах этого "восхождения" и на всех его этапах переходы от одной ступени к следующей за ней совершались однотипным способом: путем обогащения предмета исследования /предмета науки/ все новыми и новыми определениями и признаками. При этом в рамках математики и при переходе от нее к механике, а затем и в рамках механики процесс "восхождения" совершался как происходящий в нашем собственном мышлении /в нашем познании/; начиная же с физики /после перехода к ней от механики/ он совершался уже как происходящий в самой объективной действительности /в природе/. Таким образом именно в пункте перехода от математики к физике через механику сошлись оба аспекта взаимосвязи между науками /оба аспекта их классификации/: аспект объективный и аспект познавательный /субъективный/.

27. Если теперь с позиции принципа "восхождения" представить общий ряд основных наук /основных их групп/, то можно было бы в первом приближении придти к следующей схеме: логика /философия/; математика; естествознание; гуманитарные науки. Здесь эти науки показаны без связующих /переходных/ звеньев, которые существуют между каждой парой смежных групп наук. Так между философией /логикой/ и математикой стоит математическая логика; между математикой и естествознанием стоит целая группа математизированных наук, к числу которых относится не только механика /классическая/, но и квантовая механика /механика микроявлений/; термодинамика и, очевидно, кибернетика; между естествознанием и гуманитарными /общественными/ науками стоят технические науки.

28. В соответствии со сказанным выше, на основе общего принципа "восхождения" составляется следующий ряд /костяк, или скелет/ различных групп наук /в разрядку выделены основные группы наук, нормальным шрифтом - переходные, или промежуточные группы/: ф и л о с о ф и я /л о г и к а/; математическая логика; м а т е м а -

т и к а ; математизированные науки; е с т е с т в о-з н а н и е ; технические науки; г у м а н и т а р-н ы е н а у к и /включая общественные и науки о человеке и его духовной жизни/. Весь этот ряд основных и промежуточных групп наук строится полностью на основе принципа "восхождения"; благодаря последнему удается органически объединить ту часть единого ряда наук, которая базируется на познавательной /субъективной/ стороне дела, с той, которая базируется на объективной его стороне. Но этим не исчерпывается приложение принципа "восхождения" к проблеме классификации наук. С особой силой значение этого принципа обнаруживается при разборе вопроса о линейности и нелинейности конкретных схем /форм/ классификации наук.

3. ЦИКЛИЧЕСКАЯ И ВООБЩЕ ЗАМКНУТАЯ ФОРМА КЛАССИФИКАЦИИ НАУК

29. Любая линейная схема классификации наук обладает тем существенным недостатком, что переходы в каждом звене общей цепи наук существуют в обе стороны в крайних же ее звеньях - в начале и конце общего ряда наук - таких двусторонних связей нет; связи здесь сугубо односторонни. Тем самым науки, стоящие по краям общего ряда наук, оказываются в неравноправном положении по сравнению со всеми остальными науками. Эту неравноправность наук устраняет любая замкнутая схема, в том числе и циклическая. Однако самым трудным в этом случае является вопрос о том, как замкнуть линейный ряд в кольцо и как обосновать исходя из общего принципа "восхождения" подобное замыкание. Для этого необходимо тщательно проанализировать особенности составленного выше общего ряда наук и детальнее выяснить характер обоих крайних его пунктов - начального и конечного.

30. В начале ряда наук стоит логика, как часть философии. За ней следует математика, с которой логика связана через математическую логику и т.д. Завершается же общий

ряд наук областью гуманитарных наук. Но сама группа гуманитарных наук включает в себя, во-первых, все общественные науки - исторические и те, которые изучают различные стороны /или "уровни"/ жизни общества - от экономического базиса, государственно-правовой надстройки до самых высоких идеологических надстроек, вплоть до философии. Во-вторых, в число гуманитарных наук входят те, которые изучают личность самого человека, его психическую жизнь /психология/, его мышление. Такое дифференцированное рассмотрение всей группы гуманитарных наук дает возможность отыскать в заключительном звене всей цепи наук определенный контакт с ее первоначальным звеном.

31. Анализируя различные схемы классификации наук, которые выдвигались начиная с 19-го в., можно обнаружить, что в одних схемах логика стоит в самом конце общего ряда наук, после психологии, в других же, напротив, - в самом его начале, перед математикой. Из сопоставления этих схем между собой вытекает вывод, что у логики существует как бы двоякая связь с другими науками: можно допустить, что она одновременно связана и с математикой, стоящей обычно в начале общего ряда наук, открывая его собою, и с психологией, стоящей обычно в конце ряда наук, замыкая его собою. Однако в линейных схемах невозможно выразить правильно эту двустороннюю связь логики с другими науками, а потому и получается так, что в одних она стоит в начале ряда, в других - в его конце. Если допустить, что логика, как часть философии, обладает отмеченной выше двусторонностью в отношении к другим наукам, то отсюда открывается возможность именно на логике замкнуть в цикл /в круг/ общий ряд всех наук; логика в этом случае соединит в себе и начало и конец прежнего линейного ряда наук. Если же после этого "выпрямить" схему, превратив ее снова в линейный ряд, то предоставляется возможность разорвать цикл /круг/ в одном из двух мест: либо между логикой и психологией /так делает иногда Пиаже/, и тогда логика окажется перед математикой; либо между логикой и математикой, и тогда впереди будет стоять математика, а логика окажется в конце ряда, после психологии.

32. Объяснение того, почему логика обладает по меньшей мере двойной связью /и с математикой, и с психологией/ сводится к следующему: двигаясь от физики к математике /через математизированные науки, в том числе и механику/ и далее от математики /через математическую логику/ к логике, мы приходим в конце концов к области познания наиболее общих отношений между предметами /и процессами/ мира, например, их структурных отношений. Эти общие отношения абстрагируются нами сначала до их качественного, вещественного, материального содержания /таким содержанием занимается физика и все естествознание/, затем от фактора времени /который учитывается еще механикой/, затем от пространственной формы бытия /которую изучает геометрия и топология/, затем от количественной, числовой стороны вещей и их отношений, в результате чего мы приходим ко все более и более абстрактным отделам математики. В итоге всего этого движения мы приходим, наконец, к логике, которая в данном случае выступает как результат операции последовательного абстрагирования от конкретного содержания и конкретных форм изучаемых явлений и которая, по этой причине, неизбежно оказывается связанной с математикой. В этом случае логика может быть определена как логика операции абстрагирования /выделения/ субъектом определенных /общих/ сторон изучаемого объекта. Поэтому ее предметом являются в данном случае наиболее общие отношения вещей и процессов мира.

33. К той же логике /в смысле ее законов, категорий, приемов и правил/ мы приходим, двигаясь в прямо противоположном направлении - от биологии /изучающей жизнь/ к физиологии человека /изучающей его высшую нервную деятельность, его мозг, органы чувств и речи/ и, наконец, к логике /изучающей законы правильного человеческого мышления, точнее, - мышления, ведущего к истине/. К этому же мы придем, если будем двигаться от общественных наук через социальную психологию к той же логике. Все это означает, что в итоге такого движения мы приходим к логике, предметом которой оказывается процесс мышления, протекающий в мозгу человека как мыслящего существа. К этому ее предмету мы приходим, двигаясь от изучения живых существ во-

обще к изучению высших из них /с их способностью ощущать и мыслить и с материальной основой этой способности/, а также к изучению социальных условий протекания мыслительной деятельности у людей; и далее, переходя от изучения совокупности психических процессов к изучению сложнейших и высших из них, представленных процессами мышления и познания. Следовательно, логика, к которой мы приходим в итоге движения от жизни к мышлению и которая связана с психологией, может быть определена как логика познавательной деятельности человека, направленной на любой объект. Ее предметом, в таком случае, являются результаты мыслительной деятельности человека.

34. Таким образом, в первом случае, логика в ее связи с математикой, выступает субъективно, как способ исследования, способ абстрагирования определенных сторон действительного мира, а во втором случае, в ее связи с психологией, как объект исследования, как сторона самого предмета изучения - мышления как предмета нашего изучения. Значит, оба аспекта логики - объективный и субъективный - не только не разорваны между собой, но составляют внутренне единую науку, причем основой субъективной логики /логики мышления/ является объективная логика /логика вещей/. Такое совпадение субъективной логики с объективной означает, что в рассмотренных двух случаях - /а/ переход к логике от математики и /в/ переход к ней от психологии - фактически речь идет о том, что к одной и той же науке /логике/ мы приходим с двух противоположных сторон: в одном случае, двигаясь к ней от анализа связей между явлениями внешнего мира, в другом - от анализа связей между явлениями, происходящими в мышлении человека. Значит, это - лишь две разные стороны /аспекта/ одной и той же науки, которая примыкает этими своими сторонами к двум другим наукам: с одной стороны, к математике, с другой стороны, к психологии.

35. Концепция двух противоположных аспектов логики - объективного и субъективного - и их внутреннего единства объясняет возможность и необходимость определенным образом "замкнуть" в кольцо /цикл, или круг/ обычный линей-

ный ряд наук. Такая концепция была впервые изложена нами в сообщении "О современной классификации наук", прочитанном в Академии наук СССР в 1946 г. /см. нашу книгу: "Классификация наук", т. 2, Издательство "Мысль", Москва, 1965, стр. 416-417/. В случае принятия циклической схемы классификации наук чрезвычайно важно установить возможность выражения в ней специфических для современной науки процессов интеграции и синтеза научного знания. Такие процессы могут быть отражены в известной степени в любой схеме классификации наук, в том числе и линейной. Однако циклическая схема позволяет это сделать более полно, поскольку сам по себе синтез всегда есть в определенном смысле слова "замыкание" в некотором внутреннем единстве ранее разобщенного материала /расчлененных между собой элементов чего-то целого/.

36. Одним из специфических для современной науки процессов подобного синтеза является "цементирование" общего ряда или цикла наук путем заполнения пустовавших дотоле промежутков между отдельными /смежными/ его звеньями. Такое заполнение достигается в результате продолжающейся дифференциации наук, когда, во-первых, возникают новые междисциплинарные науки и отрасли науки, во-вторых, когда методы одной /нижестоящей/ науки, например, физики, начинают распространяться на изучение предмета других наук /вышестоящих по лестнице "восхождения" от низшего к высшему, от простого к сложному, от абстрактного к конкретному/. Выше уже отмечалось, что в общем ряду основных групп наук между общественными и естественными науками стоят технические науки, поскольку их назначение - использование законов природы /открываемых естествознанием/ в интересах общественно-исторической практики /составляющей предмет изучения со стороны общественных наук/. При дальнейшей своей детализации подобный процесс приводит к тому, что прежняя дискретность /разрывность/ между науками, которая служила главной основой для применения в прошлом формального принципа координации наук, ликвидируется и на ее место становятся взаимосвязи и взаимопереходы между науками, которые исключают возможность применения к их классификации принципа координации. Так

до середины 19-го в. астрономия была лишь небесной механикой /механикой небесных тел/, а физика занималась земными процессами, связанными с качественным состоянием физических тел. Создание спектрального анализа, как физического метода, дало возможность изучать не только механические, но и физические процессы, происходящие на небесных телах. Тем самым заполнилась пропасть между астрономией и физикой: между ними встала связующая их астрофизика. Так же возникла физическая химия, заполнившая собой пробел между ранее разобщенными между собой физикой и химией, каждая из которых имела до тех пор свой собственный, строго очерченный предмет. В итоге такого цементирования наук их цикл из прерывистого, составленного из дискретных звеньев, превратился в сплошной, где стираются былые резкие грани между отдельными науками, прежде всего между фундаментальными науками.

37. Другой путь взаимного связывания наук, классификация которых выражена в виде циклической схемы, состоит в их "пронизывании" более общей наукой подобно тому, как отдельные бусинки нанизываются на общую нить, завязанную затем в кольцо. Такое связывание может иметь достаточно широкий характер, охватывая все науки, как это наблюдается в случае математики, которая имеет своим предметом соответствующую сторону /количественную, пространственную или же общеструктурную/ любых объектов, включая и мысленные построения. В силу этого для математики создается двойственное положение: как особая наука она занимает вполне определенное место в общей системе научного знания; вместе с тем она пронизывает и объединяет собой все остальные науки - одни в большей, другие в меньшей степени, - служа для них методом исследования, научного поиска и открытия, а также способом изложения достигнутых результатов. Это в еще большей мере относится к математической логике, стоящей на стыке между математикой и логикой /как частью философии/. В циклической схеме такая двойная роль математики может быть показана графически в виде внутреннего кольца, проходящего через все науки, но берущего свое начало от "украшающего" это кольцо "драгоценного" камня /математики/. Связывание наук с помощью одной общей науки

может быть распространено не на все науки, а на какую-то определенную их часть. Так кибернетика пронизывает собой только те науки, которые так или иначе связаны с живыми существами и в предмет которых входят самоуправляемые процессы. Сюда относится вся область живой природы /биология/, все общественные и вообще гуманитарные науки и все технические науки, поскольку сама техника органически включает в себя момент управления человеком природными процессами, поставленными на службу обществу. Вместе с тем кибернетика неразрывно связана с математикой, являясь частью математизированных наук.

38. Кроме частных наук, таких как математика и кибернетика, которые обладают сравнительно более общим характером, нежели естественные и общественные /и вообще гуманитарные/ науки, существует одна единственная общая наука, которая может быть сопоставлена именно как общая со всеми частными науками независимо от степени их общности. Это - философия, представленная логикой и диалектикой. Изучение наиболее общих законов всякого движения, всякого развития, где бы оно ни происходило - в природе, обществе или в мышлении /предмет диалектики/, - как раз и делает философию общей наукой. Она исследует и раскрывает не какую-либо частную сторону у предметов других наук, как это мы видим у математики и кибернетики, а самую общую для них всех, которая и объединяет их друг с другом без всякого ограничения. Поэтому в циклической схеме философия /диалектика/ выступает как внутренний стержень, идущий от того места в общей системе научного знания, которое занимает философия, через все науки вообще. Сказанное касается и логики, как части философии. Поскольку логика есть наука о законах правильного, точнее ведущего к истине мышления, ее законы, принципы и приемы обладают той же самой степенью общности, как и те законы, принципы и приемы, которые составляют предмет диалектики. Независимо от специфики изучаемого предмета от глубины и полноты его изучения логика применяет свои всеобщие приемы и методы, которые лишь конкретизируются соответственно особенности изучаемого предмета.

39. Особым вариантом циклической схемы классификации наук может служить треугольная схема, основанная на том же принципе "восхождения" /развития/. Один /"верхний"/ угол этого треугольника составляет естествознание /Е/, изучающее природу, которая представлена начиная от неорганической /в самом верху/, затем органической, и кончая высшими животными, которые обладают зачатками психики. Через зоопсихологию и физиологию высшей нервной деятельности совершается переход от природы к человеку и, соответственно этому, - от естествознания к гуманитарным наукам. Но человек выступает в своей истории двояко: как существо социальное, и как существо мыслящее. Хотя оба проявления сущности человека нераздельны и взаимообуславливают друг друга, тем не менее они изучаются разными, но внутренне связанными между собой науками: социальными /С/ и науками о мышлении, составляющими часть философии /Ф/. Обе эти гуманитарные области знания представлены в треугольнике наук двумя "нижними" его углами; соединительная черта между ними образует основание треугольника. Процесс "восхождения" /развития/ может быть представлен как движение сверху вниз по нашему треугольнику.

40. Если мы впишем наш треугольник в круг, то получим новый вариант циклической схемы; при этом весь круг разделится на три дуги, соединяющие вершины вписанного треугольника. Дуга Е - С будет выражать область технических наук, стоящих между естественными и общественными науками. Дуга Е - Ф будет представлять область перехода от естествознания через математизированные науки /механику и др./ к математике /М/, а от нее через математическую логику к логике как части философии. Философия же будет представлена не только вершиной Ф /наукой о мышлении/, но и общим кругом, охватывающим или пронизывающим все науки вообще /диалектикой как основой всей философии/. Наконец, дуга С - Ф представит собой область исследования духовных явлений в жизни общества, которые изучают языкознание, социальная психология и др.

41. В особом положении оказывается в таком случае психология /П/, которая связана одновременно со всеми тремя

основными группами наук: естественными, философскими и социальными. Поэтому ее место должно быть определено в данном случае не на самом круге /цикле/ наук, скажем между общественными науками и логикой, т.е. среди остальных гуманитарных наук, но внутри общего цикла наук /внутри их треугольника/. Это вызвано тем, что психология связана с естествознанием через зоопсихологию и учение о высшей нервной деятельности, с общественными науками через социальную психологию, языкознание и другие социальные науки, и с философией через логику как учение о мышлении и его законах. Философия /логика/ изучает мышление со стороны его всеобщих законов, как оно при всех условиях протекает в человеческом мышлении, ищущем и открывающем истину. Психология же изучает тот же процесс, но не в его всеобщей форме, а в тех конкретных, сугубо частных условиях, в каких он протекает в голове отдельного человека.

42. Многочисленные дискуссии вокруг определения предмета психологии и самой психологии как науки, были в значительной степени обусловлены тем, что выбиралась одна какая-либо ее связь с той или иной группой наук и игнорировались или умалялись другие ее связи, не менее существенные и определяющие. Так долгое время психология включалась в философию, поскольку и там и здесь речь шла о мышлении человека. В противоположность этому нередко психология включалась и включается в состав общественных наук, так как человек есть существо социальное и все его поведение, вся его психика должна, дескать, детерминироваться социальными условиями. При этом стирался момент, изучаемый психологией личности. Наконец, особенно под влиянием чрезмерного подчеркивания учения о высшей нервной деятельности делались попытки растворить психологию в этом учении и тем самым стереть различие между ней как гуманитарной наукой и науками естественными /физиологией/. На наш взгляд, лишь одновременный учет всех отмеченных трех решающих связей психологии /П/ с естествознанием /Е/, социальными науками /С/ и философией /Ф/ можно правильно решить вопрос о предмете психологии и о ее месте в общей системе наук. В разбираемой циклической системе психология находится на пересечении трех хорд, идущих от вершин

Е, С и Ф через точку П, причем соответствующая точка их пересечения /П/ лежит ближе всего к точке Ф, несколько дальше отстоит от точки С и дальше всего от точки Е, поскольку, повторяем, психология есть прежде всего гуманитарная, но не естественная наука.

43. Идея треугольника наук, изложенная выше, была выдвинута нами в 1954 г. и впервые доложена на Международном конгрессе по логике, методологии и философии науки в Цюрихе в докладе "О классификации наук" /см. том Трудов этого Конгресса, пленарное заседание/. Эту идею на самом Конгрессе и впоследствии отметил Ж. Пиаже /см. нашу книгу "Классификация наук", т. 2, 1965, стр. 466-484/. Независимо от нас идею о циклической форме классификации наук выдвинул Ж. Пиаже в труде "Введение в генетическую эпистемологию" /1950/. Такая эпистемология, в отличие от обычной, рассматривает знание не со статической точки зрения, а динамически, со стороны "механизмов" его возрастания. "Замыкающее звено" всей цепи наук Пиаже видит в психологии /в ее связи с социологией/. Психология выступает у него как принципиально исходный пункт и основа всей системы наук, но не в том смысле, что она стоит в начале всего их ряда или над ним, а в том смысле, что она как бы пронизывает собой и связывает собой воедино все науки. Позднее циклическую схему классификации наук выдвинул С.Г. Струмилин в работе "Наука и развитие производительных сил" /"Вопросы философии", 1954, № 3, стр. 55/.

44. Итак рассмотрены три круга вопросов: 1/ исторический взгляд на классификацию наук и на три основные этапа ее разработки в истории человеческой мысли; 2/ различные принципы классификации наук и среди них общий принцип "восхождения", в котором воплощается основной принцип развития и субординации наук и который позволяет объединить объективный и субъективный /познавательный/ аспекты как самих наук, так и их классификации; 3/ циклическая и вообще замкнутая форма /схема/ классификации наук, позволяющая путем "замыкания" общей цепи наук поставить их в равноправное относительно друг друга положение и найти место таким наукам, которые связываются со многи-

ми другими науками или же пронизывают их. Но, разумеется, как и всякая схема, циклическая форма классификации наук имеет свои недостатки и слабые стороны, которые, в частности, обнаруживаются, когда место той или иной науки оказывается лежащим не на самом цикле /круге/, а где-то внутри него. Однако, во всяком случае такая форма классификации наук заслуживает, на наш взгляд, самого серьезного внимания, какое ей оказывают многие ученые /Ж. Пиаже и др./. Преимуществом ее, как нам кажется, является ее внутренняя цельность, обусловленная замыканием ряда наук на самого себя.

Normand Lacharité (Montréal)

Questions about "Classifiables" and "Classifyers" submitted as a commentary to various papers

1. Are the classifications within logic useful in designing a general scheme for the classification of knowledge?
(This question refers to Miss Jennifer Ashworth's paper.)

In what way do the "logical classifications" like: valid and invalid inferences, formal and informal language, formal syntax and formal semantics, or else the classification into various calculi (propositional, quantificational, etc.) affect a general scheme for the classification of knowledge? Suppose a person would not be aware of the distinctions between the various figures of argument or, even more generally, between 'true' and 'valid', would that lead him into adopting or refusing a certain division of knowledge, taxonomic categories, etc., different from those he would have otherwise adopted or refused? Some logical categories such as those mentioned above might turn out handy in classifying specific knowledge pertaining to logic and related fields. I think for instance of a "fallacies" rubric. . .; but a librarian would not have to decide whether a document is fallacious in order to classify it under that rubric; he may have to decide whether some story is "fiction" or not, but not whether it is true (even though we speak of a classification of *knowledge*!)

Indeed, what is needed to classify a knowledge content is a name for it and not a judgment on its logical form or properties. The logical properties of arguments or of expressions naming the content of a document are not "classifyers", i.e. *terms endowed with a classificatory power,* in a classification scheme for knowledge. A subject-matter *cannot* be evaluated as to its consistance, for example, and that which is said about a given subject-matter *should not* be so evaluated.

2. Theories of meaning and theories of science in their relation to classification problems.
(Question referring to Mr. William Shea's paper.)

There are, at present, various attempts at answering the question about the logical structure of a scientific theory, and, specifically, the question about 'how do *theoretical terms* have meaning in a scientific language? ' In each attempt, various epistemological theses are maintained, and in particular, theses about concept formation and conceptual frameworks. Do different theories of science ("Wissenschaftslehren") and of concept formation actually lead to different classification schemes for knowledge?

This general question can be brought down to a more concrete level and reformulated in either of the following ways:

a) given a certain classification scheme and a document on the speed of electrons in a certain situation, would the documentalist classify the document differently according to whether he adopts a "meaning postulates" solution, or a "logical reducibility to observables" solution, or an operationalist solution to the problems concerning the meaning of the term "electron"?

b) or would a classification theorist set up different schemes, say for classifying knowledge within physics, according to whether he adopts one or the other of the above mentioned solutions?

Should not a classification scheme for physical knowledge be independent of any *epistemological* interpretation concerning the nature of the electron, or the status of the term "electron" in the theory? This seems to point out an aspect of a classification scheme which is analogous to the 'logical neutrality' mentioned at the end of question 1 above: there must be a way of classifying subject-matters without passing judgment on their epistemological properties, which can only be described (and debated) in a philosophical metalanguage.

This last remark raises another question: to what extent are epistemological considerations useful and relevant to the construction of a classification scheme for knowledge? Could it be that epistemological considerations on the actual discourse of various disciplines are the least probable factor of change in the division of knowledge, and consequently the last argument to be used in justifying a particular classification scheme? Does that mean that a general conception of what knowledge is can be dispensed with or that it is *not* at work in a general classification scheme? Probably not. But to make clear how much epistemology is actually at work, and where, it would require more than otiose arguments like: "How can one classify knowledge, if one does not know what knowledge is? " It is not any, nor all knowledge of what knowledge is that is epistemology. . .

3. Questions about change and obsolescence.

Among various procedures to establish a general classification of knowledge, two have been discussed with particular attention in this conference: the division-composition of concepts and the division of knowledge into disciplines. Let us take as a prototype of the first approach: the thesaurus classification, and as a prototype of the second: the Dewey classification. In a thesaurus, the various words of language cluster into classes according to a system (more or less "systematic". . .) of meaning interrelations. It is clear that the structure of such a system is dependent upon the empirical state of a language at a given moment of its evolution.

3.1 How does the rate of obsolescence of semantic interrelations compare with the rate of obsolescence of the criteria for dividing disciplines into distinct classes? If one of those two systems would change less rapidly than the other, would that be a sufficient reason for adopting the corresponding procedure in designing a general classification scheme for knowledge?

3.2 To what extent is the dynamic of disciplines relevant to the classification theorist?

3.21 A case of socio-cultural dynamics: the striving for hegemony.

An illustration of what I mean by this expression has been pointed out in Guy Lafrance's paper: ". . . l'anthropologie se situerait nettement au carrefour des savoirs sur l'homme. Perspective qui n'est pas sans soulever de conflit avec la sociologie dont l'ambition va aussi dans le sens d'une totalité des savoirs sur l'homme, ou, selon l'expression de Mauss, dans le sens d'une connaissance des faits sociaux totaux."

Curiously enough, disciplines are cultural entities of such a kind that practicians of a discipline eventually happen to fear that their own discipline may be subjugated by another. . .

It is often the case that interdisciplinary disputes over "power" are nothing but "querelles de clocher"; and in those cases it is not very relevant for the theory of classification to describe and analyse the pronouncements of various scientists concerning the extension of their discipline, its power to synthesize other disciplines, and so forth. Could it be however that a regrouping of such phenomena, together with many others, under the perspective of the *sociology of knowledge* would lead to general results concerning the socio-cultural dynamics of disciplines, results that would help designing a more stable classification scheme? Is this a matter of anticipating historical processes? Is there, for example, anything law-like (from the point of view of a sociology of knowledge) in the creation of a quasi-disciplinary nucleus like blood research and technology? Of what use would it be for a classificationist to have a *theory* of the interactions between state intervention and discipline development, or between class-conflicts (in society) and knowledge acquisition?

It seems that the type of knowledge that sociology of knowledge can provide about the disciplines and their dynamics has a fate quite similar to the fate of *epistemological* or *logical* knowledge discussed above in questions 1. and 2.: it has no general classificatory usefulness when it does not provide the classificationist with *names* for domains of reality, or of cognitive activities; when it does not suggest reasons and/or means to name anew such domains, etc.

3.22 Cases of epistemological dynamics.

My questions (and suspected answers) concerning these cases are quite symmetrical with respect to the preceding ones. Here the change in the distribution of diciplines and in the development of a single discipline is observed from the point of view of the theoretical structure (somewhat in the styles of Bachelard, Kuhn, Heelan. . .) or from the point of view of validation in general: for instance, questions are asked about "le degré de scientificité" of a discipline. 1)

To what extent does a clarification of the formal dynamics of the successive language leading to a given state of a discipline help constructing a more adaptable (i.e. remaining efficient over long periods of time) division of knowledge into disciplines? Would such a formal dynamics be of more consequence in a composition-of-concepts approach to classification, somewhat by breaking-up discipline categories into classes of *problems* which lead to theories as they are being solved? (This applies particularly well to scientific theories; see HEELAN's article: "The Logic of Framework Transpositions", *International Philosophical Quarterly,* Vol. XI, No 3, Sept. 1971).

4. Questions relative to the approach centered on the division-composition of concepts.
(Questions referring to Mrs. S. Datta's and Mr. J.E.L. Farradane's paper.)

The proponents of a composition-of-concepts approach to the problem of a general classification of knowledge admit that they are trying to minimize the arbitrary character of classification schemes. I suppose "less arbitrary" in that context would imply, among other things: "less dependent on cultural milieux and on the contingencies of historical development"; it would further imply, by way of consequence: "more universally applicable and more stable in time, i.e. more apt to absorb changes in knowledge without changing the *basis* of classification".

The way to reach that aim: universality and stability, is to provide a rationale for *divisions into classes* (let's not use that word in a purely set-theoretical or logical sense here). To say that is to use a quite formal way of speaking since the statement holds no matter *what* is to be divided: domains of reality, the continuum of meaning (something like Hjelmslev's "substance du contenu", in general linguistics. . .), objects, beings, concepts, practices, states of affairs, discourses (e.g. disciplines), events (has anyone ever tried to classify events?), etc. But by starting in that manner one realizes more readily that the choice of *concepts* as that which is to be divided is but one particular choice.

Let us admit that composition is the reverse of division (although a mere *introspective* account or report may not corroborate that statement), both being *operations:* one may compose one concept *out of* several ones or analyse one concept *into* several ones. Now, what is it that could be nonarbitrary about a certain way of building up classes of concepts? I would summarize as follows what seems to me a possible answer of the proponents of the present approach: What is non-arbitrary is the fact that the human mind is of such a constitution that it combines sense data into certain patterns, which are the concepts, and that, in so doing, it follows certain processes of composition and division. It is the task of the investigator (is he a linguist? an epistemologist? a psychologist? . . . he can hardly be a documentalist. . .) to find out what these processes *actually* are; and this, should be possible, it is claimed, since there is empirical evidence concerning the actual "sense data patterns", and it is therefore a matter of experiment and observation to establish them as empirical facts.

My questions will concern:

1 – the nature of that empirical evidence, and

2 – the form of the grid-display which is supposed to exhibit the relationships between concepts, and thus to achieve an array of concept classes ameable to coding and suitable for the classification of any piece of knowledge whatever.

4.1 What's empirical about concept formation?

On the basis of what kind of empirical evidence are the relationships between concepts construed? That question could apply, for example, to the mutual exclusion relationships between:

- "physical" and "living" and "chemical", in the grid for "entities";
- "physical" and "mental", in the grid for "activities";
- "physical" and "symbolic" and "behavioural", in the grid for "abstracts";
- "physical" and "behavioural/physiological" and "abstract", in the grid for "properties".

Another example of the question would be: On what type of knowledge is based the division of the class of physical entities into the six following sub-classes: a) elementary particles, b) atomic (entities), c) molecular (entities), d) molecular assemblage, e) ordered molecular assemblage, f) complex physical aggregate? Surely, there is an indefinite number of other good ways of dividing the class of physical entities into sub-classes. . .

I will reformulate question 4.1 by naming certain aspects of the process of concept formation; I suggest that an empirical knowledge is obtainable about each of these aspects but that it would be vague, possibly fallacious, methodology to leave unspecified the aspect(s) which is (are) to be selected as the relevant empirical basis to justify the choise of classificatory terms (classifyers) and their order. (Would it not even be desirable, from an epistemological point of view, that each allocation of a concept to a certain sub-class of concepts be made on the basis of one uniform type of knowledge, for all concepts – the "classifiables" – and all concept-classes – the "classifyers" and conjuncts thereof? Would not that condition be necessary in order to ensure the consistency and non-redundancy of the classification scheme?) *What's empirical, then, about concept formation?*

- is it the fact that a certain discipline has been successful in forming and using a certain number of concepts? It is physics, for example, and nothing else, that has defined and used the term "elementary particle", term which would have no meaning outside physical theory. . . But so has psychoanalysis defined an entity called 'id': should it also be used as a classifyer? (as a species of "living" entity, for example. . .)
- is it the actual sequence in which human beings usually form their concepts, somewhat in the manner genetic psychology teaches us they do: a child would form the notion of an object (let us not yet say "the concept of an entity". . .) *before* the notion of an *order* between objects, or of a *causal influence* from one object to another (see Piaget's numerous studies). Could one decide on that basis whether the notion of "living" is formed before or after the notion

of "physical"? or simultaneously? Could one observe an order of acquisition of the notions composing the following name of a sub-class: "a) complex b) symbolic c) abstract", sub-class of which the concept of a sentence could be a member?

- is it the way linguistic terms are used as means of reference in the observable *speech practice?* Does that presuppose that the way terms are *used* is a reliable hint at the way concepts are formed? What would be a suitable method for studying the *use* of terms: a sort of wittgensteinian or austinian language analysis, or a distributional analysis, or a test technique with questionnaires, etc.?
- is it the law-like structure of the semantic interrelation of linguistic terms, as shown in the empirically observable lexicon? is it, for example, the fact that a certain group of concepts concur in establishing the *meaning* of a given concept? In that case, the methods of descriptive semantics would be suitable empirical approaches.
- is it the psychological processes by means of which sensory experience leads to conceptual representation? Is the link (or the transformation) between such experience and such representation empirically observable? Should one then use the methods of behavioural psychology for collecting data and controling statements about them?

Either one must point out the type of empirical knowledge on which alone the relevant parts of one's classification scheme are based, or else one must think of means to bring together the results obtained from the study of the various empirical aspects mentioned above. Do these results simply add up? Aren't they sometimes incompatible? Are they even comparable most of the time? 2) Finally, how do the data from the various types of experiments (observations) fare with time: psychological knowledge about conceptualisation processes would probably be less time-bound than lexicographical knowledge: would it be for that reason more reliable as a basis for a classification system?

4.2 Can a net catch itself?

At time I am tempted to interpret a classification scheme, such as Mr. Farradane's, as a sort of conceptual net. In that allegory the classifyers are viewed as that which catches – the mesh–, and the individuals which appear at the very bottom of the hierarchy are viewed as that which is caught.

I suspect that this allegory is one of the obstacles to an understanding of the distinction between taxon and taxonomic category (a reference to Miss Bernier's paper). The allegory of a net carries a suggestion that the lowest level of the taxonomic hierarchy has a degree of reality incommensurable with the rest, that *it is* the reality: that's where *being* would be at home. . .

A classification of "entities" in Mr. Farradane's manner strongly supports such an interpretation, since the concrete individuals are in effect confined to the last line of an inverted-tree hierarchy, or else, if one uses a grid representation, they alone ultimately *fill* the pidgeonhole of the grid and they can appear only there, whereas all the rest is but classes, classes of classes, etc., i.e. *names* for smaller or bigger, eventually empty, pidgeonholes.

But of course any term appearing in such a classification can be the subject-matter of some discourse, perhaps even of true or false discourse. And, to put it bluntly, if something factually true can be said about x, there must be some sense in which this x can be said *real;* then the simplified ontological postulate about concrete individuals is questioned all over again.

Putting aside the ontological considerations, all I want to remark is: a classification scheme purporting to classify concepts, and all the concepts, must have the property of being able to classify, among other things, all the concepts which are classificatory in the scheme itself, i.e. the classifyers. That property may be exemplified as follows:

– the adjectives used in the specification of the classes of *entities* should all reappear as individuals in the classification of *properties,* for instance the adjective "physical" in the class-name "physical entities";
– the adjectives used in the specification of the classes of *properties* should reappear as individuals in the same grid or, preferably, should be referred to the grid for *abstracts;* for instance, the adjectives "behavioural" and "pre-concept" in the class-name "behavioural pre-concept properties";
– the terms "entities", "activities", "abstracts", and "properties" should all reappear in the grid for *abstracts.*

Now, a problem concerning the form of the classification scheme thus sketched is: how can it purport to exhibit the true composition or analysis of a concept if that concept reappears at various places in the scheme, thus apparently having more than one true analysis? Let us suppose that the proponent of that scheme wishes to maintain that the names of the sub-classes directly display an analysis (although not completed) of the member concepts, and that each concept is analysable into simpler ones until one gets simple unanalysable concepts, and that this analysis is representable by one or more ordered sequences of concepts, the "direction" of the sequence corresponding to the division of classes into sub-classes. (The foregoing statement expresses what I would call a Leibnizian wish. . .) I shall hereafter distinguish two cases of reappearance of a term in more than one section of the classification scheme.

4.21 In the case when a term appears once as a classifyer and once as a classifiable, there is not much of a riddle if one wants to avail oneself of the distinction between language and metalanguage; the term "entity" appearing in the grid for "abstracts" belongs to the language, and the same term appearing as a heading in the grid for "entities" belong to the metalanguage used for classificatory purposes.

But what about the reappearance of the term "physical" in each of the four grids of that scheme? Either one has an epistemological bias in favor of that term or that term has special properties that make it fit. That remark suggests a few questions about the status of that and similar terms in relation to the main headings. Let us represent the series; "entities", "activities", "abstracts", "properties" by K1, K2, K3, K4. Would it be correct to say that the division into K1, . . ., K4 is a division into *logical kinds* (the grammar of being. . .), whereas the division into A, B, . . ., S (these are Mr. Farradane's labels for the classes immediately following the headings, namely: physical, living, chemical, etc.) is a division into *ontological kinds* (regions of being, degrees of being, etc.) Can one read the classes A, B, . . .,S as the vertical entries of a two-dimensional table, the horizontal entries of which would be K1, . . ., K4? If so, why are the relevant classes A, B, . . . for each Ki not the same? If they were the same for each Ki, what would it mean in terms of a "psychological basis for general classification"? and what would it mean in terms of ontology? Do the reappearing terms, like "physical", "living", etc., reappear everywhere with their primitive unanalyzable meaning?

4.22 Can a term appear at more than *one* place with the status of *a member of a class?*

Apparently, yes. For example, take the term "entities" or "entity":

- Occ. 1: it appears as a main heading (name for one of the four main grids);
- Occ. 2: it must appear as a member of some sub-class of abstracts, say the sub-class "symbolic abstract of higher level" also designated by "J.d.";
- Occ. 3: it should appear as a member of some sub-class of the grid "Properties" since "being an entity" could very well be a property, as in the case of a document discussing whether God, or a number, *is an entity or not* (or: whether the concept God is the concept of an entity or not).

Then, does our dichotomy language-metalanguage account (even partly) for that multiplicity, or must we distinguish three levels of language? In the former case, the term "entity" would have in the object language two different "componential analysis", namely 1° –one 'qua' abstract: a) higher level, b) symbolic, c) abstract; 2°–one 'qua' property: a) ontological, b) simple, c) abstract, d) pre-conceptual, e) property. Would this circumstance make the term amphibolous?

On the other hand, we would have reasons to consider that the three occurrences of the term "entity" are instances of different levels of language:

- *Occ. 2* would be at a first level; let us call it the ocject language L1. The document written in that language could be a piece of ontology and would be classified in the sub-class of abstracts mentioned above;
- *Occ. 1* would be at the second level: the term "entity" then belongs to the metalanguage L2; note that this is the classificatory language: therefore the documents to be classified would not be written *in* that language; in other words, the term "entity", in an *Occ. 2* context, is not used to *say* anything and nothing is said about it: the classification grid, if it is a language at all, is not a language designed to speak about something. Occurrences of the type Occ. 2 are not 'classifiables'.
- *Occ. 3* would be at a third level; the language there is a metametalanguage, say

L3, since it speaks *about* the classification scheme in discussing the positioning of certain concepts or terms in the grid; it has to talk *about* the classifyer-term "entity". I suppose a document written in that language must be classifiable: indeed that is what is meant by saying that "being an entity" is classified in a sub-class of properties.

If this language-ordering is correct, one consequence is that the relation between the classification language (L2) and the languages of the 'classifiables' is not an invariant. In other words, in the *same* sub-class of the classification, we may encounter documents written in L1 *and* documents written in L3. There is certainly a sense in which we may say that these documents, because of that difference in their language-level, do not talk about *the same thing, the same subject-matter.* How does that affect the commonality condition for being a member of that sub-class?

4.3 About the possibility of an empirical theory of knowledge.

If I understand correctly Mr. Farradane's intention, his proposed system of concepts, when brought to completion, would be both an *epistemological* ("erkenntnistheoretisch") and an empirical – based on experimental psychology – account of knowledge in general. The aim seems to be to give a philosophical elucidation of the notion of knowledge by accounting for the way it is empirically reached (formed). (Note that with respect to that aim, a presumably optimal classification scheme would be but a useful *by-product* of the proposed epistemological-empirical theory of concepts, as Mr. Farradane himself admits.)

Now, does that amount to expecting that the theory of knowledge, traditionally a branch of philosophy, will turn into an empirical science? Does one hope to convert the philosophy of mind as well into an empirically verifiable-falsifiable discourse?

Such questions remind us of the positivists' program of the twenties and thirties. The proponents of the conceptual analysis approach to classification may be reluctant to have their project likened to Leibniz's project of constructing the complex concepts of the various sciences from simple primitive ones, given in the nature of things; they may allege that Leibniz naively thought of that procedur as an "ars inveniendi", or that he actually provided nothing but a few constructions for a *calculus* of reasoning (in some restricted domain of mathematics), or that he erroneously proceeded altogether 'a priori'. . . But how would they compare their undertaking with Carnap's attempt at constructing a "system of the constitution of concepts"? (See "Der logische Aufbau der Welt", 1929)

Carnap did start, like the present-day concept-analysts we refer to, from a phenomenalist conception of knowledge, and he did use the construction of concept classes to "constitute" the various complex concepts from fundamental elements, viz. "Elementarerlebnisse", and fundamental relations, viz. "Teilgleichheit" and "Teilähnlichkeit". But Carnap did not claim that the complex concepts would be classified differently than they are in the actual language of science and he did not claim that his constructions were the results of experiments: he knew that they were "logical constructions"; indeed, the general method of philosophical analysis that he used, namely rational reconstruction ("rationale Nachkonstruktion") could not lead to anything else. . . And he later recognized that phenomenalism is but an epistemological postulate among others, that cannot be justified by "experiments".

To which difference in method between Carnap's and Mr. Farradane's analyses can we ascribe the striking difference in claims? Does anything in the psychology-based system of concept construction correspond to what Carnap calls "constitution"? Carnap recognized some problems in the use of "equivalence classes" (see also Wittgenstein's "family resemblances") as a construction device; are these problems avoided or solved in Mr. Farradane's proposal?

Could it be that the project claiming to base a general classification on psychological knowledge is a misrepresented attempt at a *logical* construction of concepts?

Footnotes

1) In that respect, I would be curious to see whether Professor Heelan's line of research (see his paper presented to the present conference) could give a formal representation of the relationship between the so-called *ideological* stage of a discipline and its *scientific* stage. The passage from ideology to science, as it is presently theorized by French épistémologists, is an instance of a break in a tradition – it is called "coupure épistémologique" – and the scientific character of some discourse consists precisely, it is maintained, in that very break: there and thus a science comes of age. The major examples offered by those theorists are: saussurian linguistics, marxian theory of social structure, and structural ethnology in Levi-Strauss' manner; those disciplines are sometimes referred to as the first "scientific" "sciences humaines".

2) Here are some cases when such questions may arise:
   a) what is *done* with the concept 'freedom'-knowledge about usage- is frequently incompatible with what is lexically meant by it -semantical knowledge;
   b) how does one compare one's lexicographical results about the term "will" with one's psycho-genetical results? How about the term "number" under the same circumstances?
   c) how does one bring the pronouncements of an empirical theory of concept formation to bear upon (to be relevant for) the empirical observations of language uses, considering that these observations hold only for a determinate language, a determinate population of language users, and perhaps for a specific cultural environment or for a discipline-bound discourse.